Our Warming Planet

Topics in Climate Dynamics

Lectures in Climate Change

Series Editors:

David Rind (*NASA Goddard Institute for Space Studies, USA & Columbia University, USA*)
Cynthia Rosenzweig (*NASA Goddard Institute for Space Studies, USA & Columbia University, USA*)

Climate change is a portending issue not only for our time, but indeed for generations to come. Informing current and future leaders regarding this issue is a vital task, so that they harbor the knowledge to devise and implement the actions that are required to forestall its consequences. Those actions must be directed towards curtailing the emissions of greenhouse gases to the greatest extent possible, and towards adapting to unavoidable changes of climate.

The goal of this series is to provide resources that can be used by scholars and students worldwide concerning topics related to climate change. Released in both print and online versions, the lectures in this series are targeted at university-level educators and their students across the globe to aid in teaching these particular topics. The series would provide the materials necessary for online learning by individuals seeking to gain a deeper understanding of relevant topics. The lecture presentations with their accompanying written introductions and notes are peer reviewed.

Published:

Lectures in Climate Change   Vol. 1

Our Warming Planet

Topics in Climate Dynamics

Editors

Cynthia Rosenzweig
NASA Goddard Institute for Space Studies, USA & Columbia University, USA

David Rind
NASA Goddard Institute for Space Studies, USA & Columbia University, USA

Andrew Lacis
NASA Goddard Institute for Space Studies, USA

Danielle Manley
Columbia University, USA

World Scientific

NEW JERSEY · LONDON · SINGAPORE · BEIJING · SHANGHAI · HONG KONG · TAIPEI · CHENNAI · TOKYO

Published by

World Scientific Publishing Co. Pte. Ltd.

5 Toh Tuck Link, Singapore 596224

USA office: 27 Warren Street, Suite 401-402, Hackensack, NJ 07601

UK office: 57 Shelton Street, Covent Garden, London WC2H 9HE

Library of Congress Cataloging-in-Publication Data

Names: Rosenzweig, Cynthia. | Rind, David. | Lacis, Andrew A. | Manley, Danielle.

Title: Our warming planet : topics in climate dynamics / [edited by] Cynthia Rosenzweig
   (NASA Goddard Institute for Space Studies, USA & Columbia University, USA),
   David Rind (NASA Goddard Institute for Space Studies, USA & Columbia University, USA),
   Andrew Lacis (NASA Goddard Institute for Space Studies, USA),
   Danielle Manley (Columbia University, USA).

Description: New Jersey : World Scientific, 2017. | Series: Lectures in climate change |
   Includes bibliographical references.

Identifiers: LCCN 2016055366| ISBN 9789813148772 (hardback : alk. paper) |
   ISBN 9789813148789 (pbk. : alk. paper)

Subjects: LCSH: Global warming. | Climatic changes. | Atmospheric circulation. | Hydrology.

Classification: LCC QC981.8.G56 O9727 2017 | DDC 551.5/2--dc23

LC record available at https://lccn.loc.gov/2016055366

British Library Cataloguing-in-Publication Data

A catalogue record for this book is available from the British Library.

Cover image credit: Night Lights 2012 – The Black Marble. NASA Visible Earth, 2012. NASA Earth Observatory image and animation by Robert Simmon, using Suomi NPP VIIRS data provided courtesy of Chris Elvidge (NOAA National Geophysical Data Center). Suomi NPP is the result of a partnership between NASA, NOAA, and the Department of Defense. Caption by Mike Carlowicz. Image can be accessed online at: http://visibleearth.nasa.gov/view.php?id=79803

For any available supplementary material, please visit
http://www.worldscientific.com/worldscibooks/10.1142/10256#t=suppl

Desk Editor: Amanda Yun

Typeset by Stallion Press
Email: enquiries@stallionpress.com

Acknowledgements

This compendium of lectures and lecture notes on climate change is the product of the work of the 20 chapter authors, who each exhibit a unique passion and knowledge for the topic which they present. We thank them very much for the time and effort spent in making this volume.

We would like to thank our team of expert reviewers: Timothy Eichler, Claire Parkinson, Dorothy Peteet, Colin Price, and Gary Russell. Their work was vital to improving the readability, accessibility, and visibility of the lecture contributions. We would also like to thank Jeff Jonas at GISS for his participation and insights.

We send our deepest gratitude to the staff at World Scientific Publishing — especially Dr. Zvi Ruder, Senior Executive Editor, and Amanda Yun, Senior Editor — for their professional expertise in publishing this unconventional work.

Finally, we would like to thank David Rind, to whom this Volume is dedicated. As the saying goes, "You embrace that which defines you." This publication proves that he has embraced deep knowledge of climate change and inspired a cohort of researchers imprinted with a devotion to this field.

Cynthia Rosenzweig and Danielle Manley
November, 2017
New York, NY, USA

Contents

Introduction

Our Warming Planet: Topics in Climate Dynamics is a new type of 'text book' on climate and climate change. It is composed of presentations by experts in their respective fields, with helpful introductory explanations that support each lecture. The goal is to provide teachers with suitable material for classroom expositions, and students with the appropriate knowledge and understanding in both textual and graphical form. As such, it is available in both a 'hard copy' book-like format and digital files suitable for downloading.

A brief glance through the topics covered in this book provides a glimmer of the heterogeneous nature of this subject. The climate system and the consequences of climate change encompass many diverse activities and fields of study, in fact too many to cover in one text. Consider these disciplines:

- Solar physics
- Atmospheric radiation
- Cloud physics
- Atmospheric aerosol physics
- Atmospheric dynamics
- Hydrologic processes
- Climatology and paleoclimatology
- Atmospheric chemistry
- Numerical modeling
- Vegetation dynamics
- Carbon cycle and other trace gas cycles
- Soil physics and ground hydrology
- Ocean dynamics
- Sea ice dynamics
- Land ice dynamics
- Climate change impacts (on health, natural and man-made ecosystems, economy, demography, and foreign policy)

All of them in one way or another are brought into play in this subject. Not only does this pose problems for students trying to understand the various component parts, but also for scientists working in the field, since these disciplines interact in the climate system. Often the individual fields have their own terminology, some of which is incomprehensible to 'outsiders'; yet researchers from these disparate fields of study must talk with one another, and learn to understand the underlying philosophy and approach their counterparts employ. Not an easy task, but one that can begin by familiarizing oneself with what might otherwise be considered 'foreign' subjects. For this text, we have grouped the presentations into the following sections: Understanding climate change, radiative processes, dynamical responses, hydrologic responses, polar responses, paleoclimate perspective, climate change impacts, and educational perspective.

Hence the desire to include in this book a wide diversity of relevant topics, each presented by experts in their field. They became experts by focusing on specific components, not by trying to embrace them all; but generally they were at least somewhat exposed to other areas that could interact with theirs, and so have been able to more fully understand the consequences of their results. Not only is that of benefit to society, it makes their research more dynamic and broadens their minds.

At what particular academic discipline is this book aimed? Since climate and climate change represents the poster child for 'interdisciplinary studies', one would suppose that only a 'Climate' department would suffice; unfortunately, no university has one. Climate is often a subset of an atmospheric science department, which itself is often part of a broader 'Earth and Environmental Sciences' faculty. The list above, however, also includes astronomy, chemistry, biology, agronomy, geology, economics, and political science. It is the clever student indeed who can manipulate the often rigid university structure to allow a course of study that incorporates the needed elements from these other departments. So, another goal of this publication is to allow students to grasp some of the relevant components of these other disciplines by him/herself. All this is to say: do not skip over topics that at first glance are unfamiliar, it might be your best chance to come in contact with them.

At what level are these presentations aimed? Since these are actual presentations often given by specialists, sometimes to other specialists, the authors have not been satisfied presenting only introductory material — they are meant to be 'state-of-the-art', and so can include quite complicated concepts. The suggestion would be: if you do not understand something, do the best you can, go on to the next slide, and get out of it whatever is possible. References are provided if you want to understand something more fully, and clearly, there is no way to master any of these subjects on the basis of one (primarily graphical) presentation alone. Hopefully, at least some of these will strike a resonant chord, and stimulate you to do further research in that area of interest.

For professors, if one of these subjects aligns itself with your particular course, so much the better. If not, or if you are teaching a more general class, the different topics offer the possibility of cherry picking (via the digital slides) from various lectures to produce the required breadth. The benefit of our flexible modern technology!

This book is nominally dedicated to me, for which I am most grateful, though actually I am simply the rallying point around which these presentations are organized. I have been fortunate to interact with these other scientists, either as students or as fellow researchers; I have been enlightened by their discoveries, and enlivened by their enthusiasms, all of which come through in these presentations. Not only is the subject matter in this field highly diverse, so are the personalities of the people involved — both of which have made it a very interesting subject to pursue! Hopefully, you will have similar experiences.

David Rind
November, 2017
New York, NY, USA

SECTION 1

Understanding Climate Change

CLIMATE LECTURE 1

Explaining Climate

Andrew Lacis

NASA Goddard Institute for Space Studies, New York, NY, USA

Andrew Lacis (PhD in Physics from University of Iowa, 1970) is a Senior Research Scientist at the NASA Goddard Institute for Space Studies (GISS) in New York City. He has been a member of the GISS climate modeling group since the mid-1970s. His area of expertise is development of fast and accurate radiative transfer techniques to model solar and thermal radiation with applications to the study of global climate change and the remote sensing of planetary atmospheres. He is the principal architect of the radiation modeling methodology that is used in the GISS climate GCM.

Throughout his professional career David Rind has been explaining global climate to audiences that have included students, teachers, policy-makers, scientists, and the general public. David's great success in explaining the complexities of climate science can be inferred from the 21 PhD candidates that he advised. Their PhD dissertations, dating from 1988 to 2011, and the resulting published papers have significantly advanced our understanding of the terrestrial climate system. Of David's 21 PhD students, all continue their top-level research in climate, with three having already been elected AGU Fellows for their outstanding research accomplishments.

Key Points and Key Components

It all begins with the Sun. The global-mean solar illumination of the Earth has been determined to be about 340 W/m^2 of which about 100 W/m^2 is reflected back to space. Hence, the Earth absorbs a global annual-mean 240 W/m^2 of solar energy. This amount of energy is just sufficient to support a global-mean temperature of 255 K. However, the global-mean surface temperature of the Earth is known to be about 288 K, which implies that the Planck emission from the ground surface must be about 390 W/m^2. It is this 'missing energy' circumstance that led Joseph Fourier to conclude back in 1824 that there must be an atmospheric greenhouse effect causing thermal heat energy to be radiated downward from the atmosphere in order to supply the additional heat energy needed at the ground surface. This is because in the absence of the greenhouse effect, 240 W/m^2 of absorbed solar energy is not sufficient to sustain a surface temperature of 288 K.

The flux difference of 150 W/m^2 between the 390 W/m^2 emitted by the ground surface and the 240 W/m^2 of longwave (LW) flux going out to space at the top of the atmosphere is a direct measure of the strength of the terrestrial greenhouse effect. The greenhouse action builds up the surface temperature and the surface-emitted flux to 390 W/m^2, and is also responsible for the ensuing reduction by 150 W/m^2 of the surface-emitted flux as it makes its way to space — all of that is accomplished by radiative energy processes (via sequential emission, absorption, and re-emission interactions). Detailed calculations of radiative flux attribution show explicitly that water vapor accounts for about 50% of the 150 W/m^2 greenhouse effect, and that LW cloud opacity accounts for 25%. Both of these radiative contributions stem from the fast feedback processes of the climate system. The remaining 25% of the terrestrial greenhouse effect comes from the radiative forcings contributed by the *non-condensing* greenhouse gases (GHGs, which coincidentally also act to sustain the terrestrial greenhouse effect at its present strength). Of these non-condensing GHG contributions, CO_2 is by far the strongest contributor accounting for about 20% of the 150 W/m^2 greenhouse effect, with the remaining 5% due to minor GHGs such as CH_4, N_2O, O_3, and CFCs (Lacis *et al.*, 2010).

The key point is that these non-condensing GHGs behave as the principal radiative-forcing agents of the climate system because of their thermodynamic, chemical, and radiative properties. CO_2 and the minor GHGs are chemically slow-reacting with atmospheric lifetimes that range from decades to many centuries. Once introduced into the atmosphere they effectively remain there indefinitely because they do not condense or precipitate at the prevailing atmospheric temperatures while continuing to exert their radiative forcing. Since CO_2 is by far the strongest and most effective of these non-condensing radiative-forcing gases, it follows that CO_2 can be identified as the principal LW control knob that governs the global climate of Earth. The fact is that the other forms of radiative climate forcing (e.g., changes in solar irradiance, surface albedo, and aerosol forcing) are small by comparison. This makes it that much more compelling to recognize CO_2 as the principal climate control knob (Lacis *et al.*, 2013).

Atmospheric water vapor, on the other hand, has its role as the principal fast-feedback process of the climate system — condensing and precipitating from the atmosphere in response to changes in local meteorological conditions (as constrained by the exponential temperature dependence of the *Clausius–Clapeyron* relation). Thus, the atmospheric distribution of water vapor (and clouds) can change rapidly in response to changing weather conditions. Radiative forcings that heat the atmosphere will cause more water vapor to evaporate, generating more LW opacity, and more radiative greenhouse effect. These changes in water vapor can produce big changes in radiative heating or cooling, but they remain limited in magnitude by how much the water vapor amount is physically able to change as it approaches its new equilibrium distribution. As a result, water vapor and clouds can only act to magnify an initial radiative perturbation, but cannot act on their own initiative to manufacture or impose a sustained warming or cooling trend on global climate, even though they may contribute more strongly to the overall atmospheric radiative structure than the radiative-forcing GHGs that actually drive and control the global temperature trend.

Cause and Effect

Global warming is basically a straightforward conservation-of-energy *cause-and-effect* problem in physics, and as such, not all that complicated. The basic 'cause' of global warming has been understood for decades, and is accurately quantified by precise measurements of atmospheric CO_2 (the Keeling curve). This reality is amply corroborated by the annual reports of fossil fuel extraction that show 10 gigatons of carbon/year (equivalent to about 10 cubic km of coal/year, which when burned, convert to nearly 5 ppm CO_2, half of which will remain in the atmosphere). The radiative effects of CO_2 and other GHGs are well understood, as is the radiative transfer modeling of atmospheric radiation. This all meshes harmoniously with the geological context made ever more clear by the precise analysis of air bubbles that have been sequestered in polar icecaps over geological time scales. Additional confirmation comes from carbon isotope analyses and direct measurements of atmospheric oxygen depletion due to the burning of fossil fuel. Still more, there is closure available from carbon cycle analyses of the ocean and biosphere uptake of about half of the CO_2 injected into the atmosphere. This leaves no room for lingering doubts that the observed increase in atmospheric CO_2 is anything but the direct result of human industrial activity. Given its radiative properties, adding CO_2 to the atmosphere restricts the natural flow of LW thermal energy to space, causing a radiative energy imbalance that requires restoration by inducing the appropriate changes in the surface and atmospheric temperature.

To be sure, it is decidedly more difficult to identify the 'effect' component of the global warming cause-and-effect problem with the same unambiguous clarity as the causal mechanism directly from the global temperature record. This is because the ongoing climate change consists of a steadily increasing global warming component (driven by continued injection of CO_2 from fossil fuel burning), on which there is superimposed a random-looking natural variability component (global temperature fluctuations about a reference point covering a broad range of time scales). These random-looking natural fluctuations obscure the global warming trend, but they are not random (as in random walk), such that if given enough time, they could divert the global climate arbitrarily far from its equilibrium reference point. Both real-world climate and global climate models (GCMs) must conserve energy, making arbitrarily large temperature departures from the equilibrium reference point impossible to sustain in the absence of externally applied forcing.

There are actually many different 'climate effects' that result from (and are thus 'caused' by) the steady increase in atmospheric CO_2. A more refined understanding of the Earth's climate system requires that all of these different CO_2 change-driven effects need to be verified and documented by accurate measurements not only for purposes of numerical closure, but also for reassurance that significant climate processes have not been overlooked. Besides the widely acknowledged increase in global land surface temperature, it is equally important to monitor the changes that take place in ocean heat content and ocean temperature, the increase in atmospheric water vapor and the decrease in stratospheric temperature. The inevitable rise in sea-level concomitant with the decrease in polar ice amount needs to be monitored closely, including also the theoretically expected radiative energy imbalance between solar and thermal energy, as more solar radiation is absorbed by the Earth than is being radiated back to space in the form of LW radiation.

Radiative Forcing and Natural Variability

Accurate measurements are also needed to characterize the other radiative-forcing perturbations of the climate system, such as changes in solar irradiance, land surface albedo, and volcanic and anthropogenic aerosol loading. Not to be overlooked are the 'virtual radiative forcings' that arise from changes in ocean circulation that spread warm surface water from the western to the eastern Pacific as happens during El Niño events, and ocean circulation changes that promote upwelling of colder water across the eastern Pacific as occurs during La Niña events. These are examples of disequilibrium changes in the ocean surface temperature that originate from basically unforced (without any direct radiative forcing) changes in ocean circulation, which then induce feedback responses that affect the other climate variables to produce changes in climate variables that can mask the response from CO_2 radiative forcing. Long-term variability in ocean circulation also occurs on decadal time scales, and produces longer lasting regional ocean temperature variations such as the Pacific Decadal Oscillation and the Atlantic Multi-decadal Oscillation.

This *unforced variability* of the climate system arises because key climate system components (water vapor and clouds for the short term, atmospheric and ocean dynamics for the long term) are not configured to respond to the radiative/temperature-forcing perturbations on a sufficiently small enough incremental scale that would allow a monotonic approach to global energy balance equilibrium. What happens is that when water vapor condenses, there is large-scale condensation and cloud formation with concomitant radiative effects that go much beyond the energy balance equilibrium point. This over-reaction, and a similar over-response by atmospheric dynamics to pressure–temperature differences, produces the quasi-chaotic weather noise that is a persistent characteristic of the terrestrial climate system. Decadal-scale variability arises from shifting ocean circulation patterns that involve much larger ocean heat capacity. The climate system equilibrium point can be identified by averaging the global variability over time periods that are significantly longer than the time scale of these 'unforced' quasi-chaotic fluctuations. The approach to energy balance equilibrium is akin to a drunken-dizzy sailor trying to step on a dime that is laying in the street — he can see the dime, but does not possess the fine-control muscle coordination needed to zero in on his objective with the necessary precision.

Global climate has undergone drastic change in the past, as is clearly evident from the geological record. The Earth was once an ice-covered snowball in pre-Cambrian times and then experienced world-wide tropical conditions during the Cretaceous period. Subsequent climate has oscillated between recurring ice-age climate and warm interglacial climate, representative of present-day conditions. During these major global climate changes, atmospheric CO_2 has been a key factor. Volcanic CO_2 is the principal forcing agent over million-year time scales. On the thousand-year interglacial time scale, CO_2 acts both as a radiative forcing and as a feedback product, changing in response to ocean chemistry–biosphere interactions. Atmospheric CO_2 caused by industrial activity is the principal driver of the current global climate warming, which, when viewed in the geological context, has produced a decadal-scale radiative forcing of climate that is a hundred times faster than prior rates of CO_2 change in the geological record.

Climate Feedback Sensitivity

All this was understood over 30 years ago when (Hansen *et al.*, 1984) published their classical paper on climate sensitivity using the GISS Model II GCM as their analysis tool. The basic conjecture of this study was that radiative forcings of similar magnitude (2% increase in solar irradiance and doubling of atmospheric CO_2) would produce a similar climate response. They also showed that the GISS GCM model physics could reproduce the ice-age global climate when operated with representative ice-sheet surface albedo and ice-age CO_2 radiative forcings. To first order, both forcings (the 4.8 W/m^2 solar irradiance and 4 W/m^2 doubled CO_2) produced roughly similar 4 K increases in surface temperature as the climate system approached its new global energy-balance equilibrium after 35 years of run time. The 4 K equilibrium climate sensitivity was the temperature difference between the forcing run and a control run that was also run for 35 years, but without the radiative forcings. The approach to equilibrium was asymptotic, but it was not monotonic, with both the radiative-forcing run and the control run exhibiting a semi-realistic inter-annual variability. For both forcings, the polar regions showed a greater climate sensitivity, warming by about 6 K, while the tropics warmed only by 3 K.

Perhaps the most significant characteristic of the climate system response to the applied radiative forcings was the pronounced positive feedback magnification that resulted from induced changes in water vapor, cloud, and snow/ice albedo distributions. The direct (no-feedback) temperature response that is just sufficient to counteract the 2% solar irradiance or the doubled CO_2 forcing is about 1.3 and 1.2 K, respectively. The threefold magnification of the direct surface warming due only to the applied radiative forcings was an important finding of this study. The increase in atmospheric water vapor as the temperature increased was expected because of the exponential temperature dependence of the *Clausius–Clapeyron* relation that stipulates that warm air can hold more water vapor than cold air. The changes in snow/ice cover were also expected to yield a positive feedback since warmer temperatures imply a reduction in the snow/ice cover with subsequent increases in absorbed solar radiation to produce the additional warming. Clouds, on the other hand, are more complicated in that optically thin high clouds are strong contributors to the greenhouse effect, while optically thick low clouds are strong reflectors of solar radiation to produce cooling. In these 1984 climate sensitivity experiments, near-surface warming tended to reduce the low-level cloud cover, while it increased the atmospheric humidification with altitude, and the cirrus cloud cover. This produced a decidedly positive cloud feedback effect that largely accounted for the equilibrium surface temperature sensitivity being 4 K for doubled CO_2.

Climate Forcing and Feedback Attribution

Because the atmospheric temperature structure is sustained primarily by radiative transfer means, it follows that attribution of the equilibrium temperature changes can be evaluated by radiative modeling. The 4 K increase in global temperature due to doubled CO_2 caused the global-mean atmospheric water vapor to increase by about 33%. Evaluating the radiative effect of this 33% increase in water vapor by off-line radiative–convective equilibrium calculation yielded a 1.6-K global-mean surface temperature change. The feedback efficiency gain factor for water vapor can be defined as $g_w = DT_w/DT_s = 1.6\ K/4.0\ K = 0.4$, where $DT_w = 1.6$ K is the temperature change due only to water vapor, and $DT_s = 4.0$ K is the total surface temperature change due to all of the contributors: i.e., the (no-feedback) $DT_0 = 1.2$ K direct heating by CO_2, the feedback temperature contributions due to cloud and snow/ice albedo changes, and the feedback warming due to water vapor itself. This temperature change attribution for the CO_2 doubling derived cloud changes yields the cloud climate feedback efficiency gain factor $g_c = \Delta T_c/\Delta T_s = 0.8\ K/4.0\ K$, or $g_c = 0.2$, and the snow/ice feedback efficiency $g_s = \Delta T_s/\Delta T_s = 0.4\ K/4.0\ K$, or $g_s = 0.1$.

The individual contributions to surface temperature are additive: $\Delta T_s = \Delta T_0 + \Delta T_w + \Delta T_c + \Delta T_s$. As a result, ΔT_s and ΔT_0 are algebraically related by $\Delta T_s = \Delta T_0/[1 - (g_w + g_c + g_s)]$. This suggests that the feedback efficiency factors are also linearly additive, but not their effects on the global surface temperature. Accordingly, doubled CO_2 with only the water vapor feedback yields 2 K global-mean surface warming. If the cloud feedback were truly zero ($g_c = 0$), the global warming would be 2.4 K. With all basic feedbacks operating, the doubled CO_2 result is 4 K warming.

This type of radiative attribution analysis was also used to deduce the latitudinal dependence of feedback sensitivity using previously unpublished supplementary material from this 1984 study. This showed that radiative forcing for doubled CO_2 is globally uniform while the solar irradiance forcing is more concentrated in the tropical regions where the snow/ice albedo feedback does not operate. This explains why the global feedback response was less efficient for solar irradiance forcing than for CO_2. The results also showed that the latitudinal dependence of the feedback responses was similar for both forcings, and that the individual feedback contributions to the equilibrium surface temperature change obtained from the 2-D radiative–convective–advective modeling analysis were basically additive, closely reproducing the GCM calculated results.

This study showed that water vapor feedback is strongly constrained by the *Clausius–Clapeyron* relation, and is the most robust of the climate system feedbacks, and that there are three principal components to the water vapor forcing: (1) a globally similar increase in column water vapor by about 33%, (2) increased low-latitude tropospheric humidification with height, which increases the water vapor greenhouse efficiency, and (3) a lapse-rate feedback due to the increased vertical transport of latent heat that warms the upper troposphere more than the surface, thus producing a substantial negative feedback in the tropical and mid-latitude regions, but a positive feedback in the polar regions. Expressed in the form of feedback efficiency g_w, the water vapor feedback was shown to have the same variability with latitude for both the solar irradiance and CO_2 forcings.

For both forcings, clouds were shown to provide similar positive feedback contribution except in the polar regions where the cloud feedback was decidedly negative. Not surprising, the snow/ice feedback is associated primarily with changes in sea-ice cover, and was strongly positive only in polar regions. Advective feedbacks, which represent changes in horizontal transports of sensible, latent, and geopotential energy as the climate changes, are specifically relevant when considering latitudinally resolved spatial scales. Globally averaged, horizontal energy transports by definition must add to zero, permitting simpler analysis for global-mean energy and feedback comparisons. Being an order of magnitude larger than the radiative feedback contributions, the sensible, latent, and geopotential energy changes tend to cancel each other but still produce positive and negative latitudinal feedback contributions, implying regional shifts in atmospheric circulation patterns.

After 32 years, 21 PhD dissertations, and dozens of research papers, this broad-brush description of global climate change as presented in the 1984 study remains remarkably accurate. What has changed is the far greater awareness of the complexity of climate system interactions between its different components, in particular the ocean, which is key to understanding natural variability and the time rate of change of the climate system. Model spatial resolution is now over an order of magnitude better than in 1984. Modeling runs with detailed solar, aerosol, GHG, and land-use changes can now be readily evaluated over century-long time scales. Current emphasis has shifted to analyzing atmospheric and ocean processes in greater detail, while focusing more on understanding the impacts of regional climate change on the environment. But *the bottom line* remains — atmospheric CO_2 is the principal driver of global warming and global climate change.

References

Hansen, J., Lacis, A., Rind, D., Russell, G., Stone, P., Fung, I., Ruedy, R. and Lerner, J. (1984). Climate sensitivity: Analysis of feedback mechanisms. In J.E. Hansen and T. Takahashi (Eds.), *Climate Processes and Climate Sensitivity* (Vol. 5, pp. 130–163). Washington DC: AGU Geophysical Monograph 29, Maurice Ewing American Geophysical Union.

Lacis, A.A., Schmidt, G.A., Rind, D. and Ruedy, R.A. (2010). Atmospheric CO_2: Principal control knob governing Earth's temperature. *Science*, 330, 356–359, doi:10.1126 /science.1190653.

Lacis, A.A., Hansen, J.E. Russell, G.L. Oinas, V. and Jonas, J. (2013). The role of long-lived greenhouse gases as principal LW control knob that governs the global surface temperature for past and future climate change. *Tellus B*, 65, 19734, doi:10.3402/tellusb.v65i0.19734.

Slide 1

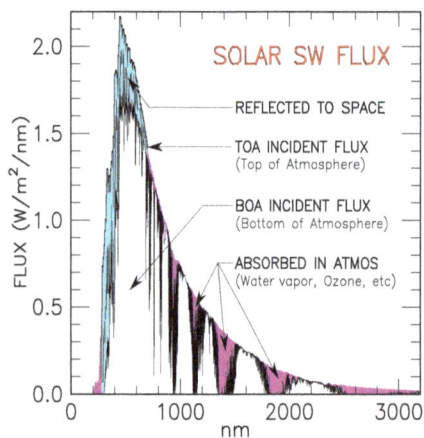

Explaining Climate

Andrew Lacis
NASA Goddard Institute for Space Studies

Our Warming Planet: Topics in Climate Dynamics
Lectures in Climate Change, Vol. 1
2017

Slide 2

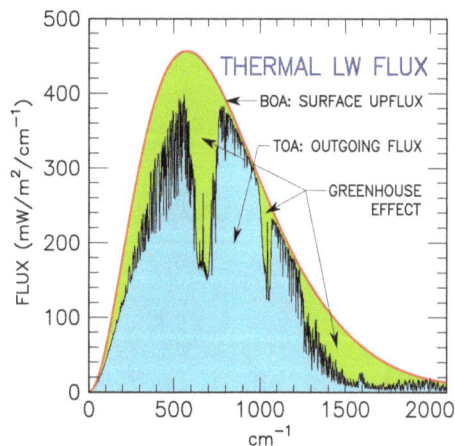

Global Energy Balance between Absorbed Solar and Emitted Thermal Energy

Spectral distribution of the incident solar radiation.

The spectral distribution of LW thermal energy emitted to space by Earth.

Slide 3

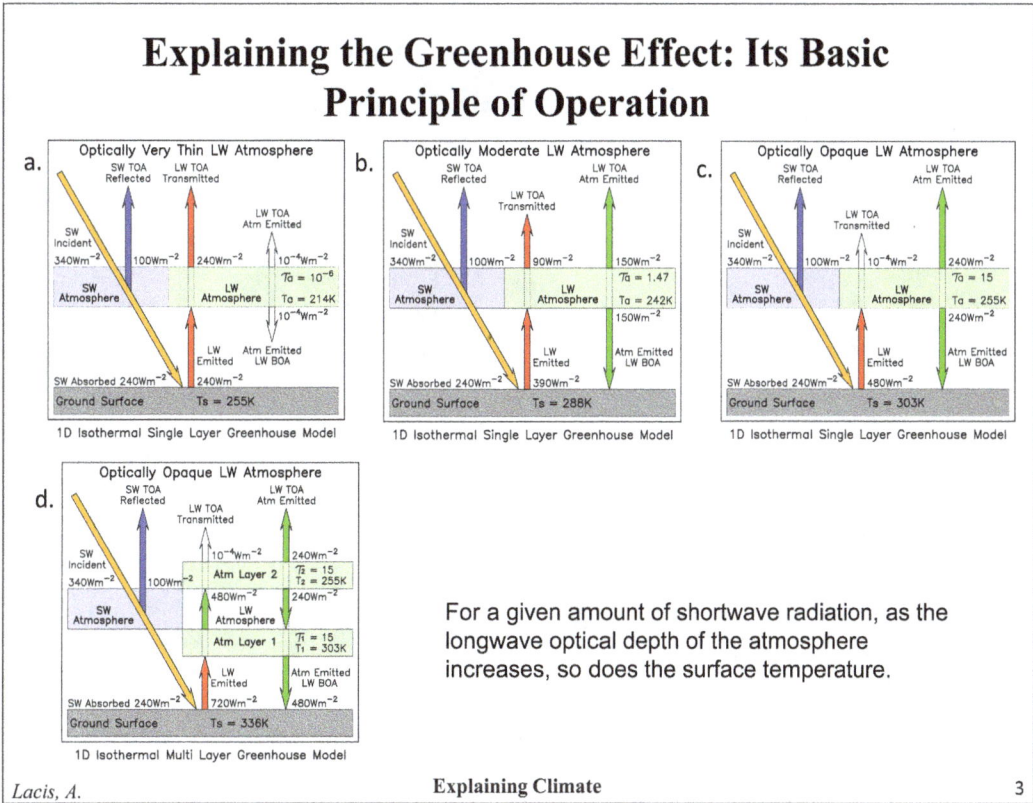

Explaining the Greenhouse Effect: Its Basic Principle of Operation

For a given amount of shortwave radiation, as the longwave optical depth of the atmosphere increases, so does the surface temperature.

Lacis, A.   **Explaining Climate**   3

Slide 4

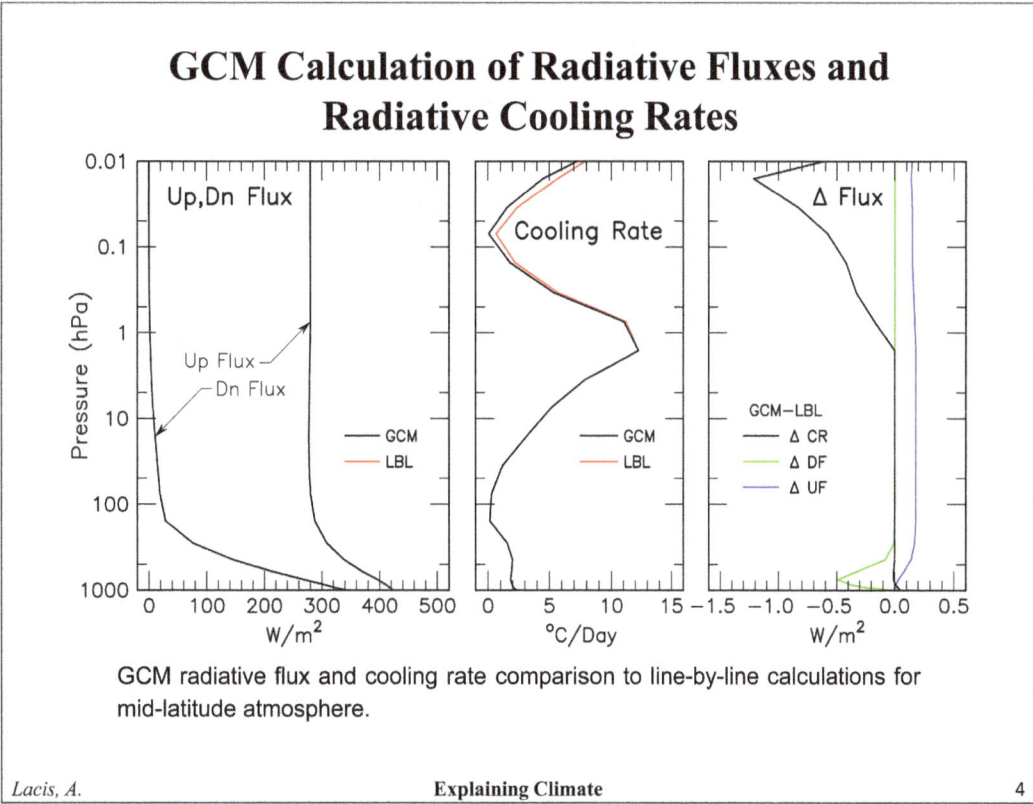

GCM Calculation of Radiative Fluxes and Radiative Cooling Rates

GCM radiative flux and cooling rate comparison to line-by-line calculations for mid-latitude atmosphere.

Lacis, A.   **Explaining Climate**   4

Slide 5

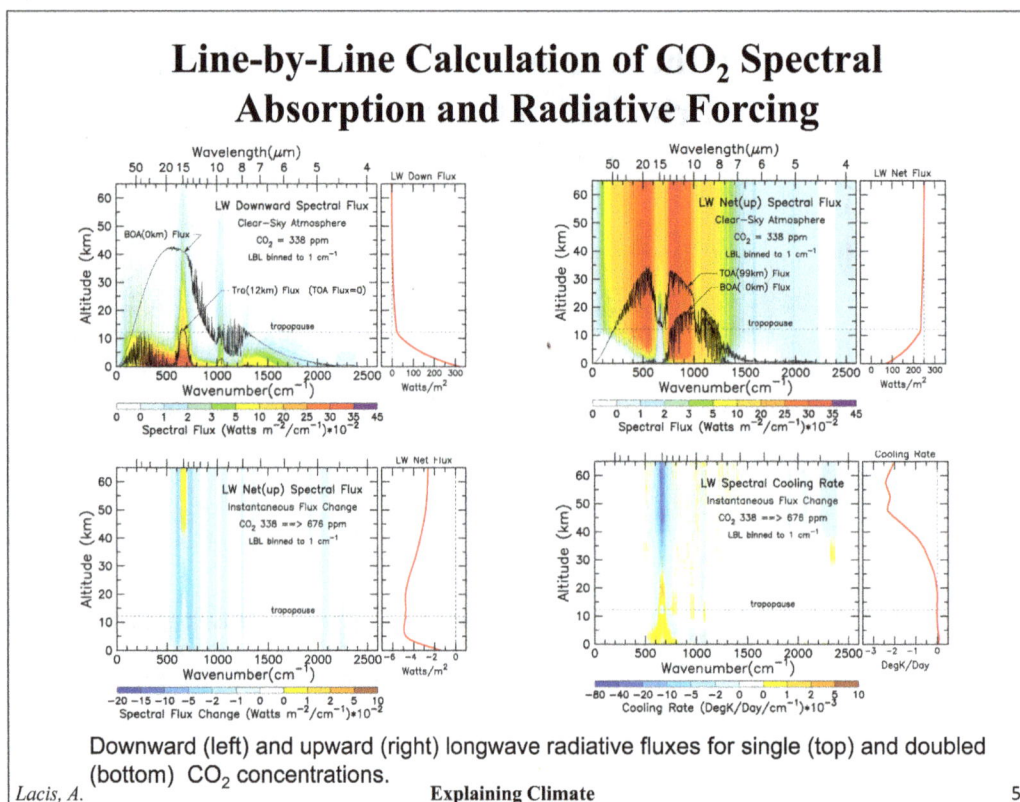

Line-by-Line Calculation of CO_2 Spectral Absorption and Radiative Forcing

Downward (left) and upward (right) longwave radiative fluxes for single (top) and doubled (bottom) CO_2 concentrations.

Lacis, A.   **Explaining Climate**   5

Slide 6

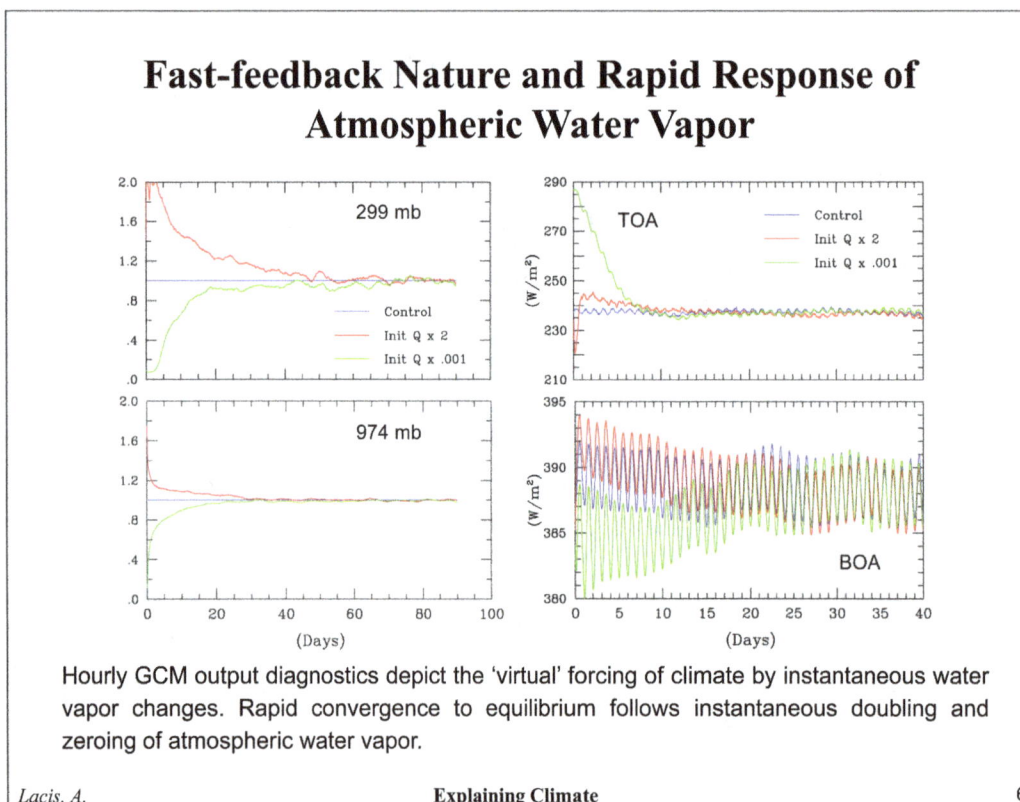

Fast-feedback Nature and Rapid Response of Atmospheric Water Vapor

Hourly GCM output diagnostics depict the 'virtual' forcing of climate by instantaneous water vapor changes. Rapid convergence to equilibrium follows instantaneous doubling and zeroing of atmospheric water vapor.

Lacis, A.   **Explaining Climate**   6

Slide 7

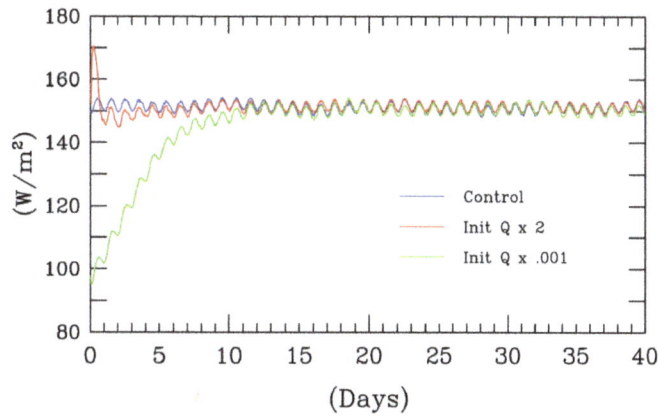

Greenhouse Strength *vs* Water Vapor and Diurnal Temperature Change

Hourly GCM diagnostics of the globally averaged greenhouse effect convergence to equilibrium following instantaneous doubling (red) and zeroing (green) of atmospheric water vapor. Diurnal oscillations in the globally averaged greenhouse strength result from land-ocean differences in diurnal surface temperature changes.

Lacis, A.   **Explaining Climate**   7

Slide 8

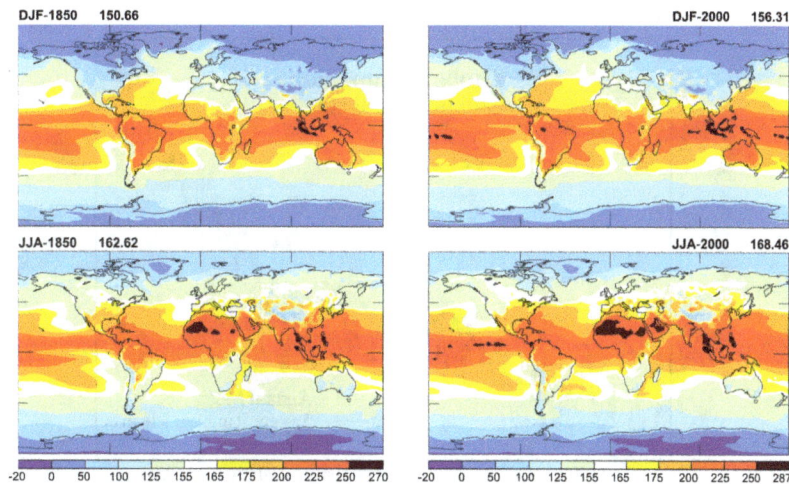

Seasonal *vs* Long-term Variability of the Greenhouse Effect Strength

Global maps of seasonal change of the greenhouse effect for 1850 (left panels) and 2000 (right panels).

Lacis, A.   **Explaining Climate**   8

Slide 9

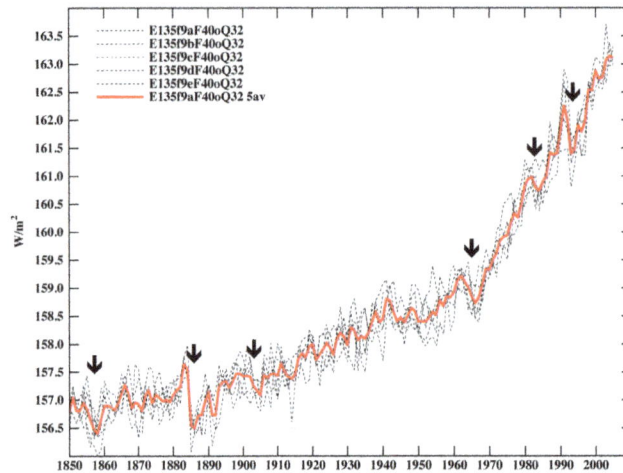

Global Mean Greenhouse Strength: Its Natural and Secular Variability

Trend of the terrestrial greenhouse effect from 1850 to 2010.

Lacis, A.   **Explaining Climate**   9

Slide 10

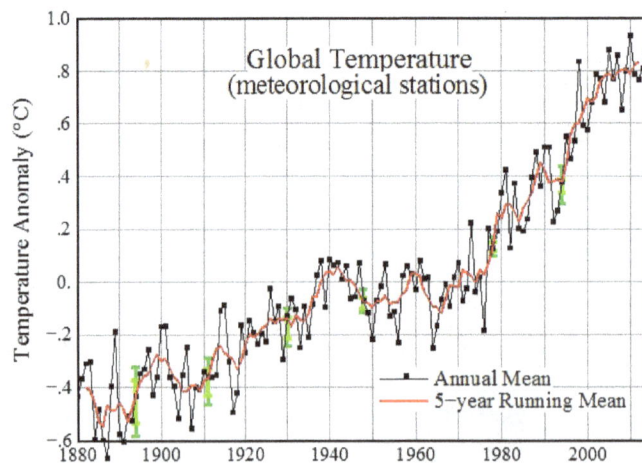

Observed Global-mean Temperature Trend from Meteorological Stations

Observed global annual-mean trend in surface air temperature relative to the base period 1951–1980.

Lacis, A.   **Explaining Climate**   10

Slide 11

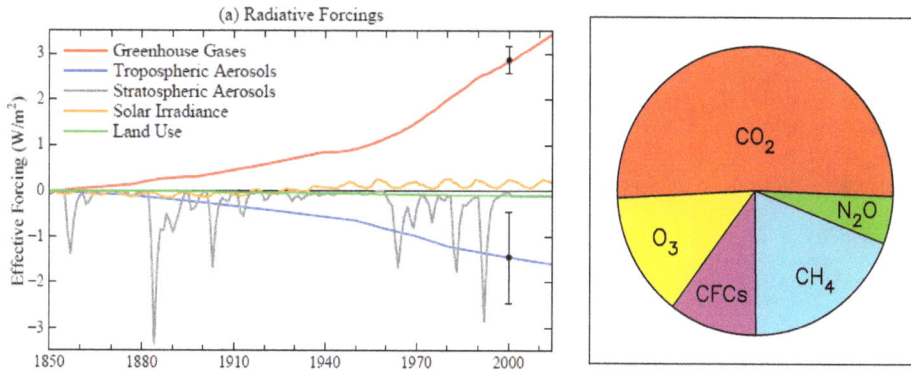

Natural and Anthropogenic Radiative Forcing of Global Climate Change

(a) Radiative Forcings

Natural and anthropogenic radiative forcing of global climate change over the industrial period.

Lacis, A.   **Explaining Climate**   11

Slide 12

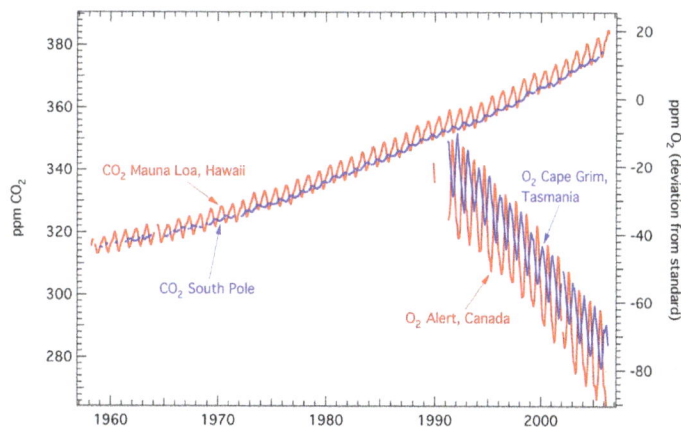

Increased anthropogenic CO_2: Principal Driver of Global Warming

Time trend of atmospheric CO_2 concentration observed at Mauna Loa (Hawaii) and the South Pole (left scale) with corresponding decrease in atmospheric oxygen (right scale, deviation from reference amount) observed at the stations Cape Grim (Tasmania) and Alert (Canada) since 1990.

Lacis, A.   **Explaining Climate**   12

Slide 13

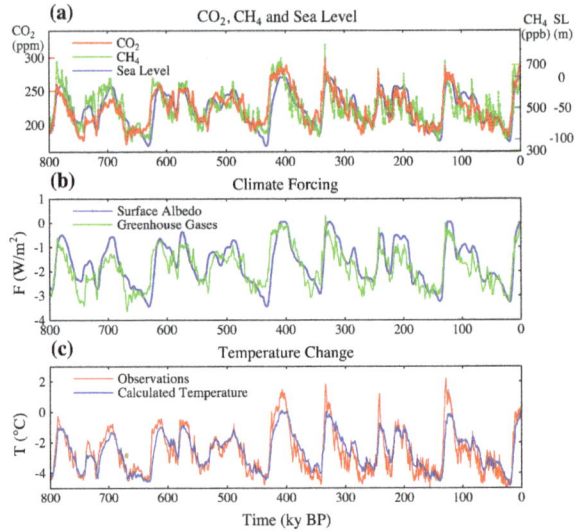

Geological Context: Temperature, CO$_2$ and Sea Level from Ice Core Analysis

Geological record of climate forcings, temperature, sea level, and surface albedo response.

Slide 14

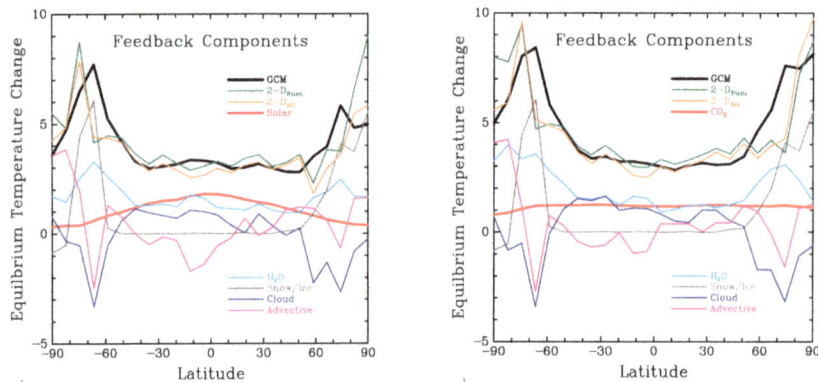

Surface Temperature Zonal Attribution in 1984 Climate Sensitivity Study

The 1984 Hansen *et al.* GCM climate sensitivity study was a defining moment in quantitative climate feedback analysis with climate forcing experiments performed for a 2% increase in solar irradiance, and doubled CO$_2$.

Slide 15

Water Vapor Feedback Component Analysis: 1984 Climate Sensitivity Study

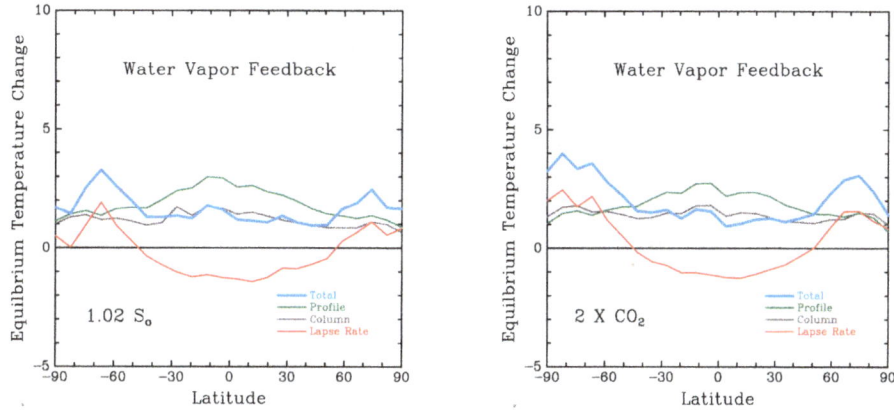

Annual-mean water vapor feedback components for 2% increase in solar irradiance and doubled atmospheric CO_2.

Slide 16

Water Vapor and Cloud Feedback Strength in 1984 Climate Sensitivity Study

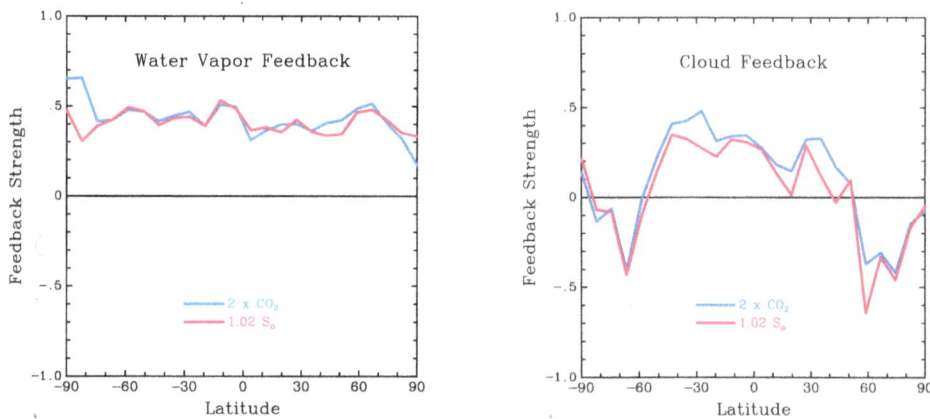

Water vapor feedback strength.　　Cloud feedback strength (or efficiency).

Slide 17

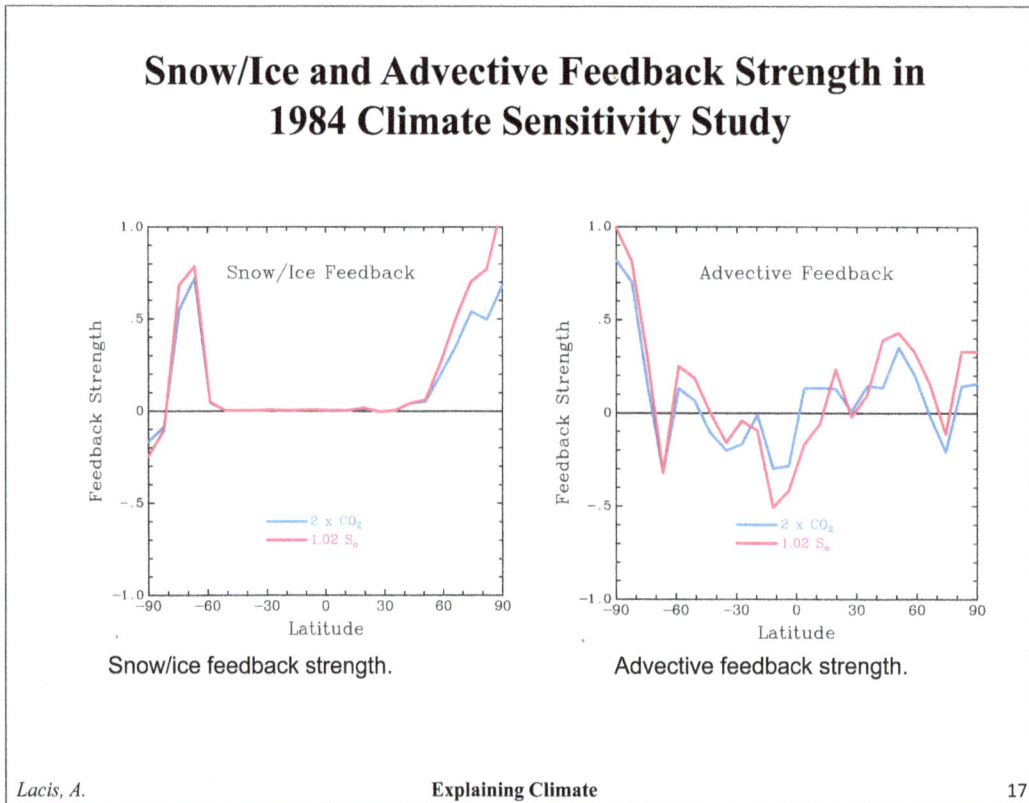

Snow/Ice and Advective Feedback Strength in 1984 Climate Sensitivity Study

Snow/ice feedback strength.

Advective feedback strength.

Slide 18

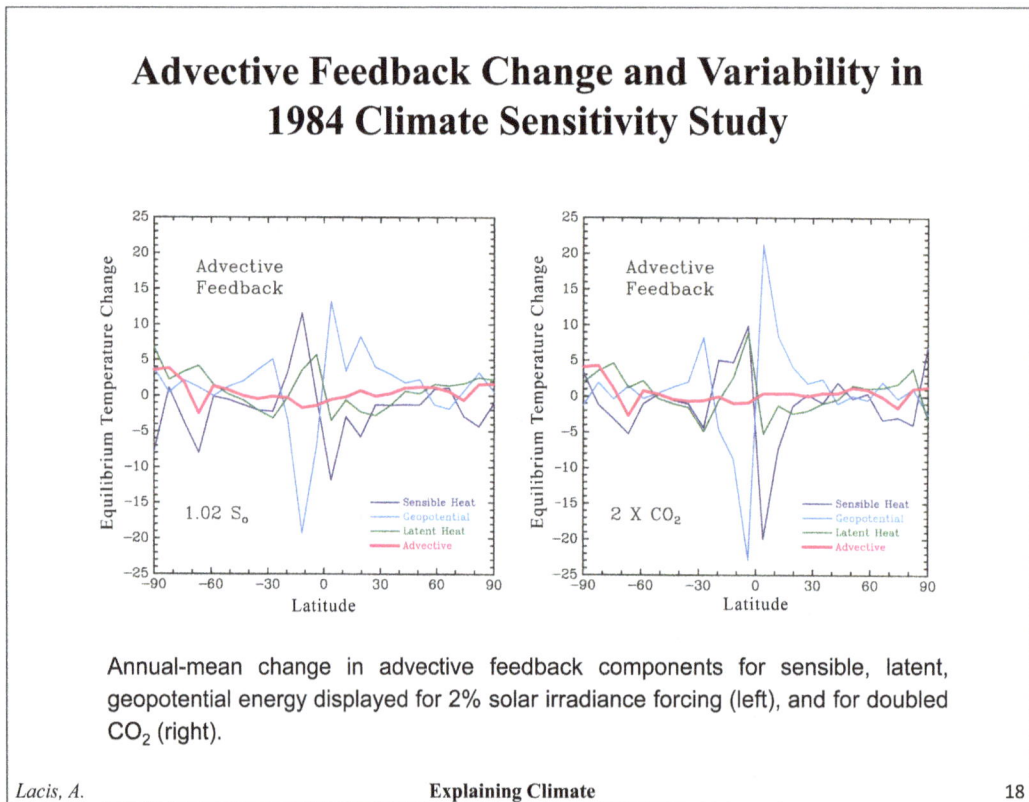

Advective Feedback Change and Variability in 1984 Climate Sensitivity Study

Annual-mean change in advective feedback components for sensible, latent, geopotential energy displayed for 2% solar irradiance forcing (left), and for doubled CO_2 (right).

Slide 19

Terrestrial Greenhouse Effect Flux Attribution

LW Absorber	Single-inclusion W/m²	Fraction	Single-exclusion W/m²	Fraction	Normalized average W/m²	Fraction
H_2O	94.7	0.458	59.8	0.534	74.7	0.490
Cloud	56.2	0.272	23.2	0.207	37.3	0.244
CO_2	40.3	0.195	24.0	0.214	31.0	0.203
O_3	4.0	0.019	1.7	0.015	2.7	0.018
N_2O	4.1	0.020	1.5	0.013	2.6	0.017
CH_4	3.5	0.017	1.2	0.011	2.2	0.014
CFCs	1.0	0.005	0.2	0.002	0.6	0.004
Aerosol	3.0	0.014	0.4	0.004	1.5	0.010
Sum	206.8	1.000	112.0	1.000	152.6	1.000

The greenhouse strength of different atmospheric constituents excluding or including overlaps between them.

Lacis, A.   **Explaining Climate**   19

Slide 20

Solar and Surface Thermal Energy Attribution for 1/8 to 256 × Current CO_2

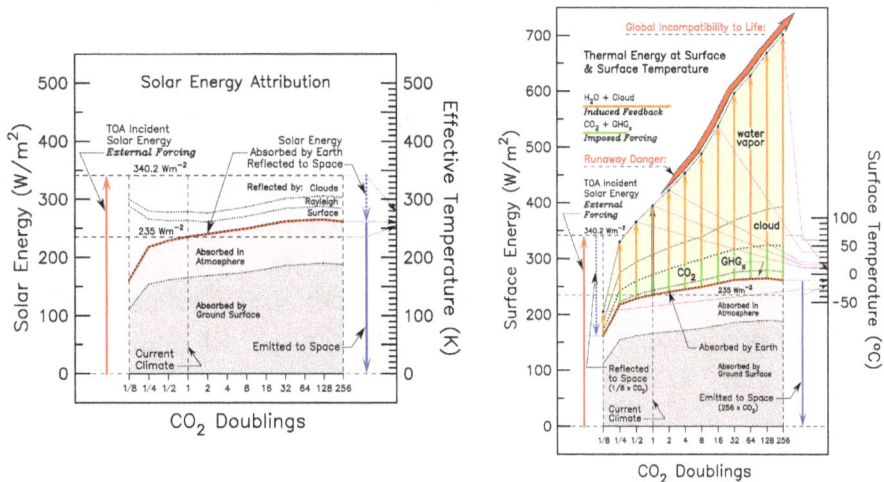

Solar energy attribution, even under extreme climate change, remains reasonably steadfast, and thus there is little expected change in the top-of-the-atmosphere SW and LW radiative fluxes for moderate shifts away from current climate conditions.

Lacis, A.   **Explaining Climate**   20

Slide Notes

Slide 2   The spectral distribution of the incident solar radiation (left-hand panel). The Sun is the ultimate source of energy that sustains the ever moderate temperature structure of the terrestrial climate system. The globally averaged annual-mean incident solar flux is about 340 W/m², of which about 30%, or 100 W/m², is reflected back to space, leaving a net 240 W/m² as the heat energy input to the terrestrial climate system. About 20% (70 W/m²) of the incident solar radiation is absorbed within the atmosphere, primarily by water vapor, but also by ozone, clouds, aerosols, O_2, CO_2 and other minor atmospheric gases. But for the 11-year sun-spot cycle variability of about 0.1%, and the seasonal variation with an amplitude of about 3.4% due to Earth's orbital eccentricity, the Sun has been a remarkably steady source of energy over the entire recorded history.

The spectral distribution of LW thermal energy emitted to space by Earth (right-hand panel). The annual-mean globally averaged LW thermal flux emitted by the ground surface, at temperature 288 K, is depicted by the solid red line. The black border of the blue area depicts the spectral distribution of the outgoing LW thermal radiation that is emitted out to space at the top-of-the-atmosphere (TOA). At global energy balance equilibrium, the 240 W/m² LW flux at TOA matches the 240 W/m² absorbed SW solar radiation. The LW flux difference (150 W/m²) between the upwelling LW flux at ground surface and outgoing LW flux at TOA is a measure of the strength of the terrestrial greenhouse effect. The greenhouse effect area (depicted in green, right) represents a sequence of absorption and re-emission events are occurring at every wavelength throughout the atmosphere.

Slide 3   Conservation of energy is the basic physics principle behind the greenhouse effect. It was well understood by Joseph Fourier in 1824, and was described mathematically by Arrhenius in 1896. The simplest example has 240 W/m² of SW energy absorbed by the ground surface. For SW–LW energy balance equilibrium, 240 W/m² of LW energy must be radiated back to space. At top left, a single layer, spectrally gray, optically exceedingly thin atmosphere absorbs a tiny amount of LW energy, and must radiate half of that absorbed energy upward and half downward. At temperature of 214 K, the atmospheric layer is in radiative equilibrium with the ground surface. As the layer's optical depth is increased, its temperature will rise. For given optical depth $\tau_a = 1.47$, the layer will absorb $[1 - e^{-1.47}]$ of the upwelling LW flux. In radiative equilibrium, the layer will emit 150 W/m² to space and to the ground, and transmit 90 W/m² from the ground directly to space. For large optical depth, the layer will emit 240 W/m² to space and to the ground. The ground will then emit 480 W/m² (at 303 K), and it will exhibit the maximum single-layer greenhouse effect of $G_F = 240$ W/m² (48 K). Adding more such layers adds additional 240 W/m² increments to G_F.

Slide 4   GCM radiative flux and cooling rate comparison to line-by-line calculations for mid-latitude atmosphere. Utilizing spectrally gray and isothermal layers, while useful for illustrative purposes, is not representative of the real atmosphere where the absorbing gas spectral characteristics, the distribution of absorbers, and the vertical temperature profile must be explicitly resolved. For example, the stratosphere cannot cool while the surface warms in a spectrally gray atmosphere when atmospheric opacity is increased. In the Earth's atmosphere, the relatively transparent LW window allows the stratosphere to cool while the surface warms as atmospheric CO_2 is increased.

In all panels, the reference line-by-line (LBL) calculations are depicted in red. GISS ModelE calculations using the correlated k-distribution methodology are plotted in black. The left-hand panel shows very close agreement for downwelling and upwelling fluxes. Cooling rates are shown in the middle panel, with flux differences plotted at right. It is shown that the 33 spectrally non-contiguous correlated k-distribution intervals in the GCM radiation model are able to closely reproduce LBL calculation results that utilize over 10^7 spectral points. For further comparison, the dotted lines depict the corresponding results for an earlier 25-k interval GCM radiation model which was used in the climate forcing and sensitivity experiments described by Hansen et al. (1984). GCM radiation modeling must be fast, yet accurate. LBL results serve as benchmark reference for GCM radiation verification.

Slide 5   Line-by-line calculations of CO_2 spectral absorption and radiative forcing are the basic means of GCM radiation model validation. The top-left panel shows LW downward flux and the top-right panel shows the net upward LW flux for 338 ppm CO_2 concentration. The bottom-left panel depicts the LW net flux change for doubled CO_2. Radiative forcing is exerted throughout the atmosphere, but it is normally referenced at the tropopause.

Slide 6   Hourly GCM output diagnostics depict the 'virtual' forcing of the climate system by instantaneous changes in water vapor. Rapid convergence to equilibrium follows instantaneous doubling and zeroing of atmospheric water vapor. Left-hand panels show the global-mean water vapor at 299 and 974 mb converging to control run equilibrium values. Right-hand panels show up-welling LW flux at top (TOA) and bottom (BOA) of the atmosphere. Diurnal oscillations in global-mean LW flux arise from diurnal surface temperature change over land areas. Red curves depict model response to doubled water vapor amounts. Green curves refer to model response to zeroed water vapor. Blue curves depict control run water vapor reference results. Water vapor changes in the left-hand panels have been normalized relative to the control run results.

Slide 7   Hourly GCM diagnostics of the globally averaged greenhouse effect convergence to equilibrium following the instantaneous doubling (red) and zeroing (green) of the atmospheric water vapor. Diurnal oscillations in the globally averaged greenhouse strength result from land-ocean differences in diurnal surface temperature changes. The convergence to equilibrium is faster for the doubled water vapor experiment since relative humidity above 100% leads to rapid condensation and rainout, whereas humidifying a dry atmosphere first requires the evaporation, then the transport and mixing through the atmosphere. The Clausius–Clapeyron relation is the principal constraint in establishing the equilibrium distribution of atmospheric water vapor. Even as a global average, the greenhouse effect strength responds to the land-ocean asymmetry in diurnal surface temperature change. The results show that water vapor and clouds converge rapidly to their equilibrium distribution (towards which they contribute radiative their feedback, but without explicit control of the equilibrium).

Slide 8 Global maps of seasonal change of the greenhouse effect for 1850 (left panels) and 2000 (right panels). The strength of the greenhouse effect has increased by about 6 W/m² since 1850, with the geographical patterns remaining unchanged. The seasonal change in greenhouse strength (12 W/m²) also remained invariant, being twice as large as the secular trend. The seasonal change in greenhouse strength is largest over land areas, and is affected by changes in surface temperature and seasonal shifts in water vapor and cloud distributions. The greenhouse strength is near zero in the polar regions, with negative values occurring over Antarctica during the winter season due to large local temperature inversions. The seasonal change in solar radiative-forcing controls the fast-feedback water vapor and cloud GH response.

Slide 9 The time trend of the terrestrial greenhouse effect from 1850 to 2010 from a 5-run ensemble average (depicted by the heavy red line) calculated using the GISS 40-layer 2°‐2.5° ModelE coupled atmosphere-ocean model. This model utilized a 3-D 26-layer ocean model that can generate El Niño-like inter-annual variability. Also evident in the greenhouse effect time trend are the radiative-forcing effects of major volcanic eruptions, which are denoted by the heavy black arrows for the major volcanoes (Shiveluch, Krakatoa, Santa Maria, Agung, El Chichon, Pinatubo). It is the continuing rise in greenhouse gases and the induced changes in atmospheric water vapor and surface temperature that account for the growing strength of the greenhouse effect. Surface cooling due to the major volcanic eruptions is sufficiently large to noticeably reduce the rising strength of the greenhouse effect. The general shape of the greenhouse strength has a strong resemblance to the time trend increase of the non-condensing greenhouse gases over the same time period. Defined as the difference between the BOA and TOA LW fluxes, the growing strength of the greenhouse effect is verifiable from observational data.

Slide 10 Observed global annual-mean trend in surface air temperature relative to the base period 1951–1980, based on the analysis of traditional meteorological station data from the meteorological station network. The red curve represents a 5-year running mean. Uncertainty bars (shown in green for both the annual and 5-year means) indicate 95% confidence limits to account for incomplete spatial sampling of data. The radiative forcings that account for most of the rise in global surface temperature change are due to the increases in atmospheric greenhouse gases, which are accurately known and monitored. There are smaller lesser known radiative forcings due to changes in tropospheric aerosols and land use. Much of the short-term variability is dominated by simulated El Niño and La Niña changes in ocean-surface temperature. Also present are global cooling events lasting for several years due to large volcanic eruptions. There is also the decadal time-scale variability associated with Pacific Decadal Oscillation and Atlantic Multi-decadal Oscillation events. It is important to understand that the observed change in global surface temperature is composed of two separate components – a steadily increasing global warming component that is driven primarily by the increase in anthropogenic greenhouse gases, superimposed upon which is a natural variability component operating on a broad range of time scales about the climate system's evolving equilibrium zero-reference point.

Slide 11   The principal source of radiative forcing over the industrial era has been the increase in the well-mixed non-condensing greenhouse gases (red curve, left panel), due primarily to human industrial activities. The non-condensing greenhouse gases have a long residence time in the atmosphere. Once injected into the atmosphere, these gases remain there virtually indefinitely while providing a steady source of radiative forcing. The right-hand pie chart shows CO_2 at 52% as the largest component of the non-condensing radiative-forcing gases, with methane second at 19%, ozone at 14%, CFCs at 10%, and nitrous oxide at 5%. There is a small negative forcing (green) attributable to anthropogenic land-use changes, and to a larger negative forcing attributed to human-made aerosols. Of the natural forcings, the change in solar irradiance is characterized by the 11-year solar sunspot cycle (orange), and by the episodic occurrence of large volcanic eruptions (gray).

At about 240 W/m² of absorbed energy, the Sun is the ultimate source of energy. There are also non-solar energy inputs to the climate system. These are geothermal, waste heat (e.g., nuclear, fossil fuel combustion), and tidal energy. Globally averaged, geothermal energy, at 0.092 W/m², is the largest of the non-solar energy sources. Waste heat generated by nuclear energy and the burning of fossil fuel contributes 0.028 W/m², while tidal energy contributes merely 0.006 W/m². While significant at a local level as sources of heat energy, they are negligible when compared to the uncertainty in the 240 W/m² of global solar radiation input.

Slide 12   Time trend of atmospheric CO_2 concentration observed at Mauna Loa (Hawaii) and the South Pole (left scale) with a corresponding decrease in atmospheric oxygen (right scale, deviation from reference amount) observed at the stations Cape Grim (Tasmania) and Alert (Canada) since 1990. Seasonal cycles of both gases reflect the seasonal biosphere activity occurring on land and in the ocean. A smaller source of anthropogenic CO_2 is associated with changes in land use and land management, foremost deforestation. During the conversion of forested areas to pasture and to agricultural fields, the organic carbon that is stored in vegetation is largely removed, burned in part and released as CO_2 into the atmosphere. Also freshly exposed soils tend to lose carbon by oxidation to the atmosphere. Only about 40% of the fossil fuel-related carbon accumulates in the atmosphere, the other 60% is taken up by the ocean and by the land biosphere. The relative distribution of CO_2 among the ocean, atmosphere, and biosphere reservoirs is governed by carbon cycle processes.

Slide 13   Geological record of climate forcings, temperature, sea level, and surface albedo response. **(a)** CO_2 and CH_4 data are derived from the Antarctic Dome C ice core analysis, sea-level record based on the analysis of Bintanja *et al.* (2005). **(b)** Greenhouse gas and surface albedo forcing are from GCM modeling studies, with ice-sheet area inferred from sea-level changes. **(c)** Dome C-derived tempera-ture change is divided by two to represent global-mean temperature change. Calculated temperature change is based on a fast-feedback climate sensitivity of 0.75°C per W/m², as per 3°C for doubled CO_2 (after Hansen *et al.*, 2008). On this inter-glacial time scale, the change in atmospheric CO_2 is driven primarily by ocean–biosphere interactions. However, the accompanying change in global temperature and sea-level responds directly and more rapidly to the ambient atmospheric CO_2 concentration, thus confirming the climate control nature of atmospheric CO_2.

Slide 14   The 1984 Hansen *et al.* GCM climate sensitivity study was a defining moment in quantitative climate feedback analysis with climate forcing experiments per-formed for a 2% increase in solar irradiance, and doubled CO_2. Both experiments produced 4 K increases in the global-mean surface temperature. Zonal-mean annually averaged model diagnostics were accumulated over the last 10 years of the experiment and control run output results, including the latitude depend-ence of atmospheric profiles of temperature, water vapor, and clouds, surface temperature and surface albedo, as well as the horizontal transports of latent, sensible and geopotential energy. Together with the annually averaged zonal-mean incident solar energy and the LW fluxes emitted to space, this provided the necessary input to apply the GCM radiation model to perform 2-D radiative–convective equilibrium calculations to reproduce the GCM surface temperature changes and enable attribution for individual radiative component contributors to the surface temperature change.

In both panels, the heavy black lines depict GCM zonal surface temperature changes for the two experiments, with light green and orange lines depicting the 2-D reconstructions, summed individually, and with all constituent changes computed simultaneously. The heavy red lines depict the radiative forcings. The other curves depict latitude-dependent feedback contributions by water vapor, snow/ice, clouds, and advection, as described in the following slide.

Slide 15   Annual-mean water vapor feedback components for 2% increase in solar irradi-ance and doubled atmospheric CO_2. For both experiments, the 4 K increase in equilibrium surface temperature generated an increase in column water vapor by about 33%. 2-D radiative–convective equilibrium calculations show the radiative impact of this water vapor increase on the zonal surface temperature, as depicted by the gray lines. The increase in atmospheric water vapor as surface and atmospheric temperature warm is an expected result as a direct consequence of the Clausius–Clapeyron relation and its exponential temperature dependence. Moreover, the added water vapor carries with it latent heat energy that is released upon condensation to warm the higher levels of the troposphere, causing a pro-portionately greater increase in water vapor amount with altitude. And, since the greenhouse effect is proportional to the temperature differential between the ground surface and the temperature of the radiating constituent, the vertical redistribution of water vapor enhances the water vapor greenhouse efficiency (green lines). Higher altitude warming also decreases the atmospheric lapse rate, which causes a negative feedback (red lines), except in polar regions. The three components illustrate how water vapor feedback materializes, not only by the increase in water vapor concentration, but also through its vertical distribution, and by its effect on the temperature lapse rate. The overall water vapor feedback is represented in its combined form by the heavy blue lines.

Slide 16 Water vapor feedback strength (left-hand panel). When expressed as a feedback efficiency (i.e., the ratio of the zonal surface temperature change attributed only to the water vapor increment, relative to the total zonal surface temperature change due to all contributors), the water vapor feedback has the same basic variability with latitude for both the CO_2 and solar forcings. This is a clear indication that the water vapor feedback response is governed primarily through the temperature-dependent constraints of the Clausius–Clapeyron relation, and that the relative interaction of evaporation, transport, and condensation processes remains basically unchanged between the two different types of radiative forcing for the two different climate equilibrium states.

Cloud feedback strength (or efficiency) for doubled CO_2 and 2% solar irradiance forcing (right-hand panel). Cloud feedback is complex in that high clouds (cirrus) contribute strongly to the greenhouse effect (positive feedback), while low clouds contribute little to the greenhouse effect, but are strong reflectors of solar radiation (negative feedback effect). Since GCM diagnostics did not provide sufficiently clear separation by cloud type, the total cloud feedback response is presented, depicting positive cloud feedback from mid to low latitudes, and negative in the polar regions with strong latitudinal similarity for both CO_2 and solar forcings. The overall cloud feedback is characterized by a small decrease in low-level cloud cover along with a small increase in high clouds.

Slide 17 Snow/ice feedback strength (left-hand panel). Snow/ice feedback is expressed as a feedback efficiency (i.e., the ratio of the zonal surface temperature change attributed only to snow/ice albedo changes, relative to the total zonal surface temperature change due to all contributors). Snow/ice feedback is strongly positive and is associated primarily with changes in sea ice, and thus effectively restricted to the polar regions. Interestingly, the snow/ice feedback is negative over Antarctica where climate warming induced more snow, and no melting. Like the water vapor and cloud feedbacks, the snow/ice feedback behaves similarly for CO_2 and solar forcings.

Advective feedback for doubled CO_2 and 2% solar irradiance forcing is displayed in the right-hand panel. Advective feedback arises as a result of changing meridional transport of sensible, latent, and geopotential energy as the global climate system responds to forcing, producing latitudinal changes in energy deposition at different latitudes by the atmospheric dynamical processes. Note that the horizontal energy transports must by definition add to zero when averaged globally (the global-mean advective feedback is zero). The latitudinal changes in convergence and divergence of advected energy are seen to be of comparable strength to the radiative feedback contributions, and exhibit similar zonal variability for both the CO_2 and solar forcings, tracking each other closely even though the forcings have a significantly different dependence on latitude.

Slide 18   Annual-mean change in advective feedback components for sensible, latent, geo-potential energy displayed for 2% solar irradiance forcing (left-hand panel), and for doubled CO_2 (right-hand panel). The advective feedbacks depict the change in annual-mean energy that is delivered and deposited at different latitudes by the atmospheric dynamical processes as global climate evolves from its reference state to its new equilibrium. The advected energy is converted to an equivalent change in surface temperature by 2-D radiative–convective calculations. Compared to radiative feedback contribution to the surface temperature change, the individual advective feedback contributions are an order of magnitude larger. However, to a large extent the advective feedbacks cancel each other, resulting in the net advective feedback depicted by the heavy red line. This compensation is accounted for in the transport of 'moist static energy', which includes all three of these energy transport terms. There is remarkable latitudinal similarity in the advective feedback, even though the solar and CO_2 forcings have a significantly different latitude dependence. There is more excess SW radiation absorbed in the tropics, and more LW radiation radiated to space in the polar regions – hence the need for pole-ward energy transport. Since the warmer atmosphere contains more water vapor, there is more latent heat energy to transport, even though the reduced equator-to-pole temperature gradient implies a reduced pole-ward transport of sensible heat. Taken together, these results emphasize the interactive nature of dynamic and radiative feedback processes.

Slide 19   The greenhouse strength is defined as the LW flux difference between the bottom (BOA) and top (TOA) of the atmosphere, shown here with $G_F = $ **152.6** W/m^2 for a sample GISS GCM 1980 atmosphere. The general-purpose GCM radiation model can calculate the radiative fluxes for all of the radiative contributors acting together, or individually, while completely accounting for the spectral and vertical overlap of absorption. In the left-hand (single-inclusion) columns, each radiative constituent is evaluated individually one-by-one. By excluding the spectral overlap of absorption, the flux difference sum (**206.8** W/m^2) is over-subscribed. In the central (single-exclusion) columns, each radiative constituent is subtracted one-by-one from the complete atmospheric composition. In this case, the flux difference sum (**112.0** W/m^2) is under-subscribed. In both cases, the fractional flux changes are similar, though not identical, indicating that there is also vertical, not just spectral overlap of absorption, The overall GH flux attribution is determined by a closeness weighted fit of the normalized single-inclusion and single-exclusion LW flux evaluations. In round numbers, current climate greenhouse strength is 75% due only to feedback effects and 25% due to radiative forcings. Water vapor accounts for 50% of the greenhouse strength, clouds 25%, CO_2 20%, with the minor GHGs comprising about 5%. Thus, if the overall temperature and absorber structure of the atmosphere were to remain proportionately invariant, as the climate system responds to radiative forcings due to an injection of greenhouse gases into the atmosphere, the initial greenhouse gas radiative forcing will be augmented by a threefold contribution to the strength of the greenhouse effect by water vapor and cloud feedback contributions.

Slide 20 Solar energy attribution, even under extreme climate change, remains reasonably steadfast (except under snowball-Earth conditions), with roughly 30% reflected (mostly by clouds and Rayleigh scattering), with 20% absorbed by the atmosphere, and 50% by the ground surface. Thus, there is little expected change in the Top-of-the-Atmosphere SW and LW radiative fluxes for moderate shifts away from current climate conditions. But the LW energy at ground surface (set by surface temperature) is a more definitive indicator of climate change. With the absorbed SW radiation remaining relatively unchanged near 240 W/m^2, the LW energy at ground surface will increase due to the GHG forcing and water vapor feedback effects. In the right-hand panel, the green section represents the radiative forcing contributed by the non-condensing GHGs and CO_2, while the yellow section depicts the feedback contributions due to water vapor and clouds. This makes it very clear that water vapor is the single largest feedback contributor to the global surface warming. This demonstrates the exponential temperature dependence of the water vapor feedback effect, as imposed by the Clausius–Clapeyron relation.

CLIMATE LECTURE 2

Global Change in Earth's Atmosphere: Natural and Anthropogenic Factors

Judith L. Lean

Space Science Division, Naval Research Laboratory, Washington DC, USA

Judith Lean is the Senior Scientist for Sun-Earth System Research at the U.S. Naval Research Laboratory. Her research focuses on the mechanisms, measurements, modeling, and forecasting of variations in the Sun's radiative output at all wavelengths, and responses to this variability of the Earth's global climate, middle atmosphere, and space climate and weather.

Among many challenges, climate change research seeks to understand:

- how and why Earth's atmosphere is changing,
- the extent to which observed changes are the consequence of human activity, such as the emission of carbon dioxide and other greenhouse gases, or of natural variations driven by, for example, the Sun or volcanoes,
- why Earth's surface warmed barely, if at all, in the last decade,
- why the atmosphere just 20 km above the surface is cooling instead of warming,
- how Earth's temperature might evolve in the 21st century,
- when — and whether — the ozone layer will recover from its two-decade decline due to chlorofluorocarbon depletion.

David Rind's research has advanced the scientific basis of each and all these topics, in unique, creative, and diverse ways. For more than three decades, he has developed, utilized, validated, and interpreted numerical models that parameterize Earth's physical processes and their responses to changing natural and anthropogenic influences. In particular, David's trademark curiosity and open mindedness motivated him to participate in the early 1990s in an NRC study of the Sun's Role in Global change (NRC, 1994), at a time of considerable skepticism about the reality of such an effect. After 20 years of subsequent research, David is the world's foremost expert on modeling and interpreting the Sun's influence on global change in the past, present, and future, at the surface and throughout the atmosphere (Balachandran and Rind, 1995; Rind and Balachandran, 1995; Lean and Rind, 1998, 1999, 2001; Shindell *et al.*, 1999; Rind, 2002; Rind *et al.*, 1999, 2004, 2008, 2014a).

Observations of the Earth made during the last three decades elucidate how the environment is changing: globally, the surface, lower atmosphere and upper ocean are warming, the middle atmosphere is cooling and the ozone layer possibly recovering from a two-decade long decline. Measurements of Earth's surface temperature indicate that the trend in the last three decades is part of a larger warming trend of 0.75°C during the past century, a warming that has proceeded more rapidly than at any time during the past millennium (Bradley and Jones, 1995; Jones *et al.*, 1998; Mann *et al.*, 1998; Marcott *et al.*, 2013). A controversial suggestion is that anthropogenic-driven climate change may have actually begun 10,000 years ago with the introduction of agriculture, and that this 'anthropocene' warming may have alleviated orbit-related 'Milankovitch' cooling (Ruddiman, 2010).

Associations of climate change with changing greenhouse gas concentrations are well established in paleo and contemporary climate change research. But at the same time that greenhouse gases were rising from pre-industrial levels of 270 ppm to current values near 400 ppm, the Sun's activity was rebounding from the 17th century Maunder minimum of anomalously low activity, leading to questions about the possible contribution of speculated solar irradiance changes to global warming (e.g., Rind *et al.*, 1999; Robertson *et al.*, 2001).

Variations in the Sun's radiative output, like changing greenhouse gas concentrations (and other radiative forcings), disrupt Earth's energy balance, forcing a new equilibrium temperature that causes climate to change. The Sun provides radiant 'short' wavelength energy at near UV, visible, and near IR wavelengths that heats the Earth, which then radiates part of this energy back to space, at 'long' wavelengths near 10 µm (Lean, 2005). Greenhouse gases in the atmosphere warm Earth's surface further because they trap infrared energy (Trenberth *et al.*, 2009). Aerosols cool Earth's surface because they reflect incoming solar radiation, and orbital motions (which act on much longer time scales) alter both the distance of the Earth from the Sun and its orientation to incoming radiation. Whether from changing solar radiation, volcanic aerosols, anthropogenic gases or Earth's orbital motions, David's achievements have advanced understanding of dynamical responses of the entire Earth system to radiative forcing on all time scales (e.g., Rind 2002, 2008; Rind *et al.*, 1998, 2014b).

There are many causes of climate change (IPCC AR5, 2013). Natural influences include changes in the Sun's brightness (Kopp and Lean, 2011), episodic injections of volcanic aerosols into the atmosphere, as well as semi-regular fluctuations associated with the El Niño Southern Oscillation (ENSO) near the surface and the Quasi Biennial Oscillation in the atmosphere. Anthropogenic influences include increasing concentrations of atmospheric greenhouse gases, especially CO_2, industrial aerosols, ozone-depleting substances (ODSs, primarily chlorofluorocarbons), and land-use changes. Changes in radiative forcing initiate radiative and dynamical feedbacks, each with characteristic latitudinal and altitudinal patterns. Radiative feedbacks involve changes in water vapor, clouds, and sea ice. Dynamical feedbacks involve redistribution of energy at the Earth's surface and in the atmosphere (e.g., Rind, 2008) that can alter, for example, the Hadley and Walker circulations (see, e.g., Hartmann, 1994).

Models that appropriately combine natural and anthropogenic influences, determined statistically from the measurements, reproduce both the multi-year fluctuations and the overall trends in temperature at the surface and in the atmosphere in recent decades (Lean and

Rind, 2008), successfully capturing the lull in surface warming from 2001 to 2010. On time scales of a few years, El Niños and volcanic eruptions produce episodic global surface warming of as much as 0.3°C; a 'super' El Niño in 1998 caused that year to be the warmest on record until recently, whereas the El Chichón volcanic eruption mitigated the impact of a comparable El Niño event in 1982. Also extracted from the measurements are changes of the order of 0.1°C at the surface due to increasing solar brightness from the minimum to maximum of the Sun's 11-year activity cycle.

Statistical analysis of observations suggests that a combination of more La Niña (rather than El Niño) conditions and declining solar brightness countered much of the anthropogenic warming in the decade from 2001 to 2010. So in terms of its statistical representation, the climate of the past decade is not exceptional at all. However, physical climate models are generally unable to reproduce climate change on decadal time scales, including during the past decade, leading to the suggestion of a recent 'hiatus' in global warming. Simulations made with the Community Climate System Model (CCSM4) model (Meehl *et al.*, 2012), for example, overestimate both the warming due to increasing greenhouse gases and cooling due to aerosols, possibly underestimate the influence of the solar irradiance cycle, and do not reproduce observed El Niño variability. This suggests that as the physical models improve their ability to more realistically simulate climate's responses to shorter term natural processes, especially volcanic aerosol loading, the solar irradiance cycle and ENSO fluctuations, so too will their decadal-scale simulations and forecasts improve. This should lead to greater appreciation of episodes such as the so-called hiatus as normal climate fluctuations, rather than a supposed stalling of anthropogenic global warming.

The natural and anthropogenic factors that simultaneously change Earth's surface and atmosphere each have distinct temporal, geographical, and altitudinal signatures. Regional responses to individual influences can also be extracted statistically from observational datasets and compared with simulations made using physical general circulation models (such as NASA GISS Model 3, Rind *et al.*, 2008). Such comparisons facilitate quantitative cross-validation of the physical models and observations. Both the statistical and physical model simulations indicate overall warming in response to increasing solar irradiance and greenhouse gas concentrations, especially at mid- to high northern latitudes, and cooling in response to volcanic activity. There are, however, distinct regional differences, such as the larger solar-induced warming over the USA and larger volcano-induced cooling over northern Eurasia in Model 3. Since a pattern of longitudinal warming and cooling at mid- to high northern latitudes is typically associated with the North Atlantic Oscillation, these differences may reflect dynamical parameterizations in the physical model. Knowledge of the regional responses of climate to individual forcings can also help interpret paleo time series by providing a global framework for reconciling multiple, seemingly disparate, datasets obtained at individual locations around the world (Lean, 2010).

In Earth's lower stratosphere, as at the surface, a combination of natural and anthropogenic influences accounts for much of the measured variations in temperature (near 20 km) in recent decades (Lean, 2010), including its distinctly different temporal evolution compared with simultaneous variations at the surface. This is the result of the quite different relative strengths and characteristic regional signatures of individual natural and anthropogenic influences in the atmosphere near 20 km compared with at the surface. And as for surface temperature variations, the statistical and physical models differ somewhat in the relative distributions and strengths of these effects. Whereas increasing greenhouse gases warm the surface, they cool the stratosphere because the much thinner overlying atmosphere allows the heat to escape. NASA GISS Model 3 suggests somewhat larger anthropogenic cooling at high latitudes in the lower stratosphere than does the statistical model. While aerosols injected into the stratosphere during a volcanic eruption cool the surface, they warm the stratosphere. Both Model 3 and the statistical model indicate significant warming at low to mid-latitudes in response to volcanic aerosols but Model 3 suggests that the warming is somewhat larger. Increases in the Sun's brightness warm Earth's atmosphere, throughout, with a cycle of 0.3°C at 20 km due to increasing solar brightness from the minimum to the maximum of the Sun's 11-year activity cycle. Model 3 indicates more solar-related lower stratospheric warming at high northern latitudes and less warming over Antarctica than does the statistical model. Since a key aspect of stratospheric variability is ozone heating by absorption of solar UV radiation, and subsequent dynamical responses, the differences between the physical and statistical models may relate to the specification and parameterization of ozone and its variability in Model 3. The decline in 20-km temperature reflects, as well, contributions of changing ozone concentrations, affected by ozone-depleting substances.

As concentrations of greenhouse gases increase, Earth's surface warms. This is the expectation of physical models that simulate global warming in the range 2–4.5°C for doubled CO_2 concentrations (4 W/m^2 radiative forcing). Uncertainties in model projections reflect incomplete understanding of climate sensitivity (which ranges over a factor of 2, IPCC AR5) as well as in future anthropogenic gas concentrations. The expectation of future warming also follows naturally from the relationship established statistically of a positive anthropogenic influence and global surface warming (Lean and Rind, 2009). A projection using a mid-range anthropogenic scenario suggests that global surface temperatures may rise ~1°C from 2015 to 2095, and 2.3°C for doubled CO_2 concentrations (Lean, 2015), which is quite consistent with the physical model projections. Simulations and projections of regional surface temperature changes are less certain than for global values, in part because of difficulty in modeling latitudinal thermal gradients which control the dynamic motions that produce regional inhomogeneities (Rind, 2008).

Northern hemisphere mid- and high latitudes over land are warming faster than are tropical and southern latitudes, and are likely to warm more in the future. This is evident in the regional patterns of the surface temperature anomalies in 2005 compared with those in 1955, and it is evident in both the statistical projections of the measurements and physical model simulations of surface temperatures in 2095. In simulations of regional surface temperature changes for doubled CO_2 concentrations, made by NASA GISS Model 3, increasing greenhouse gas concentrations produce enhanced warming at high latitudes and over land, consistent with analysis of the measurements. In the North Atlantic region, however, the modeled warming is minimal because a slower Atlantic Meridional Overturning Circulation (AMOC) reduces ocean transport of warm tropical water to higher latitudes. Causes of the AMOC reduction in the models include increased freshwater from melting glaciers and sea ice and precipitation, and changing winds and ocean gyre circulation (Zickfeld *et al.*, 2007). Statistical projections based on the surface temperature measurements foresee more Northern Atlantic warming than do the physical models, although less than in the surrounding continental regions; this emphasizes the importance of better understanding projections of ocean circulation changes.

The projection of global surface temperature that includes a simulated solar brightness cycle modulates the overall warming trajectory very slightly (Lean and Rind, 2009). Should the Sun's activity subside over the next century into a prolonged period of inactivity, such as occurred during the 17th-century Maunder minimum, solar brightness may decrease by more than has been measured in the past three decades. Current understanding from both statistical and physical models nevertheless suggests that the resultant global surface cooling would be less than a few tenths °C, which is an order of magnitude smaller than the projected anthropogenic warming. Of course, it is entirely uncertain whether or not the Sun's activity will subside into a prolonged minimum; analysis of cosmogenic proxies of solar activity indicate that such events occur approximately every 2400 years; since the Maunder minimum occurred in the mid-17th century (less than 400 years ago), this suggests that anomalously low solar activity — a new Maunder minimum — is statistically unlikely for a few thousand years or so (McCracken and Beer, 2007).

Statistical and physical models are also used to project future levels of ozone in the atmosphere, especially the time of its recover from depletion by ODSs following the phasing out of many of these gases. Statistical models of the Merged Ozone Dataset (MOD V8) total ozone observations (McPeters *et al.*, 2007) that account individually for anthropogenic ODSs and GHGs as well as natural influences (Lean, 2014) suggest that global and tropical total ozone may recover to 1980 levels as early as 2025 and increase to levels that in 2100 are the highest in two centuries, in the range 11–29 DU above 1980 levels globally, and 6–18 DU in the tropics (depending on GHG emissions). This increase exceeds the projections by WMO's chemistry climate models (World Meteorological Association [WMO], 2011), according to which total tropical ozone will be 3±3 DU lower (not higher) in 2100 than in 1980 (Lean, 2014). Might chemistry climate models overestimate the impact of GHG-induced changes in the stratosphere, exaggerating trends in tropical upwelling, and over-accelerating the Brewer-Dobson circulation? This scenario is consistent with recent observations that the age of stratospheric air has changed little in the past three decades, even with greenhouse gas forcing of 1 W/m^2 and net global surface warming of 0.3°C from 1980 to 2011 (Engel *et al.*, 2009), and that modeled changes in tropical ozone and upper troposphere temperature in recent decades may overestimate the observed changes (Mitchell *et al.*, 2013). Improvements may therefore be needed in chemistry climate model transport, for example in the parameterizations of dynamic and convective processes that connect the tropical troposphere and stratosphere.

Improved observations in combination with physical model parameterizations, simulations and validation are essential for future progress in specifying and forecasting earth's atmosphere. David Rind's expertise and creativity remains strongly in demand for securing the most rapid and reliable advances in understanding how and why Earth's atmosphere, surface and oceans are changing, globally and regionally.

Based in part on AGU Bjerknes Lecture, San Francisco, December 2013. Support by NASA and ONR is acknowledged.

References

Balachandran, N.K. and Rind, D. (1995). Modeling the effects of UV variability and the QBO on the troposphere–stratosphere system. Part I: The middle atmosphere. *Journal of Climate*, 8, 2058–2079.

Bradley, R.S. and Jones, P.D. (1995). *Climate Since 1500 AD*. London: Routledge.

Engel, A., Möbius, T., Bönisch, H., Schmidt, U., Heinz, R., Levin, I., Atlas, E., Aoki, S., Nakazawa, T., Sugawara, S., Moore, F., Hurst, D., Elkins, J., Schauffler, S., Andrews, A. and Boering, K. (2009). Age of stratospheric air unchanged within uncertainties over the past 30 years. *Nature Geoscience*, 2. doi:10.1038/NGEO388.

Hartmann, D.L. (1994). *Global Physical Climatology. International Geophysics Series*, (Vol. 56). Academic Press.

Intergovernmental Panel on Climate Change (IPCC) (2013). Fifth Assessment Report (AR5), 2013: The Physical Science Basis.

Jones, P.D., Briffa, K.R., Barnett, T.P. and Tett, S.F.B. (1998). High-resolution palaeoclimatic records for the last millennium: interpretation, integration and comparison with General Circulation Model control-run temperatures. *The Holocene*, 8(4), 455–471. doi:10.1191/095968398667194956.

Kopp, G. and Lean, J.L. (2011). A new low value of total solar Irradiance: evidence and climate significance. *Geophysical Research Letters*, 38, L01706, doi:10.1029/2010GL045777.

Lean, J.L. (2005). Living with a Variable Sun. *Physics Today*, June, 32–38.

Lean, J.L. (2010). Cycles and trends in solar Irradiance and climate. *Wiley Interdisciplinary Reviews, Climate Change*, 1. doi:10.1002/wcc.018.

Lean, J.L. (2014). Evolution of total atmospheric ozone from 1900 to 2100 estimated with statistical models. *Journal of Atmospheric Sciences*, 71, 1956–1984, doi:10.1175/JAS-D-13-052.1.

Lean, J. (2015). *Measurements, What and Why? In Climate 2020, Facing the Future*. London: Witan Media Ltd.

Lean, J. and Rind, D. (1998). Climate forcing by changing solar radiation. *Journal of Climate*, 11, 3069–3094.

Lean, J. and Rind, D. (1999). Evaluating Sun–climate relationships since the Little Ice Age. *Journal of Atmospheric and Solar Terrestrial Physics*, 61, 25–36.

Lean, J. and Rind, D. (2001). Earth's response to a variable Sun. *Science*, 292, 234–236.

Lean, J. and Rind, D. (2008). How natural and anthropogenic influences alter global and regional surface temperatures: 1889 to 2006. *Geophysical Research Letters*, 35, L18701. doi:10.1029/2008GL034864.

Lean, J. and Rind, D. (2009). How will Earth's surface temperature change in future decades? *Geophysical Research Letters*, 36, L15708.

Mann, M.E., Bradley, R.S. and Hughes, M. K. (1998). Global-scale temperature patterns and climate forcing over the past six centuries. *Nature*, 392, 779–787. doi:10.1038/33859.

Marcott, S.A., Shakun, J.D., Clark, P.U. and Mix, A.C. (2013). A reconstruction of regional and global temperature for the Past 11,300 Years. *Science*, 339(6124), 1198–1201. doi:10.1126/science.1228026.

McCracken, K.G. and Beer, J. (2007). Long-term changes in the cosmic ray intensity at Earth, 1428–2005. *Journal of Geophysical. Research*, 112, A10101. doi:10.1029/2006JA012117.

McPeters, R.D., Labow, G.J. and Logan, J.A. (2007). Ozone climatological profiles for satellite retrieval algorithms. *Journal of Geophysical Research*, 112, D05308. doi:10.1029/2005JD006823.

Meehl, G.A., Washington, W.M., Arblaster, J.M., Hu, A., Teng, H., Tebaldi, C., Sanderson, B.N., Lamarque, J-F., Conley, A., Strand, W.G. and White III, J.B. (2012). Climate system response to external forcings and climate change projections in CCSM4. *Journal of Climate*, doi:10.1175/JCLI-D-11-00240.1

Mitchell, D.M., Thorne, P.W., Stott, P.A. and Gray, L.J. (2013). Revisiting the controversial issue of tropical tropospheric temperature trends. *Geophysical Research Letters*, 40, 1–6. doi:10.1002/grl.50465.

National Research Council (NRC) and Board on Global Change. (1994). Judith Lean (Chair), David Rind and Working Group Members. *Solar Influences on Global Change*.

Rind, D. (2002). The Sun's role in climate variations. *Science*, 296, 673–677.

Rind, D. (2008). The consequences of not knowing low- and high-latitude climate sensitivity. *BAMS*. doi:10.1175/2007BAMS2520.1.

Rind, D. and Balachandran, N.K. (1995). Modeling the effects of UV variability and the QBO on the troposphere–stratosphere system. Part II: The troposphere. *Journal of Climate*, 8, 2080–2095.

Rind, D., Shindell, D., Lonergan, P. and Balachandran, N.K. (1998). Climate change and the middle atmosphere. Part III: The doubled CO_2 climate revisited. *Journal of Climate*, 11, 876–894.

Rind, D., Lean, J. and Healy, R. (1999). Simulated time-dependent climate response to solar radiative forcing since 1600. *Journal of Geophysical Research*, 104, 1973–1990.

Rind, D., Lonergan, P., Lean, J., Shindell, D., Perlwitz, J. Lerner, J. and McLinden, C. (2004). On the relative importance of solar and anthropogenic forcing of climate change between the Maunder Minimum and the present. *Journal of Climate*, 17, 906–929.

Rind, D., Lean, J., Lerner, J., Lonergan, P., McLinden, C. and Leboissitier, A. (2008). Exploring the stratospheric/tropospheric response to solar forcing. *Journal of Geophysical Research*, 113, D24103. doi:10.1029/2008JD010114.

Rind, D.H., Lean, J.L. and Jonas, J. (2014a). The impact of different absolute solar irradiance values on current climate model simulations. *Journal of Climate*, 27, 1100–1120. doi:10.1175/JCLI-D-13-00136.1.

Rind, D., Jonas, J., Balachandran, N.K., Schmidt, G.A. and Lean, J. (2014b). The QBO in two GISS global climate models: 1. Generation of the QBO. *Journal of Geophysical Research Atmospheres*, 119, 8798–8824. doi:10.1002/2014JD0216781.

Robertson, A., Overpeck, J., Rind, D., Mosley-Thompson, E., Zielinski, G., Lean, J., Koch, D., Penner, J., Tegen, I. and Healy, R. (2001). Hypothesized climate forcing time series for the last 500 years. *Journal Geophysical Research*, 106, 14783–14804.

Ruddiman, W.F. (2010). *Plows, Plagues, and Petroleum: How Humans Took Control of Climate*. Princeton University Press.

Shindell, D., Rind, D., Balachandran, N., Lean, J. and Lonergan, P. (1999). Solar cycle variability, ozone and climate. *Science*, 284, 305–308.

Trenberth, K.E., Fasullo, J.T. and Kiehl, J. (2009). Earth's global energy budget. *Bulletin American Meteorological Society*, 90, 311–323. doi:10.1175/2008BAMS2634.1.

World Meteorological Association (2011). Scientific Assessment of Ozone Depletion, Global Ozone Research and Monitoring Project — Report No. 52 Geneva, Switzerland.

Zickfeld, K., Levermann, A., Morgan, M.G., Kuhlbrodt, T., Rahmstorf, S. and Keith, D.W. (2007). Expert judgments on the response of the Atlantic meridional overturning circulation to climate change. *Climatic Change*, 82/3–4, 235–265.

Slide 1

Slide 2

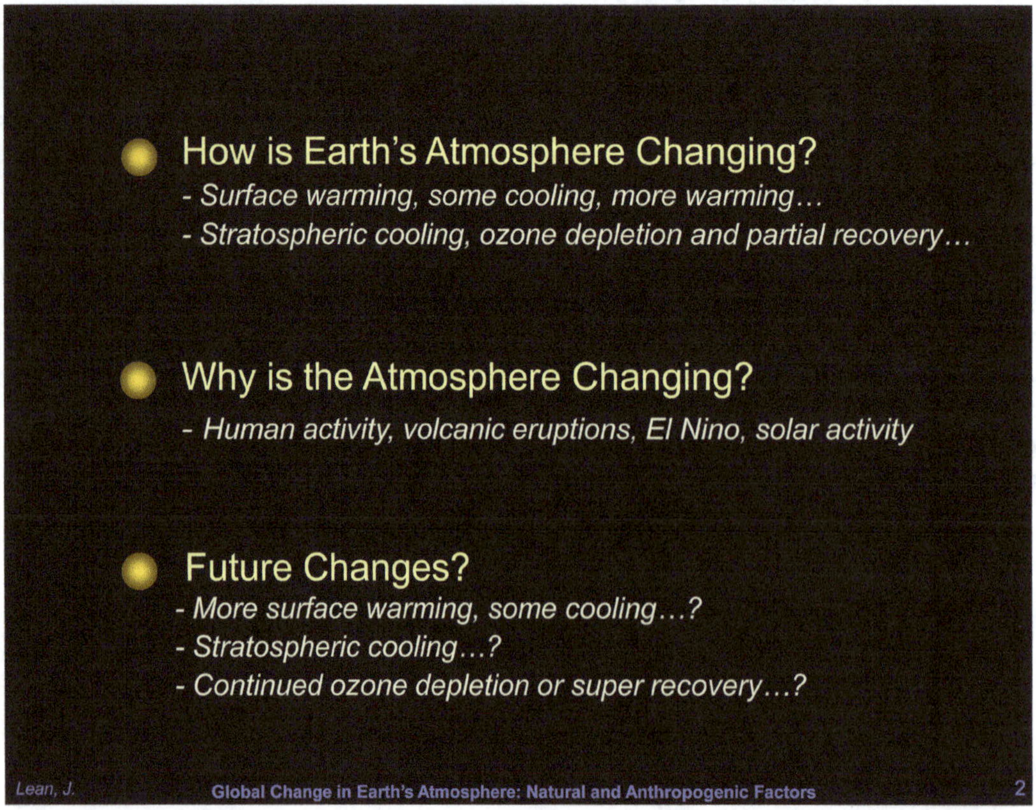

Slide 3

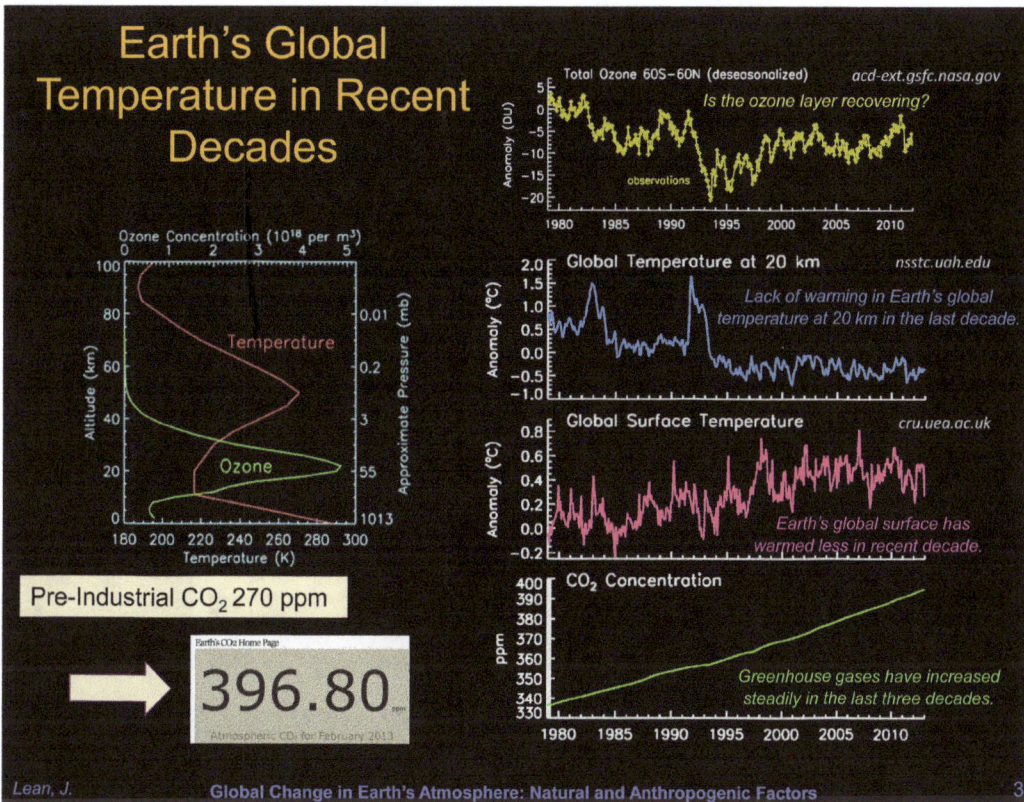

Slide 4

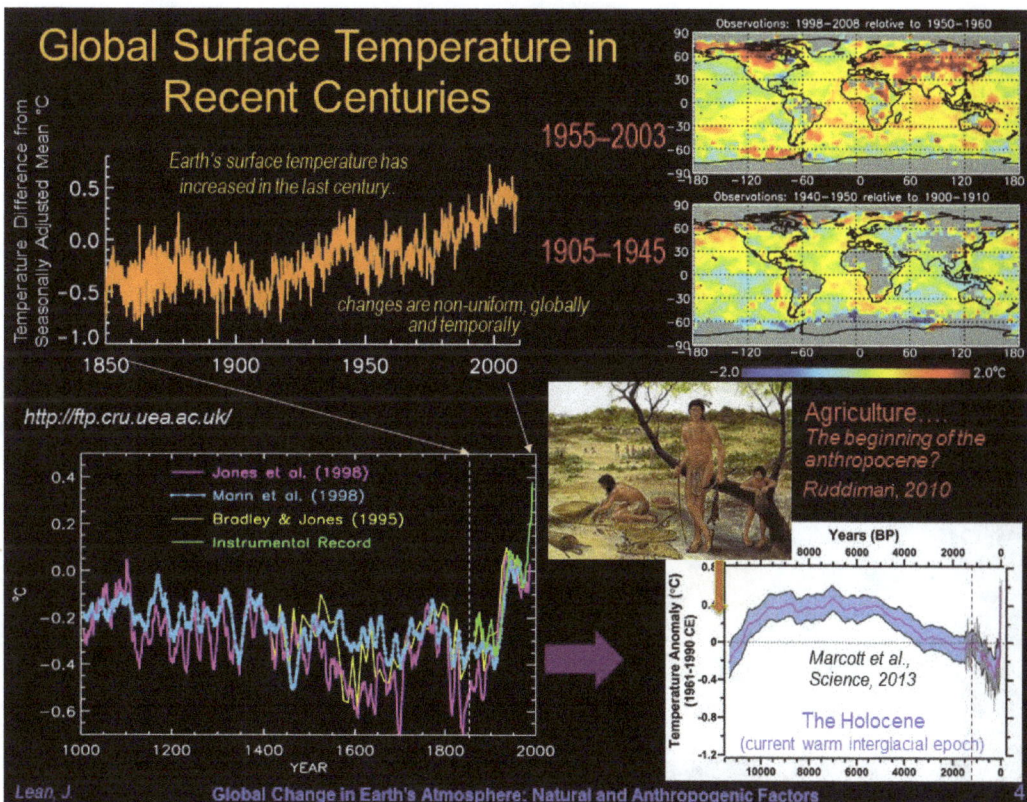

Slide 5

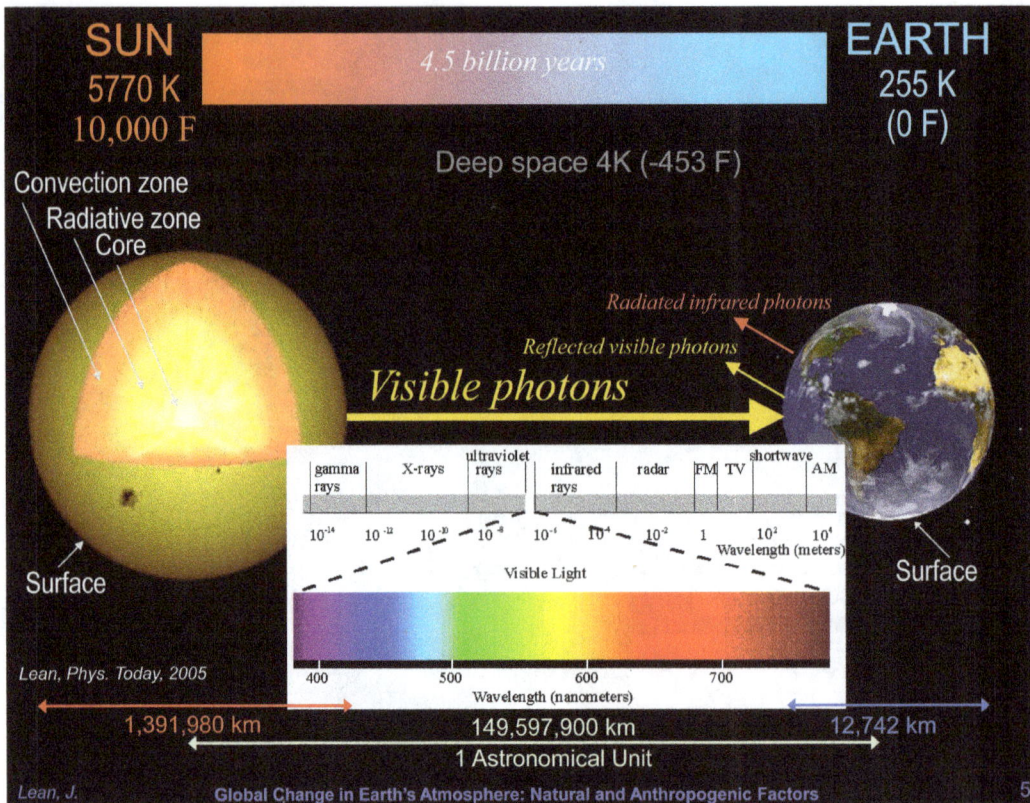

Slide 6

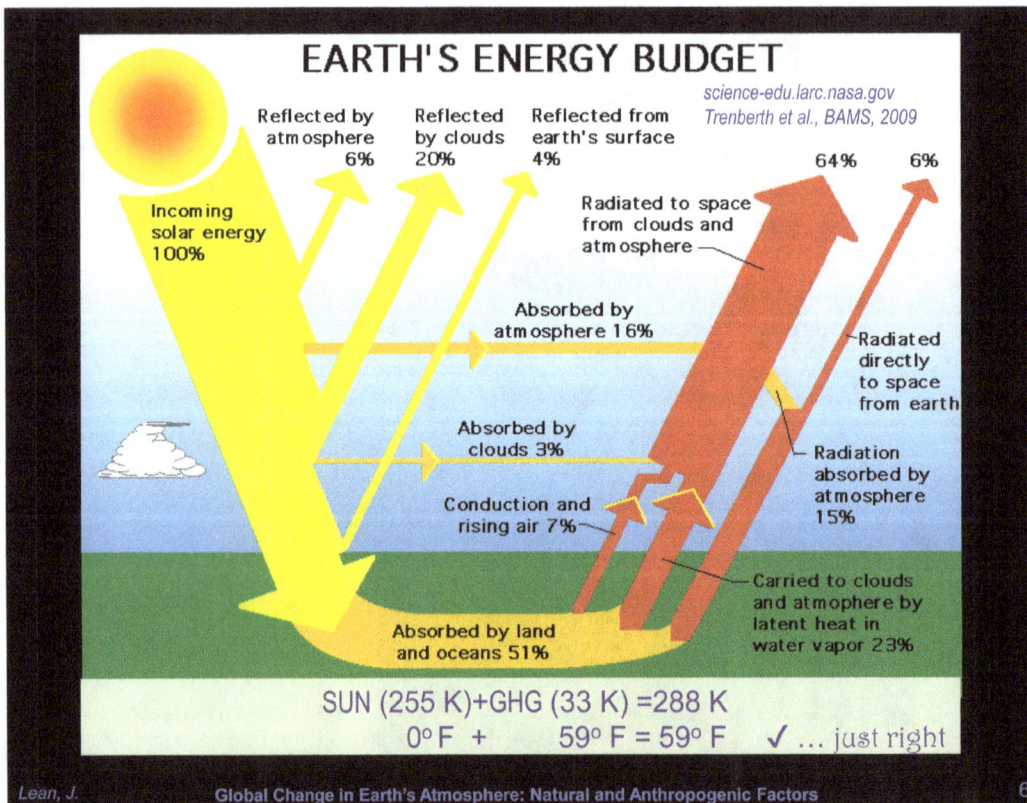

Slide 7

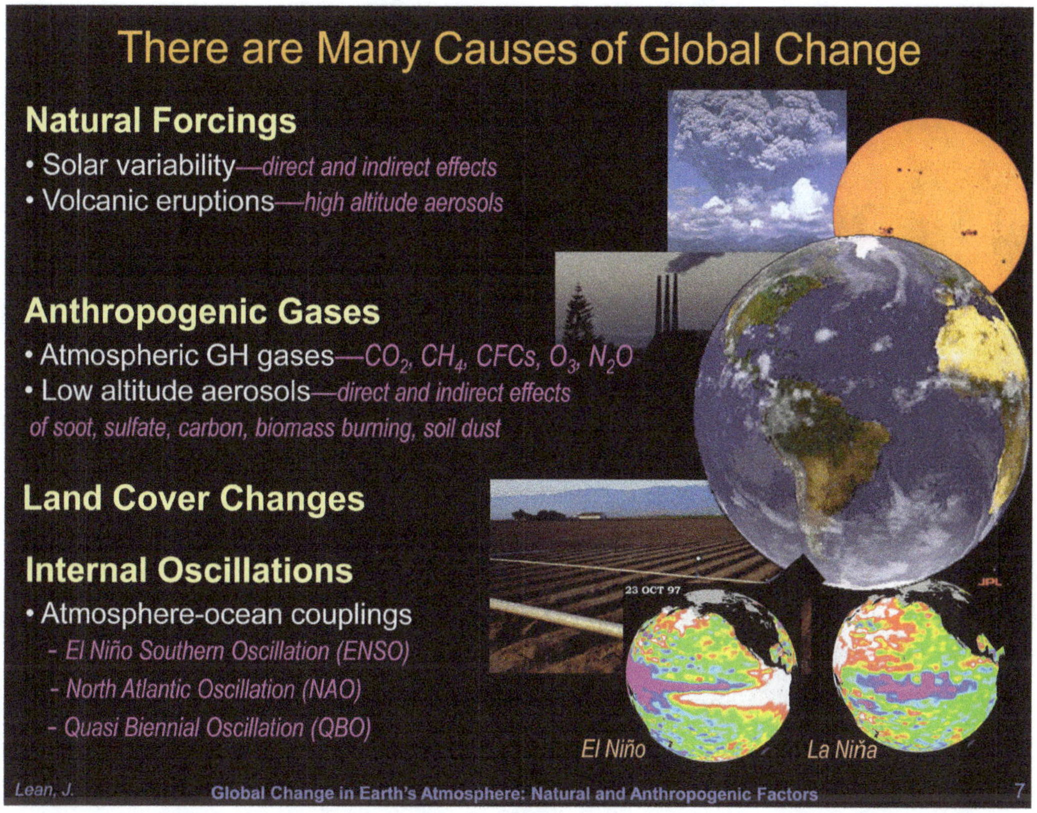

Slide 8

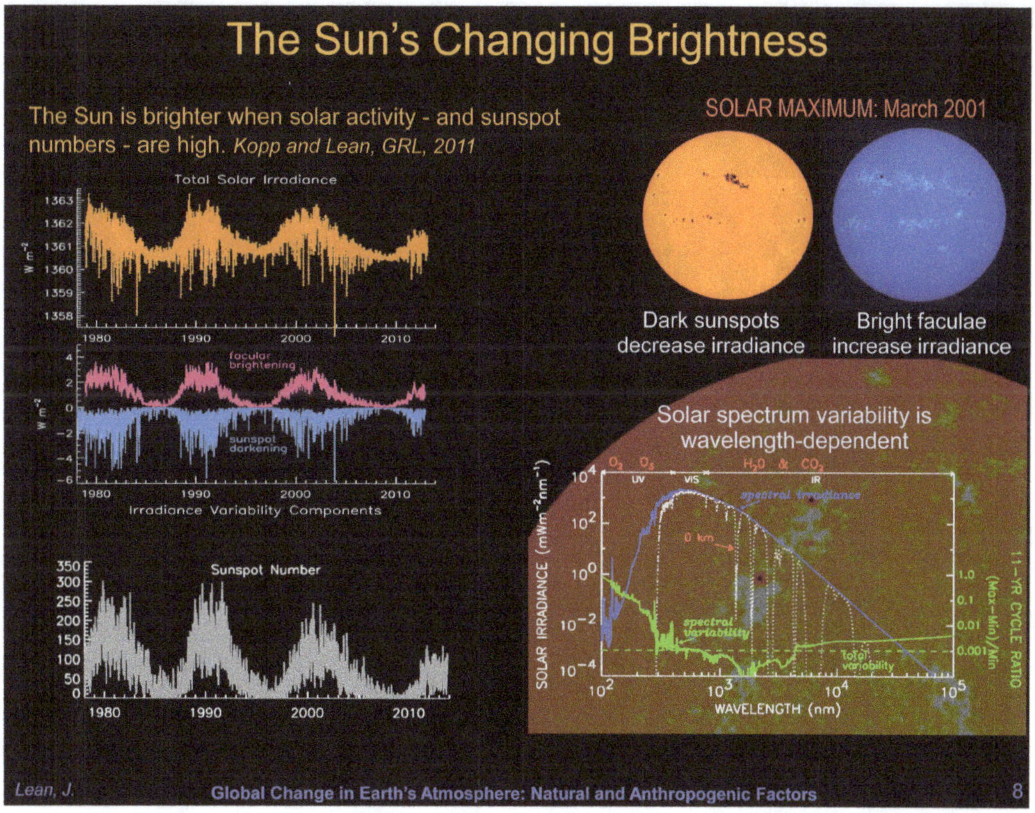

Slide 9

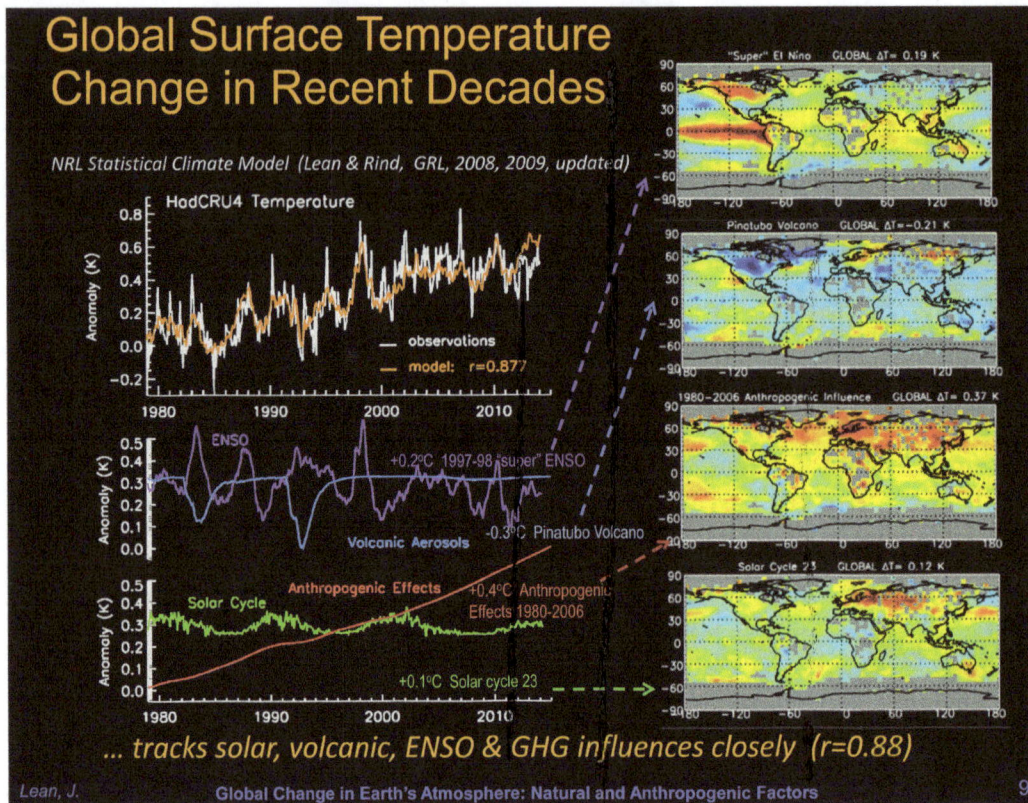

Slide 10

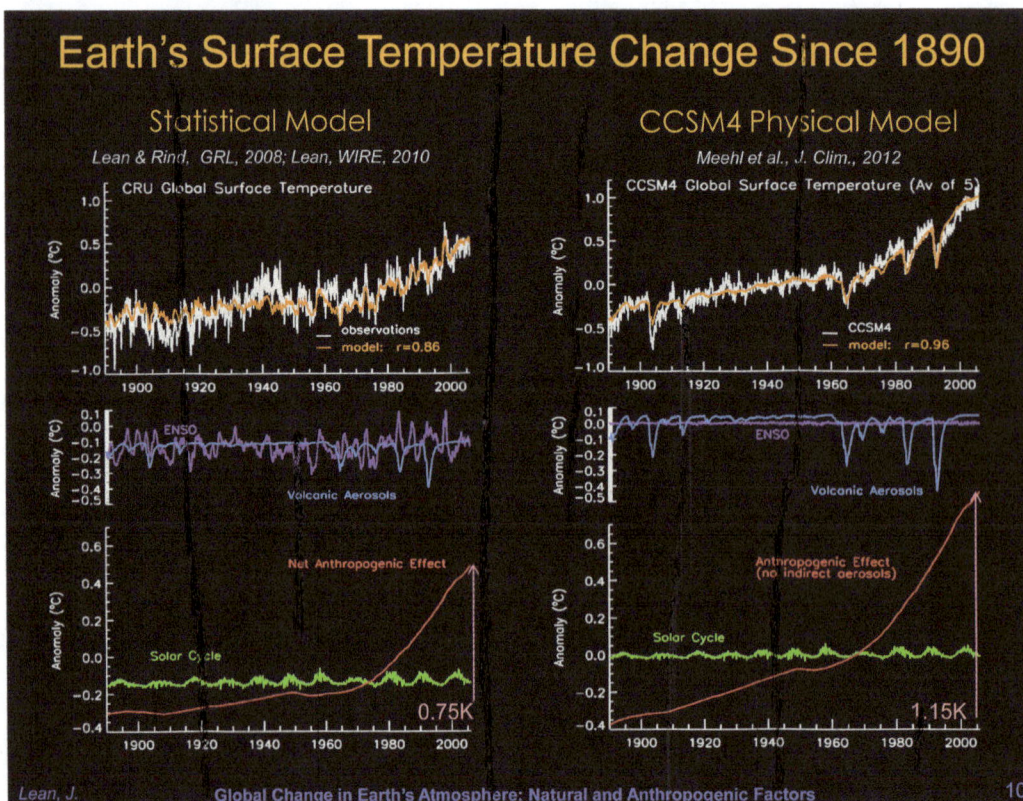

Slide 11

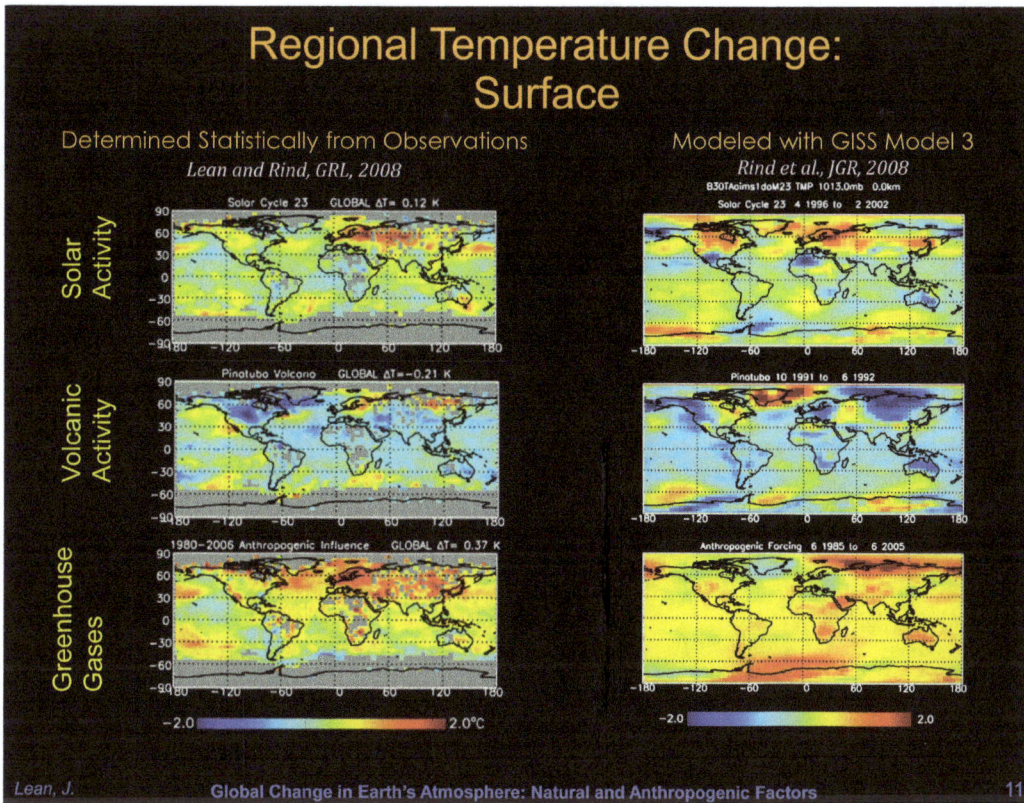

Slide 12

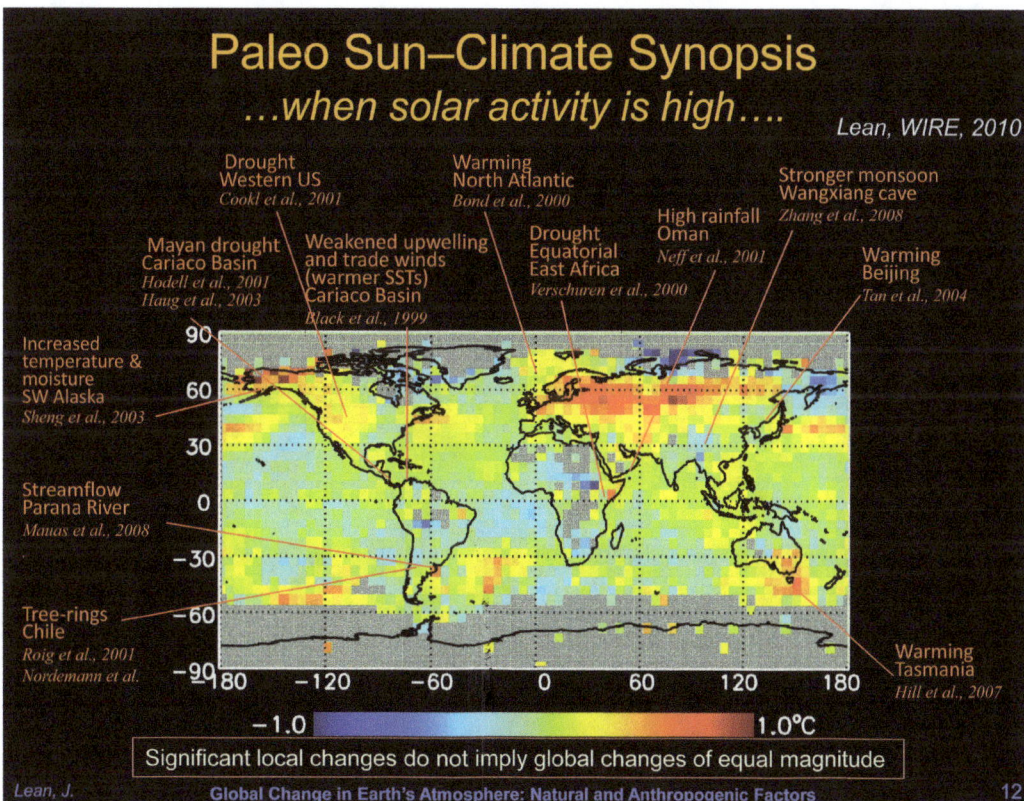

Slide 13

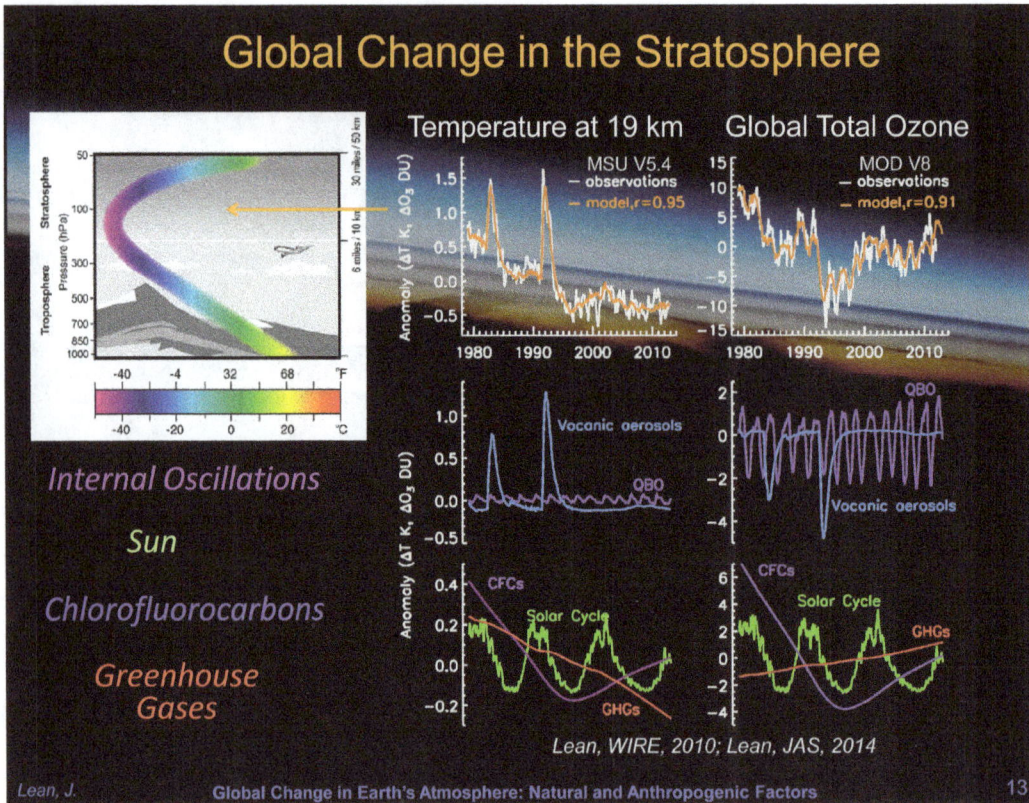

Slide 14

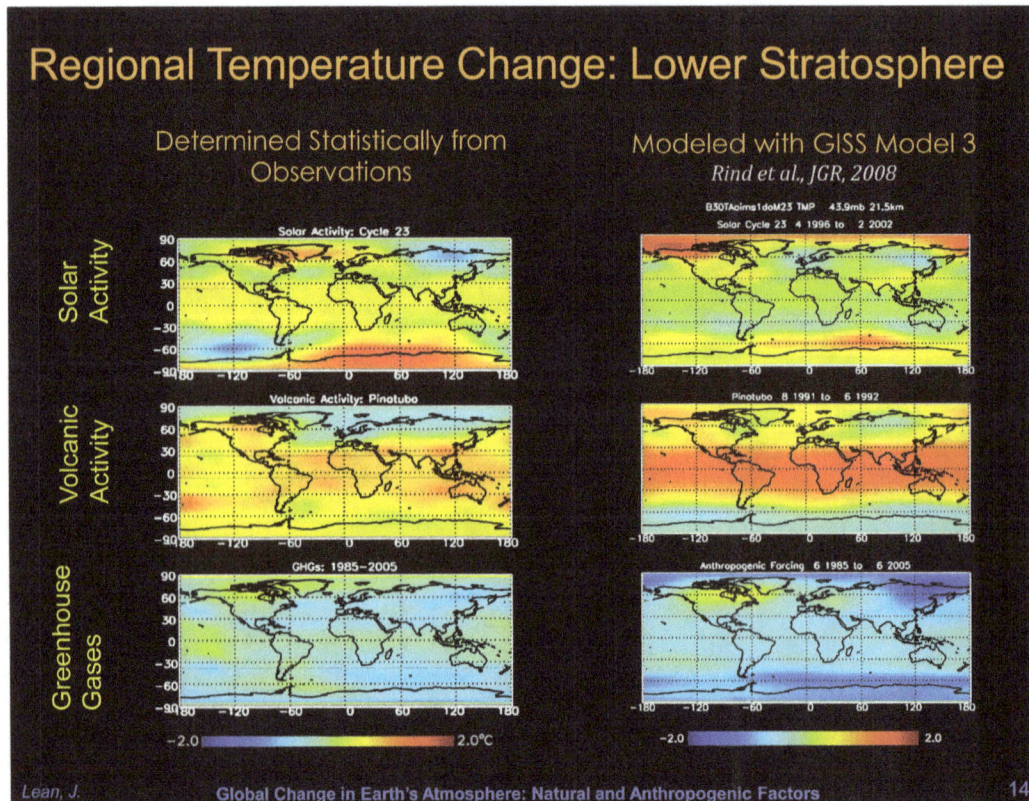

Slide 15

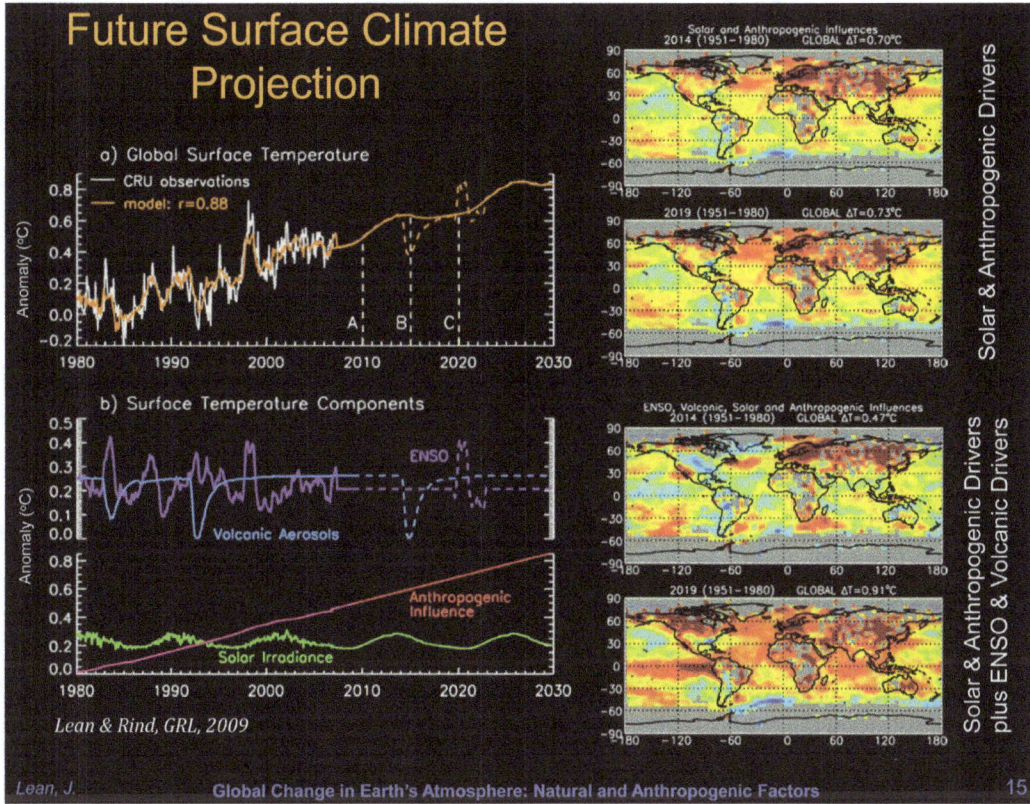

Slide 16

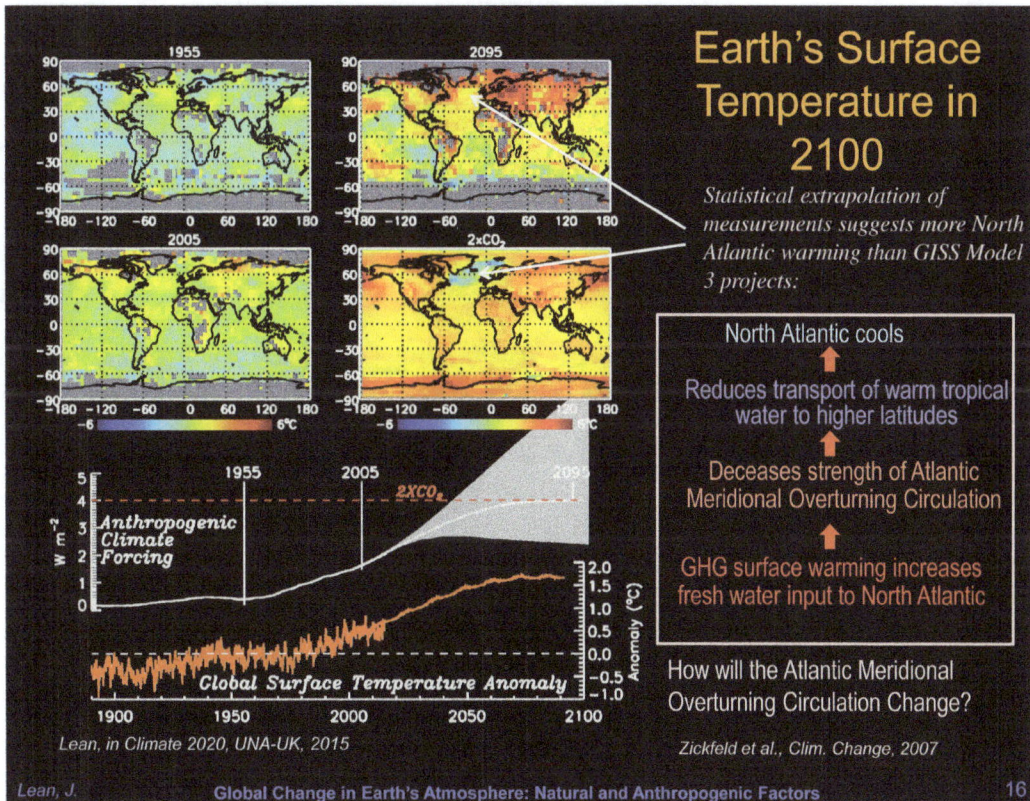

Slide 17

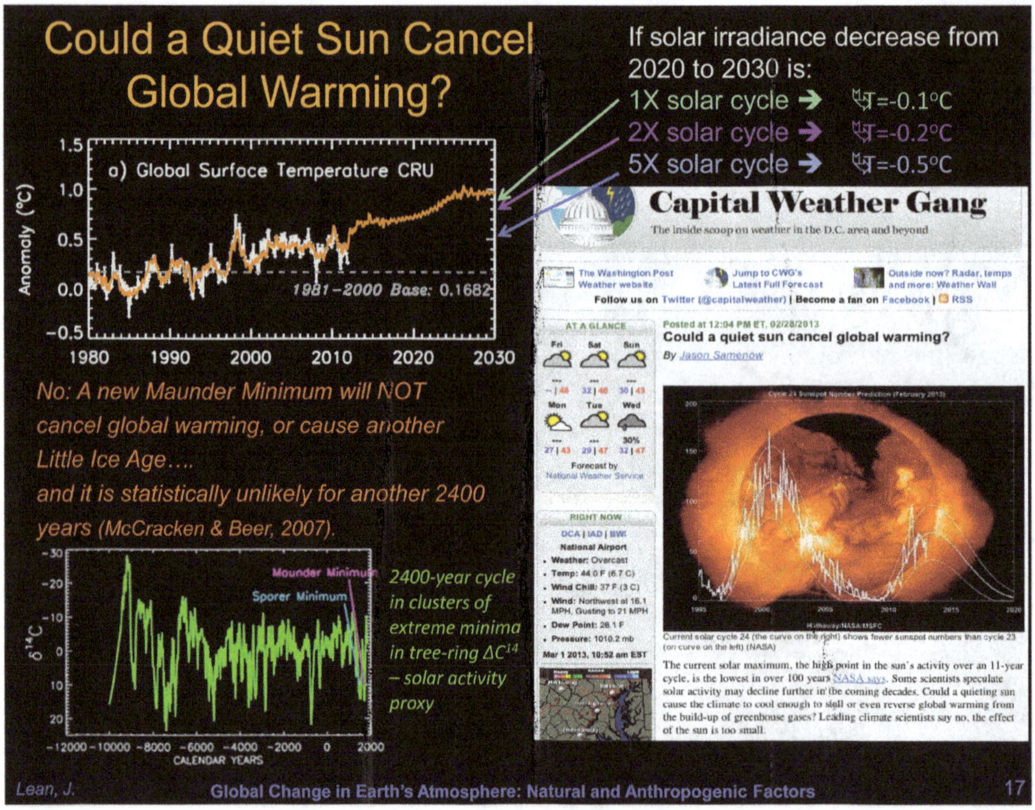

Could a Quiet Sun Cancel Global Warming?
If solar irradiance decrease from 2020 to 2030 is:
1X solar cycle → ΔT=-0.1°C
2X solar cycle → ΔT=-0.2°C
5X solar cycle → ΔT=-0.5°C
a) Global Surface Temperature CRU
Anomaly (°C)
1981–2000 Base: 0.1682
Capital Weather Gang
The inside scoop on weather in the D.C. area and beyond
Could a quiet sun cancel global warming?
By Jason Samenow
No: A new Maunder Minimum will NOT cancel global warming, or cause another Little Ice Age….
and it is statistically unlikely for another 2400 years (McCracken & Beer, 2007).
Maunder Minimum
Sporer Minimum
2400-year cycle in clusters of extreme minima in tree-ring ΔC14 – solar activity proxy
CALENDAR YEARS
The current solar maximum, the high point in the sun's activity over an 11-year cycle, is the lowest in over 100 years NASA says. Some scientists speculate solar activity may decline further in the coming decades. Could a quieting sun cause the climate to cool enough to stall or even reverse global warming from the build-up of greenhouse gases? Leading climate scientists say no, the effect of the sun is too small.
Lean, J.
Global Change in Earth's Atmosphere: Natural and Anthropogenic Factors
17

Slide 18

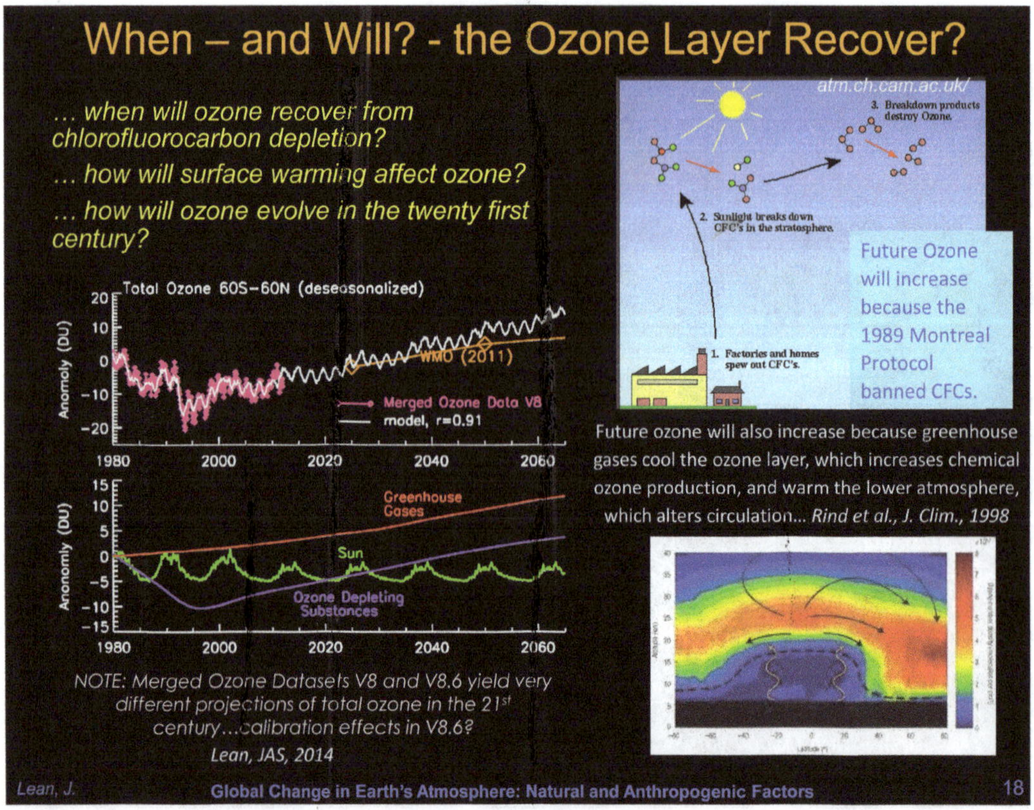

When – and Will? - the Ozone Layer Recover?
… when will ozone recover from chlorofluorocarbon depletion?
… how will surface warming affect ozone?
… how will ozone evolve in the twenty first century?
atm.ch.cam.ac.uk/
3. Breakdown products destroy Ozone.
2. Sunlight breaks down CFC's in the stratosphere.
1. Factories and homes spew out CFC's.
Future Ozone will increase because the 1989 Montreal Protocol banned CFCs.
Total Ozone 60S–60N (deseasonalized)
Anomaly (DU)
WMO (2011)
Merged Ozone Data V8
model, r=0.91
Greenhouse Gases
Sun
Ozone Depleting Substances
Future ozone will also increase because greenhouse gases cool the ozone layer, which increases chemical ozone production, and warm the lower atmosphere, which alters circulation… Rind et al., J. Clim., 1998
NOTE: Merged Ozone Datasets V8 and V8.6 yield very different projections of total ozone in the 21st century…calibration effects in V8.6?
Lean, JAS, 2014
Lean, J.
Global Change in Earth's Atmosphere: Natural and Anthropogenic Factors
18

Slide 19

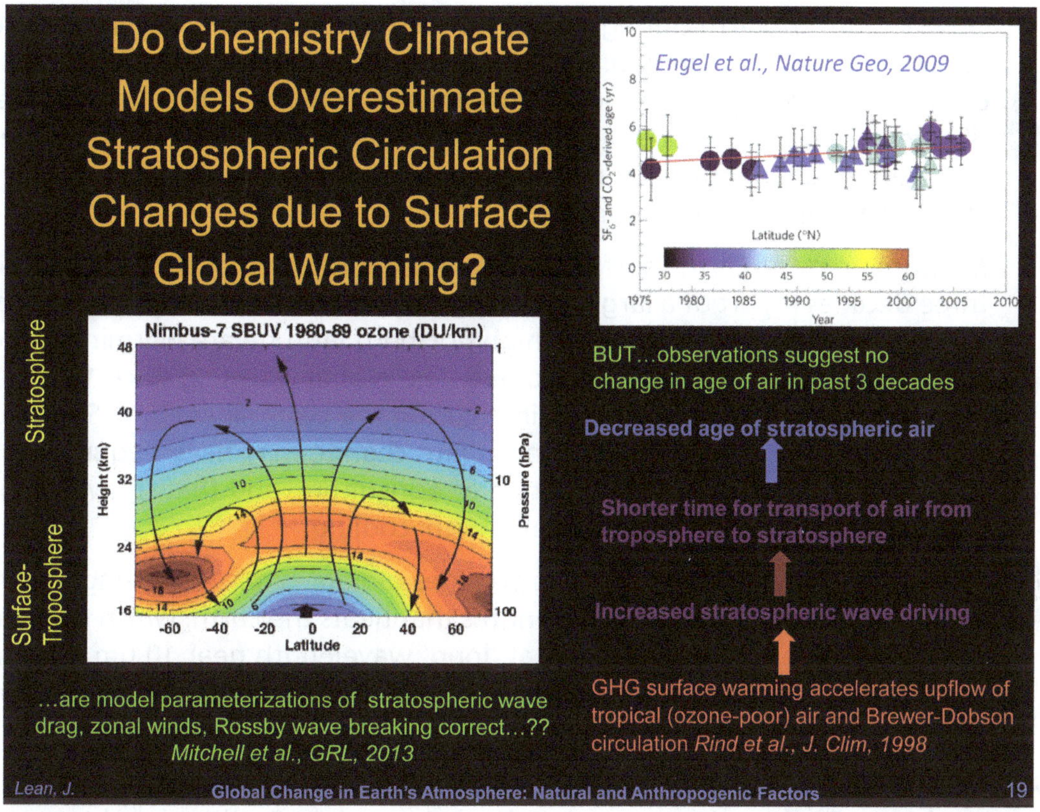

Slide 20

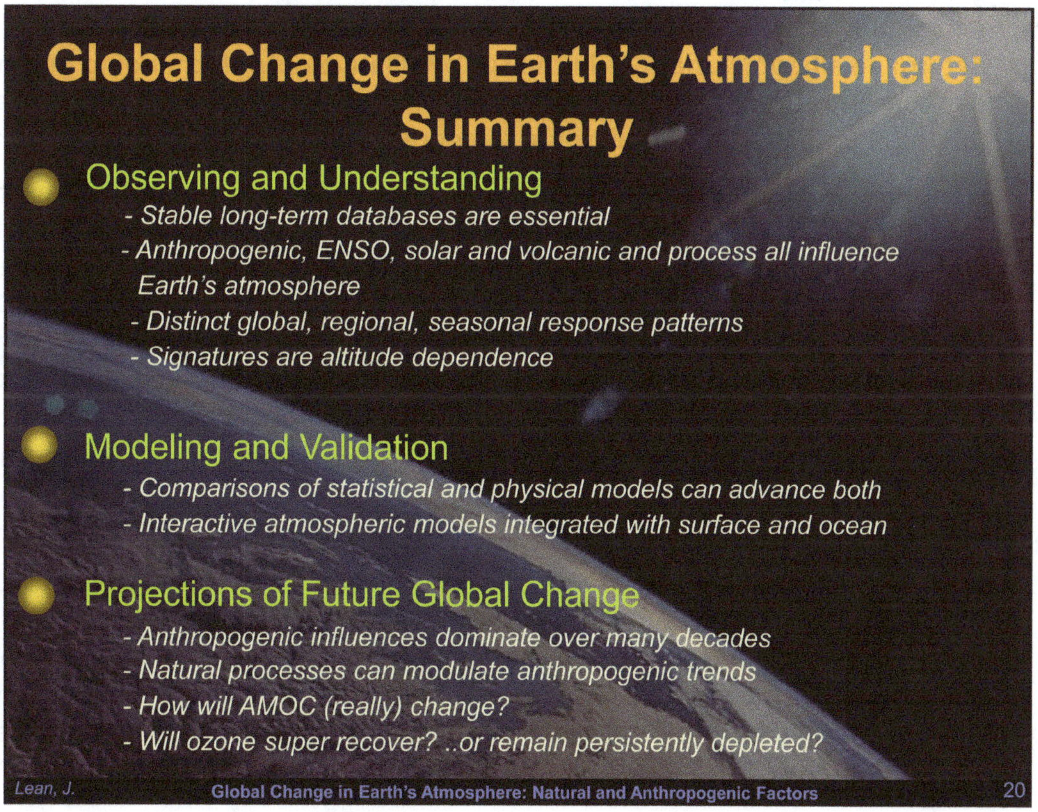

Slide Notes

Slide 3   Observations of Earth during the last three decades collectively synthesize how our environment is changing: globally, the surface, lower atmosphere (and upper ocean) are warming, the middle atmosphere is cooling and the ozone layer possibly recovering from a protracted decline from 1989 to 1995. These changes have occurred simultaneously during a period when the concentrations of CO_2 in the atmosphere have steadily increased.

Slide 4   Measurements of Earth's surface temperature indicate that the trend in the past three decades is part of a larger warming trend of 0.75°C during the past century. Regional surface temperatures at high northern latitude are warming more than in the tropics and southern hemisphere. The global warming since 1850 has proceeded more rapidly than at any time during the past millennium. Some suggest that, in fact, anthropogenic-driven climate change may have begun as far back as 10,000 years, with the introduction of agriculture, and that this warming may have alleviated orbit-related 'Milankovitch' cooling.

Slide 5   The Sun is Earth's energy source. It provides radiant 'short' wavelength energy at near UV, visible and near IR wavelengths that heats the Earth, which then radiates part of this energy back to space, at 'long' wavelength near 10 μm. Greenhouse gases in the atmosphere warm Earth's surface further because they trap infrared energy.

Slide 6   The balance of incoming energy (from the Sun) and outgoing energy (radiated from Earth's solar-warmed surface) determines Earth's global temperature. Myriad processes can alter this energy balance, including the concentrations of greenhouse and other gases in the atmosphere, volcanic and industrial aerosols, and clouds and surface features such as ice, desert, and forest coverage which determine albedo. Variations in the Sun's radiative output, changing greenhouse gas concentrations, and other radiative forcings disrupt Earth's energy balance, forcing it to seek a new equilibrium temperature, thus causing climate to change.

Slide 7   There are, thus, many causes of climate change, of both natural and anthropogenic origin. Natural influences in addition to changes in the Sun's brightness and episodic injections of volcanic aerosols into the atmosphere include semi-regular fluctuations associated with the El Niño Southern Oscillation (ENSO) near the surface, the Quasi Biennial Oscillation (QBO) in the atmosphere, and the Atlantic Meridional Overturning Circulation (AMOC) in the ocean. Anthropogenic influences include increasing concentrations of atmospheric 'greenhouse' gases, especially CO_2, industrial aerosols, and also ozone-depleting substances, primarily chlorofluorocarbons.

Slide 8   The total irradiance from the Sun, once called the solar 'constant', has been measured with space-based radiometers for more than three decades, and found to vary in phase with the Sun's 11-year activity cycle. The reason for the variability of total solar irradiance is the presence of magnetic features on the solar disk. Sunspots are compact regions of intense magnetic fields that inhibit the upward flow of energy from beneath the Sun's surface, and are therefore cooler and darker than the surrounding solar surface; their presence decreases the Sun's total irradiance. Bright faculae also occur on the Sun's disk and these more dispersed, bright regions enhance total irradiance. Observed total solar irradiance variability is the net effect of competing sunspot darkening and facular brightening, each of which increases during times of high solar activity. Because sunspot darkening and faculae brightening are wavelength dependent, the solar spectrum varies by different amounts at different wavelengths during the 11-year activity cycle.

Slide 9 Appropriate combinations of natural and anthropogenic influences (determined statistically from the measurements) reproduce both the multi-year fluctuations and the overall trends in temperature at the surface in recent decades, successfully capturing the lull in surface warming from 2001 to 2010. On time scales of a few years, El Niños and volcanic eruptions produce episodic global surface warming of as much as 0.3°C; a 'super' El Niño in 1998 caused this year to be the warmest on record until recently, whereas the El Chichón volcanic eruption mitigated the impact of a comparable El Niño event in 1982. Also extracted from the measurements are changes of order 0.1°C at the surface due to increasing solar brightness from the minimum to maximum of the Sun's 11-year activity cycle.

A combination of more La Niña (rather than El Niño) conditions and declining solar brightness countered much of the anthropogenic warming in the decade from 2001 to 2010 so that in terms of statistical analysis of measurements, the so-called hiatus is not exceptional at all, even though physical models generally failed to reproduce it.

The natural and anthropogenic factors that are simultaneously changing Earth's surface and atmosphere, each have distinct temporal, geographical and altitudinal signatures.

Slide 10 The instrumental record of Earth's surface temperature extends from 1850 to the present. The combination of natural and anthropogenic components that account for the observed changes in global surface temperature in the past three decade also simulates the changes over the past 100+ years. Neither ENSO nor volcanic aerosols exhibit trends that can account for the observed global warming of 0.75°C. The adopted solar irradiance record is a reconstruction using a flux transport model of the Sun that estimates an accumulation of magnetic flux — and hence irradiance — with increasing solar cycle amplitude. The estimated long-term irradiance change is 0.5 W/m^2 from the Maunder minimum to current solar minima. These solar irradiance changes, combined with the ENSO and volcanic aerosol changes, account for less than 15% of the warming since 1890. The most likely cause of the warming is anthropogenic not natural, due to increasing concentrations of greenhouse gases (partially mitigated by increasing industrial tropospheric aerosols).

Slide 11 The natural and anthropogenic factors that are simultaneously changing Earth's surface and atmosphere, each have distinct temporal, geographical and altitudinal signatures. Comparing regional responses extracted statistically from multiple datasets with simulations made using physical general circulation models such as GISS Model 3 facilitate unique cross-validation of the physical models and observations.

Slide 12   Knowledge of the responses of climate to individual forcings can help interpret and coagulate paleo time series from regional sites. For example, the response pattern of surface temperature to solar activity obtained from statistical analysis of the contemporary observations provides a framework for reconciling multiple, seemingly disparate reports of solar-related climate change at individual locations around the world. At some sites, especially in northern hemisphere continental regions, solar-related changes can reach 1°C but because the response pattern is regionally inhomogeneous, significant local changes do not imply equally large global changes.

Slide 13   A combination of natural and anthropogenic influences also accounts for much of the measured variations in Earth's lower stratospheric temperature (near 20 km) in recent decades, including its distinctly different behavior compared with simultaneous variations at the surface. This results from their different relative strengths at the surface and in the atmosphere near 20 km. Whereas increasing greenhouse gases warm the surface, they cool the stratosphere because the much thinner overlying atmosphere allows the heat to escape. Aerosols injected into the stratosphere during a volcanic eruption warm the stratosphere but cool the surface. Increases in the Sun's brightness warm Earth's atmosphere, throughout, with a cycle of 0.3°C at 20 km due to increasing solar brightness from the minimum to maximum of the Sun's 11-year activity cycle (compared with 0.1°C at the surface). Global cooling in the atmosphere near 20 km from 1979 to 2014 simultaneously with warming at the surface is therefore consistent with increasing concentrations of greenhouse gases. The decline in 20-km temperature reflects, as well, contributions of changing ozone concentrations, affected by ozone-depleting substances.

Slide 14   GISS Model 3 extends up to 80 km and is therefore able to simulate climate change occurring in the troposphere and stratosphere, as well as at the surface. Comparing regional responses extracted statistically from multiple datasets with simulations made using GISS Model 3 facilitate unique cross-validation of the physical models and observations.

Slide 15   As concentrations of greenhouse gases increase in the future, Earth's surface warms. This follows naturally from the relationship established statistically of a positive anthropogenic influence and global surface warming, with largest warming at mid to high northern latitudes. In upcoming decades natural influences will continue to manifest. A solar cycle-related modulation of order 0.1°C (depending on the strength of future solar activity) superimposed on the projected warming may mitigate to some extent the anthropogenic warming for a few years, causing apparent lulls in the overall warming trajectory. While future ENSO and volcanic activity are not predictable on time scales of decades, their occurrence may also modify the future global surface temperature changes significantly, both globally and regionally.

Slide 16   A projection of the statistically determined anthropogenic component of observed global surface temperatures using a mid-range anthropogenic scenario suggests that global surface temperatures may rise ~1°C from 2015 to 2095, and 2.3°C for doubled CO_2 concentrations. For comparison, physical models simulate global warming in the range 2–4.5°C for doubled CO_2 concentrations (4 W/m^2 radiative forcing). The projections include a simulated solar brightness cycle modulates the overall warming trajectory very slightly.

Regionally, northern hemisphere mid- and high latitudes over land are warming faster than are tropical and southern latitudes, and are likely to warm more in the future. This is evident in the regional patterns of the surface temperature anomalies in 2005 compared with those in 1955 and it is evident in both the statistical projections of the measurements and physical model simulations of surface temperatures in 2095.

In simulations of regional surface temperature changes for double CO_2 concentrations, made by NASA GISS Model 3, increasing greenhouse gas concentrations produce enhanced warming at high latitudes and over land, consistent with analysis of the measurements. In the North Atlantic region, however, warming is minimal because a slower Atlantic Meridional Overturning Circulation (AMOC) reduces ocean transport of warm tropical water to higher latitudes. In contrast, statistical projections based on the surface temperature measurements, foresee more Northern Atlantic warming; this emphasizes the importance of better understanding projections of ocean circulation changes.

Slide 17   Some suggest that the Sun's activity may subside over the next century into a prolonged period of inactivity, such as occurred during the 17th-century Maunder minimum. Should this transpire, solar brightness may decrease by more than has been measured in the past three decades. Knowledge of the expected magnitude of such an irradiance decrease is uncertain but current understanding limits it to a range of 0.05–0.25% compared with solar cycle changes of order 0.1%. The resultant global surface cooling would then be less than a few tenths °C, which is an order of magnitude smaller than the projected anthropogenic warming. Even in the highly unlikely event that solar irradiance decreases five times more than it has been observed thus far to change, the corresponding global surface cooling of 0.5°C will still be too small to counter the expected anthropogenic warming. The occurrence of anomalously low solar activity epochs appears to cluster in groups of five or six that reoccur approximately every 2400 years. The Spörer, Maunder, and Dalton minima that coincided with the Little Ice Age comprise the most recent such grouping; the next is thus not anticipated for a few thousand years.

Slide 18   Ozone is expected to increase in the 21st century because of the declining levels of chlorofluorocarbons (following the Montreal Protocol) and increasing concentrations of greenhouse gases. Depending on these two influences, levels of global total ozone by 2100 may exceed those in 1980. The increase in global total ozone projected by statistical models of the MOD V8 dataset exceeds the projections by WMO's chemistry climate models.

Slide 19   Might chemistry climate models overestimate the impact of GHG-induced changes in the stratosphere, exaggerating trends in tropical upwelling, and over-accelerating the Brewer-Dobson circulation? This scenario is consistent with recent observations that the age of stratospheric air has changed little in the past three decades, even with greenhouse gas forcing of 1 W/m² and net global surface warming of 0.3°C from 1980 to 2011, and that modeled changes in tropical ozone and upper troposphere temperature in recent decades may overestimate the observed changes. Improvements may therefore be needed in chemistry climate model transport, for example, in the parameterizations of dynamic and convective processes that connect the tropical troposphere and stratosphere.

CLIMATE LECTURE 3

Building a Climate Model

Gary Russell

NASA Goddard Institute for Space Studies, New York, NY, USA

Gary Russell (PhD in Mathematics from Columbia University, 1976) is a Senior Research Scientist at the NASA Goddard Institute for Space Studies (GISS) in New York City. He has been a key member of the GISS climate modeling group since the mid-1970s. His principal area of expertise is development and implementation of mathematically sound numerical techniques for modeling global climate change. As principal architect of the GISS climate GCM, he developed on-line diagnostics and a physics-based ocean model that have revolutionized the numerical modeling approach to study global climate change.

Introduction

Climate, or the average of day-to-day weather, can be very different at various points on Earth. The local climate in the Arabian Desert is hot and dry, while that in the Amazon River basin is hot and humid with frequent rain. In upstate New York, the climate changes from being warm in the summer with sporadic rain to cold in the winter with sporadic snow. Hawaii, on the other hand, has a pleasant climate all year long. However, the day-to-day weather at all of these locations is much more variable. There can be dry days in the Amazon jungle, and rainy days in the Arabian Desert. There are some days in winter that are warmer than some days in summer. For further contrast, daylight in Antarctica lasts up to six months at a time with freezing cold day-in day-out. Can a climate model be built that can reproduce all of this complex behavior?

Model Variables

Climate models require that the climate system be described in terms of physical variables such as temperature, pressure, clouds, wind, and solar radiation, variables that change with time and interact with each other and with the evolving properties of the land, ocean, and atmosphere. Of these, water is the most active, ubiquitous, and most significant substance in the entire climate system, appearing in and transforming among solid, liquid, and vapor forms. It is the principal source of the energy released in destructive storms, as well as being the indispensable enabler in sustaining an efficiently functioning biosphere. There are also tiny amounts of other gases such as ozone and carbon dioxide distributed throughout the atmosphere. It is ozone that keeps the biosphere from being destroyed by

51

extreme ultra-violet radiation. Meanwhile, carbon dioxide acts to maintain the strength of the terrestrial greenhouse effect at its present temperate level.

All of these different climate system variables are specified numerically to assure a quantitative mathematical description of the state of the land, ocean, and atmosphere at all points on the Earth, at all heights in the atmosphere and at all depths of the ocean, and for every arbitrary time of day. This imposes an enormous amount of information handling, not to mention keeping track of the interactions that take place between the different climate system variables, and requires very large and fast computers to handle the mathematical models that have been specifically designed to simulate the basic physical structure and manner of operation of the climate system. There are basically three different types of climate variables: *prescribed*, *prognostic*, and *diagnostic*.

Prescribed variables include the basic parameters that define the terrestrial climate system such as the radius and total mass of the Earth, the Earth's orbital parameters, positioning of the Sun and Moon, atmospheric composition, and land–ocean distribution and topography. There are also a number of numerical coefficients that have been empirically derived, and can be effectively grouped with the physical constants that are an integral part of the model physics. Prescribed variables can interact with all other climate components, but they themselves are not transformed by these interactions. Simpler climate models may include ozone with its seasonal variability, vegetation albedo, and aerosols, or use specified ocean surface temperature. In a Q-flux climate model, seasonal ocean heat transports are preselected from separate offline calculations. In such Q-flux simulations, ocean currents and salinity stay fixed, but the prescribed heat transports are applied to ocean heat content, so that ocean temperature and sea-ice act as prognostic variables.

Prognostic variables are the basic variables in terms of which the climate system components interact with each other, undergo change, and evolve with time. They are continuously defined at each three-dimensional model grid-box, and include temperature, dry-air mass, water substance, wind components, and changes in clouds, snow/ice cover, and subsurface reservoirs. In coupled atmosphere–ocean climate models, ocean mass, salt, heat content, and currents are all prognostic variables. For some simulations, tracer quantities for chemistry, aerosols, or water isotopes may be prognostic. Scalar quantities are defined at the grid-box center and fluxes at grid-box edges, but velocity locations for atmosphere, ocean, and sea-ice depend on the differencing scheme used (of which there are several). At each model time step, the horizontal and vertical fluxes of mass, momentum, and energy of the three-dimensional grid-boxes are acted upon according to the established laws of physics, thermodynamics, and fluid mechanics, all ensuring that mass, momentum, and energy are globally conserved at every time step (Hansen *et al.*, 1983).

Diagnostic variables are continually being derived from prognostic variables. Thus, pressure at any level is computed from the mass of air, ice, and water above that level multiplied by gravity. Other examples are surface albedo, which includes the effects of changing snow accumulation; also surface temperature, which is determined from the net balance of energy contributions from changes in radiative heating and cooling, sensible heat, cooling by evaporation, and net heating or cooling by falling precipitation. Potential, kinetic, latent, and radiative energy components, angular momentum, vertical wind, are diagnostic, although they eventually affect the prognostic variables. Other diagnostic variables are used only for model output, including global maps of reflected solar radiation and emitted thermal radiation that arise from radiative transfer modeling. Such maps serve to compare model performance against observational data. Model diagnostics also provide illustrative explanations of the physical functioning of the climate system.

Serving as the central nervous system of the climate model are the *fundamental equations* that describe the atmosphere and the physical interactions between the different constituents of the climate system. These insure that the conservation of mass, momentum, and energy are strictly enforced at all times and at all points of the climate system. The *equation of motion* is written for a rotating frame of reference with a fixed rotation rate. It accurately represents the gravitational, pressure-gradient, and Coriolis force control over the dynamic motions of the atmosphere. The *thermodynamic equation* describes the work performed upon a unit mass in compressing it, and includes heating and cooling by all other processes. Except for topography, the Earth is treated as a sphere with the atmosphere being thin relative to the planetary radius. The vertical component of the equation of motion is replaced with the *hydrostatic assumption* to filter out vertical sound waves, and permit longer time steps. The Earth's *orbital parameters* are model specifications that enable paleo-climate simulations. They define the precise amount of solar radiation incident on each model grid column at each time step, in accord with the local time of day and season.

GISS Climate Model Beginnings

Climate modeling at the Goddard Institute for Space Studies (GISS) began in the mid-1970s. Of note was the IBM 360/model-95 computer that occupied half of the second floor of the GISS building and had 5 mb of then impressive internal memory. David Rind, a knowledgeable meteorologist, was one of the key developers of the GISS climate model. Using a coarse $8° \times 10°$ latitude–longitude grid with nine vertical layers, the model employed an alternating dynamics/source term 60-minute time-marching strategy with shorter dynamics time steps that utilized a leap-frog formulation. At the roughly 1000-km spatial resolution, most storm systems went unresolved. Climate system variables needed to be specified at some 7500 grid-boxes in order to numerically describe the pressure and temperature structure of the atmosphere. Global distributions of winds, clouds, and water vapor required the use of numerical *parameterizations* to represent the essence of the unresolved sub-grid physical interactions in terms of their grid-box-mean values.

It is noteworthy that one year's worth of model-simulated time contains hundreds of millions of pressure, temperature, wind, cloud, and humidity data points. In 1970s climate modeling, it was common practice for the instantaneous prognostic and diagnostic variables to be loaded onto *history tapes* with a daily time-sampling frequency. Then, at some later date, statistical analyses would be performed to extract the climate relevant information from the accumulated data as if the model-generated data constituted a detailed collection of real meteorological measurements.

This cumbersome modeling approach was a distinct impediment to rapid model development and assessment of GCM performance, but it was resolved with the implementation of comprehensive *online model diagnostics* in the GISS climate model. The numerical information required to formulate model diagnostics is always readily available as the model runs its course. Assembling and packaging this information online was the GISS climate modeling innovation that enabled literally hundreds of model simulations to be rapidly conducted and evaluated, screening and testing the model physics and parameterizations for validity, accuracy, and reliability. The earlier approach of analyzing the model-generated history tapes would have required much more time to achieve the same model development objectives. Because climate model simulations calculate the full repertoire of weather fluctuations, model diagnostics can be programmed to also tabulate the frequency of occurrence as well as the severity of extreme weather events such as prolonged heat waves, droughts, and onsets of damaging frosts. Thus, online diagnostics can provide many options for rapidly assessing and evaluating climate model output. However, with the advent of ever more powerful computers and readily available mass storage, the 'history tape' approach has renewed important applications for inter-model comparisons and other statistical analyses.

The overarching principle in constructing a successful climate model is achieving explicit and detailed conservation of mass, energy, momentum, and water substance. It is taken for granted that explicit conservation of all of these quantities takes place in real-world climate, even though the observational verification of that is well beyond measurement capabilities. As the ensuing corollary, in order for any climate model to be considered credible, its model diagnostics must first demonstrate precise conservation of mass, energy, momentum, etc., even though some other model diagnostics might suggest good agreement with observations.

Climate vs. Weather Models

In essence, a climate model is intrinsically a weather model. The key difference is that a weather model operates in a mode of what is called an *initial value problem* in physics. This means that in weather forecasting models, to the extent feasible, pressure, temperature, wind, and humidity fields are all *initialized* with assimilated data from current meteorological observations. As the weather model is then stepped forward in time, it is actively predicting future weather.

The operational strategy in running a climate model is different from that of a weather model. Here, with the same mathematical model operated as a climate model, we are actually working to solve what is called a *boundary value problem* in physics. The computational procedure may appear similar to that of a weather model, since the climate model is also started from an initial state, and then time marched forward into future time. The key difference is that in a weather model, the initial conditions define the sought-after output, whereas in a climate model, the initial atmospheric state is of no particular relevance. The objective of a climate model is to determine the '*climatic mean*' of weather variability over a specified time interval (month, season, year, decade, or century), and study how the climate system responds to the *radiative forcings* that act to shift the existing average state of the climate system away from its previously established equilibrium to a new quasi-equilibrium point. It is this 'shift' in the average climate state that defines the climate system response to the applied radiative forcing, and it is independent of the initial state of the model. Small sustained radiative forcing changes have a cumulative effect that is magnified by the climate system feedback effects. They must all be modeled accurately.

Computer limitations and the model spatial and time-stepping resolution come into play when simulating unresolved physical processes on a sub-grid scale. For example, physical processes that operate on a sub-grid scale cannot be rigorously evaluated, and instead must be replaced by *parameterizations* that are only approximately correct. Also, in both climate models and the real world, the climate system is striving towards its energy balance equilibrium. However, neither the climate model nor the real world is able to approach the equilibrium point in small enough incremental steps so as to reach equilibrium smoothly. Because of this characteristic response, the climate system's physical processes (condensation, cloud formation, precipitation) to the disequilibrium forcing are so strong, they systematically overshoot the equilibrium point, thus producing the familiar quasi-chaotic behavior (or *natural variability*) of terrestrial weather.

Thus, in a way, initializing the weather model with current meteorological data is an attempt to co-align the otherwise unrelated space–time phasing of the quasi-chaotic behavior of the model with that of the real world. A limitation of weather models is that they accumulate errors arising from their sub-grid parameterizations and unresolved physical process, resulting in a substantial decline in forecast accuracy beyond lead times of only a few days. Climate models, however, are not attempting to solve an initial value problem; their focus is on how realistic their modeling is of the external radiative forcings and of the corresponding climate feedback response.

GISS Model Evolution

To be considered trustworthy for analyzing climate change, the model must first demonstrate that it can successfully reproduce the basic seasonal and geographic variability of current climate — based more on realistic model physics and less on process parameterizations that might have been tuned for current climate conditions. To this end, as computers have become more powerful and much faster, the GISS climate model has evolved to the point that many prescribed variables have now been reconstituted as prognostic variables. In particular, the space–time variability and distribution of ozone and aerosols are now being calculated online using atmospheric chemistry calculations in conjunction with interactive dynamical transport. Prescribed vegetation properties have also been transformed into prognostic variables with model vegetation cover responding to cumulative changes in precipitation and temperature. Nevertheless, it still remains convenient to utilize prescribed variables, such as the atmospheric change in carbon dioxide, which is known accurately, without incurring the added computational burden of modeling the carbon cycle, when that is not the specific point of interest.

More importantly, the ocean heat transport (Q-flux) is no longer prescribed, having been made fully interactive in what is called a *coupled atmosphere–ocean model*. Like the atmosphere, the ocean is a fluid subject to the same gravity, Coriolis force, and basic laws of fluid mechanics. The principal differences are that the ocean is a thousand times denser and more massive than the atmosphere, and much less compressible. Space–time changes in temperature and salinity are the driving factors that affect ocean circulation, constrained by the ocean-bottom topography. The ocean, with its enormous heat capacity, interacts strongly with the atmosphere by exchanging heat radiatively, and by means of sensible and latent heat. Since the ocean absorbs nearly half of the incident solar radiation, mostly in the tropical regions, this requires a redistribution of heat toward the polar regions. A key aspect of the atmosphere–ocean interaction is the wind influence on ocean surface currents, and its effect on heat redistribution within the ocean mixed layer.

As a result, present-day climate models are invariably coupled atmosphere–ocean models. This is because the question being posed to climate models is: how does the increase in greenhouse gases affect both global and regional climate? The direct global warming due to the increasing greenhouse gases is easy enough to calculate. It is more difficult to determine the precise rate of heat uptake by the ocean, which determines the speed at which the surface temperature warms. As an added complication, the ocean is responsible for the climate system's *natural variability*, which is characterized by radiatively unforced fluctuations occurring on inter-annual and decadal time scales such as the El Niño and La Niña events, or the Pacific Decadal Oscillations.

Current State of the Art

Present-day climate models with their growing realism have become indispensible research tools (Schmidt *et al.*, 2006, 2014). Reliable prediction of future climate change will continue to be the principal objective. This includes understanding how the climate will change on a regional basis, including the impact of global warming on regional precipitation patterns, fresh water resources, sea level rise, severity of storms, and ecological stability. Other areas of active research include reconstructing climate on geological time scales from when Earth was an ice-covered snowball 650 million years ago, to global tropical conditions that prevailed during the Cretaceous period. Also of great interest is modeling the regional climate change since the last ice age, in particular what sustained the lush vegetation that existed 6000 years ago in what is now the Sahara desert.

The recent discovery of the existence of thousands of extra-terrestrial planets has opened a new area of investigation where a general-purpose climate model is needed to assess and evaluate the prospects of habitability over a wide range of orbital parameters, parent star type, and planetary mass, size, and atmospheric composition. This has renewed interest for inter-planetary climate comparisons with neighbor planets Mars and Venus, which have atmospheres composed almost entirely of carbon dioxide. Mars, with an atmosphere that is only about 1% as massive as that of the Earth, has a surface temperature scarcely 5 K warmer than direct solar heating can support, whereas the surface temperature on Venus is 500 K warmer than the absorbed solar energy would imply. Exploring how planetary climate responds to extreme changes in radiative forcing is one example of the types of research studies that can be investigated with a climate model. Other examples include the study of isotope fractionation used in calibrating the temperature-dependent precipitation rate of water vapor for deuterium and oxygen-18 isotopes to determine changes in global temperature from geological ice-core records (Schmidt et al., 2007).

The current emphasis in climate model development is to upgrade the ocean model component. With its large heat capacity, and as the principal instigator of the natural variability component in the ongoing climate record, modeling this behavior is a major challenge. A significant part of the problem is that extensive measurements of ocean temperature, salinity, and circulation have not been available, as compared to the observational scrutiny of the atmosphere. The impact of lunar and solar tides on ocean circulation has been successfully included. Work continues to upgrade the model treatment of sea-ice formation, degradation, and transport by ocean currents. A better performing ocean model is key to developing a reliable climate forecast model.

Significant progress has been made to upgrade the capability of the numerical methodology used in model calculations, such as replacing the traditional latitude–longitude grid with a cube-sphere grid (Putman and Lin, 2009). A few years in the future, the two horizontal velocity components will be replaced with three symmetric horizontal components of angular momentum on an icosahedral grid (Russell and Rind, 2017). Interactive computations are also being revised with interactive carbon cycle chemistry modeling having been recently added to the expanding climate modeling capabilities (Romanou et al., 2014).

References

Hansen, J., Russell, G., Rind, D., Stone, P., Lacis, A., Lebedeff, S., Ruedy, R. and Travis, L. (1983). Efficient three-dimensional global models for climate studies: Models I and II. *Monthly Weather Review*, 111, 609–662, doi:10.1175/1520-0493(1983).

Putman, W.M. and Lin, S.-J. (2009). A finite-volume dynamical core on the cubed-sphere grid. *Preprints, Numerical Modeling of Space Plasma Flows: ASTRONUM-2008, Astronomical Society of the Pacific Conference Series*, 406, 268–276.

Romanou, A., Romanski, J. and Gregg, W.W. (2014). Natural ocean carbon cycle sensitivity to parameterizations of the recycling in a climate model. *Biogeosciences*, 11, 1137–1154, doi:10.5194/bg-11-1137-2014.

Russell, G.L. and Rind, D.H. (2017). Symmetric equations on the surface of a sphere. *Monthly Weather Review*.

Schmidt, G.A., Ruedy, R., Hansen, J.E., Aleinov, I., Bell, N., Bauer, M., . . . Yao, M.-S. (2006). Present day atmospheric simulations using GISS ModelE: Comparison to in-situ, satellite and reanalysis data. *Journal of Climate*, 19, 153–192, doi:10.1175/JCLI3612.1.

Schmidt, G.A., LeGrande, A.N. and Hoffmann, G. (2007). Water isotope expressions of intrinsic and forced variability in a coupled ocean-atmosphere model. *Journal of Geophysical Research*, 112, D10103, doi:10.1029/2006JD007781.

Schmidt, G.A., Kelley, M., Nazarenko, L., Ruedy, R., Russell, G.L., Aleinov, I., . . . Zhang, J. (2014). Configuration and assessment of the GISS ModelE2 contributions to the CMIP5 archive. *Journal of Advances in Modeling Earth Systems*, 6(1), 141–184, doi:10.1002/2013MS000265.

Slide 1

Building a Climate Model

Gary Russell
NASA Goddard Institute for Space Studies

Our Warming Planet: Topics in Climate Dynamics
Lectures in Climate Change, Vol. 1
2017

Slide 2

GISS General Circulation Climate Model: The Early (1983) Version

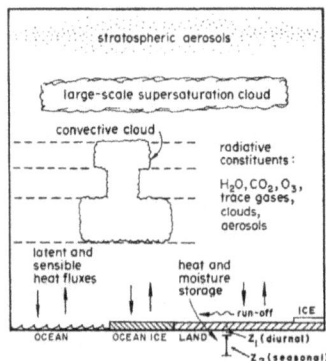

Schematic structure of the terrestrial climate system as depicted in a 1983 version of the GISS climate model.

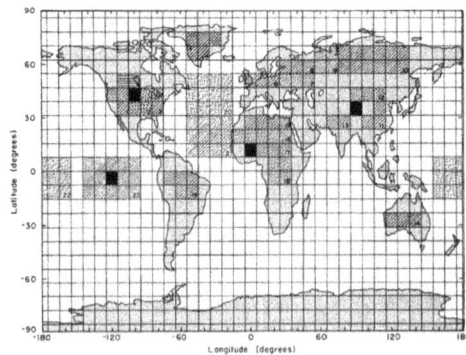

The model's horizontal resolution and surface topography were described on a 8° × 10° latitude-longitude grid.

Slide 3

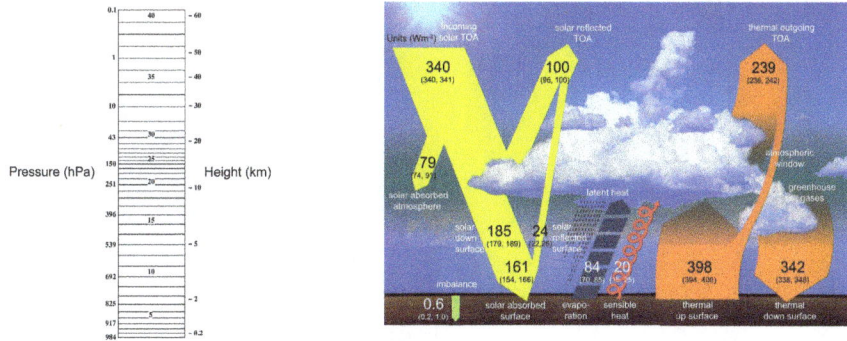

Radiative, thermodynamic, and dynamic processes are included in a model which is broken up into various vertical levels in the atmosphere.

Slide 4

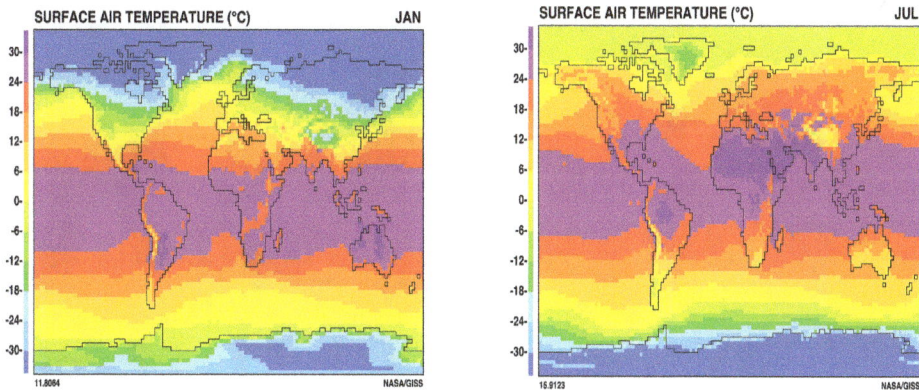

Seasonal variability of the surface temperature, between January (left) and July (right).

Slide 5

Seasonal Air Temperature as Function of Latitude and Pressure (mb)

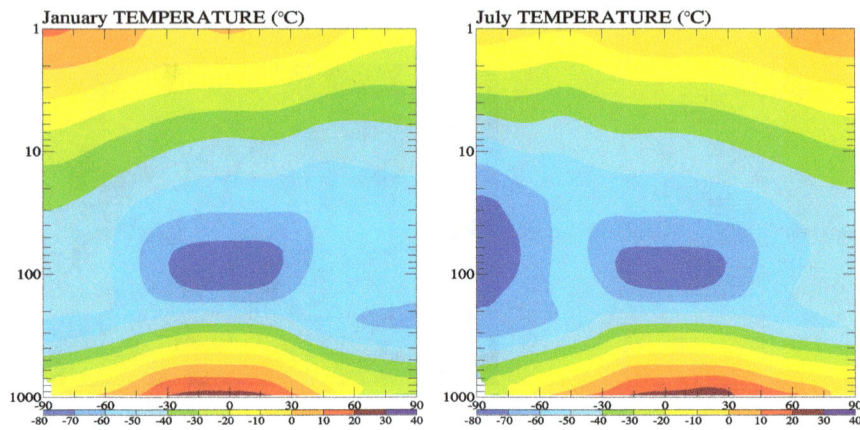

January TEMPERATURE (°C)   July TEMPERATURE (°C)

Atmospheric temperature decreases steadily with height, from the surface until the minimum temperature tropopause which ranges in altitude from 12 to 20 km. Above the tropopause, temperature increases with height until the stratopause near 50 km, then (beyond the graph) decreasing until the mesopause near 80 km, and finally increasing again into the thermosphere.

Russell, G.   **Building a Climate Model**   5

Slide 6

Seasonal Variability of the Atmospheric Stream Function

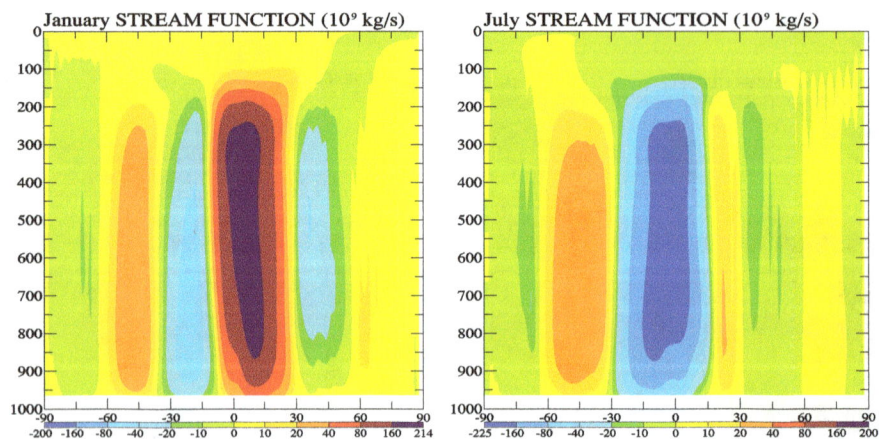

January STREAM FUNCTION (10⁹ kg/s)   July STREAM FUNCTION (10⁹ kg/s)

The chief feature of the atmospheric streamfunction, the Hadley Cell, changes its circulation from clockwise (in the plane of the figure) in January to counter-clockwise in July.

Russell, G.   **Building a Climate Model**   6

Slide 7

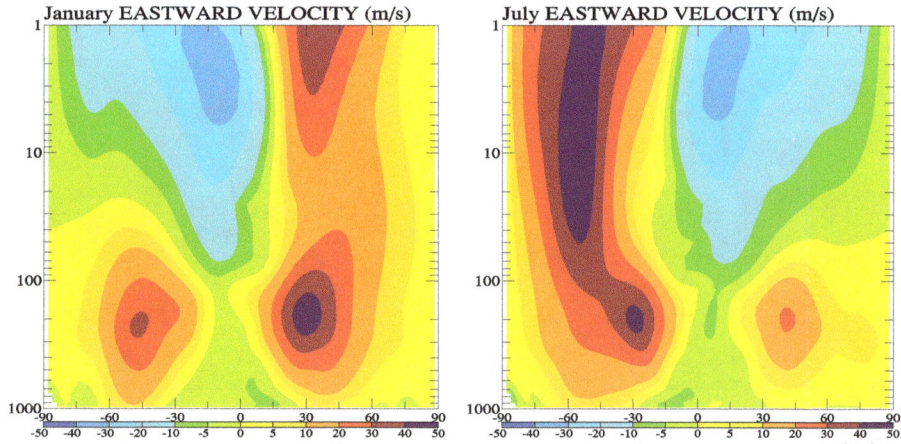

Seasonal Variability of the Zonal Wind Velocity

January EASTWARD VELOCITY (m/s)   July EASTWARD VELOCITY (m/s)

Conservation of angular momentum causes surface winds from the Hadley cell heading to ward the equator to be redirected toward the west; this causes the tropical trade winds. Similarly, pole-ward heading surface winds from the Ferrel cell are redirected eastward. Angular momentum discrepancies between tropics and mid-latitudes also drive the mid-latitude jets.

Russell, G.   **Building a Climate Model**   7

Slide 8

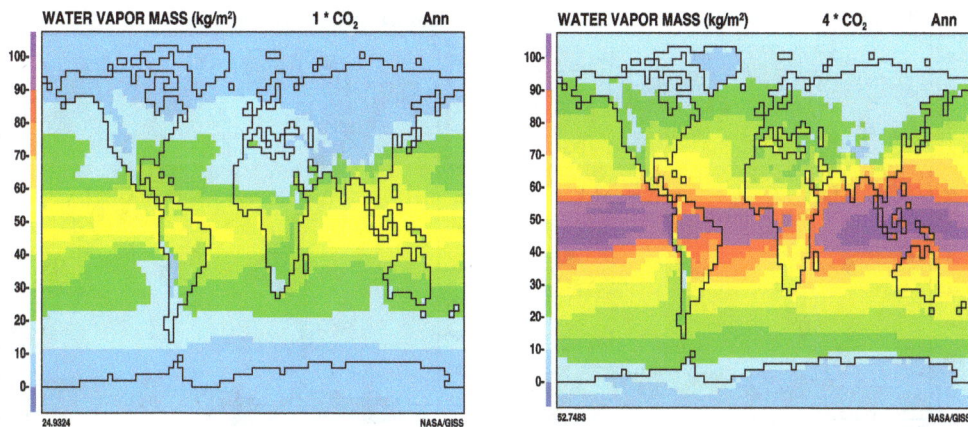

Water Vapor Mass (kg/m^2), 1950 CO$_2$ Concentration and 4 times CO$_2$

WATER VAPOR MASS (kg/m^2)   1 * CO$_2$ Ann   WATER VAPOR MASS (kg/m^2)   4 * CO$_2$ Ann

The amount of water vapor that air can hold is an exponential function of temperature, increasing as temperature increases. Thus, as the planet warms, the atmosphere can, and will, retain more water vapor.

Russell, G.   **Building a Climate Model**   8

Slide 9

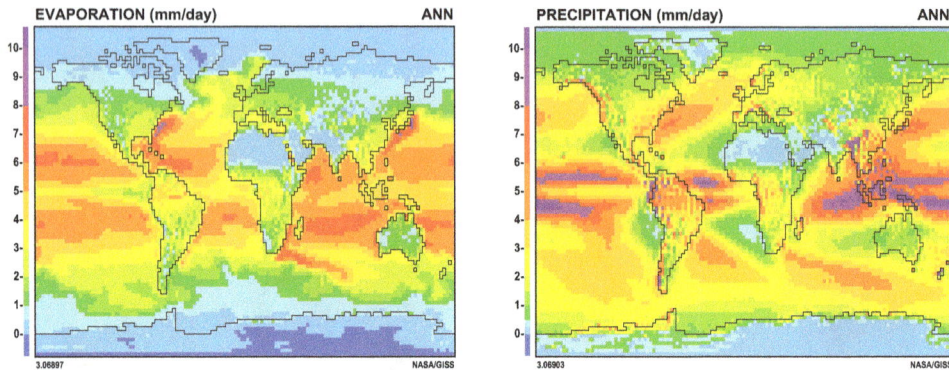

Evaporation and Precipitation, Global Annual Average is 3 (mm/day)

Evaporation is greater in the tropics than over high latitudes, and greater over oceans or lakes than over ground or ice surfaces.

Precipitation occurs when moist air cools. The amount of water vapor that the air can hold is inversely proportional to pressure, and is strongly governed by the exponential temperature dependence as set by the Clausius-Clapeyron formula.

Russell, G. **Building a Climate Model** 9

Slide 10

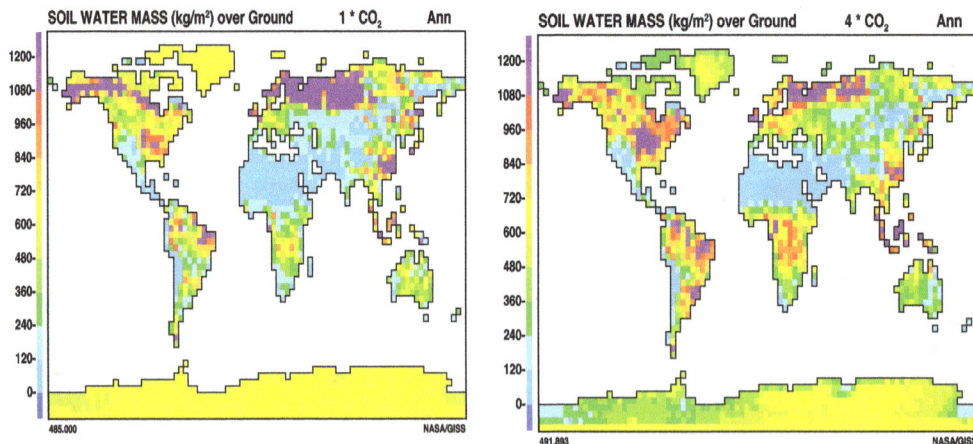

Soil Water Mass (kg/m^2), 1950 CO$_2$ Concentration and 4 times CO$_2$

With severe climate change, the distribution of ground water on Earth will be altered.

Russell, G. **Building a Climate Model** 10

Slide 11

Sensible and Latent Heat Fluxes from Surface to Atmosphere

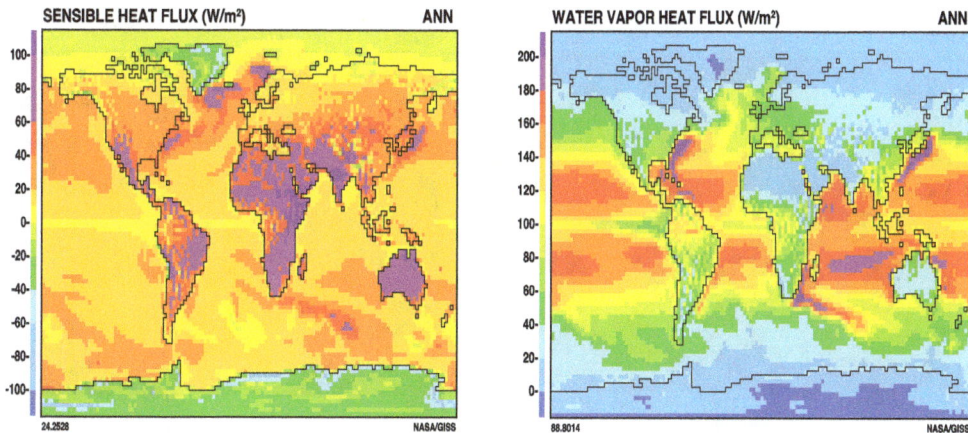

Fluxes of sensible heat (larger over land) and latent heat (larger over the oceans) help distribute heat from the surface into the atmosphere.

Russell, G.   **Building a Climate Model**   11

Slide 12

High and Low Cloud Global Annual-mean Distribution

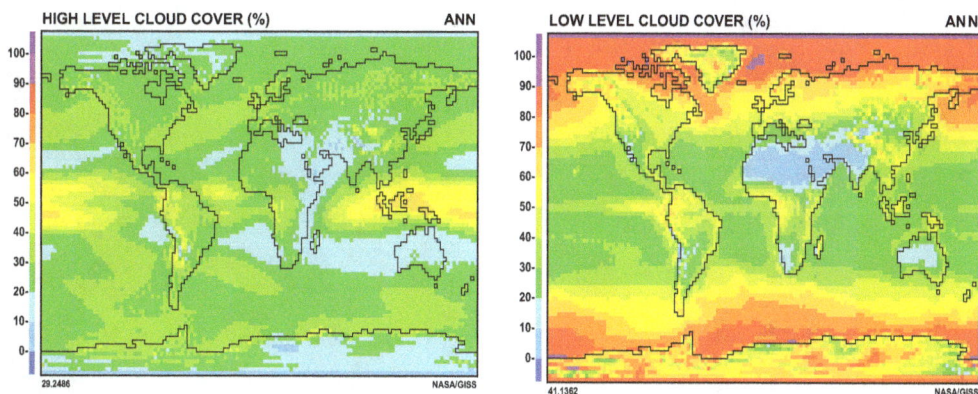

Clouds are important regulators of the Earth's global energy balance. During the day, clouds reflect sun-light, thus cooling the planet. Clouds at night -timekeep the planet warmer by reducing the escape of thermal radiation to space.

Russell, G.   **Building a Climate Model**   12

Slide 13

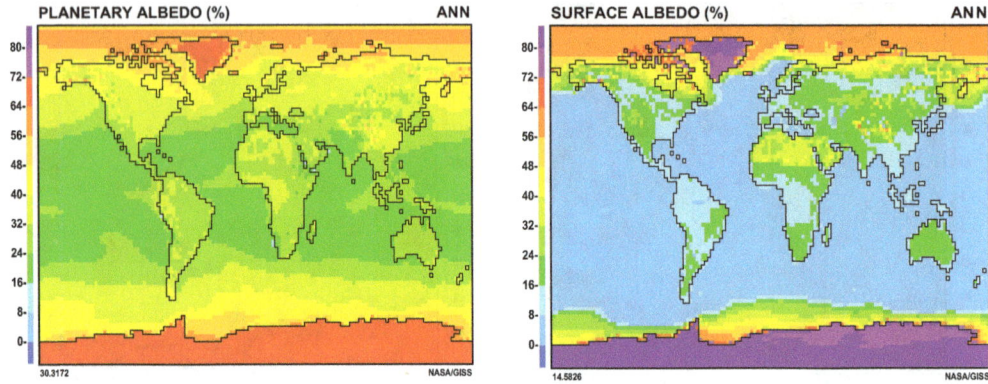

Planetary (Top of Atmosphere) and Surface Albedos

Albedo is the ratio of the (spectrally integrated) reflected sun-light divided by incident sun-light.

Slide 14

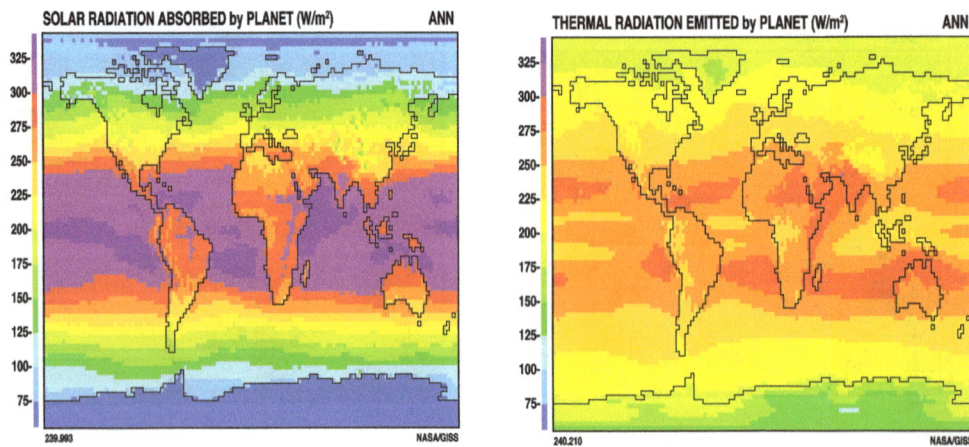

Solar Radiation Absorbed and Thermal Radiation Emitted by Planet

The Earth absorbs shortwave energy from the sun (left) and emits longwave energy back to space (right). When they are equal, the earth's temperature is in equilibrium.

Slide 15

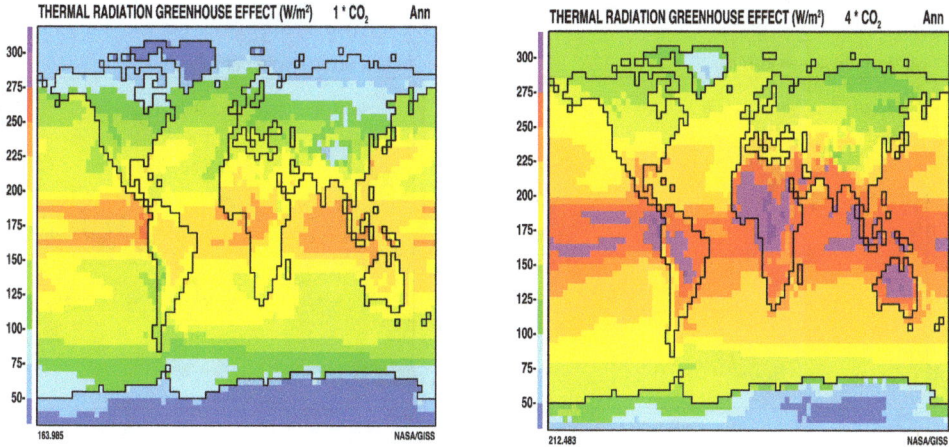

Atmospheric Greenhouse Effect, 1950 CO$_2$ Concentration and 4 times CO$_2$

The greenhouse effect (W/m^2) is defined as the difference between the upward thermal radiation emitted from the surface minus thermal radiation emitted at top-of-atmosphere out to space. The strength of the greenhouse effect is a measure of the heat trapping efficiency of spectral absorption by greenhouse gases and by longwave cloud opacity.

Russell, G.   **Building a Climate Model**   15

Slide 16

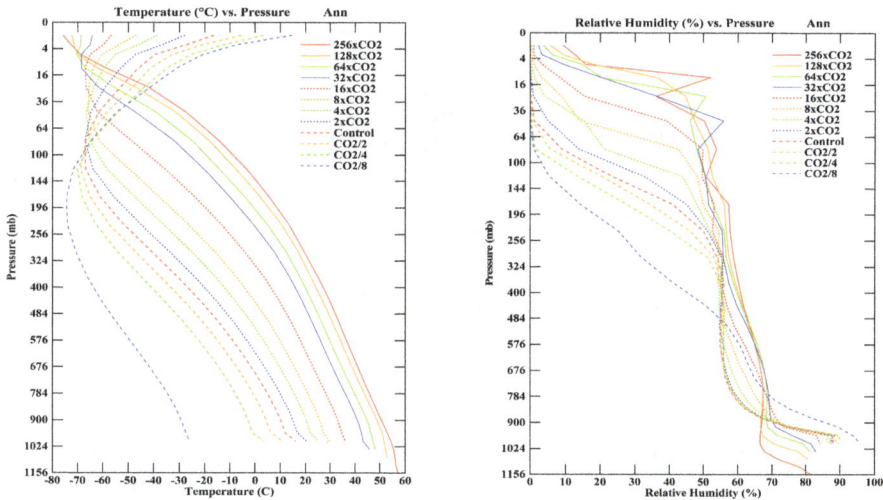

Climate Change Experiments: Temperature and Relative Humidity

The climate of Earth is strongly controlled by the amount of atmospheric CO$_2$.

Russell, G.   **Building a Climate Model**   16

Slide 17

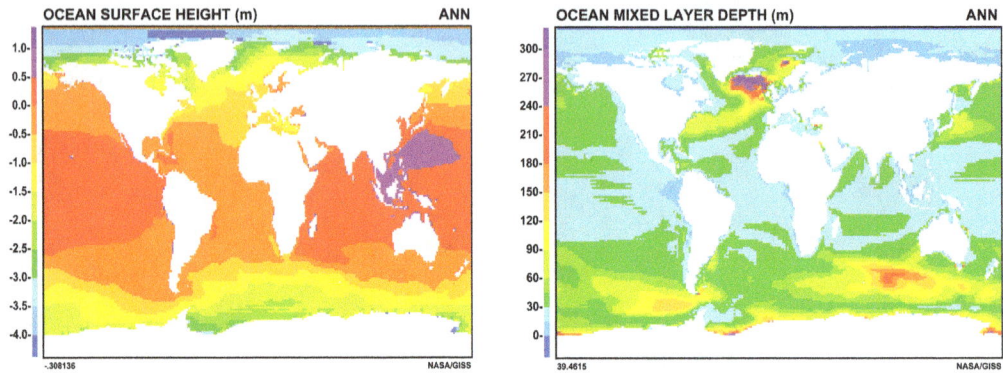

Ocean Surface Height (m) and Ocean Mixed Layer Depth (m)

The reference geoid for the ocean surface is a level where the surface pressure gradient force is zero. This includes effects due to variations of both gravity and of the Earth's rotation.

Slide 18

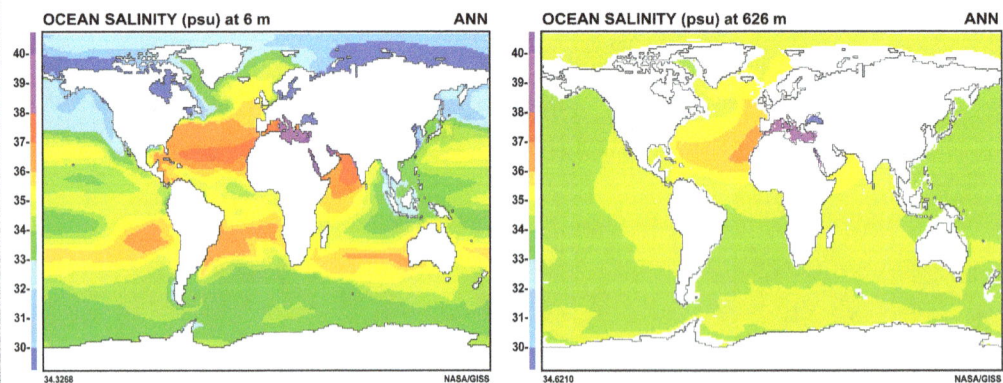

Ocean Salinity at the Surface and at Ocean Depth of 626 (m)

Salinity, because it directly affects the density of sea water, along with the water temperature, are the principal controlling factors that affect the global circulation of the ocean, and thus the transport of heat energy pole-ward from the tropics.

Slide 19

Downward Mass Flux in Atmosphere and Downward Velocity in Ocean

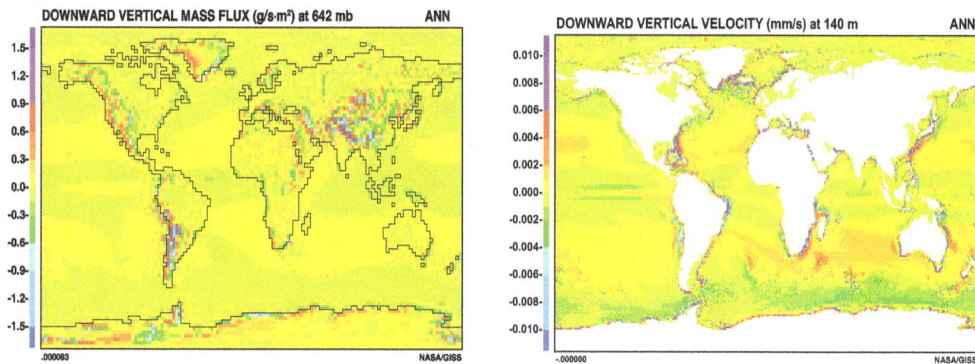

Vertical velocities in both the atmosphere (left) and the ocean (right) are subject to numerical instabilities in the model, especially along mountain ranges and coastlines.

Slide 20

Realistic Lunar and Solar Tides that Accelerate Horizontal Ocean Velocities

Click 'play' button on the video to see tidal shift over the month of January 1912.

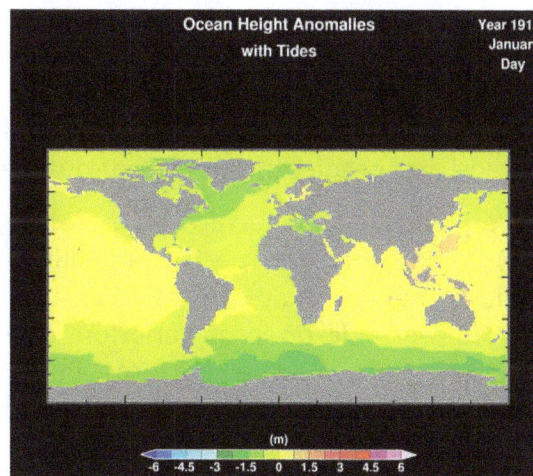

The side of the Earth facing the Moon is more strongly pulled by the Moon's gravity and the opposite side is pulled more weakly.

Slide Notes

Slide 2 Schematic structure of the terrestrial climate system (left panel) as depicted in a 1983 version of the GISS climate model. Atmospheric composition included the principal radiatively active gases: water vapor, carbon dioxide, ozone; as well as the minor absorbers such as methane, nitrous oxide, and the major CFCs. Also included were stratospheric and tropospheric aerosols, and multi-layered water and ice clouds. Solar irradiance was programmed with explicit diurnal and seasonal variability. The model surface used four horizontal surface types: ocean or lake, ocean-ice, land, and land-ice, and nine dynamically active vertical layers up to 10 mb, topped off by three radiative equilibrium layers. The model included latent, sensible, and precipitation heat exchange between the surface and the atmosphere; liquid water run-off and snow accumulation; and full accounting of water and heat stored in atmosphere, ocean, ground, and lake reservoirs.

The model's horizontal resolution and surface topography were described on a $8° \times 10°$ latitude–longitude grid (right panel). Special geographic areas (labeled 1–23) were selected for collecting regional on-line monthly diagnostics for more focused analysis. The four grid-boxes (shown in black) were designated to collect hourly weather-type information to assess and evaluate model variability.

Slide 3 The standard version of GISS ModelE2 coupled atmosphere–ocean climate model (Schmidt *et al.*, 2014) operates on a $2° \times 2.5°$ latitude–longitude grid, but can work at higher or lower resolution, if required. It utilizes a vertical resolution of 40 layers with the model top at 0.1 hPa, with adjustable layer spacing (left panel) to provide higher vertical resolution in the more critical near-surface and tropopause regions of the atmosphere. ModelE2 includes all significant radiative, thermodynamic, and dynamic physical processes and interactions, as schematically depicted in the right panel (IPCC-AR5, Fig. 2.11). Numerous model improvements since 1983 include a dynamical ocean, river flow, cloud physics, stratospheric and tropospheric aerosols, and accurate radiative transfer calculations. Non-standard versions of ModelE2 include interactive chemistry of aerosols and ozone, dynamic vegetation, a carbon cycle, and stable water isotopes with fractionation – all important aspects of the increasing physical realism in climate model development. ModelE2 is an indispensible research tool to study the physical processes of the climate system, to reconstruct historical and geological climate changes, and to understand and predict the climate of our ever evolving planet.

Slide 4 Seasonal variability of the surface temperature, TS, between January (left) and July (right). TS best characterizes the prevailing habitability of the biosphere; animal and plant life is strongly constrained by TS even if it lives in the ocean. On a daily basis, TS can undergo large variations due to solar heating by day, and the effects of storms and weather patterns. In general, the Sun heats the ground which, in turn, heats the atmosphere via sensible, latent, and thermal radiation fluxes. TS is established as the balance between these competing fluxes and is buffered by the large heat capacity of the ocean or the smaller capacities of the land and atmosphere. Note that the tropical regions remain warm all year long. The oceans undergo relatively small changes in temperature, while continental areas have large seasonal changes. The global mean TS is 4°C higher in July than in January, even though the July solar irradiance is 6.5% less in January.

Slide 5   Atmospheric temperature decreases steadily with height, from the surface until the minimum temperature tropopause which ranges in altitude from 12 to 20 km. Above the tropopause, temperature increases with height until the stratopause near 50 km, then (beyond the graph) decreasing until the mesopause near 80 km, and finally increasing again into the exosphere. Radiative cooling is the principal cause for temperature decreasing with height. Were it not for ultra-violet absorbing ozone concentrated in the stratosphere, temperature would decrease monotonically toward space. Radiatively, the larger the thermal opacity, the steeper the temperature gradient. Dry convective instability limits the temperature gradient to be no steeper than about 10°C/km, but latent heat released by condensing water vapor in the troposphere limits the most-convective temperature gradient to approximately 5–6°C/km. Tropospheric temperatures are driven by insolation with convective and advective redistribution of energy. Stratospheric temperatures are close to radiative equilibrium, balanced by solar ozone heating and thermal CO_2 cooling.

Slide 6   At any latitude and pressure, the mass stream function is computed as the northward mass flux integrated over longitude and from top of atmosphere down to the pressure level. Negative (counter-clockwise flow on the graph) and positive values of the stream function closest to the equator are described as branches of the Hadley cell. The rising warm air between the branches indicates position of the ITCZ (inter-tropical convergence zone) which varies on either side of the equator and is driven by the seasonal variation of maximum insolation. The ITCZ is where the largest rain storms occur, when rising moist air condenses below the tropopause; this causes the summer Asian monsoons. Just outside branches of the Hadley cell are the weaker Ferrel cells, in which air flows in the opposite direction from the adjacent Hadley cell branch. Sinking dry air between the Hadley and Ferrel cells tends to cause arid and dessert conditions on the surface. Outside the Ferrel cells are the polar cells which are weaker still.

Slide 7   Conservation of angular momentum causes surface winds from the Hadley cell heading toward the equator to be redirected toward the west; this causes the tropical trade winds. Similarly, poleward heading surface winds from the Ferrel cell are redirected eastward. Angular momentum discrepancies between tropics and mid-latitudes also drive the mid-latitude jets. The right-hand plot above is an average of 10 Januarys; the jets meander, but are much more confined on an individual day. A property of fluid dynamics is that to transport fluid through a medium, it is most efficient to transport the fluid by means of a narrow jet. The stratospheric zonal wind velocity reverses sign with season. However, the tropospheric jet streams maintain their west to east direction, but are substantially stronger during the winter season. The logarithmic vertical coordinate in the plot is pressure (mb).

Slide 8   The amount of water vapor that air can hold is an exponential function of temperature, increasing as temperature increases. This dependence is governed by the

Clausius–Clapeyron formula such that for every increase in temperature by 10°C, the water vapor holding capacity by the atmosphere doubles. Thus, as the planet warms, the atmosphere can, and will, retain more water vapor. Water vapor is an important greenhouse gas, and is radiatively a stronger contributor than CO_2. This makes water vapor a very important climate feedback effect in response to the imposed forcing of CO_2. In the four-time CO_2 scenario illustrated above, water vapor contribution to the total greenhouse warming effect far exceeds the CO_2 contribution.

Slide 9   Evaporation is greater in the tropics than over high latitudes, and greater over oceans or lakes than over ground or ice surfaces. Some of the water on continents runs into rivers and lakes and eventually into the ocean as opposed to evaporating. Vaporization of liquid water or ice cools the ground and the latent energy of water vapor escapes into the atmosphere; this occurs more readily when the air is warm or dry. Moist air is transported upward into the atmosphere and also toward the poles and to drier continents. Precipitation occurs when moist air cools. The amount of water vapor that the air can hold is inversely proportional to pressure, and is strongly governed by the exponential temperature dependence as set by the Clausius–Clapeyron formula.

Slide 10   With severe climate change, the distribution of ground water on Earth will be altered. Some lakes may dry up while other may expand. The present climate model predicts that with four times CO_2, the eastern half of the United States will become wetter and the western half will be drier; the Amazon and African rain forests will be wetter, but Greenland and Antarctica will be drier. Soil moisture is determined largely by the balance between precipitation and evaporation. In general, the Clausius–Clapeyron formula states that in a warmer climate the atmosphere can hold more water vapor, implying the likelihood of increased precipitation. But a warmer climate also increases the efficiency of evaporation. Shifting patterns in the atmospheric circulation also play an important role.

Slide 11   Sensible heat flux is proportional to the difference between the ground temperature and the surface air temperature. Greater wind speed increases this flux and greater atmospheric stability decreases it. Greater instability leads to convection cells and eventually to turbulence in which sinking cold air encounters the ground that increases the sensible heat flux. Water vapor heat flux includes both the sensible and latent heat content of the vaporized ground water and ice that escapes into the atmosphere. Negative water vapor flux indicates that dew exceeds evaporation; moist air condenses onto the cold ground, warming it because latent heat is released. The sensible and latent heat fluxes are important mechanisms, along with radiation, in establishing the vertical as well as the horizontal temperature structure of the atmosphere.

Slide 12   Clouds are minute particles of liquid water or ice suspended in the air. Clouds are important regulators of the Earth's global energy balance. During the day, clouds reflect sun-light, thus cooling the planet. Clouds at night-time keep the planet warmer by reducing the escape of thermal radiation to space. The high clouds are typically cirrus clouds made up of ice particles. They reflect some sun-light, but their greater affect is to reduce the outward going thermal radiation by absorbing some of it and reemitting some of it back downward toward to warm the surface. For the optically thicker low clouds, reflection of solar radiation is more important. Since thermal emission depends on the cloud temperature and the temperature of low clouds is closer to that of the ground than for cirrus clouds, low clouds are less able to reduce the outgoing thermal radiation.

Slide 13   Albedo is the ratio of the (spectrally integrated) reflected sun-light divided by incident sun-light. The global albedo cannot be computed by area-weighting the local albedo. Instead, both the numerator and the denominator need to be area-weighted separately, and then the ratio will be an accurate representation of the global albedo. The main difference between planetary (top-of-atmosphere) albedo and ground albedo is the intervening large influence of the atmosphere. Clouds, aerosols, and Rayleigh scattering by air molecules are the principal atmospheric contributors to the planetary albedo. After large volcanic eruptions, sulfuric acid particles in the stratosphere can also reflect sun-light to increase the planetary albedo and thus cool the Earth. For the most part, the atmosphere acts to increase the albedo, although in some locations such as Greenland and Antarctica, the surface albedo is higher than the planetary albedo.

Slide 14   The global annual insolation on the Earth is about 340 W/m^2. About 100 W/m^2 is reflected back to space leaving 240 W/m^2 is absorbed by the Earth. Of this, about 68 W/m^2 is absorbed by the atmosphere and 172 W/m^2 by the ground surface. There is an additional 1 W/m^2 globally that is received by the ground from warm rain. To maintain global energy balance equilibrium, 173 W/m^2 must be extracted from the surface by sensible, latent, and thermal radiation heat fluxes. Most of the solar radiation is absorbed in the tropical regions, with progressively less solar radiation absorbed at higher latitudes. By comparison, thermal radiation is more uniformly distributed with latitude. This implies that there is atmospheric and ocean transport of heat horizontally, moving heat energy from the tropics to the poles. For the Earth as a whole to maintain energy equilibrium, 240 W/m^2 must be emitted out to space via thermal radiation.

Slide 15   The greenhouse effect (W/m^2) is defined as the difference between the upward thermal radiation emitted from the surface minus thermal radiation emitted at top-of-atmosphere out to space. The strength of the greenhouse effect is a measure of the heat trapping efficiency of spectral absorption by greenhouse gases and by longwave cloud opacity. In the present model, contributions to the greenhouse effect are roughly 55% by water vapor, 22% by clouds, 18% by CO_2, and 5% by other minor (non-condensing) greenhouse gases such as methane, nitrous oxide, and ozone. Although water vapor and clouds are by far the largest contributors to the greenhouse effect, their contributions are feedback effects in response to changing temperature. The Clausius–Clapeyron formula causes water vapor to increase more rapidly than the applied radiative forcing due to the increase in CO_2, and other non-condensing greenhouse gases.

Slide 16   The greenhouse effect (W/m^2) is defined as the difference between the upward thermal radiation emitted from the surface minus thermal radiation emitted at top-of-atmosphere out to space. The strength of the greenhouse effect is a measure of the heat trapping efficiency of spectral absorption by greenhouse gases and by longwave cloud opacity. In the present model, contributions to the greenhouse effect are roughly 55% by water vapor, 22% by clouds, 18% by CO_2, and 5% by other minor (non-condensing) greenhouse gases such as methane, nitrous oxide, and ozone. Though water vapor and clouds are by far the largest contributors to the greenhouse effect, their contributions are feedback effects in response to changing temperature. The Clausius–Clapeyron formula causes water vapor to increase more rapidly than the applied radiative forcing due to the increase in CO_2, and other non-condensing greenhouse gases.

Slide 17   The reference geoid for the ocean surface is a level where the surface pressure gradient force is zero. This includes effects due to variations of both gravity and the Earth's rotation. The ocean surface height differs (left-hand panel) from this geoid because of wind stresses and local steric expansion. The reductions in stability of the water column and momentum stresses at the top cause mixing in the upper ocean layers, from the ocean surface down to the mixed layer depth. There is a deficiency in the present ocean model in that there is insufficient mixing in the Labrador Sea, and too much mixing in the Greenland Sea (right-hand panel). These deficiencies arise from the complex interactions of the ocean circulation with sea-ice formation, melting, and transport, and with changing fresh water inputs and radiative environment. Improved ocean modeling is a key factor in improved climate model reliability.

Slide 18   Surface salinity is low in the Arctic Ocean because of the significant fresh water river input and low evaporation. Salinity is also low adjacent to California because the ocean temperature there is cold which reduces evaporation. On the other hand, surface air crossing the Sahara and Arabian deserts is very dry. When this air encounters the ocean, it causes enhanced evaporation resulting in greater salinity. High salinity water, escaping at lower depths from the Mediterranean Sea at Gibraltar, spreads all the way across the Atlantic Ocean. Salinity, because it directly affects the density of sea water, along with the water temperature are the principal is the principal controlling factor that affect the global circulation of the ocean, and thus the transport of heat energy pole-ward from the tropics.

Slide 19   In both the atmosphere and the ocean, horizontal velocity is prognostic whereas vertical velocity is computed diagnostically from the convergence of horizontal mass fluxes. This computation is sensitive to numerical instabilities and checkerboarding which is reduced by numerical filters, although it is still noticeable in the atmosphere. Most of the large vertical velocities occur near mountainous regions in the atmosphere (left panel) and next to coast lines in the ocean (right panel). Both water vapor and latent heat energy, which are generated at the surface by evaporation, must first be transported vertically before they are dispersed horizontally by atmospheric winds. In the oceans, near-surface and deep water circulation patterns are quite different, and must be bridged in key locations by vertical velocities.

Slide 20   The side of the Earth facing the Moon is more strongly pulled by the Moon's gravity and the opposite side is pulled more weakly. These three-dimensional variations of gravity, by the Moon and Sun, cause the Earth's tides. The maximum horizontal tidal acceleration is about 10^{-7} (m/s^2). Over three hours, the change in velocity can be as large as 1 (mm/s) at time-varying locations. Tides were ignored when the video starts, ocean surface height was relatively calm, but then tides are invoked. Notice the counter-clockwise motion in the North Atlantic. (Click on power-point figure to activate lunar tide movie.)

SECTION 2

Radiative Processes

CLIMATE LECTURE 4

Atmospheric Radiation

Valdar Oinas

NASA Goddard Institute for Space Studies, New York, NY, USA

Valdar Oinas The climate system is well known for its great complexity and complex interactions that involve dynamic, thermodynamic, radiative, chemical, biological and human-driven processes. This view of the climate system has emerged from detailed measurements, meticulous record keeping, and theoretical analyses arising from, and made possible by the science and technology revolution that greatly advanced our understanding of the role of physical processes that operate in the global climate system. These measurements also show very clearly that the global surface temperature has been rising over the past century, and that this is a consequence of human industrial activity.

Key Points

Most basic are the near-daily measurements of surface air temperature at meteorological stations around the globe, some dating back to the 18th century. There are also precise measurements of the key atmospheric greenhouse gases (CO_2, CH_4, N_2O, O_3, CFCs) that are associated with fossil fuel burning and with other human industrial activities. Accurate monitoring of changes in solar irradiance and other radiative forcings that impact the climate system is likewise important. But these latter effects are small compared to the anthropogenic greenhouse gas forcings. The other important components of the scientific documentation of global climate change include field campaigns, theoretical studies, and laboratory analysis of climate-constituent radiative properties.

Given the limited length of recorded climate data and the complexity of the natural variability that is superimposed on the steadily increasing global warming component, statistical analyses of the measured data alone are not sufficient to establish a convincing cause-and-effect relationship between increasing atmospheric greenhouse gases and the increase in global surface temperature. Comprehensive global climate models (GCMs) are needed to quantify, attribute, and completely understand the relative importance of the individual components of the climate system, and their role and contribution to the ongoing global warming and climate change.

Fortunately, a basic understanding of the ongoing global climate change does not require complete knowledge of all physical processes that operate in the climate system. Conservation of energy is the overarching principle that constrains all of the changes that take place in the climate system. Absorbed solar energy (240 W/m² out of 340 W/m² global annual-mean incident solar energy) is virtually the entire source of energy input to the climate system. Other non-solar energy inputs to the climate system come from geothermal and tidal friction energy sources, and waste heat from nuclear energy and fossil fuel burning, together contribute approximately 0.1 W/m² to the global annual-mean energy budget, and are therefore of little consequence. This suggests then that atmospheric radiation is the key physical process that best describes the global energy balance of the Earth. Thus, the global-mean surface temperature of the Earth can be explained as being sustained by the steady input of absorbed solar shortwave (SW) heat energy, warmed further by the greenhouse effect, and kept in energy balance by longwave (LW) thermal cooling to space.

The spectral distribution of the incident and reflected solar energy, and of the emitted LW thermal energy to space, provides a full description of the global radiative energy balance. Given the spatial distribution of water vapor, clouds, aerosols, and other atmospheric gases, radiative analysis explains the surface temperature and thermal structure of the atmosphere. Convection and advection of heat also help define the atmospheric temperature structure, but such dynamic transports of energy are orders of magnitude slower than the radiative transports, and thus, do not directly interfere with, or otherwise impede, the radiative analysis of global energy balance.

Nature of Atmospheric Radiation

There are multiple aspects of electromagnetic radiation that are exceedingly complex. However, much of this complexity is not directly relevant to the radiative transfer modeling that is required for global climate studies. The key concepts that are needed to understand the nature of global warming and climate change are a working knowledge of the atmospheric greenhouse effect and of the radiative transfer modeling of heating by solar radiation and cooling by thermal emission. Explicit knowledge of the speed of light, particle-wave duality, or the probabilistic nature of quantum transitions is not specifically required in climate-related applications (Petty, 2006).

Nevertheless, explicit numerical modeling of solar and thermal radiation is needed to determine the radiative heating and cooling rates at all points of the atmosphere and at the ground surface. This requires the differencing of height-resolved spectrally integrated solar and thermal fluxes, which need to be computed accurately. It is also important that the wavelength dependence of the spectral absorption and scattering by atmospheric constituents, and the angle dependence of the scattered and thermally emitted radiative fluxes be calculated accurately. For the absorbing gases in the atmosphere (e.g., water vapor, CO_2, and ozone), their spectral absorption consists of many thousands of individual absorption lines that need to be explicitly taken into account. Of equal importance is the angle dependence of both solar and thermal radiation since the amount of solar radiation heating the atmosphere depends on solar zenith angle, while atmospheric temperature gradients cause thermal radiation to depend on emission angle. Then too, solar radiation reflected by clouds, aerosols, and by the ground surface is characteristically angle dependent.

Solar Radiation

Solar radiation illuminates the Earth in the form of an effectively parallel beam of radiative flux that delivers energy at the rate of 1360 W/m^2 referenced at the annual-mean Sun–Earth distance. This energy, averaged globally, is 340 W/m^2 of incident solar radiation. It is distributed over the spectrum from the UV to microwave wavelengths, with much of the energy residing at visible wavelengths. Solar energy arrives in minute energy packets (photons) containing energy $E = h\nu$, where h is Planck's constant, ν is the frequency, which is related to the speed of light by $c = \lambda\nu$, where λ is expressed in wavelength units. Basically, there are three things that can happen to incoming solar energy photons when they encounter an atmospheric constituent — they can be scattered, or absorbed, or continue forward in the original direction. Photons scattered upward into the upper hemisphere constitute *reflection*, while the photons scattered downward constitute *diffuse transmission*. Photons continuing in their original direction constitute *direct transmission*. These options impart a probabilistic nature to atmospheric radiation modeling. The interaction strength with which atmospheric constituents interact with radiation is expressed by their spectral *optical depth* τ (a dimensionless ratio of constituent optical cross-section area relative to the area covered by the incident radiation). An optical depth of $\tau = 1$ implies an absorber (or a scattering agent) with a cross-section area that is just equal to the incident beam area (with the constituent compacted into a thin opaque layer perpendicular to the beam direction, without overlapping).

Radiative constituents are both exceedingly tiny and exceedingly numerous, and are randomly distributed along the light-beam path. In the mathematical limit as randomly distributed particles become ever smaller, convergence takes the form of the exponential function $e^{-\tau}$, which accounts for the shadowing due to random particle overlap. This expression is recognized as the *direct beam transmission*. Without the random overlap, an absorber amount of $\tau = 1$ would have totally extinguished the incoming radiation beam. But, for randomly overlapping small particles of optical depth $\tau = 1$, significant transmission remains, and is given by $e^{-1} = 0.368$.

Mie Scattering

The wavelength and viewing-geometry dependence of atmospheric radiation can be complicated. But this dependence is actually well understood, and it can be included accurately and efficiently in climate modeling and remote sensing applications. Although requiring substantial computing effort, Mie theory is a very precise recipe for computing scattering of electromagnetic radiation by homogeneous spheres. Derived by Gustav Mie in 1908, it explains why scattering of light by atmospheric particulates is far more important for solar radiation than it is for thermal radiation. Because clouds are non-absorbing at visible wavelengths, they scatter visible light. But clouds are strongly absorbing at thermal wavelengths. The ratio of particle size to the wavelength of incident light determines the magnitude and wavelength dependence of particle scattering, and also the degree to which radiation is scattered in the forward direction. The key input parameters that define Mie scattering are the *particle size* and (wavelength-dependent) *refractive index*.

For particles that are very small relative to the wavelength of light, Mie scattering efficiency is inversely proportional to the fourth power of the wavelength of the incident light. This scattering is also largely isotropic and is known as Rayleigh scattering. The prime example is scattering by air molecules that cause the characteristic blue sky color, since the shorter wavelengths are more strongly scattered relative to longer wavelengths. Cloud droplets (10 micrometers in radius) are significantly larger than the wavelength of visible light, so their scattering efficiency is largely independent of the wavelength, causing clouds to appear white. Another key characteristic of Mie scattering is that large particles tend to scatter light more efficiently in the forward direction compared to smaller particles. Aerosols (typically smaller than the wavelength) have scattering characteristics that are intermediate between Rayleigh scattering and scattering by clouds.

Mie scattering is computationally too intense to be performed directly in climate model radiative calculations. Instead, off-line calculations are made for a wide range of cloud particle sizes using laboratory refractive indices that cover the entire spectrum from the UV to the microwave. It is variations in the imaginary part of the refractive index that determine the strength of cloud and aerosol wavelength dependent absorptivity. Absorptivity is near-zero at visible wavelengths for water, ice, and most aerosol compositions. Hence, these substances scatter visible light without absorbing it. Water, ice, and aerosols have several absorption bands at near-IR wavelengths, and are strongly absorbing at most thermal wavelengths. Accurate modeling of the angle dependence of scattering by clouds and aerosols is the main concern when modeling solar radiation, while spectral absorption is the principal cloud and aerosol effect at thermal wavelengths.

Since cloud and aerosol particles interact with radiation at all wavelengths of the spectrum, Mie scattering calculations are tabulated to define the radiative parameters at all solar and thermal wavelengths. Specifically, the wavelength-dependent radiative parameters are: (1) the *extinction efficiency factor*, Qx, which is the ratio of the Mie extinction cross-section relative to the particle geometric cross-section and can range from being much smaller than unity to as large as a factor of four; (2) the *single scattering albedo*, ω_o, which is the non-absorbed fraction of light in a single scattering event ($\omega_o = 1$ for conservative scattering, $\omega_o = 0$ for total absorption); and (3) the *asymmetry parameter*, g; it defines the degree of forward scattering ($g = 1$, forward directed; $g = -1$, back directed; and $g = 0$, isotropically scattered). Derived from an exact electromagnetic theory and laboratory refractive indices, these radiative parameters serve as input to a radiative transfer model to calculate *multiple scattering* by atmospheric constituents and reflection by the ground surface. In climate GCMs, these calculations determine where solar radiation heats the atmosphere, how much is absorbed by the ground, and what fraction is reflected back to space.

Multiple Scattering of Solar Radiation

An effective way to calculate multiply scattered atmospheric radiation is the energy-conserving doubling-and-adding algorithm described by Lacis and Hansen (1974). This algorithm accurately reproduces the spectral and angular dependence of Mie scattering results for clouds and aerosols for any specified optical depth, atmospheric distribution, and solar zenith angle geometry. The model also accounts for the absorption of solar radiation by atmospheric absorbing gases such as water vapor, ozone, and CO_2. Furthermore, the radiation model has the built-in capability to include or exclude *individual* scattering and absorbing constituents in prescribed combination. This enables the evaluation of each constituent's contribution to the total radiation budget by systematically evaluating the non-linear radiative interactions between the different radiative components. In performing such attribution for an annual-mean 1980 atmosphere, it is found that, about 30% of the incident solar radiation is reflected back to space, and of this, about 18% is by clouds, 7% by the ground surface, 4% by Rayleigh scattering, and about 1% by aerosols. Of the 70% of incident solar radiation that is absorbed by the Earth, about 50% is absorbed by the ground surface. About 20% is absorbed within the atmosphere, of which about 13% is absorbed by water vapor, 3% by clouds, 2% by ozone, and roughly 1% by oxygen and 1% by aerosols.

Longwave Thermal Radiation and Greenhouse Effect

Similar flux attribution can be performed for thermal LW radiation. The land–ocean surface, water vapor, clouds, CO_2, and ozone are the principal emitters and absorbers of thermal radiation. Because of strong absorption by clouds and aerosols at thermal wavelengths, thermal radiation does not have to deal with the complexities of multiple scattering that are characteristic for solar radiation. With the global mean surface temperature being close to 288 K, the Planck radiation law requires that the LW thermal flux emitted by the land–ocean surface must be about 390 W/m². For energy balance equilibrium, the global-mean outgoing LW flux at the top of the atmosphere (TOA) must be 240 W/m² in order to balance the absorbed solar radiation. The 150 W/m² flux difference between the 390 W/m² emitted upward by the land–ocean surface and the TOA 240 W/m² going out to space is generated through the absorption and re-emission of thermal radiation by the atmospheric constituents. This 150 W/m² flux difference is a direct measure of the strength of the terrestrial greenhouse effect.

In attributing the 150 W/m², water vapor accounts for about 50% of the total greenhouse effect, LW cloud opacity contributes about 25%, while the *non-condensing* greenhouse gases (CO_2, CH_4, N_2O, O_3, CFCs) account for the remaining 25%, with the strongest contributor CO_2, accounting for 20% (Lacis *et al.*, 2010). If the LW gaseous absorption and cloud opacity were to be removed from the atmosphere, strong surface cooling would ensue. Initially, the outgoing LW flux would be 390 W/m², a value not sustainable by the 240 W/m² of absorbed solar radiation. The global surface temperature would then drop rapidly to 255 K in order to maintain global energy balance. By partitioning of the LW flux contributors into *condensing* and *non-condensing* species, it becomes workable to infer a climate feedback sensitivity based only on the radiative structure of the atmosphere. This assessment follows because the atmospheric distribution of water vapor is established by fast feedback processes in which the condensation of water vapor is constrained by the temperature dependence of the Clauius–Clapeyron relationship. Accordingly, about 75% of the greenhouse effect is contributed by the fast feedback processes, while only 25% is due to radiative forcing by the non-condensing greenhouse gases. Hence, the direct radiative heating caused by the increase in atmospheric CO_2 is magnified by nearly a factor of three by the water vapor and cloud feedback effects. This makes atmospheric CO_2 as the effective control knob that governs the global surface temperature of Earth (Lacis *et al.*, 2013).

Technical Issues

The basic role of radiation in climate modeling is to calculate SW and LW radiative fluxes from which global energy balance and radiative heating and cooling of the atmosphere are determined. There are also important climate process details that add to the complexity of radiative modeling. Hygroscopic aerosols (e.g., sulfates, nitrates, and sea salt) change size in response to changes in relative humidity. Similarly, cloud particle size, optical depth, and water/ice phase also change as meteorological conditions change. All of these changes have radiative consequences, and they affect the intensity and angle dependence of the scattered radiation. These effects are included in parameterized form in GCM radiation calculations. While the radiative effects of such structural changes may be small, they all need to be fully taken into account since the resulting changes in particle size and refractive index cause the Mie scattering parameters to change, thus altering their cumulative radiative contributions to climate change forcing (Hansen and Travis, 1974).

Computational speed is a key constraint when performing climate model radiation calculations. The principal atmospheric absorbing gases (H_2O, CO_2, O_3) each have literally tens of thousands of spectral absorption lines spread over the spectrum. These absorption lines each have different strengths, with line-widths that are strongly dependent on atmospheric pressure and temperature. Line-by-line calculations are needed for accurate determination of their radiative contributions but are prohibitively costly for climate modeling use. However, since only the spectral integral is actually needed, the absorption lines can be regrouped (off-line) according to absorption strength by using the correlated *k*-distribution approach (Lacis and Oinas, 1991). This approach, using angle-dependent thermal emission, retains line-by-line precision at much-reduced computing cost (Lacis *et al.*, 2013). Similarly, precise multiple scattering calculations of the solar zenith angle dependence of scattered radiation are prohibitively expensive to employ within a climate model. A single-quadrature version of the doubling-and-adding method was adapted to enable accurate calculation solar radiation multiple scattering in a climate modeling setting (Hansen *et al.*, 1983).

Remote Sensing Applications

Besides global energy balance and radiative heating and cooling rates, atmospheric radiation has another key role in remote sensing applications. Satellite observations of reflected solar radiation and emitted thermal radiation, including lidar, radar, and narrow spectral channel measurements, are used to determine the radiation balance of Earth, along with the retrieval of cloud and aerosol radiative properties, surface and atmospheric temperature, vegetation cover, water vapor, ozone, CO_2, and other absorbing gas distributions. All of this radiation-based satellite-measured climate system information has been used extensively in a comparative and diagnostic sense to test and evaluate climate model performance.

Typically, satellite remote sensing measurements utilize the variability of the *spectral intensity* (sometimes with very high spectral resolution) of the reflected solar radiation or emitted thermal radiation. This approach based on intensity-only measurements produces ambiguity (for aerosol radiative properties) because detectors cannot readily distinguish radiation scattered by aerosols from reflectivity by the ground surface or due to sub-pixel cloud contamination. *Polarimetric* measurements provide an enhanced retrieval discrimination capability since radiation reflected from the ground surface tends to be unpolarized, while cloud and aerosol particles have distinct polarimetric signatures. This enabled Hansen and Hovenier (1974) to unambiguously identify the visible Venus cloud deck composition as 1 micrometer in radius sulfuric acid droplets. Similar polarimetric measurements were planned to retrieve quantitative aerosol information for Earth (Mishchenko *et al.*, 2007), but the 2011 NASA *Glory* mission launch failure set back this plan.

References

Hansen, J.E. and Hovenier, J.W. (1974). Interpretation of the polarization of Venus. *Journal of Atmospheric Sciences*, 31, 1137–1160, doi:10.1175/1520-0469(1974).

Hansen, J., Russell, G., Rind, D., Stone, P., Lacis, A., Lebedeff, S., Ruedy, R. and Travis, L. (1983). Efficient three-dimensional global models for climate studies: Models I and II. *Monthly Weather Review*, 111, 609–662, doi:10.1175/1520-0493(1983).

Hansen, J.E. and Travis, L.D. (1974). Light scattering in planetary atmospheres. *Space Science Reviews*, 16, 527–610, doi:10.1007/BF00168069.

Kopp, G., and Lean, J.L. (2011). A new, lower value of total solar irradiance: Evidence and climate significance, Geophys. Res. Lett., 38, *L01706, doi:*10.1029/2010GL045777.

Lacis, A.A. and Hansen, J.E. (1974). A parameterization for the absorption of solar radiation in the Earth's atmosphere. *Journal of the Atmospheric Science*, 31, 118–133, doi:10.1175/1520-0469(1974).

Lacis, A.A., Hansen, J.E., Russell, G.L., Oinas, V. and Jonas, J. (2013). The role of long-lived greenhouse gases as principal LW control knob that governs the global surface temperature for past and future climate change. *Tellus B*, 65, 19734, doi:10.3402/tellusb.v65i0.19734.

Lacis, A.A. and Oinas, V. (1991). A description of the correlated k distributed method for modeling nongray gaseous absorption, thermal emission, and multiple scattering in vertically inhomogeneous atmospheres. *Journal of Geophysical Research*, 96, 9027–9063, doi:10.1029/90JD01945.

Lacis, A.A., Schmidt, G.A., Rind, D. and Ruedy, R.A. (2010). Atmospheric CO_2: Principal control knob governing Earth's temperature. *Science*, 330, 356–359, doi:10.1126/science.1190653.

Mishchenko, M.I. (2009). Gustav Mie and the fundamental concept of electromagnetic scattering by particles: A perspective. *Journal of Quantitative Spectroscopy Radiative Transfer*, 110, 1210–1222, doi:10.1016/j.jqsrt.2009.02.002.

Mishchenko, M.I., Cairns, B., Kopp, G., Schueler, C.F., Fafaul, B.A., Hansen, J.E., Hooker, R.J., Itchkawich, T., Maring, H.B. and Travis, L.D. (2007). Accurate monitoring of terrestrial aerosols and total solar irradiance: Introducing the Glory mission. *Bulletin of the American Meteorological Society*, 88, 677–691, doi:10.1175/BAMS-88-5-677.

Petty, G.W. (2006). *A First Course in Atmospheric Radiation* (459 pp). Madison, WI: Sundog Publishing.

Tang, I.N. and Munkelwitz, H.R. (1991). Simultaneous determination of refractive index and density of an evaporating aqueous solution droplet. *Aerosol Science and Technology*, 15, 201–207.

Tang, I.N. and Munkelwitz, H.R. (1994). Water activities, densities, and refractive indices of aqueous sulfates and sodium nitrate droplets of atmospheric importance. *Journal of Geophysical Research*, 99, 18801–18808.

Slide 1

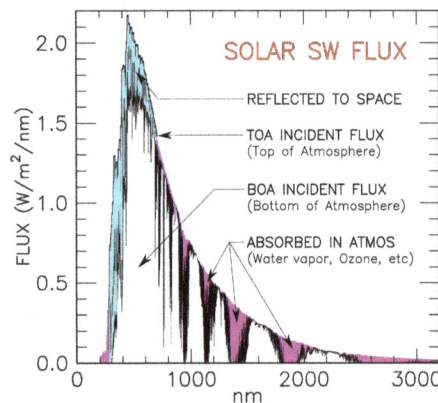

Atmospheric Radiation

Valdar Oinas

NASA Goddard Institute for Space Studies

Our Warming Planet: Topics in Climate Dynamics
Lectures in Climate Change, Vol. 1
2017

Slide 2

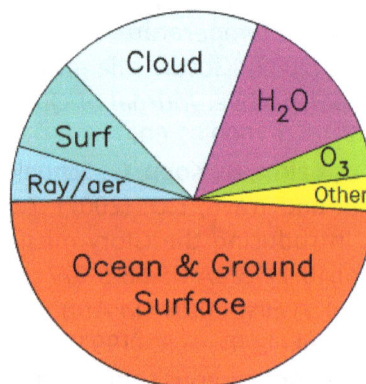

Spectral Distribution and Global Attribution of Solar SW Radiation

Solar radiation is the ultimate source of energy that supports the temperature structure of the terrestrial climate system.

Slide 3

Spectral Distribution and Forcing/Feedback Attribution of Thermal Radiation

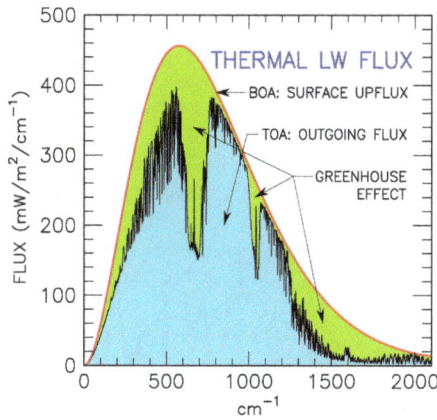

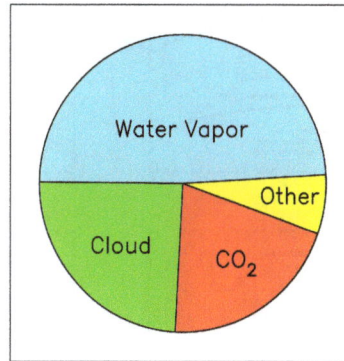

Top of atmosphere (TOA) spectral distribution of the outgoing thermal radiation emitted by Earth (left panel). By means of radiative modeling attribution (right-hand pie chart), it can be shown that water vapor accounts for nearly 50% of the terrestrial greenhouse effect, with clouds contributing about 25%, CO_2 about 20% and the minor GHGs (CH_4, N_2O, O_3, CFCs) the remaining 5%.

Oinas, V.   **Atmospheric Radiation**   3

Slide 4

Mie Scattering Size Dependence and Relative Spectral Extinction

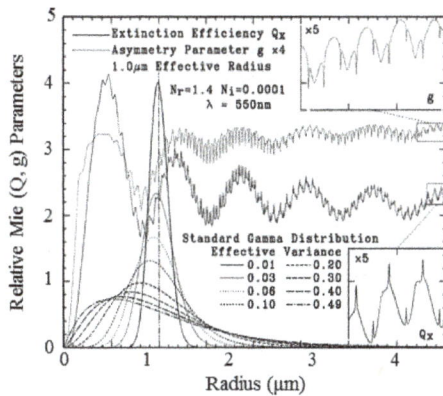

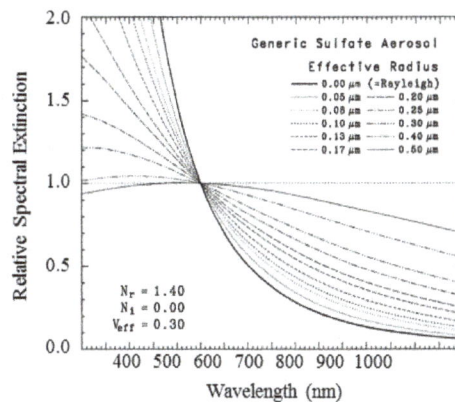

Optical properties (scattering, absorption) of atmospheric aerosols are a function of the size of the particle and the wavelength of the radiation.

Oinas, V.   **Atmospheric Radiation**   4

Slide 5

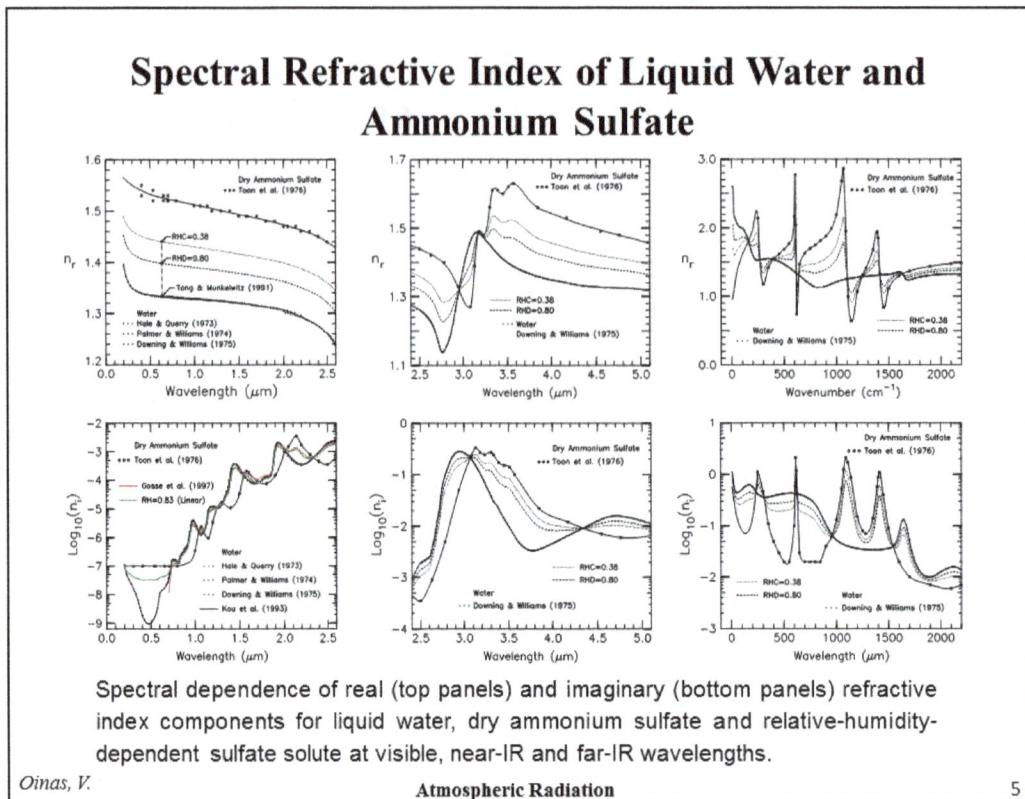

Spectral Refractive Index of Liquid Water and Ammonium Sulfate

Spectral dependence of real (top panels) and imaginary (bottom panels) refractive index components for liquid water, dry ammonium sulfate and relative-humidity-dependent sulfate solute at visible, near-IR and far-IR wavelengths.

Oinas, V.   **Atmospheric Radiation**   5

Slide 6

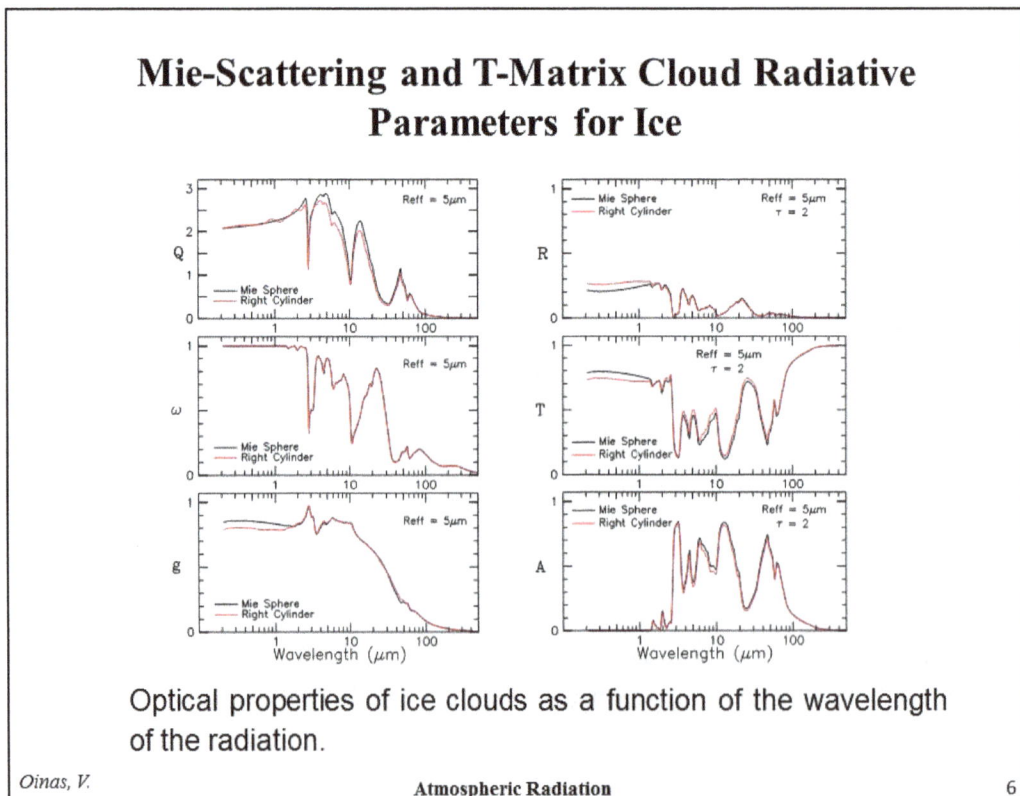

Mie-Scattering and T-Matrix Cloud Radiative Parameters for Ice

Optical properties of ice clouds as a function of the wavelength of the radiation.

Oinas, V.   **Atmospheric Radiation**   6

Slide 7

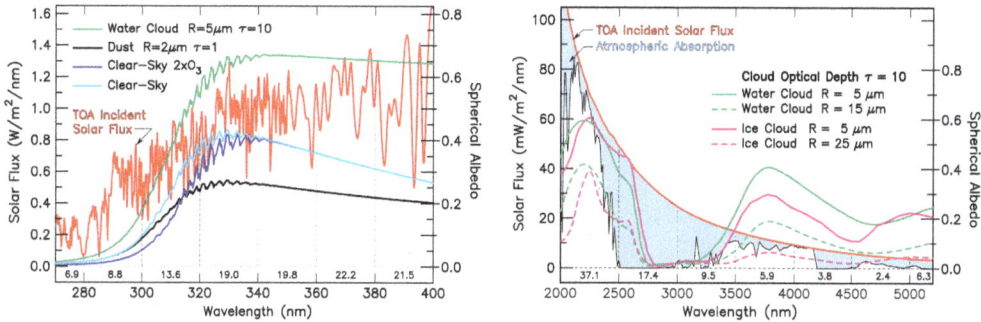

Cloud, Ozone, and Dust Radiative Effect on Solar SW Spectral Reflectivity

Radiative effects of cloud, ozone and mineral dust on the spectral dependence of reflected solar SW radiation.

Oinas, V.   **Atmospheric Radiation**   7

Slide 8

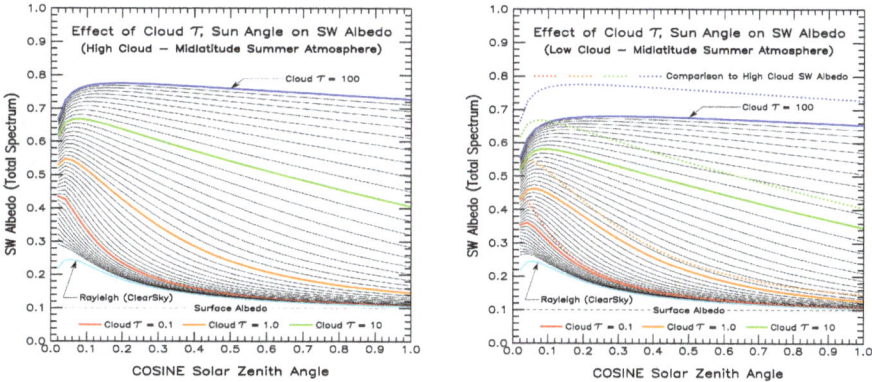

Cloud Optical Depth and Sun-Angle Effect on Total-Spectrum SW Albedo

Total-spectrum SW albedo dependence on cloud optical depth, solar zenith angle and cloud height.

Oinas, V.   **Atmospheric Radiation**   8

Slide 9

Relative Humidity Dependence of Hygroscopic Aerosol Radiative Parameters

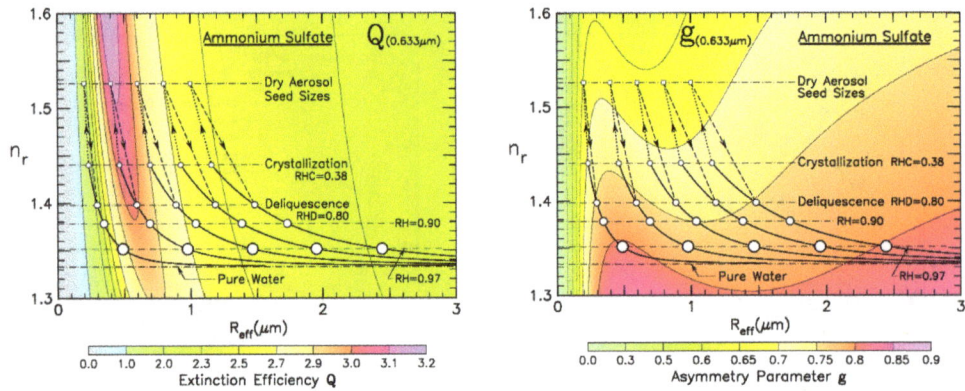

Hygroscopic aerosols absorb moisture and so their radiative properties depend upon the amount of moisture available.

Slide 10

Modeling Relative-Humidity-Dependent Properties of Hygroscopic Aerosols

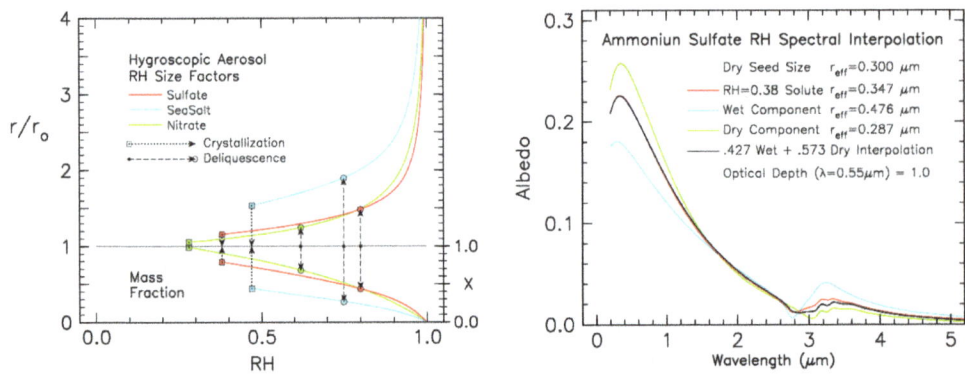

Further size and radiative dependencies of hygroscopic aerosols on relative humidity.

Slide 11

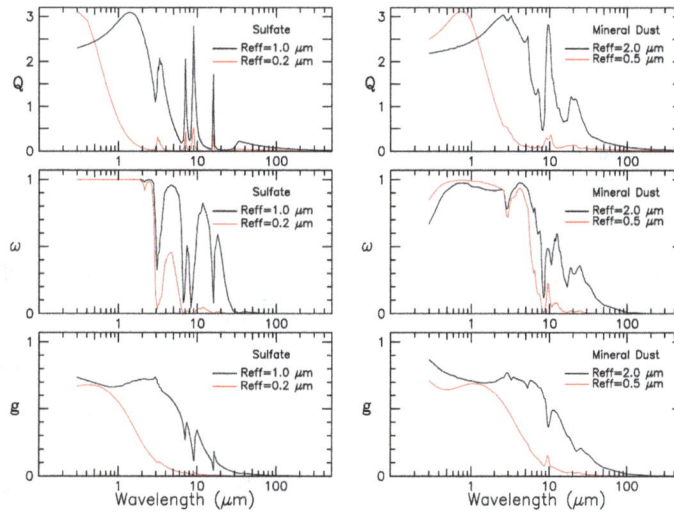

Mie Scattering Parameters for Sulfate and Mineral Dust Aerosol

Mie scattering of aerosols as a function of their size displays a complex relationship.

Oinas, V.   **Atmospheric Radiation**   11

Slide 12

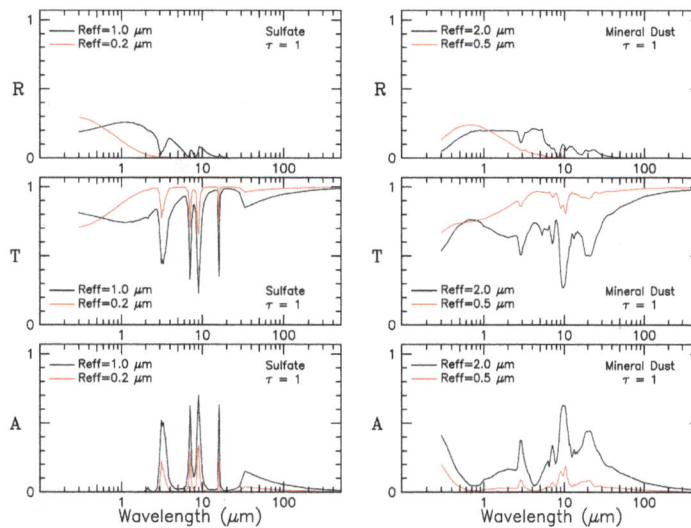

Reflection, Transmission, Absorption for Sulfate and Mineral Dust

Similarly, the impact of aerosols on atmospheric radiation is a complicated function of aerosol size and radiative wavelength.

Oinas, V.   **Atmospheric Radiation**   12

Slide 13

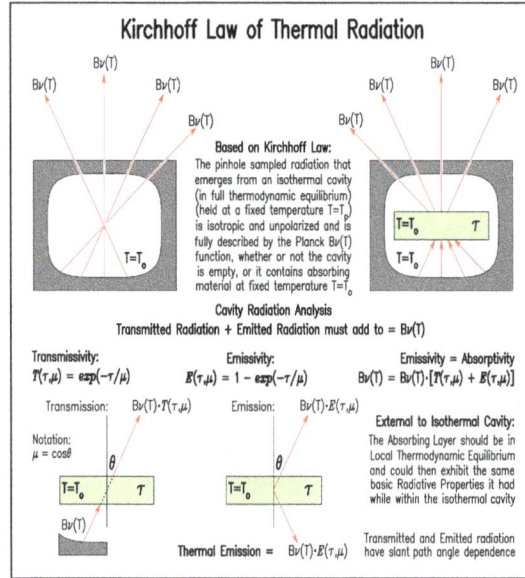

Physical Basis for Thermal Radiation Emission, Absorption, Transmission

Kirchhoff's Law and the isothermal cavity concept provide a convenient starting point to develop the radiative transfer formulations needed to calculate and model the emission, absorption and transmission of thermal radiation.

Oinas, V.   **Atmospheric Radiation**   13

Slide 14

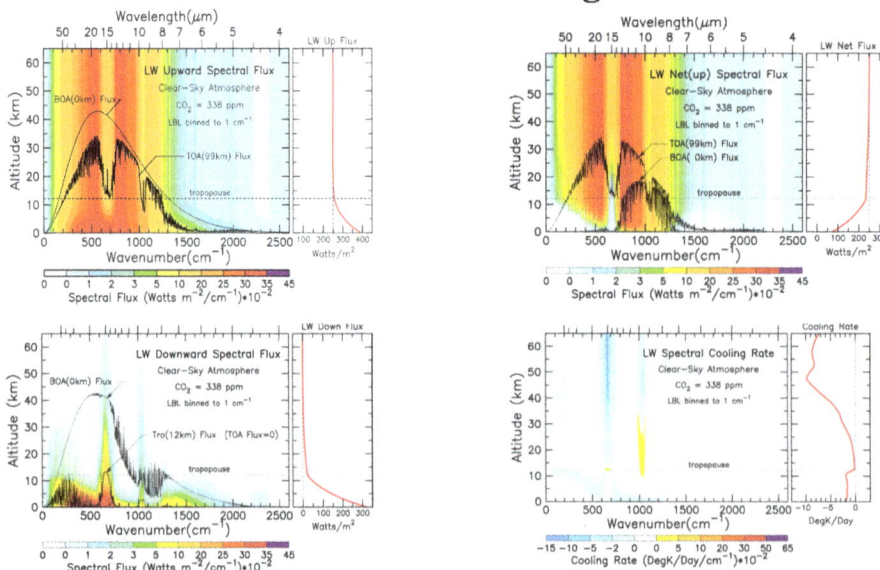

Line-by-Line Calculation of LW Spectral Fluxes and Radiative Cooling Rate

Line-by-Line CO_2 spectral fluxes and radiative and radiative cooling rate profiles.

Oinas, V.   **Atmospheric Radiation**   14

Slide 15

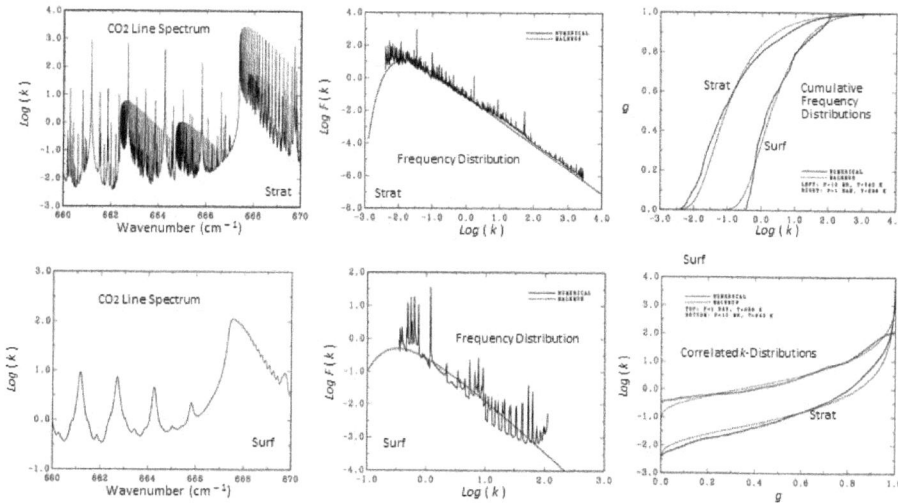

Converting CO$_2$ Line Spectrum into Correlated *k*-Distribution Form

The line-by-line absorption pattern for CO$_2$ has to be parameterized for use in the GCM.

Oinas, V.   **Atmospheric Radiation**   15

Slide 16

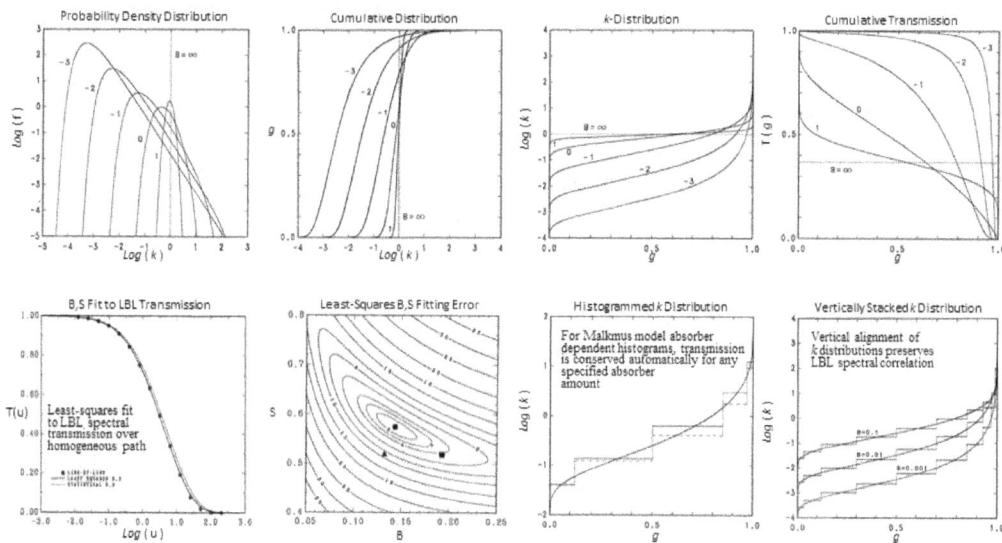

Malkmus Model Utilization in Formulating the Correlated *k*-Distribution

The Malkmus Model is used for this parameterization.

Oinas, V.   **Atmospheric Radiation**   16

Slide 17

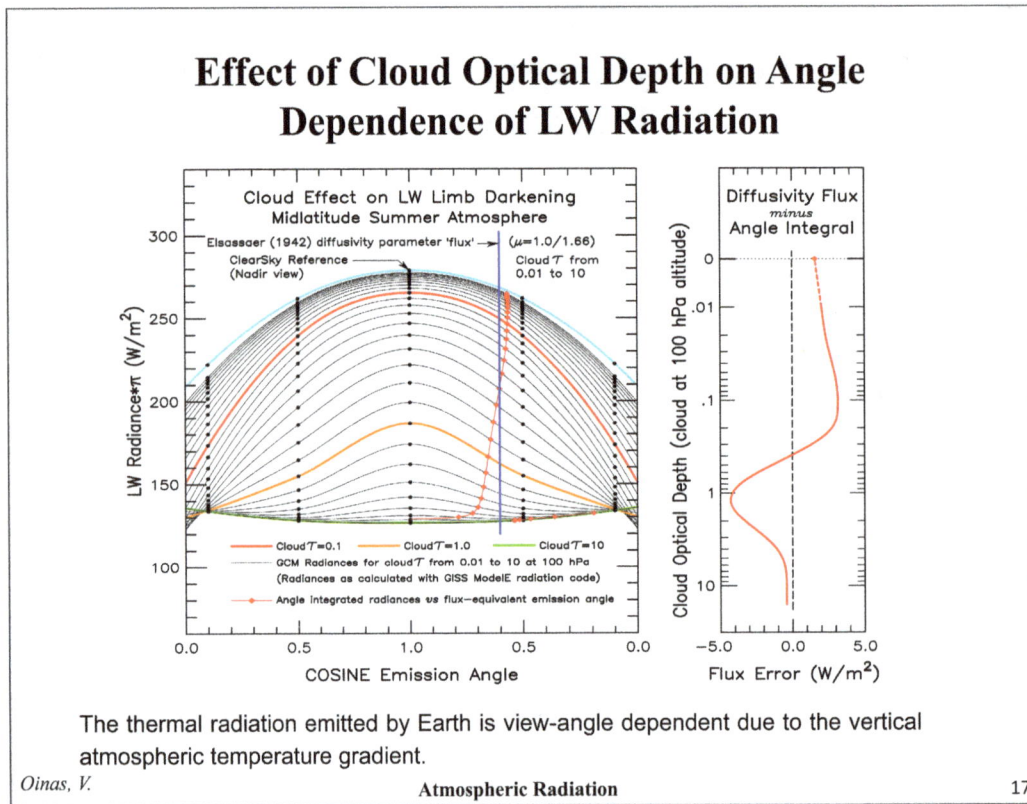

Effect of Cloud Optical Depth on Angle Dependence of LW Radiation

The thermal radiation emitted by Earth is view-angle dependent due to the vertical atmospheric temperature gradient.

Slide 18

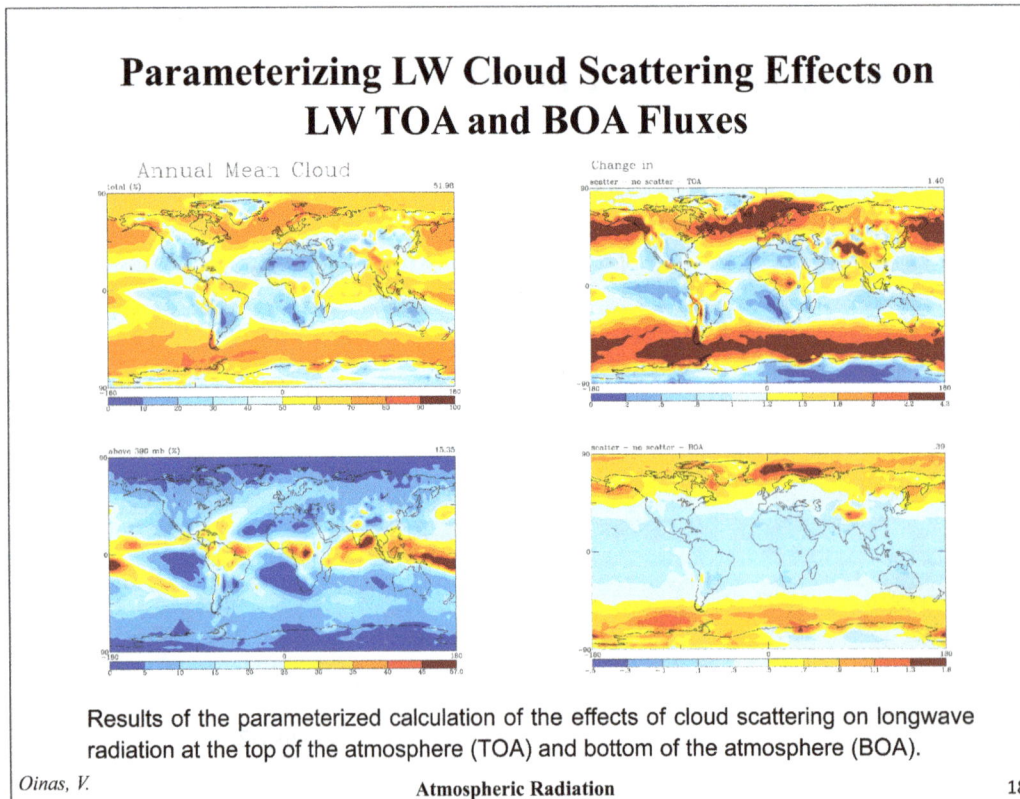

Parameterizing LW Cloud Scattering Effects on LW TOA and BOA Fluxes

Results of the parameterized calculation of the effects of cloud scattering on longwave radiation at the top of the atmosphere (TOA) and bottom of the atmosphere (BOA).

Slide 19

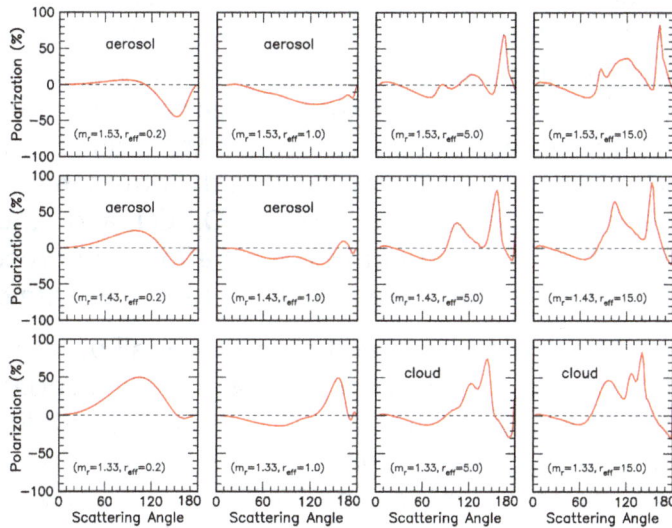

Polarimetric Scattering-Angle Signature: Set by Size and Refractive Index

The polarization of light by atmospheric aerosols and clouds is a function of particle size and refraction index.

Oinas, V.   Atmospheric Radiation   19

Slide 20

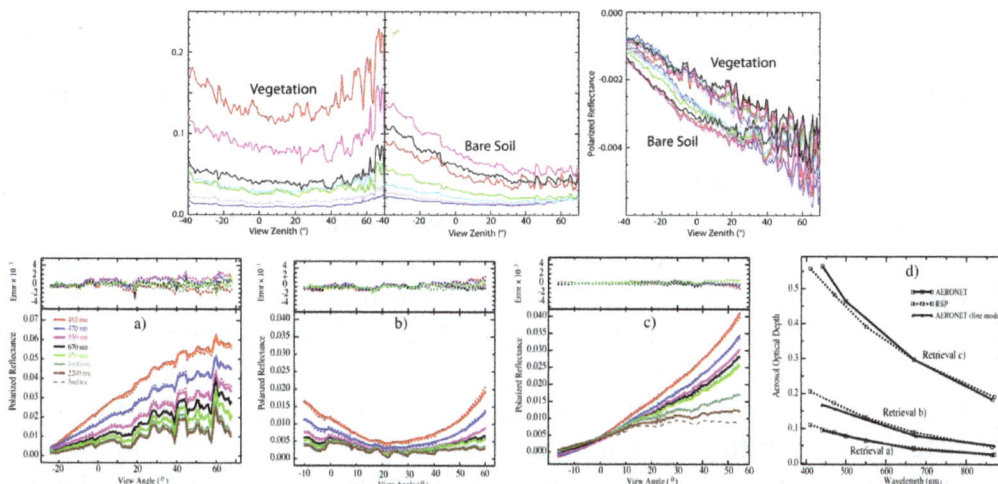

Performance Capability of Polarimetry in Remote Sensing Applications

Observations of the polarization properties of reflected light can help determine characteristics of the reflecting features.

Oinas, V.   Atmospheric Radiation   20

Slide Notes

Slide 2   Spectral distribution of incident solar radiation (left-hand panel) with indicated clear-sky spectral reflection due to atmospheric Rayleigh scattering (blue) and absorption (magenta) by water vapor, oxygen and ozone. Solar radiation is the ultimate source of energy that supports the temperature structure of the terrestrial climate system. Globally-averaged, the annual-mean incident solar radiation is about 340 W/m². As shown in the (right-hand) pie-chart, about 30%, or 100 W/m², is reflected directly back to space primarily by clouds, Rayleigh scattering, aerosols and by the ocean, land, and snow/ice surfaces, leaving a net 240 W/m² as the heat energy input to the terrestrial climate system. Approximately 20% (70 W/m²) of the incident solar radiation is absorbed within the atmosphere, primarily by water vapor, but also by ozone, clouds, aerosols, O_2, CO_2 and other minor atmospheric gases. Roughly half of the incident solar radiation is absorbed by the ocean and land surface. The results shown are from a reference 1980 atmosphere.

The Sun has been a remarkably steady source of energy over all of recorded history. Decades of precise measurements show a 0.1% variability associated with the 11-year sun-spot cycle. There is also a seasonal cycle in solar irradiance due to Earth's orbital eccentricity such that the incident solar energy in January is about 3.4% greater than the annual mean, and 3.4% less in July, even though the global temperature is about 4°C warmer in July than in January. The normal-incident solar radiation at the mean Sun-Earth distance is 1360.8 W/m² (Kopp and Lean, 2011).

Slide 3   Top of atmosphere (TOA) spectral distribution of the outgoing thermal radiation emitted by Earth (left panel). The annual-mean globally-mean LW thermal flux emitted upward by the ground (surface temperature 288 K) is about 390 W/m² (red line). The blue area depicts the spectral distribution of the outgoing LW thermal radiation emitted to space by the Earth (about 240 W/m²). This is expected for global energy balance with the 240 W/m² of absorbed SW solar radiation. The flux difference of 150 W/m² (green area) between the upwelling LW flux at ground surface and outgoing LW flux at TOA is a measure of the strength of the terrestrial greenhouse effect. It is not a simple atmospheric transmission effect, arising instead from a series of absorption and emission events which depend on the spectral absorption strength and occur throughout the atmosphere.

This 150 W/m² greenhouse effect is due entirely to the LW spectral absorption by the atmospheric greenhouse gases and by the spectral cloud opacity. By means of radiative modeling attribution (right-hand pie chart), it can be shown that water vapor accounts for nearly 50% of the terrestrial greenhouse effect, with clouds contributing about 25%, CO_2 about 20%, and the minor GHGs (CH_4, N_2O, O_3, CFCs) the remaining 5%. This also shows that CO_2 and the minor GHGs represent the radiative forcings of the climate system (since they remain in the atmosphere virtually indefinitely once injected), while water vapor and clouds are the fast feedback effects of the climate system changing in response to local meteorological conditions. Thus, in the terrestrial atmosphere, feedback effects multiply the direct CO_2 or GHG greenhouse radiative forcing by about a factor of three.

Slide 4   Mie scattering extinction efficiency (**Qx**) and asymmetry (**g**) parameter dependence on particle size. Scattering of electromagnetic radiation by homogenous spherical particles is remarkably complex. As the particle changes minutely in size, there is changing interference between light transmitted though the particle, refracted around the particle, and light reflected internally within the particle, to produce the low-frequency and high-frequency fluctuations in **Qx** and **g** (shown magnified by a factor of 5 in the left-hand inset). Extinction efficiency **Qx** is the ratio of the Mie scattering cross-section relative to the particle's geometric area, ranging from zero in the small particle limit to as large as a factor of 4. The asymmetry parameter **g** is similarly near-zero in the small particle limit, approaching unity in the large particle limit (multiplied by a factor of 4 in the figure). In Mie scattering, the particle size (R) and wavelength (λ) of the scattered light are reciprocally related through the ratio $X = 2\pi R/\lambda$. In practice, cloud and aerosol particles are never found to be mono-modal, being described instead by a particle size distribution (e.g., standard Gamma) characterized by an effective radius Reff and effective variance Veff.

The right-hand panel shows the relative spectral dependence of the Mie extinction efficiency **Qx** for typical sulfate aerosol particles ranging from near-zero to 0.5 μm in size, with a moderately broad effective variance of Veff = 0.3. At near-zero size, aerosol particles exhibit the strong λ^{-4} (blue-sky) Rayleigh spectral dependence. As particle size increases beyond Reff = 0.5 μm, spectral extinction becomes increasingly flatter (and whiter).

Slide 5   Spectral dependence of real (top panels) and imaginary (bottom panels) refractive index components for liquid water, dry ammonium sulfate and relative-humidity-dependent sulfate solute at visible, near-IR and far-IR wavelengths. Radiative properties of solids and liquids are defined by their complex refractive index. The real part of the refractive index **nr** refers to overall substance refractivity, while the imaginary part **ni** describes the degree of absorption. For typical substances, **nr** tends to vary slowly across visible wavelengths, exhibiting spectral features that are associated with absorption bands in the near and far-IR spectral regions. The imaginary refractive index is small in the visible for both water and sulfate, increasing in strength and variability toward the infrared. Laboratory data are the basic source of refractive index information. Particle size and the complex refractive index are the key input parameters to Mie scattering calculations needed to compute cloud and aerosol radiative properties.

Slide 6   Spectral dependence of Mie scattering (spherical) and T-matrix (non-spherical) radiative parameters (Q_x, w_o, g) for 5 mm ice-cloud particles (left-hand panels). Mie scattering is an exact theory for spherical particles. T-matrix is an exact theory for randomly oriented non-spherical particles and is depicted here for 5 mm effective-radius equivalent-sphere right cylinders with a 1:1.5 aspect ratio. The principal effect of particle non-sphericity is a decrease in forward scattering, manifested by decreased asymmetry parameter g for wavelengths shortward of 2 mm, resulting in enhanced reflectivity and decreased transmission of SW radiation (right-hand panels). This is also the spectral region of essentially zero absorption by cloud particles. A doubling/adding radiative transfer model is used to convert the input (Q_x, ω_o, g) radiative parameters into the physically more relevant (R, T, A) reflection, transmission and absorption information.

Slide 7 Radiative effects of cloud, ozone and mineral dust on the spectral dependence of reflected solar SW radiation. The left-hand scale of the left panel depicts the spectral variation of solar flux in the UV part of the spectrum (red curve). The Spherical Albedo scales on the right side of the figures depict the fraction of the incident solar radiation as function of wavelength (integrated over solar zenith angle) that is reflected back to space. Most evident in the UV is the strong absorption by ozone. UVA radiation (340–400 nm) is scarcely affected by ozone, while the UVC radiation (<290 nm) is so strongly absorbed that scarcely any reaches the ground. On the other hand, UVB radiation is modulated by ozone amount as indicated by the dark-blue curve for doubled O3 relative to the clear-sky Rayleigh scattering. The green curve depicts the spectrally steady reflectivity (>340 nm) of a typical ($\tau = 10$) water cloud, with little dependence on particle size or cloud height. By comparison, absorption by ($\tau = 1$) desert dust results in less reflection than the clear-sky Rayleigh scattering case.

In the near-IR (right-hand panel), the spectral dependence of reflected radiation is more complicated. Within the gray areas, there is strong absorption by water vapor and CO_2, so little of the incident radiation is reflected. But there are significant 'window' regions (2000–2500 nm and 3500–4000 nm) where clouds have opportunity to reflect some of the incident solar radiation. However, because clouds exhibit some absorptivity in the near-IR spectral region, the cloud radiative parameters become more sensitive to particles size since for larger particles there is more opportunity for light traversing the particle to be absorbed, than is the case for smaller particles. Also, in the near-IR, there is a substantial dependence of cloud reflectivity on water/ice phase. In the visible spectrum where cloud absorption is small, the light path difference has a much smaller radiative effect.

Slide 8 Total-spectrum SW albedo dependence on cloud optical depth, solar zenith angle and cloud height. The left-hand panel shows the reflected spectrally integrated SW flux at the top of the atmosphere (plane albedo) as a function of the solar zenith angle (sun-rise to direct over-head), calculated with the GISS ModelE radiation code for high-cloud (cirrus) optical depth ranging from near-zero ($\tau = 0.01$) to optically extreme ($\tau = 100$). For reference, the blue-green line at the bottom is for clear-sky Rayleigh scattering. The cloud optical depth is logarithmically incremented with the decadal values ($\tau = 0.1, 1, 10, 100$) identified by the red, orange, green and blue lines, respectively. What is evident is the strong solar zenith angel dependence for small optical depths which gradually flattens out as cloud optical depth increases. Of note is the marked decrease in reflectivity at low sun angles, caused by increased ozone absorption along an increasing atmospheric path.

The right-hand panel displays the same basic results as the left-hand panel, except that the cloud is located at low altitude near ground level. The difference now is that the tropospheric gases, primarily water vapor, are located above the cloud, and are able to absorb the incident radiation on its way down, as well as the reflected radiation on its way up. The colored dotted lines corresponding to the decadal optical depth values quantify the amount of this absorption.

Slide 9   Relative humidity dependence of ammonium sulfate Mie scattering (**Qx,g**) radiative parameters according to laboratory measurements (at 0.633 μm) by Tang and Munkelwitz (1991; 1994). The left-hand panel is a color-contour grid of the Mie scattering extinction efficiency parameter (**Qx**) calculated for refractive indices (**nr**) and particle sizes (**Reff**) as shown. Superimposed on this grid are parametric curves relating the refractive index and particle size measured for a suspended ammonium sulfate droplet as relative humidity undergoes change. The sulfate aerosol exhibits hysteresis behavior by remaining 'dry' until a deliquescence relative humidity is reached (RHD = 0.80), whereupon it then rapidly takes on moisture to become a solute, and moves along the equilibrium curve, changing in size as relative humidity changes. If the relative humidity falls below the critical crystallization value (RHC = 0.38), the solute rapidly loses its remaining moisture to become a dry particle again.

Similarly, the right-hand panel is the Mie scattering asymmetry parameter (**g**) grid, superimposed on which are the parametric curves that relate the refractive index and particle size dependence on relative humidity for the sulfate aerosol droplet. In this way, the Mie scattering **Qx(nr,Reff)** and **g(nr,Reff)** grids serve as look-up tables for the Tang and Munkelwitz (**nr,Reff,RH**) parametric formulations to enable rapid determination of the (**Qx,g**) radiative parameters for any dry-size sulfate aerosol as a function of changing relative humidity.

Slide 10   Hygroscopic aerosol size and (dry) mass fraction as functions of relative humidity. Size, density and refractive index of hygroscopic aerosols follows a hysteresis loop as has been shown by laboratory measurements for sulfate, sea salt and nitrate aerosols (Tang and Munkelwitz, 1991; 1994; 1996). As shown in the left-hand panel, when relative humidity (RH) reaches the deliquescence point (filled circles), the dry aerosol particles rapidly take on moisture to become solute particles. They then follow the equilibrium curves for the respective species, changing in size as RH changes. If RH decreases below the crystallization point (filled squares), the solute droplets rapidly lose their remaining moisture and drop from the equilibrium curve.

The right-hand panel demonstrates how RH-dependent radiative parameters for a hygroscopic aerosol can be parameterized. The (reference) red curve utilizes Mie scattering parameters computed for the appropriate solute spectral refractive index at RH = 0.38, clearly an *internally* mixed aerosol composition. The black line depicts the spectral albedo computed for an *externally* mixed aerosol comprised of *dry sulfate* and *pure water* components. Thus, the radiative properties for an originally 0.300 μm dry sulfate aerosol (which becomes a 0.347 μm solute particle at RH = 0.38) can be accurately represented as the weighted average of a pure dry 0.287 μm dry sulfate particle and a 0.476 μm pure water aerosol, with the dry sulfate and pure water sizes and fractional amounts selected to match the (**Qx,g**) of the solute sulfate aerosol. This parameterization simplifies the tabulated input data needed for computing the radiative effects of hygroscopic aerosols.

Slide 11   Spectral dependence of Mie scattering parameters (Q, ω, g) for representative 0.2 and 1.0 mm sulfate aerosol (left-hand panels) and for 0.5 and 2.0 mm mineral dust aerosols (right-hand panels). The extinction efficiency parameter Q is the ratio of the Mie scattering cross-section relative to the particle's geometric cross-section. It varies with wavelength because the sulfate refractive index varies with wavelength. The spectral features are co-aligned in wavelength, but notably, the Mie size dependence is non-linear. The single scattering albedo ω is a measure of aerosol absorptivity with $\omega = 1.0$ indicating no absorption, and $\omega = 0$ indicating total absorption. The asymmetry parameter g is a measure of the degree of forward scattering. The right-hand panel illustrates similar Mie scattering (Q, ω, g) parameters for mineral dust particles. The differing spectral refractive indices are the means for aerosol composition retrieval.

Slide 12   Reflection, transmission, and absorption for unit optical depth based on the appropriate Mie scattering parameters (Q, ω, g) for 0.2 and 1.0 mm sulfate aerosol (left-hand panels), and 0.5 and 2.0 mm mineral dust aerosol (right-hand panels). The spectral radiative transfer calculations are performed using the doubling/adding method. The notable characteristics are that sulfate aerosol is basically non-absorbing for wavelengths shortward of 3 mm, while mineral dust exhibits substantial SW absorption, significantly stronger for the larger particle size and increasing toward the blue. Also notable are the four prominent sulfate aerosol absorption features in the mid-thermal spectral region. By comparison, the LW absorption by mineral dust has significantly broader absorption bands. Also of further note is the substantial reflectivity by the 2.0 μm mineral dust in the near-IR, and extending longward to the 10–30 μm region.

Slide 13   Kirchhoff's Law and the isothermal cavity concept provide a convenient starting point to develop the radiative transfer formulations needed to calculate and model the emission, absorption and transmission of thermal radiation. Basically, the radiation field inside an isothermal cavity is Planck radiation at temperature To of the isothermal cavity. Sampling the cavity radiation by means of arbitrarily positioned pinholes will all yield the same Planck intensity B(To). If an absorbing slab of optical depth τ and temperature T = To is inserted into the cavity, Kirchhoff's Law maintains that the radiation sampled through the pinholes will yield the same Planck intensity B(To). Since the thermal emission from the inner surface of the isothermal cavity must be B(To), it follows that transmissivity, emissivity and absorptivity must be related as shown in the figure in order that the pinhole radiation maintain its Planck intensity B(To). The radiation field inside the isothermal cavity being in thermodynamic equilibrium serves also to define the absorption coefficients and the optical depth consistent with the prevailing pressure-temperature conditions in the cavity. In the outside world, under slowly changing temperature gradients, local thermodynamic equilibrium closely approximates thermodynamic equilibrium.

Slide 14   Line-by-line CO_2 spectral fluxes and radiative and radiative cooling rate profiles. Top-left panel shows LW up flux, and bottom-left panel shows the down flux for 338 ppm CO_2 concentration. The top-right panel depicts the net upward LW flux. The incremental change in the net flux profile with height defines the vertical profile of the radiative cooling rate shown in the bottom-left panel. Water vapor is the principal cooling gas in the troposphere. CO_2 cooling dominates in the stratosphere, but water vapor and ozone also contribute.

Slide 15 Top left panel depicts the line-by-line (LBL) absorption coefficient spectrum from near-center (660–670 cm^{-1}) of the 15 μm CO_2 band for a pressure of 10 hPa and temperature 240 K. The bottom-left panel displays the same spectral region but for a pressure of 1000 hPa and temperature 296 K. Pressure broadening accounts for the marked difference between the two spectra. Central panels depict frequency distributions **F(k)** of corresponding LBL spectra. Superimposed are Malkmus band model frequency distributions that closely reproduce the LBL spectral transmission. The top right-hand panel depicts the cumulative frequency distributions **g** of both LBL and Malkmus model frequency distributions displayed in the central panels. Inverting the axis of the cumulative frequency distributions produces the corresponding k-distributions shown in the bottom right-hand panel. There are no substantial approximations in converting the LBL spectra to k-distributions. For a homogeneous path, both methods reproduce the same 660–670 cm^{-1} spectrally integrated transmission. Since line positions do not change with pressure, stratospheric and surface k-distributions remain strongly correlated as functions of **g**.

Slide 16 The Malkmus band model is a convenient analysis tool well-suited for numerical manipulation of the correlated k-distribution. This is because analytic 2-parameter formulations exist for the probability density **f**, cumulative fraction **g**, inverse **k** distributions, and the corresponding transmission (top-row panels), derived specifically for randomly overlapping Lorentz-line spectra with exponential line strength distribution. The Malkmus **B** = b/d, and **S** = s/d, parameters (where b,s,d are spectral line mean half-width, strength, and line spacing) provide a close statistical rendering of spectral transmission (dashed line at bottom-left). Statistical B and S coefficients are refined (solid circles) by least-squares fitting the Malkmus transmission to line-by-line calculated transmission over a broad (**log u**) absorber range. Bottom-right panels show how Malkmus k-distribution can be analytically converted to histogram form for any absorber amount, stacked vertically (as in a multi-layered atmosphere) to preserve the implicit spectral information. Results are compiled as a pressure, temperature, absorber amount look-up table that closely reproduces line-by-line accuracy at a small fraction of the computing cost.

Slide 17 The thermal radiation emitted by Earth is view-angle dependent due to the vertical atmospheric temperature gradient. Clear-sky satellite nadir-view sees deepest into the atmosphere, and thus encounters the warmest along-site temperatures. Off-nadir radiances originate from higher altitudes and proportionately cooler temperatures. In the above, a cloud is placed at high near-tropopause altitude, and its optical depth is varied logarithmically from near-zero ($\tau = 0.01$) to saturation ($\tau = 10$). In the GISS GCM radiation code, LW radiances are calculated at three emission angle quadrature points (black dots). A numerical quadrature is then applied to obtain the angle-integrated LW flux (red diamonds). Because of angle-

dependent transmission, clouds exert significant control over the angle dependence of outgoing radiation. The angle dependence (limb-darkening) is steepest
for optically thin clouds (τ = 0.1), converting to limb-brightening for optically
dense clouds.

Climate GCMs typically utilize a diffusivity approximation (heavy blue vertical line)
when calculating LW fluxes. Overall, this a good approximation, but as shown in
the right-hand panel, compared to angle-integrated flux, the diffusivity approximation produces LW flux biases of 3–4 W/m² for clouds in the cirrus τ = 0.1–1.0
range.

Slide 18   Cloud scattering effects on LW fluxes are small, and they are usually ignored
in climate modeling applications. However, given the three-quadrature point
approach in LW flux calculation in the GISS GCM, a parameterized correction
based on off-line calculations of the LW spectral cloud reflectivity and emissivity
is possible and has been implemented. The parameterization includes spectral
emissivity corrections for cloud-emitted radiances and reflected radiance contributions at cloud top and at cloud bottom. The net effect of LW cloud scattering
is a reduction in the outgoing (TOA) LW flux (top-right panel) by 1.4 W/m² on a
global annual-mean basis, and an increase in the downwelling flux (bottom-right
panel) at the ground surface (BOA) by 0.39 W/m².

LW scattering is more efficient for ice clouds because of their higher reflectivity in
the 20–30 μm region. It is more pervasive at higher latitudes, coinciding broadly
with the global cloud cover (top-left panel) and with high clouds (bottom-left
panel). The downwelling flux correction is more evident in the drier polar regions.
The LW cloud scattering enhances the cloud greenhouse effect by reducing the
outgoing LW flux.

Slide 19   Mie scattering imparts a unique polarimetric signature to every refractive index
and particle size combination. First-order scattering constitutes the 'polarization
angle spectrum' as a function of the Sun-particle-detector scattering angle that is
uniquely characteristic for each (**nr,Reff**) combination. Successive orders of scattering tend to add mostly unpolarized intensity, so that the polarization imparted
by the first-order scattering remains readily identifiable in the observed radiance.
When several (**nr,Reff**) combinations are present in the observed radiance, each
contributor can be identified by its polarization angle spectrum similar to identifying gases by their spectral signatures. In the above Mie scattering polarization grid, the refractive index increases vertically (**nr** = 1.33, 1.43, 1.53 for water,
sulfate, dust), and particle size increases horizontally (**Reff** = 0.2, 1.0 μm for aerosol, and 5, 15 μm for cloud).

Slide 20   Airborne Research Scanning Polarimeter (RSP) makes polarimetric and radiance
intensity measurements at 410, 470, 555, 670, 865, 1590 and 2250 nm (blue, mauve,
turquoise, green, red, magenta, black). Viewing geometry in the top-row panels
is for a relative solar azimuth of 205°, and solar zenith of 62° for two different
surface types. The spectral reflectances for the two surface types show significant
variations in color and directionality. However, the *polarized* surface reflectance

is smaller by two orders of magnitude, and it tends to be spectrally gray, being dependent instead on viewing geometry. As shown in the bottom-row panels, this circumstance makes it possible for combined polarimetry-intensity based analyses to retrieve accurate cloud and aerosol radiative properties compared to co-located AERONET ground-truth measurements, a capability that is not possible using intensity-only observational data.

Current intensity-only satellite data are unable to retrieve aerosol properties commensurate with AERONET ground-truth measurements because of their inability to differentiate light intensity from the ground surface, aerosol or from sub-pixel cloud contamination. Polarimetry offers an additional capability to help resolve this retrieval ambiguity (Mishchenko et al., 2007).

CLIMATE LECTURE 5

The Role of Clouds in Climate

Anthony D. Del Genio

NASA Goddard Institute for Space Studies, New York, NY, USA

Anthony Del Genio is a Senior Research Scientist at NASA Goddard Institute for Space Studies. He develops the cumulus and stratiform cloud parameterizations for the GISS climate models. His research focuses on climate feedbacks, the hydrologic cycle, interactions with the general circulation, and comparative dynamics of planetary atmospheres.

Introduction

David Rind has played a central role in the science of the modeling of climate change. He was the scientific driving force behind the development and evaluation of the first Goddard Institute for Space Studies (GISS) global climate model (GCM), 'Model II' (Hansen *et al.*, 1983). Model II was one of the three original GCMs whose projections of climate change in response to a doubling of CO_2 concentration were the basis for the influential Charney Report that produced the first assessment of global climate sensitivity. David used Model II to pioneer the scientific field of climate dynamics, performing a broad range of investigations of processes controlling individual elements of the general circulation and how they changed over a wide range of past and potential future climates (e.g., Rind and Rossow, 1984; Rind, 1986, 1987). The defining characteristic of David's papers is his unique talent for tracking down the myriad links and causal chains among different parts of the nonlinear climate system. Rather than viewing climate using a simple forcing-and-response paradigm, David showed that the global energy, water, and even momentum cycles are coupled via the general circulation and its transports.

Clouds are among the most important agents by which the climate system accomplishes this coupling. Clouds usually form when moist air rises to lower pressure, adiabatically expands and cools, and saturates. The resulting condensation of water vapor releases latent heat, and the clouds that form reflect sunlight and absorb thermal infrared radiation. For the planet as a whole, clouds are the major contributor to Earth's planetary albedo and thus the primary regulator of the energy input to the Earth system. The greenhouse effect of clouds is second only to water vapor as the means by which thermal cooling of Earth to space is limited.

Cloud formation occurs in two fundamentally different ways. Moist convective clouds require conditional instability, that is, a lapse rate that exceeds the moist adiabatic lapse

103

rate. When instability is present and the atmosphere is sufficiently humid, convective clouds are triggered when (1) lifting due to surface turbulence or other sources saturate air, (2) the latent heat release is sufficient to overcome a stable inversion layer above, making the air parcel buoyant relative to its surroundings and allowing it to rise on its own. When the atmosphere is not convectively unstable, stratiform clouds can occur, usually due to lifting by other phenomena (e.g., frontal uplift, other wave motions, boundary layer turbulence) that produces saturated but not buoyant air or occasionally by other sources of cooling of moist air (radiative or contact with a cooler surface). Thus in satellite images, clouds are tracers of the different character of the general circulation in different regions and the way in which the circulation links the global energy and water cycles—a visible reminder of David Rind's life's work. Convective and stratiform clouds play largely different, though partly overlapping, roles in the climate system. Convective cloud systems produce much of the precipitation on Earth and transport heat and moisture upward. Stratiform clouds produce less precipitation but most of the cloud radiative effect on Earth.

Deep convective clouds cool the surface and boundary layer, warm and dry the mid-troposphere, and moisten the upper troposphere. Shallow fair weather cumulus clouds tend to produce little or no precipitation but are important because they transport water evaporated from the ocean surface into the free troposphere. Shallow cumulus clouds are frequent enough to reflect a non-negligible amount of sunlight. Deep convective clouds are infrequent and radiatively unimportant by comparison, but they inject large amounts of water vapor and ice into the upper troposphere, forming extensive shields of stratiform cirrostratus 'anvil' clouds and ubiquitous thinner cirrus clouds (Liao, Rossow and Rind, 1995a, 1995b) that are very important radiatively. These extensive cloud decks, convectively generated but not convective themselves, produce the visible appearance of the Intertropical Convergence Zone (ITCZ), an irregular band of cloudiness that circles the equatorial region, in satellite images. This comprises the rising branch of the Hadley and Walker cells. The sinking branch of the Hadley and Walker cells creates an inversion layer at the top of the boundary layer, suppressing the development of deep clouds. Over land this is responsible for the major subtropical deserts, while over ocean, it leads to broken fields of shallow cumulus over the open subtropical oceans and solid decks of stratocumulus in the eastern oceans, where boundary layer turbulence lifts air to its condensation level but not above the inversion.

Shallow cumulus and stratocumulus clouds also occur in the mid-latitudes, in 'cold air outbreaks' behind the cold fronts associated with synoptic baroclinic storms. The major mid-latitude cloud features, though, are the stratiform 'comma clouds' that form along cold fronts and warm fronts and in the cold 'conveyor belt' of rising motion that wraps around the north side of low-pressure centers. These spiral-like features are made of cirrus, altostratus, altocumulus, and precipitating nimbostratus clouds, and they tend to occur in a wave-like sequence of 5–6 storms around the latitude belt. At even higher latitudes, the water transported poleward in frontal regions or evaporated from the open ocean moistens the polar low-level atmosphere and forms stratus cloud decks that emit longwave (LW) radiation down to the surface and play a role in the seasonal melting of sea ice. In the absence of such clouds, the sea-ice surface radiates 40–50 W/m^2 of heat to the atmosphere to maintain itself against melting.

Once formed, clouds evolve according to the microphysical processes that occur on the scale of individual liquid droplets or ice crystals. The pathways are somewhat different for liquid and ice but they share some things in common. We consider the simpler liquid case first. Excess vapor, usually formed in rising air at a rate τ_*^{-1}, condenses liquid water onto small particles called cloud condensation nuclei to form small cloud droplets at a rate τ_{cond}^{-1} (where τ represents a characteristic time scale of a given process). Droplets may be supplied or removed by transport from or to other locations, or by local evaporation when dynamical processes cause dry air to enter the region. Cloud droplets grow by coalescence, that is, gravitational settling of larger droplets relative to smaller droplets that produces collisional growth into raindrop sizes (10s–100s of microns) at a rate τ_{growth}^{-1}. Drops that reach a size such that their fall speeds exceed the vertical velocity of the updraft fall from the cloud as precipitation at a rate τ_{fall}^{-1} and eventually evaporate at the rate τ_{evap}^{-1} into smaller droplets and then vapor again, completing the lifecycle. Ice crystals display a similar lifecycle but in a more complex fashion, since they can form and grow not only by vapor deposition onto ice nuclei but also by contact of ice nuclei with supercooled droplets, freezing of small liquid haze droplets at very cold temperatures, and by the diffusion of water vapor from nearby evaporating liquid droplets. Growth by diffusion (the Bergeron–Findeisen process) is responsible for producing many of the large particles that lead to heavy precipitation (snow, or 'cold' rain from snow crystals that melt as they fall).

The net effect of the fractional coverage, frequency of occurrence, top altitude, and optical thickness (the combined result of the amount of liquid and ice, the physical thickness, and the droplet or crystal size) determines the magnitude of the effect of clouds on the Earth's planetary energy balance (i.e., the balance at the top of the atmosphere between sunlight absorbed and heat emitted to space). This is quantified by the concept of cloud forcing, the difference between how much radiation heats the Earth and how much it would heat a hypothetical Earth with no clouds. The shortwave (SW, sunlight) component of cloud forcing is negative, because clouds are brighter than most of the rest of the Earth and reflect more sunlight, cooling the Earth. The longwave (LW, Earth's radiated heat) component is positive because the greenhouse effect of clouds (mostly high altitude clouds) adds to the warming caused by greenhouse gases. This occurs because in the absence of clouds, the Earth emits heat to space primarily from the moderately warm middle troposphere, where the water vapor concentration starts to become small enough for LW photons to escape to space. High clouds prevent that from occurring, and radiation to space occurs instead from near the colder top of the cloud, and thus less heat is emitted.

The net cloud forcing varies geographically because different cloud types occur in different regions. Overall, net cloud forcing is negative because the SW cooling exceeds the LW warming, that is, on balance, the Earth is cooler with than without clouds. The largest negative SW cloud forcing occurs in the places where optically thick precipitating clouds are prevalent over a dark ocean: the ITCZ and the mid-latitude storm tracks. The largest positive LW cloud forcing also occurs over the ITCZ because of high altitude tropical anvils and cirrus clouds, and to a lesser extent in the nimbostratus clouds of the mid-latitude storm tracks. The SW and LW forcing largely compensate each other in the ITCZ and to a lesser extent in the storm tracks. The largest negative net cloud forcing occurs in the stratocumulus decks over the cold eastern tropical/subtropical oceans, where SW forcing is large and LW forcing is almost non-existent because of these clouds' low altitude.

By itself cloud forcing tells us nothing about the role of clouds in climate change. How clouds respond to increases in greenhouse gas concentrations or any other external forcing agent (e.g., changing solar luminosity, volcanic, or anthropogenic aerosols) is the biggest scientific uncertainty in projections of future climate change. It has often been assumed that because more water evaporates into the atmosphere as the oceans warm, cloud cover will increase with warming and clouds will have more liquid water and become optically thicker, providing a negative feedback that would mitigate much of the climate change. These assumptions are however inconsistent with what GCMs actually predict and/or what is observed, and these feedbacks in any case tell only part of the story. Cloud feedbacks consist of five components:

1. *Cloud cover feedback*: Changes in cloud cover depend on changes in both water vapor concentration (specific humidity) and temperature, since the latter determines the saturation vapor pressure, that is, the upper limit for water vapor in equilibrium over a flat surface of liquid water at a given temperature. Thus, *relative* humidity, the ratio of specific to saturation humidity, is the controlling factor for cloud cover, and relative humidity itself is controlled by the dynamics of the atmosphere on large and small scales.
2. *Cloud height feedback*: Higher cloud tops reduce outgoing LW radiation relative to the emission change without this cloud effect. An early one-dimensional radiative–convective model study in which David Rind participated (Hansen et al., 1981) demonstrated this feedback for the case in which cloud top temperature is fixed with warming but rises to higher altitude. In this case, the initial reduction in emission due to the higher colder cloud top causes the new emission level to warm to the temperature of the previous emission level to re-establish radiative balance. Extrapolating this temperature along the convectively adjusted atmospheric lapse rate to the surface implies a warmer surface temperature because of the greater distance from the emission level to the surface in the warmer climate.
3. *Cloud optical thickness feedback*: The canonical wisdom that clouds should become brighter with warming due to increased condensation of water vapor was shown to be inconsistent with satellite observations for much of the world by Tselioudis, Rossow and Rind (1992), because processes that reduce the physical thickness of clouds and/or their liquid water content oppose this effect in low and middle latitudes.
4. *Cloud phase feedback*: Fewer aerosol particles serve as effective ice nuclei than as cloud condensation nuclei for liquid clouds. Consequently, liquid clouds have more and smaller droplets than ice clouds have crystals, and the bigger ice crystals precipitate more quickly. Both effects cause ice clouds to be optically thinner than equivalent liquid clouds. Thus, a change in the relative amounts of liquid and ice as climate warms will induce a cloud feedback.
5. *Cloud location feedback*: Since the SW feedback of clouds depends not only on the cloud properties but on the amount of sunlight available for them to reflect, a shift in clouds to higher solar zenith angles (e.g., to higher latitudes) or to a location over a brighter surface will result in less SW cloud forcing even if the cloud properties do not change.

The current generation of climate models that participated in the Coupled Model Intercomparison Project Phase 5 (CMIP5) generally agree on some aspects of cloud feedback and disagree on others. Climate model responses to increasing greenhouse gas concentrations exhibit four basic behaviors:

1. An increase in cloud top altitude (decrease in cloud top pressure) in the tropics and extratropics such that cloud top temperature remains nearly constant with warming, producing a positive LW feedback. This behavior, anticipated in David Rind's early one-dimensional study (Hansen *et al.*, 1981), occurs in all models. It is well understood via two mechanisms. In the tropics, the 'fixed anvil temperature' hypothesis states that convective clouds inject ice into the atmosphere at levels at which they lose buoyancy and their compensating subsidence no longer warms the atmosphere, and this occurs at the level at which water vapor concentration (which approximately follows temperature) becomes too small to effectively radiatively cool the atmosphere. In mid-latitudes, the rising tropopause altitude due to CO_2 warming of the upper troposphere and cooling of the lower stratosphere has the same effect. In both cases, since convective and/or large-scale dynamical mixing controls the lapse rate below the higher cloud top emission-to-space level, greenhouse warming at high altitude is communicated to the surface as a higher temperature there even though temperature at the emission level remains constant.

2. A poleward shift in the mid-latitude storm tracks, which takes place in all models and is also seen in the observational record. This is associated with the poleward expansion of the Hadley cell, for which several explanations exist, for example the suppression of baroclinic instability at the equatorward edge of the storm track due to the reduced lapse rate in the warmer climate. The shift to higher latitudes produces a positive cloud location feedback, as long as the latitudes vacated by the shifting storm tracks are replaced by decreased cloudiness.

3. A general decrease in cloud cover in the subtropics and tropics, except for the equatorial central and east Pacific in some models (due to a weakening of the Walker cell and eastward shift of deep convection). Most GCMs predict this to varying degrees, but the magnitude and spatial extent of this decrease are probably the greatest source of disagreement among models and the single largest contributor to the spread in global climate sensitivity in the CMIP5 models. The response of low clouds to warming is the net result of several opposing processes. Increased surface turbulent fluxes in the warmer climate deepen the boundary layer, causing stratocumulus clouds that lie at its top to thicken and reflect more sunlight — a negative feedback. At the same time, turbulent entrainment of dry air above the cloud top inversion into the cloud evaporates liquid water and dissipates cloud, a positive feedback. In some models shallow cumulus clouds develop and penetrate the inversion top, and the dry air that descends to compensate this also dissipates the cloud. GCMs with low climate sensitivity tend to be dominated by the boundary layer deepening effect, especially in the stratocumulus regions of the eastern oceans. GCMs with high sensitivity tend to be dominated by the drying effects, e.g., within and at the edges of the stratocumulus decks, but perhaps also over the open subtropical oceans. Optical thickness also decreases somewhat with warming in these models, as originally predicted by Tselioudis, Rossow and Rind (1992).

4. An increase in high latitude cloud cover and optical depth, especially at low altitudes. Possible reasons for this include the increased poleward transport of water vapor by synoptic storms, increased evaporation of water from regions in which sea ice has melted and been replaced by open ocean, and boundary layer deepening. Much of it is microphysical, though, a result of the increased ratio of liquid to ice in the mixed-phase stratus that are ubiquitous in this region. The magnitude of this cloud phase feedback is highly uncertain, due to poor knowledge of ice nucleus concentrations and how they will change, and poor understanding of the many pathways by which ice is nucleated and supercooled liquid is maintained in these clouds.

Most of the focus in cloud feedback studies has been on low-level clouds, which continue to be a great uncertainty because of the competing mechanisms that control them. High cloud feedbacks have received less attention, in part because cancellation between their LW and SW forcing mutes their net cloud forcing in the current climate. Caution should be exercised, however, because at least two aspects of high clouds are more or less missing from today's GCMs. Thin cirrus, the only clouds whose net cloud forcing is positive in the current climate, are unresolved by the coarse vertical resolution of GCMs, and thus the physics that maintains them is largely absent from the models. Likewise, organized mesoscale convection, which occurs when favorable environmental conditions (e.g., high wind shear and/or humidity) cause a transition from individual short-lived convective cells into longer lived clusters of cells that develop an areally extensive stratiform rain region and anvil clouds, is not parameterized at all by most GCMs. In addition, most GCM cumulus parameterizations tend to produce a transition from shallow to deep convection too readily, due to insufficient sensitivity of convection depth to dry tropospheric conditions, a process that is controlled by entrainment into the clouds. Gradually, these weaknesses are giving way as more models begin to be evaluated against a broader range of satellite and field experiment data and in hindcasts of real-world events. However, convergence of model predictions of cloud feedback is still most likely to require a decade or more of further research.

Clouds also affect climate change through their interactions with aerosols. The *first indirect effect* occurs because clouds nucleate onto aerosol particles. With more aerosols, the available liquid water is partitioned among more, but smaller, droplets, and the greater collective surface area of the larger number of droplets causes more sunlight to be reflected, cooling the climate. The *second indirect effect* is proposed to occur because smaller droplets fall more slowly and thus are less likely to precipitate, increasing the liquid water that remains in the clouds and thus also causing more solar reflection. However this effect is highly uncertain. A *semi-direct* effect can also occur when absorbing aerosols (black carbon and dust) heat a cloud and cause liquid water to evaporate, or when they increase or decrease the convective stability of the atmosphere and thus change the amount or vertical development of clouds. One of the 'solar radiation management' proposals for geoengineering the climate to offset greenhouse warming involves adding aerosol particles to low stratocumulus cloud decks, making them more reflective. Studies to date indicate that while reductions in temperature may be possible via such techniques, it is much harder to predict the resulting feedbacks on the circulation and effects on precipitation patterns. Aerosols also interact with precipitation. Precipitation *scavenging* of aerosols from the atmosphere (also called wet deposition) is the primary removal mechanism for aerosols. Convective *invigoration* by aerosols has been proposed to limit growth of cloud droplets and allow them to interact more with ice crystals to increase precipitation, but this mechanism is highly uncertain.

References

Hansen, J., Johnson, D., Lacis, A., Lebedeff, S., Lee, P., Rind, D. and Russell, G. (1981). Climate impact of increasing atmospheric carbon dioxide. *Science*, 213, 957–966.

Hansen, J., Russell, G. Rind, D., Stone, P., Lacis, A., Lebedeff, S., Ruedy, R. and Travis, L. (1983). Efficient three-dimensional global models for climate studies: Models I and II. *Monthly Weather Review*, 111, 609–662.

Liao, X., Rossow, W.B. and Rind, D. (1995a). Comparison between SAGE II and ISCCP high-level clouds: 1. Global and zonal mean cloud amounts. *Journal of Geophysical Research*, 100, 1121–1135.

Liao, X., Rossow, W.B. and Rind, D. (1995b). Comparison between SAGE II and ISCCP high-level clouds: 2. Locating cloud tops. *Journal of Geophysical Research*, 100, 1137–1147.

Rind, D. and Rossow, W.B. (1984). The effects of physical processes on the Hadley circulation. *Journal of the Atmospheric Sciences*, 41, 479–507.

Rind, D. (1986). The dynamics of warm and cold climates. *Journal of the Atmospheric Sciences*, 43, 3–43.

Rind, D. (1987). Components of the ice age circulation. *Journal of Geophysical Research*, 92, 4241–4281.

Tselioudis, G., Rossow, W.B. and Rind, D. (1992). Global patterns of cloud optical thickness variation with temperature. *Journal of Climate*, 5, 1484–1495.

Slide 1

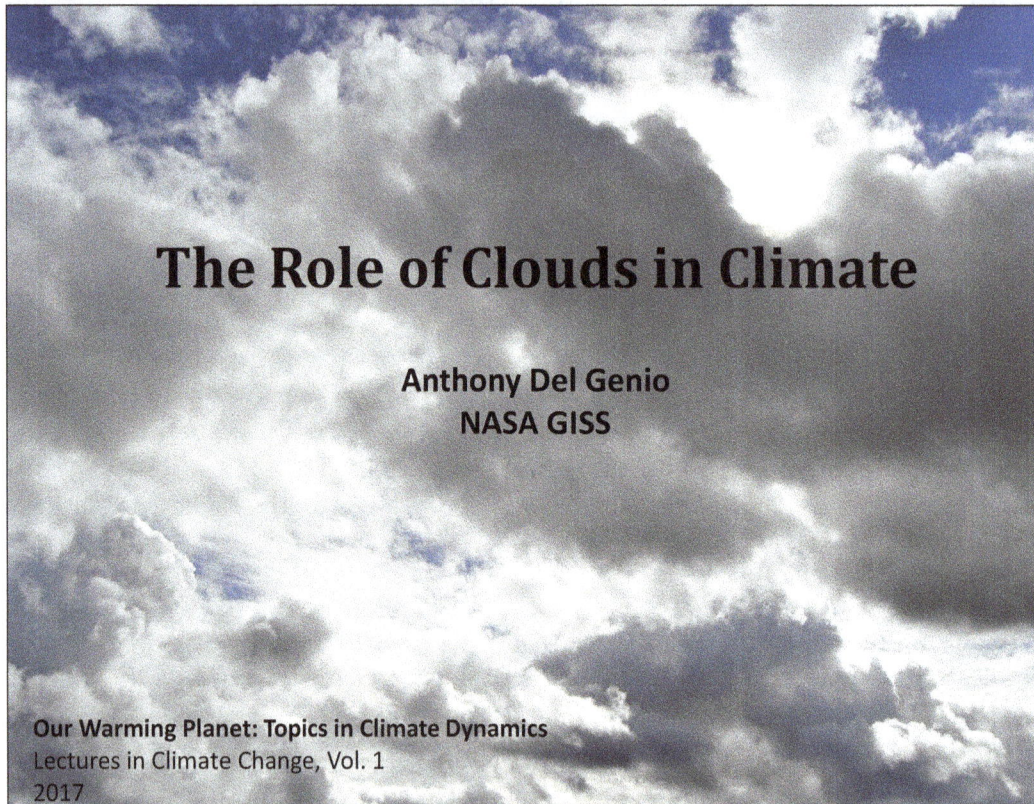

The Role of Clouds in Climate

Anthony Del Genio
NASA GISS

Our Warming Planet: Topics in Climate Dynamics
Lectures in Climate Change, Vol. 1
2017

Slide 2

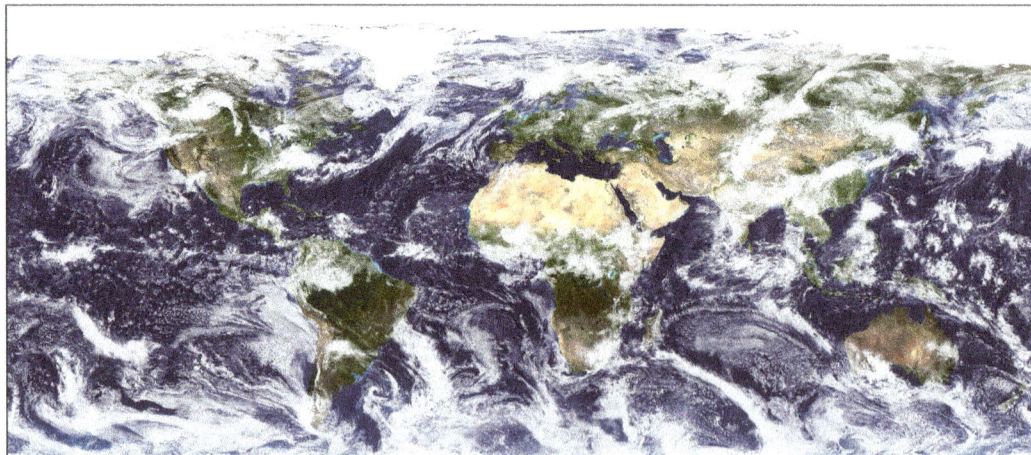

(image courtesy NASA)

Clouds form when parcels of air that contain moisture rise to higher altitude and lower pressure. The adiabatic expansion and cooling they experience brings water vapor to its saturation point, initiating condensation. Thus, clouds are tracers of the general circulation of the troposphere.

Slide 3

Two fundamentally different types of clouds exist, with different roles in the climate

- Convective clouds: Formed by rising due to buoyant instability of a column of air when air is humid and the lapse rate exceeds the moist adiabatic lapse rate

- Stratiform clouds: Formed by other sources of upward vertical motion such as lifting of warm air masses over cold air masses or boundary layer turbulence

Slide 4

Convective clouds

- Shallow (fair weather) cumulus
  - No or little precipitation
  - Moisten the lower troposphere by transporting water evaporated from the surface upward
  - Frequent enough to have a radiative impact

- Deep cumulonimbus
  - Responsible for the majority of Earth's precipitation
  - Warm and dry the atmosphere
  - Source of ~50% of cirrus clouds

Slide 5

Stratiform clouds

- Low/middle clouds
  - Mostly liquid, but can be mixed-phase or ice at low temperatures
  - No or little precipitation (stratus, stratocumulus) or significant precipitation (nimbostratus)
  - Important for reflection of sunlight
  - Little greenhouse effect because they lie below the clear sky emission level

- High clouds
  - Almost exclusively ice
  - Optically thin with little reflection of sunlight (cirrus) or thick convective anvil clouds (cirrostratus) with significant reflection
  - Always produce snow which does not reach the ground; thick anvils produce rain at the ground
  - Strong greenhouse effect because they lie above the clear sky emission level

(images courtesy NASA)

Del Genio, A.   The Role of Clouds in Climate   5

Slide 6

(a)   (b)   Cirrus   Altostratus   Nimbostratus   Stratus   WARM   COLD   Polar (mixed phase) Stratus   Mid-Latitudes   High Latitudes   10 km

(c)   Thin Cirrus   Convective Anvils   Large-scale Subsidence   Land/Sea Circulation   Melting Level   WARM SUBSIDING REGIONS   Shallow Cumulus   Stratocumulus   Trade Winds   WARM OCEAN   COLD OCEAN   Deep Tropics   Subtropics   17 km

(figure courtesy IPCC)

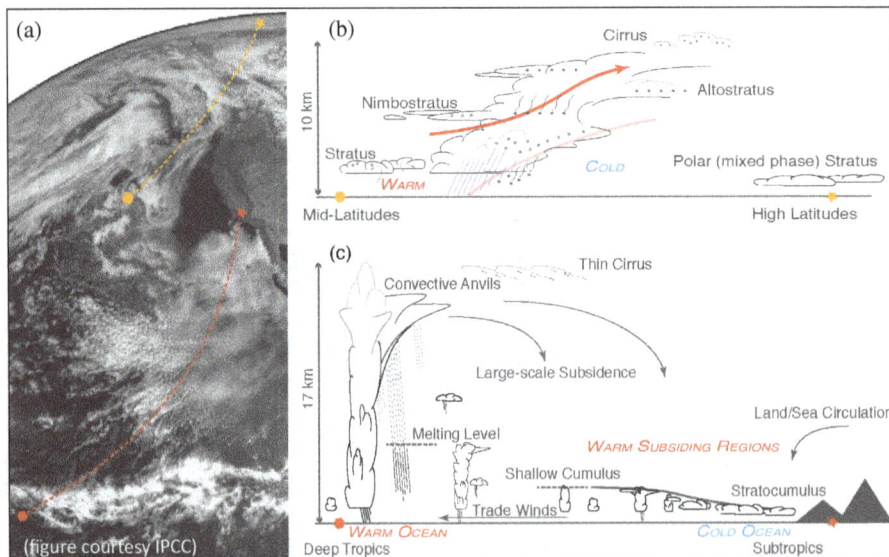

In the tropics, clouds are controlled by the Hadley/Walker circulation; in the mid-latitudes by synoptic storms; and in the polar regions by boundary layer turbulence and the snow/ice or ocean surface

Del Genio, A.   The Role of Clouds in Climate   6

Slide 7

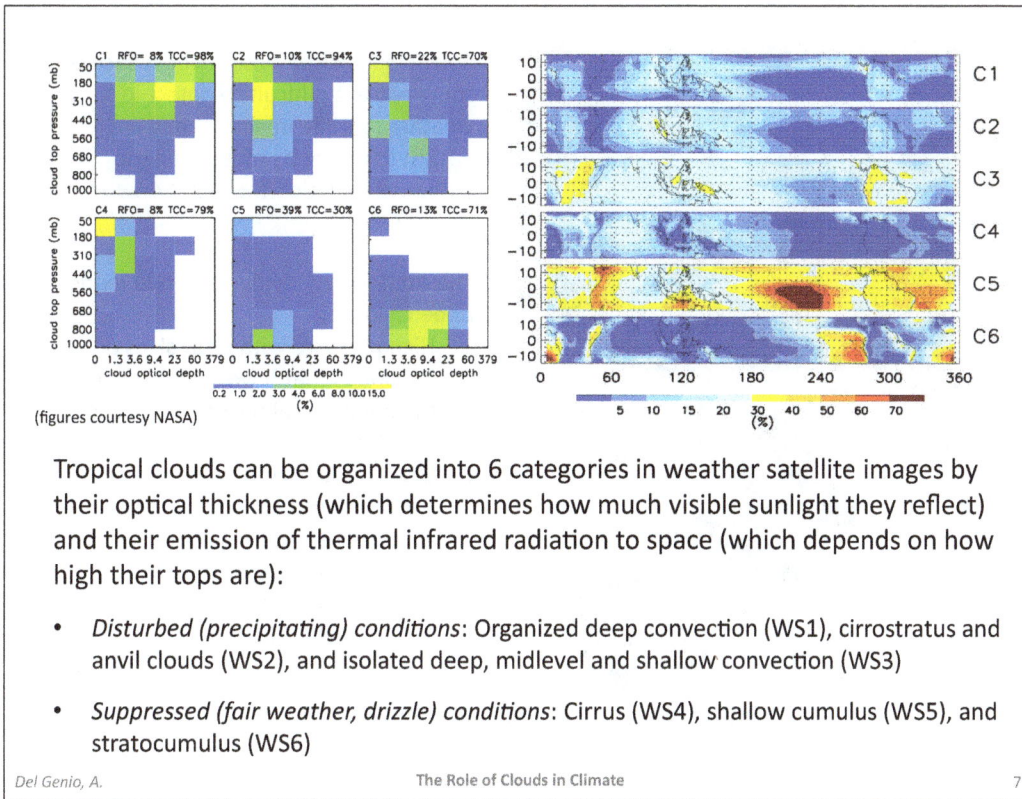

(figures courtesy NASA)

Tropical clouds can be organized into 6 categories in weather satellite images by their optical thickness (which determines how much visible sunlight they reflect) and their emission of thermal infrared radiation to space (which depends on how high their tops are):

- *Disturbed (precipitating) conditions*: Organized deep convection (WS1), cirrostratus and anvil clouds (WS2), and isolated deep, midlevel and shallow convection (WS3)

- *Suppressed (fair weather, drizzle) conditions*: Cirrus (WS4), shallow cumulus (WS5), and stratocumulus (WS6)

Del Genio, A.   The Role of Clouds in Climate   7

Slide 8

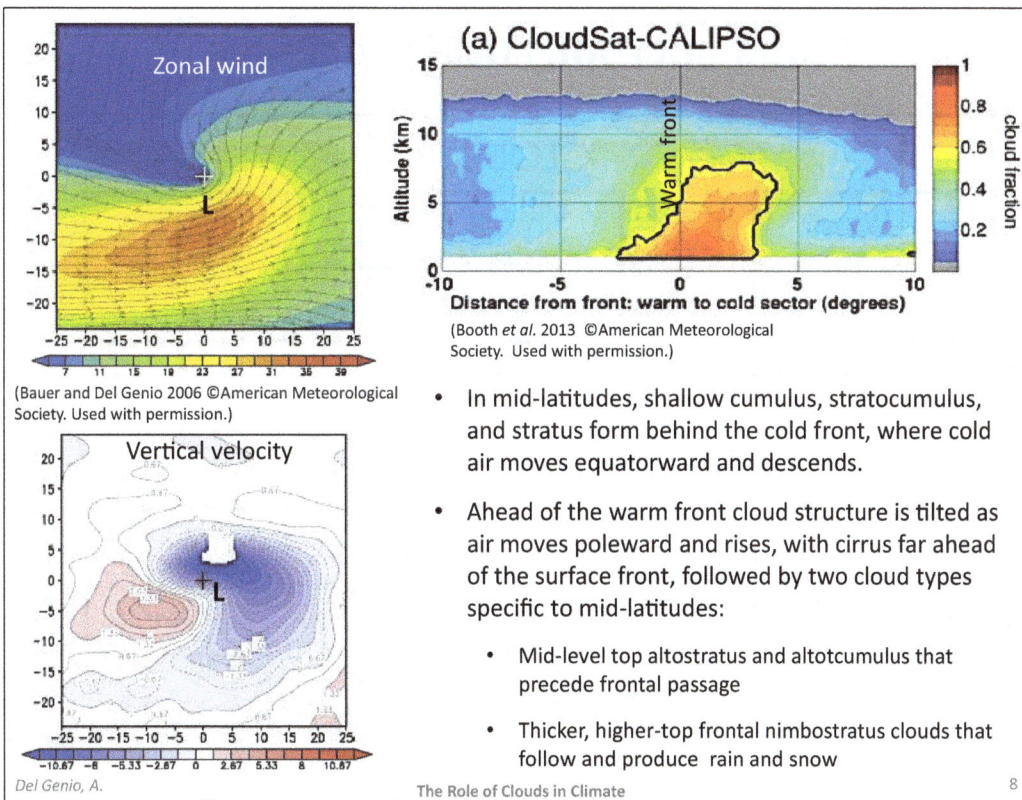

(Bauer and Del Genio 2006 ©American Meteorological Society. Used with permission.)

(Booth *et al.* 2013 ©American Meteorological Society. Used with permission.)

- In mid-latitudes, shallow cumulus, stratocumulus, and stratus form behind the cold front, where cold air moves equatorward and descends.

- Ahead of the warm front cloud structure is tilted as air moves poleward and rises, with cirrus far ahead of the surface front, followed by two cloud types specific to mid-latitudes:

  - Mid-level top altostratus and altotcumulus that precede frontal passage

  - Thicker, higher-top frontal nimbostratus clouds that follow and produce rain and snow

Del Genio, A.   The Role of Clouds in Climate   8

Slide 9

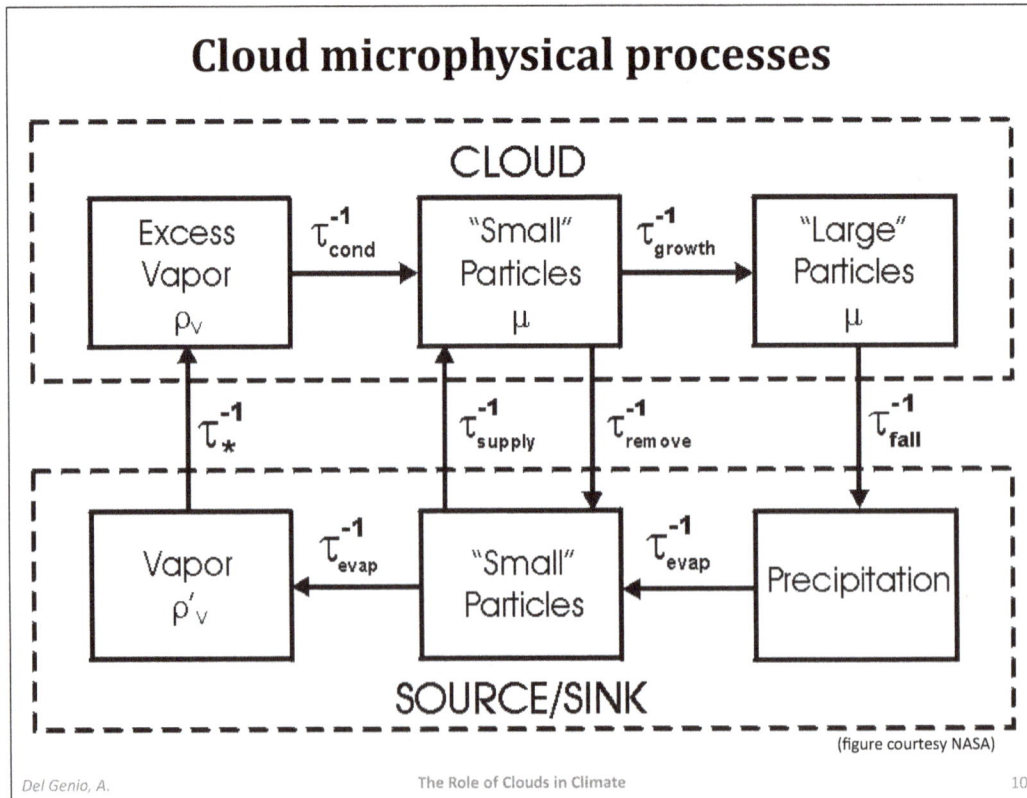

(image courtesy NASA)

Classic Transitions

Polar clouds often occur over sea ice or snow and are difficult to detect in weather satellite images. They are easiest to notice by measuring the longwave radiation they emit down to the surface.

The Arctic often divides into two modes of behavior:

1. High pressure with clear skies or thin ice clouds that have little effect on radiation and allow sea ice to cool radiatively;
2. Low pressure with thicker liquid water clouds formed from water vapor transported by storms, which emit enough longwave radiation down to balance the upward heat loss from the surface.

(Stramler *et al.* 2011 ©American Meteorological Society. Used with permission)

Slide 10

Cloud microphysical processes

CLOUD

| Excess Vapor ρ_v | $\xrightarrow{\tau_{cond}^{-1}}$ | "Small" Particles μ | $\xrightarrow{\tau_{growth}^{-1}}$ | "Large" Particles μ |

τ_*^{-1} τ_{supply}^{-1} τ_{remove}^{-1} τ_{fall}^{-1}

| Vapor ρ'_v | $\xleftarrow{\tau_{evap}^{-1}}$ | "Small" Particles | $\xleftarrow{\tau_{evap}^{-1}}$ | Precipitation |

SOURCE/SINK

(figure courtesy NASA)

Slide 11

Clouds modulate the geographic pattern of solar absorption and thermal infrared emission

Cloud forcing is a measure of the radiative effect of clouds

Defined as the actual absorbed (SW) or emitted (LW) flux
(including clouds) minus the flux in clear sky

(up/down fluxes are defined as –/+)

CERES EBAF Annual

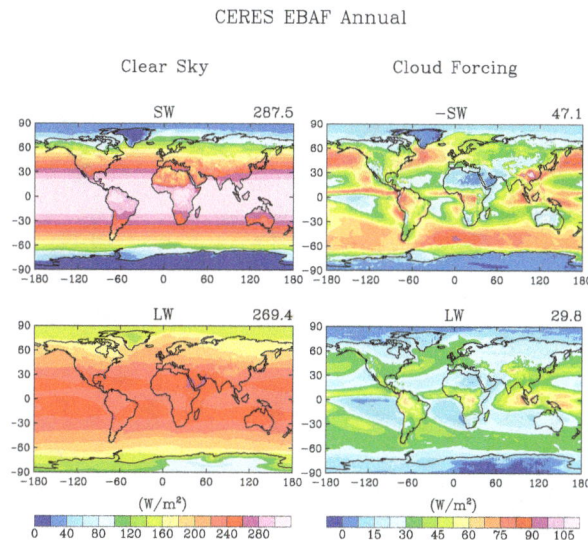

(DelGenio 2013 © The Arizona Board of Regents)

Del Genio, A.   The Role of Clouds in Climate   11

Slide 12

Components of cloud feedback

- *Cloud cover feedback:* Depends on how *relative* humidity responds to climate warming, the result of competing effects of changes in water vapor concentration and temperature, and whether the clouds that increase or decrease are those that primarily warm or cool the planet. Primarily a response to dynamics.

- *Cloud height feedback:* Higher cloud tops reduce emission to space relative to the change in emission without this change, and convective/dynamic mixing communicates this to the surface.

- *Cloud optical depth feedback:* Clouds become brighter or darker depending on how cloud water content, physical thickness, and particle size change with warming.

- *Cloud phase feedback:* An increase in the amount of liquid relative to ice causes a negative feedback due to the smaller liquid droplets, but this is reduced for clouds over snow/ice.

- *Cloud location feedback:* A shift in cloud locations to higher solar zenith angle (poleward or away from noon) or to a location with a brighter surface reduces SW cloud forcing.

Del Genio, A.   The Role of Clouds in Climate   12

Slide 13

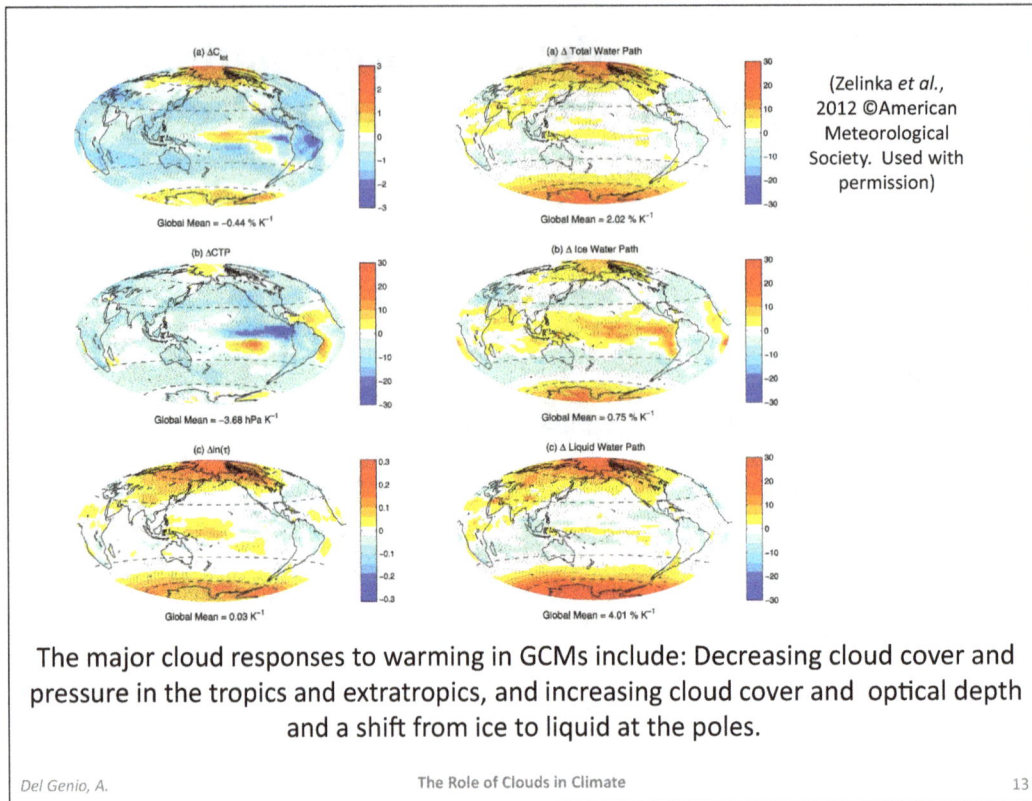

The major cloud responses to warming in GCMs include: Decreasing cloud cover and pressure in the tropics and extratropics, and increasing cloud cover and optical depth and a shift from ice to liquid at the poles.

Slide 14

Some features of the predicted cloud feedback are considered high-confidence predictions

- The positive LW feedback due to increasing cloud altitude is understood via the "fixed anvil temperature" mechanism whereby the altitude at which convective clouds detrain moisture and ice shifts upward, following the shift of the vertical profile of radiative cooling by increased water vapor.

- The positive SW feedback due to the expansion of the Hadley cell and poleward shift of the storm tracks (moving the storm clouds to latitudes with less insolation) is supported by observations, although the physical mechanism(s) responsible for the Hadley cell expansion are still debated

Slide 15

But there is still substantial disagreement among GCMs about tropical low cloud feedback

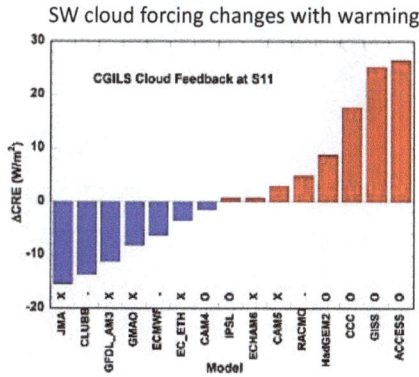

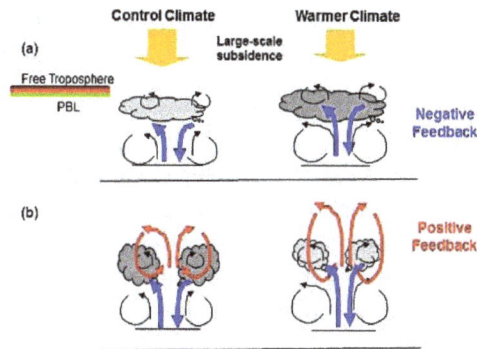

SW cloud forcing changes with warming

(Zhang *et al.* 2013; © John Wiley and Sons)

Low-sensitivity models: Increased surface evaporation deepens the boundary layer and thus stratocumulus clouds with warming.

High-sensitivity models: Drying by shallow convective subsidence and cloud top entrainment dissipates stratocumulus with warming.

Slide 16

Other sources of uncertainty in cloud feedback

- The cloud phase contribution to cloud feedback (less reflective ice to more reflective liquid) depends on the unknown concentration of ice nuclei and uncertainty in the multiple pathways by which ice nucleates and grows in different conditions

- Thin cirrus, which have a net positive LW cloud forcing, are unresolved by the vertical layering of GCMs and the mechanisms of their production and maintenance are poorly understood

- Optically thick convective anvil clouds are produced by the mesoscale organization of convection, which is not represented in any GCM

Slide 17

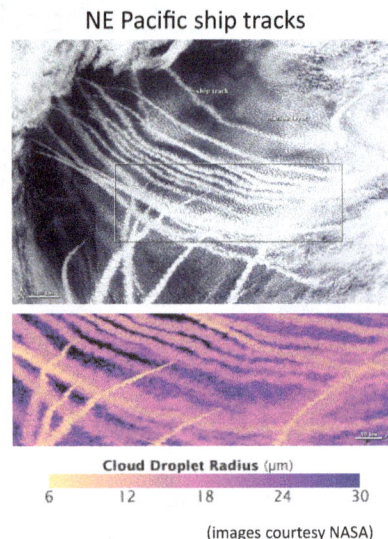

Cloud–aerosol interactions

- *First indirect effect*: Higher aerosol concentration produces more but smaller cloud droplets whose collective surface area is greater, making the cloud more reflective; a potentially significant negative climate forcing.

- *Second indirect effect*: Smaller droplets precipitate less, increasing cloud water content and lifetime; this has not yet been observationally verified and may be offset by faster evaporation of smaller droplets when dry air is entrained into the cloud.

- *Semi-direct effect*: Warming of air by solar absorption by lower albedo aerosols (e.g., black carbon) evaporates cloud droplets or affects atmospheric stability; cloud may increase or decrease depending on the type of cloud and altitude of the warming.

NE Pacific ship tracks

Cloud Droplet Radius (µm)
6   12   18   24   30

(images courtesy NASA)

Del Genio, A.   The Role of Clouds in Climate   17

Slide 18

Precipitation–aerosol interactions

- *Wet deposition (scavenging)*: Precipitating droplets collect aerosol particles and remove them from the atmosphere (below-cloud scavenging), or cloud droplets that form by nucleating onto an aerosol remove the aerosol when they rain out (in-cloud scavenging); this is the primary mechanism for removing aerosols from the atmosphere

- *Invigoration of convection*: Higher aerosol concentration limits coalescence growth of cloud droplets into rain drops in a convective updraft, allowing more liquid to be carried up to colder temperatures; the resulting increase in ice formation releases additional latent heat that strengthens the convection; this is currently a controversial hypothesis

Del Genio, A.   The Role of Clouds in Climate   18

Slide 19

Take-away messages

- Clouds form either because of convective instability or because of stable lifting of air masses; they are tracers of Earth's general circulation
- Convective clouds transport heat, moisture, momentum and produce much of Earth's precipitation; stratiform clouds produce some precipitation but dominate radiative effects
- Clouds with low and high tops play different roles in the climate system via how much precipitation they form and/or whether they influence primarily SW or LW radiation
- Cloud feedbacks in a changing climate are due to changes in cloud cover, height, optical thickness, phase, and location; some are well understood, others are not
- Interactions of aerosols with clouds affect cloud optical thickness, and potentially cloud cover and precipitation; many of these interactions are uncertain

Del Genio, A. The Role of Clouds in Climate 19

Slide Notes

Slide 2 Clouds link the energy and water cycles via the atmospheric general circulation. This composite global satellite image identifies the primary atmospheric circulation regimes by the characteristic cloud patterns associated with them. The equatorial region is characterized by an irregular but zonally oriented band of cloudiness associated with the Intertropical Convergence Zone (ITCZ). The ITCZ marks the rising branch of the tropical Hadley circulation, in which moisture evaporated in the subtropics converges and rises, creating an environment suitable for deep convection and the highest climatological precipitation rates on Earth. Actual convective clouds cover only a small fraction of the tropics. The clumps of equatorial cloudiness that collectively make up the ITCZ are the result of organization of convection into clusters with extensive stratiform rain and anvil cloud regions. Poleward of the ITCZ in the subtropics, the descending branch of the Hadley cell advects dry air down from high altitudes. Over land, where the surface source of moisture is limited, this suppresses clouds and rain and is responsible for Earth's deserts. Over ocean, surface evaporation provides water vapor that is advected upward to the top of the low-level boundary layer inversion by small-scale turbulence. Depending on the sea surface temperature, this cloudiness can take two distinct forms. Over the colder ocean waters of the eastern oceans (particularly visible off the coast of Peru and Chile, but to a lesser extent off the coasts of Namibia and California), overcast stratocumulus decks form. Over the somewhat warmer open oceans, scattered fields of shallow trade cumulus form instead. The mid-latitudes of both hemispheres are dominated by the spiral, 'comma' cloud and precipitation patterns associated with synoptic low-pressure centers generated by the baroclinic instability of the atmosphere's latitudinal temperature gradient. The clouds themselves are the result of rising motion along the cold front equatorward of the low (the tail of the comma), along the warm front east of the low, and in the 'cold conveyor belt' that wraps around the north side of the low.

Slide 3 Clouds are usually the result of upward motions that cause adiabatic expansion and cooling that brings moist air to its saturation point. Global climate models (GCMs) typically divide clouds into two general categories: (1) Convective clouds that result when air is displaced upward in a conditionally unstable environment (i.e., a lapse rate in excess of the moist adiabatic lapse rate) and reaches its level of free convection (positive buoyancy). (2) Stratiform clouds that form in moist convectively stable conditions due to other sources of lifting—this includes motions along fronts formed by baroclinic instability processes in an area that is not unstable vertically, clouds that form at the tops of the turbulent boundary layer, clouds that form as convective updrafts lose buoyancy and diverge, injecting ice into the upper troposphere, clouds that form due to other wave motions, and orographic clouds that form in flow over topography.

Slide 4 Convective clouds sometimes penetrate only a short vertical distance but at other times all the way through the troposphere, and there is actually a continuum of convective cloud depths with occurrence peaks of tops in the lower and upper troposphere and a weak tertiary peak in the mid-troposphere. Predicting the depth of a convective cloud is important because the effect of convection on the environment changes character as convection deepens. Shallow cumulus slightly heat the lower troposphere, but their primary role is to transport moisture out of the boundary layer into the lower free troposphere. This has several implications for the climate. It helps maintain sub-saturated conditions in the boundary layer, allowing surface evaporation to continue; it facilitates the transport of subtropical water vapor toward the equator by the return branch of the Hadley circulation; and by moistening the lower troposphere it provides more favorable conditions for future cumulus clouds to rise to greater altitude without being diluted by turbulent entrainment of dry air. Shallow cumulus often does not precipitate, or else precipitate only weakly. They produce scattered cloudiness but are frequent enough to significantly affect the cloud cover of the subtropics and mid-latitudes. Deep cumulonimbus produces most of the Earth's precipitation. They create a deep profile of convective heating that peaks in the mid-troposphere early in the lifecycle and then shifts upward as convection organizes on the mesoscale and produces upper troposphere stratiform rain regions. Deep convection dries the atmosphere and thus stabilizes it. The cloudiness due to deep convection itself is minimal, but the stratiform rain and anvil regions created by deep convection cover a large area. Injection of ice into the upper troposphere by deep convective updrafts accounts for about 50% of the cirrus in the tropics. (The remainder of cirrus occurs when large-scale upward motions in the atmosphere, e.g., due to waves, cause air to become supersaturated with respect to ice.)

Slide 5 Stratiform clouds are important primarily for their radiative effects. The global mean temperature is determined by the balance between absorbed sunlight and the Earth's emitted heat radiation. Clouds affect the former by their influence on Earth's planetary albedo, with ~2/3 of the reflected sunlight being due to clouds. Clouds affect the latter because like greenhouse gases, they too absorb thermal infrared radiation and re-emit it at a lower temperature, thus reducing overall emission to space. Stratiform clouds, which cover a much larger area than convective clouds, are responsible for most of clouds' radiative effect on the climate. Low-altitude stratiform clouds such as stratus and stratocumulus tend to be made of liquid and are optically thick and thus reflect significant sunlight. At the same time, they do not trap much more thermal radiation than the clear atmosphere does because they exist at altitudes where most of the atmosphere's water vapor (the most important greenhouse gas) resides. Thus these clouds have a net cooling effect on the current climate. High stratiform clouds are almost always ice and can either be optically thin (cirrus) or optically thick (cirrostratus). Cirrus clouds reflect relatively little sunlight but strongly absorb thermal infrared radiation and re-emit it to space at the very low temperatures of the upper troposphere, and they thus have a net warming effect on the climate. Thick cirrostratus clouds both strongly reflect sunlight and strongly absorb thermal infrared radiation and thus have an approximately neutral effect on the planetary energy balance. Optically and physically thick stratiform clouds such as nimbostratus can produce significant precipitation (rain or snow) in mid-latitude synoptic storms but usually not as much as convective storms do. Stratus and stratocumulus either do not precipitate at all or drizzle. Ice crystals in cirrus and cirrostratus clouds are typically much bigger than liquid droplets in low clouds and therefore precipitate more easily. Thus, technically all high clouds produce snow. However, for most high clouds, the snow sublimes long before reaching the surface, visible as wispy virga. Thick cirrostratus anvils associated with convective storms, though, produce considerable snow that melts upon reaching the 0°C line and ultimately reaches the surface as stratiform rain.

Slide 6   Tropical cloud types and locations are dictated primarily by the Hadley and Walker circulations. In the rising branch of these circulations (near the equator and over warm oceans and land), moisture converges and produces unstable conditions, giving rise to heavily precipitating cumulonimbus clouds, sometimes with stratiform rain regions and anvils, and often produces thin cirrus as well. As air drifts poleward or toward longitudes with cooler sea surface temperatures, it radiatively cools and begins to subside. This subsidence dries the column and suppresses cloud formation in the free troposphere. If the subsidence is over ocean, it is opposed at low altitudes by turbulence in the boundary layer, and where the two meet, an inversion layer forms. Water evaporated from the ocean is transported upward by the turbulence and saturates beneath the inversion, forming a stratocumulus deck, typically off the west coasts of the continents where the ocean surface is cool. If the subsidence is over land, there is no significant surface source of water vapor and skies remain clear; these are locations of the world's deserts. After air subsides, it begins its return flow toward the equator or warm ocean longitudes. As the ocean beneath these 'trade winds' begins to warm, the boundary layer deepens and shallow cumulus clouds that can penetrate the inversion begin to occur, reducing the cloud cover and injecting water vapor into the lower free troposphere, the source of the water that ultimately converges in the ITCZ. Mid-latitude cloud types are regulated primarily by the synoptic low- and high-pressure centers that develop due to baroclinic instability of the equator-pole temperature gradient. Warm moist air flows poleward east of the low and rides up and over denser colder air at the warm front, producing cirrus, altocumulus or altostratus, and eventually thicker stratus and nimbostratus near the surface warm front. This air continues to flow around the north side of the low, producing an extensive nimbostratus cloud shield there. West of the low-pressure center, cold dry air flows equatorward. It meets warm humid air at the cold front and lifts that air, often producing convective clouds and precipitation. Behind the cold front, the colder drier air produces clear skies and/or low stratus clouds. If the cold air outbreak occurs over the ocean, evaporation from the warmer ocean surface can sometimes produce extensive areas of cellular shallow convective clouds. In the polar regions, the dominant cloud types are stratus and stratocumulus and are heavily influenced by boundary layer turbulence and by the nature of the surface beneath them (open ocean, or snow/ ice). Figure reproduced with permission from IPCC from Boucher, O., Randall, D., Artaxo, P., Bretherton, C., Feingold, G., Forster, P., . . . Zhang, X.Y. (2013). Clouds and Aerosols. In: *Climate Change 2013: The Physical Science Basis. Contribution of*

Working Group I to the Fifth Assessment Report of the Intergovernmental Panel on Climate Change [Stocker, T.F., Qin, D., Plattner, G.-K., Tignor, M., Allen, S.K., Boschung, J., . . . Midgley, P.M. (eds.)]. Cambridge University Press, Cambridge, UK and New York, NY, pp. 571–658, doi:10.1017/CBO9781107415324.016.

Slide 7 The different types of clouds that are prevalent in the tropics can be objectively identified by their combination of visible optical thickness and cloud top pressure (or temperature) that is retrieved from visible and thermal infrared weather satellite images. The International Satellite Cloud Climatology Project (ISCCP) produces two-dimensional frequency histograms of these quantities for 2.5 × 2.5 degree regions and defines six dominant cloud types from these distributions. Organized deep convection is characterized by high optical thickness and low cloud top pressure and occurs mostly in the ITCZ north of the equator and the corresponding South Pacific Convergence Zone (SPCZ) south of the equator. Cirrostratus and anvil clouds, which occur in association with organized convection, have similar cloud top pressures, somewhat lower optical thicknesses, and have a similar geographic distribution. Both these cloud types occur where and when the atmosphere is very humid. In moderately humid conditions, a mix of individual deep convective clouds, shallow cumulus, and mid-level top 'congestus' clouds are common, identified by moderate to low optical thickness and a variety of cloud top pressures. These situations also occur in the ITCZ and SPCZ, and they are more likely to exist over the tropical continents. In drier conditions skies are sometimes dominated by thin cirrus (low cloud top pressure and low optical thickness), primarily over the tropical warm pool and central Africa, where deep convection has previously humidified the upper troposphere. In the cooler sea surface temperature regions of the tropics, low clouds dominate. Over the open oceans on either side of the ITCZ or SPCZ, where air subsides, shallow cumulus is prevalent. These are identified as high cloud top pressure, low optical thickness clouds, but the latter is in part an artifact of the fact that individual clouds are smaller than the resolution of weather satellite images, and thus clear sky mixes with cloud in one pixel. Over the colder eastern oceans off the west coasts of the continents, where ocean upwelling is present, denser stratocumulus decks are prevalent, with high cloud top pressure and moderate optical thickness.

Slide 8 In mid-latitudes, the flow around low-pressure centers (L in the ECMWF reanalysis storm composites above) brings cold, subsiding air equator-ward west of the low, giving rise to shallow cumulus, stratocumulus, and stratus clouds. East of the low, poleward moving warm air ascends as it meets cold air at the warm front. Cirrus occurs when surface high pressure is in place but the warm front is advancing from the west and the elevated part of the warm front is overhead. With time the altitude of the front lowers at a given location. Altostratus and altocumulus, mid-level cloud types unique to the mid-latitudes (moderate optical thickness, moderate cloud top pressure) commonly set in after the cirrus, heralding a degradation of weather conditions. As the front approaches the surface, nimbostratus (fairly high optical thickness and fairly low cloud top pressure, but not to the extent seen in cumulonimbus), another cloud type unique to the mid-latitudes, move in. A composite cross section of cloud occurrence across the warm front thus shows a tilted structure. Figures from Bauer, M. and Del Genio, A.D. (2006). Composite analysis of winter cyclones in a GCM: Influence on climatological humidity. *Journal of Climate*, 19, 1652–1672, doi:10.1175/JCLI3690.1, and Booth, J.F., Naud, C.M. and Del Genio, A.D. (2013). Diagnosing warm frontal cloud formation in a GCM: A novel approach using conditional subsetting. *Journal of Climate*, 26, 5827–5845, doi:10.1175/JCLI-D-12-00637.1.

Slide 9   Polar cloudiness is dominated by low stratus and stratocumulus clouds. More so than in other parts of the world, these low clouds can occur in either phase of water or a mixture of the two. Polar clouds are difficult to detect in weather satellite images, because in the visible they are about as reflective as the underlying sea ice or snow, and in the thermal infrared, they are often as warm or warmer than the underlying surface because low-level temperature inversions are common. They are much easier to detect from the ground via their effect on thermal infrared radiation. When skies are clear or when optically thin ice clouds are present, the surface radiates thermal infrared (LW) radiation upward, while the atmosphere emits little downward, producing a negative (upward) net LW radiative flux that cools the surface and maintains the sea ice against heat conducted up through the ice from the warmer ocean below. When liquid or mixed-phase clouds are present, they emit about as much LW radiation down as the surface emits up, producing a near-zero net radiation balance at the surface. These two contrasting conditions tend to persist for long times in the Arctic relative to other latitudes, with sudden transitions from the 'radiatively clear' state to the 'opaque cloud' state when low pressure moves in, usually accompanied by poleward transport of water vapor from lower latitudes. The lower figure shows individual observations of surface pressure and net LW radiation at the surface for an Arctic sea ice location (black crosses). The colored lines connect consecutive points in time and thus show several examples of how the two parameters jointly evolve on particular days. When low pressure departs and high pressure moves in, an equally sudden transition from near-zero net LW flux (caused by cloudy skies that prevent the surface from cooling) to strongly negative LW flux (from upward radiation in clear skies that cools the surface freely when clouds are not present) occurs. The net result is a strongly bimodal distribution of surface longwave radiation. Figure from Stramler, K., Del Genio, A.D. and Rossow, W.B. (2011). Synoptically driven Arctic winter states. *Journal of Climate*, 24, 1747–1762, doi:10.1175/2010JCLI3817.1.

Slide 10   The evolution of a cloud from birth to death is due to a combination of macrophysical dynamical processes that produce an environment favorable or hostile to the presence of clouds, and microphysical processes on the scale of water molecules or individual cloud droplets or ice crystals that cause the cloud to grow or decay. Consider first the simpler liquid water case. Excess (supersaturated) vapor concentration initiates condensation (at the rate τ^{-1}_{cond} in units of inverse time) in the presence of a suitable cloud condensation nucleus (soluble aerosol particles such as sulfate or sea salt, or insoluble but hygroscopic particles). Cloud droplets may also be supplied (at the rate τ^{-1}_{supply}) by advection from another location and can be removed (at the rate τ^{-1}_{remove}) by advection to another location or by erosion of the cloud as dry air is turbulently entrained into the cloud and evaporates cloud liquid water. Condensation can produce droplets of size ~10 μm, similar to that observed in typical clouds, but condensation slows down as droplet size increases and thus cannot explain the onset of precipitation in some clouds. Instead, clouds grow to precipitation size (at the rate τ_{growth}^{-1}) by gravitational coalescence, the process by which large droplets falling faster than nearby small droplets collide with them and incorporate them to make larger drops whose fall speeds eventually become large enough to allow them to fall out of the cloud (at the 'sedimentation' rate τ_{fall}^{-1}). The observed size of cloud droplets is the size at which coalescence becomes faster than condensation. Once the large drops fall through the base of the cloud and leave the saturated conditions of the cloud, they begin to evaporate in the sub-saturated air below (at the rate τ_{evap}^{-1}). If the sedimentation rate exceeds the evaporation rate (e.g., if the sub-cloud air is fairly humid or the cloud base is close to the surface), precipitation reaches the ground. If not, the drops completely evaporate and return water vapor and aerosols to the atmosphere. This description, for liquid clouds, applies as well to ice clouds. However, ice clouds 'heterogeneously' nucleate onto different types of aerosol (black carbon, biogenic, perhaps dust) and can also 'homogeneously' nucleate in the absence of an ice nucleus at temperatures colder than ~−35°C by the freezing of a previously nucleated liquid haze droplet. Furthermore, in addition to the coalescence mechanism of growth (called 'aggregation' for ice clouds becoming snow), ice crystals can also grow in a mixed phase cloud when water vapor is supersaturated with respect to ice but sub-saturated with respect to liquid. In these conditions, water evaporated from the liquid droplets diffuses toward ice crystals and is deposited onto the surface, causing snowflakes to grow rapidly via what is called the Bergeron–Findeisen process. This process is responsible for much of the precipitation (rain and snow) that occurs in cold seasons.

Slide 11   Earth planetary radiation balance (at the top of the atmosphere) demands that in equilibrium, the sunlight (SW radiation) absorbed by the Earth equal the thermal radiation (LW) that Earth emits to space. The role of clouds in maintaining this balance is often described in terms of the concept of 'cloud forcing' (sometimes called 'cloud radiative effect'). For energy balance purposes, radiative fluxes are designated as positive downward (into the Earth system, i.e., warming) and negative upward (out of the Earth system, i.e., cooling). The cloud forcing is then defined as the difference between the actual absorbed or emitted flux (which includes both cloudy and clear regions of the atmosphere) and the flux in clear skies (the latter being a rough proxy for what the flux would be if there were no clouds). The figures above show the clear sky SW and LW flux to space (left panels) and the SW and LW cloud forcing, i.e., the amount by which clouds alter the flux relative to clear skies (right panels). In the absence of clouds, the geographic distribution of reflected sunlight would be determined primarily by the decrease in incident sunlight from equator to pole caused by Earth's spherical shape, modulated by the seasonality due to Earth's obliquity and the different surface albedos of ocean, land, sea ice and snow. Likewise, in clear skies outgoing thermal infrared (LW) radiation would be determined primarily by the decrease in temperature from equator to pole and modulated by the mostly latitudinal variation in middle/upper troposphere water vapor concentration associated with both temperature (through the Clausius–Clapeyron equation) and the general circulation (through its effect on relative humidity). (The maximum clear sky LW flux thus occurs in the subtropics of each hemisphere, where water vapor concentrations are small and radiation to space occurs from a lower, warmer altitude.) Thus, SW cloud forcing is negative, because clouds are brighter than most of the Earth's surface and increase the planetary albedo. The largest SW cloud forcing occurs in regions of significant optically thick cloud, including the ITCZ and SPCZ, the mid-latitude storm tracks, and the eastern subtropical ocean stratocumulus decks. It is larger over ocean than land in general because the ocean surface is darker and thus the presence of cloud has more of an effect on the albedo. LW cloud forcing is positive, because the greenhouse effect of clouds reduces LW cooling to space. The largest LW cloud forcing occurs in regions with significant high cloud, i.e., regions of rising motion in the general circulation such as the ITCZ, SPCZ, and mid-latitude storm tracks, because in these places clouds lie above the altitude (and below the temperature) at which the clear atmosphere emits to space. (It is often mistakenly assumed that clear areas emit to space from Earth's surface, but only a small fraction of the emission to space comes from the surface because of the opacity of water vapor and other greenhouse gases. Applying the Stefan-Boltzmann equation flux = σT^4 to the observed clear sky flux of 269.4 W/m^2, one finds that clear regions emit to space on average at a temperature $T \sim$ 270 K, much colder than Earth's mean surface temperature of ~288 K.) The geographic pattern of LW cloud forcing is similar to that of SW cloud forcing, except that there is little LW cloud forcing in the stratocumulus regions, where almost all cloud is at low altitude and emission is dominated by the overlying water vapor and other greenhouse gases. Globally, SW cloud forcing (–47.1 W/m^2) exceeds LW cloud forcing (+29.8 W/m^2), giving rise to the common statement that 'clouds cool the Earth". This is to be interpreted as a statement about the current climate: Earth is cooler because it has clouds than it would be if it had no clouds. It is not a statement about how clouds will affect climate change. Figure from 'Physical Processes Controlling Earth's Climate' by Anthony D. Del Genio, from *Comparative Climatology of Terrestrial Planets*, edited by Stephen J. Mackwell, Amy A. Simon-Miller, Jerald W. Harter, and Mark A. Bullock © 2014 The Arizona Board of Regents. Reprinted by permission of the University of Arizona Press.

Slide 12 The long-term effect of clouds on climate change is referred to as 'cloud feedback'. It is the combined result of how changes in temperature caused by climate forcings (such as greenhouse gases, aerosols, and solar luminosity) directly affect cloud properties, and how changes in other parts of the climate system caused by changes in temperature (such as water vapor and the circulation) affect cloud properties. It has often simplistically been assumed that clouds will offset greenhouse gas-induced climate change, based on the logic that warming evaporates more water from the ocean, which causes more clouds to form, which increases the albedo, which offsets the warming. However, this is largely erroneous for three reasons. First, cloud cover does not respond to the absolute amount of water vapor in the atmosphere (the specific humidity), but rather to the relative humidity, i.e., the amount of water vapor in the atmosphere relative to its upper limit over a flat surface such as the ocean (the saturation humidity). As the climate warms, water vapor does increase, but temperature increases too, raising the saturation humidity, and so relative humidity changes only by small amounts, rising slightly in some places and decreasing slightly in other places, determined mostly by interaction with the general circulation. Thus in global climate models (GCMs), some places see a slight increase in cloudiness, others a slight decrease in cloudiness as climate warms, and the global cloud feedback depends on which of these has the greater effect. Second, the simplistic argument considers only the SW effect of clouds, ignoring the warming LW effect. So, one must know whether the clouds that increase or decrease are those that warm or cool the Earth to know the sign of the feedback. Third, not only cloud cover changes when the climate warms. Higher cloud tops with warming reduce emission to space, a positive feedback that appears as surface warming due to mixing by convection or large-scale dynamics. Increases or decreases in cloud optical thickness change reflection of sunlight and sometimes emission of heat to space. Changes from less reflective ice clouds with bigger particles to more reflective liquid clouds with smaller droplets increase the optical depth as climate warms, but much less so if the clouds are over a bright snow/ice surface. Finally, shifts in cloud locations to a higher solar zenith angle where there is less sunlight to reflect, or to a location with a brighter surface underneath, with no change in the cloud itself, reduce SW cloud forcing and produce a positive cloud feedback.

Slide 13   The current generation of GCMs produces these primary cloud feedbacks in response to warming: (1) Cloud cover generally decreases equatorward of about 50° latitude, with the exception of the equatorial central-east Pacific, where some models predict an increase in cloud cover instead. The decrease in cloud cover occurs mostly in low- and middle-level clouds and is thus a positive cloud feedback. (2) Cloud height increases in the tropics and extratropics, another positive feedback. (3) Cloud cover and optical depth increase in the polar regions with warming. (4) There is a shift from ice to liquid, especially in the polar regions, as climate warms; this constitutes part but not all of the optical depth feedback there. The stippled areas in the figures indicate where at least 75% of the GCMs agree, which gives a sense of where there is model consensus and inter-model disagreement. Overall, the total cloud feedback predicted by today's GCMs ranges from near-neutral to strongly positive. Figure from Zelinka, M.D., Klein, S.A., and Hartmann, D.L. (2012). Computing and Partitioning Cloud Feedbacks Using Cloud Property Histograms. Part II: Attribution to Changes in Cloud Amount, Altitude, and Optical Depth. *Journal of Climate*, 25, 3736–3754. doi:10.1175/JCLI-D-11-00249.1.

Slide 14   Several aspects of cloud feedback are thought to be well understood and are agreed upon by all models, lending confidence in the predictions: (1) The positive LW feedback due to increasing cloud top altitude with warming can be understood by invoking the 'fixed anvil temperature' (FAT) hypothesis (or a more recent variant, the proportionately higher anvil temperature hypothesis). The idea is based on the fact that relative humidity remains nearly constant with warming, and so increasing water vapor concentration increases the altitude at which the water vapor concentration is so low that it no longer can effectively radiate to space. At this altitude there must be a divergence of the subsidence mass flux to compensate this cooling. The divergence is compensated by an inflow of mass at this altitude where convective systems detrain water vapor and ice to create anvils and cirrus clouds. The level of this detrainment follows the level at which water vapor radiative cooling begins to decrease as climate warms. The feedback occurs because despite the fact that the emission temperature remains constant, the emission level rises in altitude. Extrapolating from the new emission level over a greater distance along the atmospheric lapse rate down to the surface, a warmer surface temperature results. This positive feedback has been predicted by all GCMs for as long as climate simulations have been performed, and there is observational evidence for FAT in current climate variability over El Niño cycles. This is a tropical feedback, but cloud height increases in the extratropics with warming as well. This can be understood by the increase in tropopause height as climate warms, a well-founded feature of climate associated with the fact that increased upper troposphere radiative warming and increased lower stratosphere cooling by increased CO_2 shifts the tropopause upward. (2) The positive SW feedback due to the poleward shift of the extratropical storm tracks (i.e., a cloud location feedback due to the higher solar zenith angle) is also agreed upon by GCMs and occurs because of Hadley cell expansion with warming, which shifts the latitude of maximum latitudinal temperature gradient required for baroclinic instability poleward. There are several hypotheses for Hadley cell expansion, which is observed over recent decades, the most promising being the fact that the lapse rate of the atmosphere decreases with warming (because of the temperature dependence of the moist adiabatic lapse rate), stabilizing the southern edge of the storm track against baroclinic instability.

Slide 15   The major difference between low and high climate sensitivity climate models is in their prediction of how low clouds respond to warming. The plot on the left shows the change in cloud radiative effect (i.e., the change in cloud forcing, which can be considered a rough measure of the cloud feedback for the clouds in question here) between a warmer climate and the current climate for a large number of climate models that simulated a stratocumulus cloud regime in the subtropical Pacific. The models range from ones with a strong negative cloud feedback (ΔCRE < 0, implying stronger negative SW cloud forcing with warming) to ones with a strong positive cloud feedback (ΔCRE > 0, implying weaker negative SW cloud forcing with warming). At least two competing mechanisms determine the sign of this feedback in GCMs. As climate warms, more water evaporates from the oceans and more sensible heat is transferred to the atmosphere, while subtropical free troposphere subsidence weakens slightly, the net effect being a deeper boundary layer. By itself, this would cause the lifting condensation level to occur farther below the inversion, creating deeper, thicker stratocumulus clouds that are more resistant to breakup by the entrainment of dry air from above the inversion. However, in some GCMs entrainment is strong enough to break up the cloud in the warmer climate, and in others, shallow cumulus clouds begin to form or increase in frequency or depth in the warmer climate and penetrate the inversion. The subsidence that compensates the upward mass flux of these cumulus clouds dries and breaks up the stratocumulus layer. This effect is limited to the stratocumulus regions and does not extend to the open subtropical oceans, where most GCMs predict decreasing low cloud cover due to the fact that drier air subsides into the boundary layer as the climate warms.

Slide 16   Regardless of GCM consensus or lack thereof, physics that is not included in any GCM must be kept in mind in assessing the true uncertainty in cloud feedback, which in turn is the single biggest contributor to the uncertainty in Earth's climate sensitivity to a given external forcing. The transition from more ice to more liquid as climate warms is not simply a function of temperature. It also depends on the nature and availability of ice nuclei, which are poorly understood and poorly constrained by observations, and on the many possible pathways by which ice can nucleate and how this depends on its interactions with the dynamics within the cloud and at cloud top. None of these is thought to be represented well in GCMs, as evidenced by the fact that the fraction of ice vs. liquid in clouds varies with temperature in very different ways in different GCMs. Thin cirrus, the only cloud type that has a net warming effect on the current climate, exist in thin layers that are unresolved by the vertical layering of today's GCMs. Thus, subtle radiation–microphysics–turbulence interactions that depend on heating gradients across thin layers are absent in GCMs. In addition, the confidence with which GCM cumulus parameterizations transfer water vapor and ice upward into the near-tropopause atmosphere is low, and even the dynamical vertical transport of water vapor in a discretized model across a gradient of several orders of magnitude can be problematic. Finally, most tropical precipitation and the extensive cloudiness associated with it occurs in organized mesoscale convective systems. The processes of mesoscale convective organization and the lifecycle and propagation of such systems are absent from today's GCMs.

Slide 17   Clouds also interact with the aerosol environment in several ways as it changes with time. The interaction depends on the type of aerosol, the concentration of the aerosol, and the type of cloud. For liquid clouds, aerosols that are soluble in water are the most efficient cloud condensation nuclei, and if small concentrations of these aerosols exist, cloud droplet formation is limited by aerosol concentration and thus an increase in aerosol leads to an increase in cloud droplet number, which for a given amount of available vapor implies smaller droplets. The increased total surface area of the smaller, but greater in number, droplets makes the cloud more reflective. This is called the first indirect effect or 'Twomey' effect. The Twomey effect has been observed in ship tracks, local concentrations of aerosols that produce more reflective cloud with smaller droplet sizes under the right conditions. The magnitude of the first indirect effect for industrial, transportation-generated, or biomass burning aerosols advected out to the more pristine oceans is less certain, since it is difficult to unravel the relative influence of aerosol and meteorological effects on observed clouds. Smaller droplets also potentially produce a microphysical second indirect effect, since smaller droplets precipitate from the cloud more slowly. Thus, cloud liquid water content should increase with reduced precipitation loss, and it has been hypothesized that this can increase cloud lifetime and thus time-averaged cloud cover. The second indirect effect has not been observationally verified. The hypothesized liquid water content increase may be offset by the entrainment of dry air into some clouds, which may evaporate smaller droplets more easily than larger droplets in pristine non-polluted clouds. In addition, any cloud lifetime effect must compete with the time scales on which circulation changes dissipate clouds dynamically. Another aerosol influence on clouds, the semi-direct effect, occurs by using the atmosphere as an intermediary. In the presence of an absorbing aerosol such as black carbon, air warms by SW absorption due to the aerosols. If the aerosol is within the cloud the heating of the air may evaporate cloud liquid water. If it is above a stratocumulus deck and strengthens an inversion, it may reduce the entrainment of dry air into the cloud and make the cloud optically thicker. If it occurs beneath the base of a convective cloud, it may destabilize the column and promote convective growth. The magnitudes of all these effects are highly uncertain.

Slide 18   Clouds also interact with aerosols via precipitation. The most obvious and important process is scavenging (wet deposition), by which precipitation removes aerosols from the atmosphere. Scavenging can occur when precipitating particles collect aerosols below-cloud base (below-cloud scavenging), or since liquid droplets nucleate onto aerosols, any droplet that precipitates by definition removes aerosol from the atmosphere (in-cloud scavenging). A more controversial hypothesis is the invigoration of deep convection by aerosols. In deep convection, liquid droplets are carried by updrafts above the 0°C level and become super-cooled. When they collide with ice crystals and freeze (glaciate), the additional latent heat release strengthens the updraft and allows it to penetrate higher. It has been proposed that with a higher aerosol concentration, the smaller droplets that form are more easily carried upward to the colder regions of the cloud and promote additional freezing, invigorating the convection. The extent to which this occurs in varying environmental conditions is poorly understood, and whether any such effect is transient or affects convection over its entire lifecycle is not yet determined.

SECTION 3
Dynamical Responses

CLIMATE LECTURE 6

How will Storms and the Storm Track Change: Extratropical Cyclones on a Warmer Earth

Walter A. Robinson[1] *and James F. Booth*[2]

[1] *Department of Marine, Earth, and Atmospheric Sciences, North Carolina State University, Raleigh, NC, USA*
[2] *Department of Earth and Atmospheric Sciences, City College of New York, New York, NY, USA*

Walter Robinson is a Professor at North Carolina State University. His research is focused on large-scale atmospheric dynamics and the dynamics of Earth's climate, including climate variability and climate change.

James Booth is an Assistant Professor in the Earth and Atmospheric Sciences Department at the City College of New York. His research interests include mid-latitude atmospheric dynamics, air–sea interactions, and moist processes in extratropical cyclones.

Storms, Storm Tracks, and Expected Changes with Global Warming

Extratropical cyclones are the familiar low-pressure systems that dominate day-to-day weather over much of the planet. These storms follow typical paths denoted as *storm tracks*. They are important weather makers, bringing rain, snow, and sometimes damaging winds, and they are the engines of the mid-latitude climate system, carrying heat, moisture, and momentum from the subtropics towards the poles. Rainfall in mid-latitudes concentrates along the storm tracks. How these storms will change with a changing climate is, therefore, a leading question for how the weather and climate will be different on a warmer Earth.

Pioneering Norwegian meteorologists (Bjerknes and Solberg, 1922) exposed the fundamental structure of extratropical cyclones early in the 20th century. They elucidated their frontal structure and they recognized that these storms gain their energy from the contrast in temperatures between the warm subtropics and the cold Polar Regions. One of the most robust projections of climate models is that t surface temperatures will warm more near the poles than in lower latitudes as the Earth warms — this denotes the *polar amplification* of global warming. From this, it is expected that mid-latitude cyclones will be *weaker* on a warmer Earth.

There is, however, a competing effect that could drive changes towards *stronger* storms. Cyclones, depending on their strength and position (say, over the ocean versus over land), can gain a significant fraction of their overall energy from the condensation of water vapor into cloud droplets and rain (also the freezing of liquid drops into ice and snow) — that is, from the release of *latent heat*. In a computer model, it is possible to compare how a storm develops with and without the effects of latent. In a classic numerical modeling study in the 1980s (Kuo and Reed, 1988), an intense Pacific cyclone was simulated including and excluding the effects of moisture. The storm modeled with moisture became much stronger, with a deeper low-pressure system and sharper frontal zones, a result that subsequently has been reproduced many times.

The saturation vapor pressure of water increases, roughly 7% per °C of increase in temperature (colloquially we say a warmer atmosphere 'holds' more vapor). We expect, therefore, that on a warmer Earth, there will be more water vapor in the atmosphere, unless relative humidity decreases. A robust projection of climate models, however, is that the relative humidity changes little with global warming. Therefore, we expect the atmosphere will become 'juicier', providing more latent heat to fuel future storms. This should make storms *stronger* over ocean regions.

That competing effects might drive opposing changes in cyclones makes projections of these changes both uncertain and complex. It is unlikely that there will be a single simple answer to the question of how the storms and storm tracks will change with global warming. Instead, changes are likely to vary with location, with the season, and with the nature of the storm. For example, it is possible that most storms will weaken but that the strongest storms will become stronger still. The limitations of current climate models in representing the dynamics of storms, particularly their release of latent heat, add to uncertainty in projecting future changes in their strength and distribution.

This problem typifies many in climate change. While the globally averaged picture may be clear — warming, melting ice, and rising seas — projections of changes in the phenomena that are of greatest interest to people where they live — will storms become more severe? Will rainfall increase or decrease? Will there be more droughts or more floods? — invariably bring into play the full, rich dynamics of the climate system. Prof. David Rind, to whom these lectures are dedicated, has never shied away from the complexity of the weather and climate. It is a hallmark of his research that he eschewed the oversimplifications often encapsulated in neat theories and instead embraced the full complexity of climate and climate change. This topic, how mid-latitude cyclones will change with global warming, is, therefore, a fitting one for a volume in his honor.

Combined Effects of Moisture and Resolution on Modeled Storms and the Storm Track

If storms strengthen in the future, it will almost certainly be a result of the increase, described above, in the release of latent heat. Idealized studies of cyclones under a range of humidities (Booth, Wang and Povani, 2013) show the expected effect that storms develop more rapidly when more moisture is present. These idealized storms also display a strong sensitivity to the model grid scale. Weather and climate models treat the atmosphere on a mesh of *grid points*. Constraints on the size and speed of computers limit the numbers of grid points in a model, so that the typical separation between grid points in a global climate model (GCM) is 100 km or greater. (The model grid scale is often referred to as the model *resolution*, with a smaller grid scale corresponding to a higher resolution.) Many important features in cyclones, such as the *warm conveyor belts* that carry moisture into storm centers, have smaller spatial extents than this, and so are not well represented on GCM grid scales.

The modeled representation of moist processes in cyclones, and the resulting influence of the latent heating on the storm circulation, is sensitive to the model's grid scale. In a recent study (Willison, Robinson and Lackmann, 2013), a generic, North Atlantic cyclone is simulated on a GCM grid scale (120 km) and a much finer, *storm-scale* grid (20 km). The distribution of heat released by precipitation is unsurprisingly much sharper at the higher resolution. This leads to rapid deepening of the storm. The contribution of latent heating to the deepening of the storm is also greater at the higher resolution — again, not surprising, because the storm is stronger. More significantly, the release of latent heat is *proportionally* more important for the storms on the finer grid. In short, the latent heating plays a greater role in the growth of the storm when it is more accurately modeled.

Moving from individual storms to the storm track, when multiple winters are simulated in a model that covers the North Atlantic basin, the energy of the storm track and its pole-ward transport of heat exhibit strong sensitivity to grid scale, increasing by tens of per cent between GCM-scale and storm-scale simulations (Willison *et al.* 2013). These limited area simulations, however, lack the self-limiting negative feedbacks on the storm-track activity found in global models and, presumably, the real climate system, so their results are an upper bound on the sensitivity of the storm track to model grid scale.

Changes Simulated in GCMs

Whether their projections are accurate or suffer from insufficient grid resolution, GCMs typically project only modest changes in cyclones with climate change. An oft-cited study on this topic (Bengtsson, Hodges and Keenlyside, 2009) uses a state-of-the-art GCM from the Max Plank Institute for Meteorology in Hamburg, Germany. The model has a grid scale in middle latitudes of about 130 km. Lower atmospheric winds in the very strongest storms are projected to increase by only a few meters/second (up to around 5 miles per hour) between the simulated late-20th-century climate and the projected late-21st-century climate. Changes in precipitation in the strongest storms are more significant, about 1–2 mm per hour.

Not all GCMs produce such modest changes in storms with warming, but any interpretation of the results is complicated by the differences among studies in the weather variables considered and in the methods of analysis. For example, Lambert and Fyfe (2006) analyzed GCM results from 11 different models prepared for the Fourth Assessment of the Intergovernmental Panel on Climate Change (IPCC) under a range of greenhouse gas emission scenarios. In contrast to Bengtsson *et al.*'s results, they found that with global warming the GCMs produced a pronounced increase in the frequency of the most intense cyclones, as measured by their central pressures. Changes in pressure, however, can convolve changes in the strengths of cyclones with changes in their positions. Because sea-level pressure decreases towards the poles, storms tend to have lower central pressures if their tracks shift pole-ward. GCMs consistently project such a shift, as first reported in a model intercomparison study by Yin (2005) at the National Center for Atmospheric Research. Similarly, Gastineau and Soden (2009) found increases in the occurrence of the strongest lower atmospheric winds in sub-polar latitudes of both hemispheres with warming. This, again, may represent more a poleward shift and, especially in the Southern Hemisphere, an intensification of the mean near-surface winds than an intensification of storms.

Changes Simulated in Limited-area Models

As described above, there is evidence that the strengths of cyclones are under-represented in models at GCM grid scales. Computers are not yet powerful enough to allow multiyear climate-change simulations to be run in global models, complete with coupled dynamical oceans and cryospheres, on storm-scale grids. A 'workaround' is to embed limited-area models with storm-scale resolution (i.e., regional climate models), within the projections of climate change produced by GCMs.

In results recently published by Willison, Robinson, and Lackmann (2015) using this approach, at both GCM and storm-scale resolution the energy of the storm track and its pole-ward transport of heat increase with global warming, especially on the eastern side of the Atlantic basin. The changes are substantially (about 20%) stronger at the higher resolution. These results are consistent with the hypothesis that the moistening of the atmosphere with warming is responsible for the changes. First, the changes are greatest in the eastern Atlantic basin where temperature gradients are relatively weak and, therefore, latent heating is relatively more important as a source of cyclone energy. Second, the amplification of the climate-induced changes at higher resolution is consistent with the importance of resolution for the dynamics through which latent heating amplifies cyclones. This situation is, however, made more complex by dynamical feedbacks that may function differently in a global model. In particular, in *this* model, the stronger cyclones in a warmer world and at higher resolution feed more momentum and energy into the westerly jet aloft, producing an atmospheric state that is more conducive to cyclone growth.

Another way to study changes in storms with warming is to consider the development of individual storms over short periods of time. Marciano, Lackmann, and Robinson (2015) performed such a study on 10 strong US East coast cyclones, comparing how these storms develop in current and future climate conditions. In the future simulations, the strengthening of upper-level westerly winds that accompanies global warming causes the paths of the storms to shift eastward. All but one of the storms deepens more rapidly, but these are small changes. More interesting are changes in storm dynamics brought about by the increased moisture content. The future storms show a marked increase in precipitation. This is not surprising, but it leads to a strong increase and tightening in the distribution of a dynamical quantity called the *potential vorticity*, which captures the tendency of the atmosphere to rotate cyclonically. For these storms, the increased strengthening effect of latent heat release more than compensates for a weaker north–south temperature gradient.

Complexity, Variability, and Uncertainty

The science presented so far demonstrates significant progress in understanding the processes that could make storms and the storm track different on a warmer Earth, but the results offer little confidence that we can make a reliable projection of whether storms will get weaker or stronger, more or less numerous, or shift in their predominant locations and tracks. Why is this problem so hard?

First as described earlier, there is a competition between changes in the temperature conditions that promote the formation of cyclones and the increased availability of moisture. *In general*, the former effects are expected to favor fewer and/or weaker cyclones in the future, while the latter — the 'juicier' atmosphere — favors stronger storms. This competition increases the uncertainty in the outcome (like the errors that results from taking a small difference between two large numbers). Perhaps more importantly, however, a *general* result may not be useful to a person who lives in a particular place and at a particular time. A group of GCMs — those that produce the best simulations of storms and the storm track in the current climate — show an increase in the numbers of strong cyclones along and inland from the East Coast of the United States between the early and mid-21st century (Colle *et al.* 2013), yet this change is projected to weaken later in the century. Moreover, while projections from models that *cannot* simulate the present-day storm and storm track are clearly not credible, there is no guarantee that models that *can* simulate present-day conditions will do equally well-simulating future changes (and there is no way to test this until the change has happened).

A second source of uncertainty for any *place-based* projection of change at any specific time is the large intrinsic variability in the climate system. An example of such variability is the El Niño Southern Oscillation (ENSO) phenomenon. El Niño events occur irregularly, with a typical frequency of one every three to four years. ENSO has substantial influences on the Northern Hemisphere storm tracks, and these are roughly as strong as those expected from climate change (Seager *et al.* 2010). In an El Niño year, the jets and storm tracks strengthen and shift towards the equator.

Finally, humans influence Earth's climate in ways other than the emission of greenhouse gases. An example is the erosion of the stratospheric ozone layer by man-made chemicals, such as chlorofluorocarbons, manifested most dramatically in the Antarctic ozone hole. In the Southern Hemisphere, the effects of the ozone hole on the storm track are stronger than those of global warming, according to climate model experiments (Grise, Son, Correa, and Polvani, 2014), in which these two effects are applied separately. In projecting *future* changes in the Southern Hemisphere, it is, therefore, necessary to include the expected *recovery* of the ozone layer, which should reverse the influences ozone depletion, at the same time that the effects of global warming are expected to amplify.

Summary

In recent years Norfolk, Virginia, a city on the United States East Coast has experienced severe floods during *nor'easters*, the local name for strong extratropical cyclones. Increased risk of flooding is, in part, associated with rising sea level (from ocean warming and melting ice sheets and glaciers) increases the overall risk of flooding, but acute floods occur in storms. A resident of Norfolk might want to know, therefore, if there will be more or fewer nor'easters in the future and whether they will be stronger or weaker than today. Climate scientists try to answer such questions starting from our basic understanding of the climate system — the physics of the greenhouse effect — and then applying expectations for global-scale warming to estimate the likely effects on storms. We are quite certain that Earth will continue to warm, and we expect that the pattern of warming will, in general, lead to fewer and weaker cyclones, while the increase in moisture — another confident projection — should make storms stronger. To tell the people of Norfolk what to expect, we must project the outcomes of these competing effects to a specific location and season. We are, at present, very far from being able to do this with any confidence. Through ever-more comprehensive modeling and improved direct observations, our understanding of storm processes and dynamics continues to grow, but we will make confident projections for specific places — will winter floods in Norfolk, Virginia get worse? — only when our suites of different global and regional models begin to converge on an answer that is consistent with our dynamical understanding. Will this come before the change happens and can be observed by 'looking out the window'? Probably yes, but much hard work remains to be done.

References

Bengtsson, L., Hodges, K.I. and Keenlyside, N. (2009). Will extratropical storms intensify in a warmer climate? *Journal of Climate*, 22, 2276–2301.

Bjerknes, J. and Solberg, H. (1922). Life cycle of cyclones and the polar front theory of atmospheric circulation. *Geophysics Publication*, 3, 3–18.

Booth, J.F., Wang, S. and Povani, L. (2013). Midlatitude storms in a moister world: lessons from idealized baroclinic life cycle experiments. *Climate Dynamics*, 41, 787–802.

Colle, B.A., Zhang, Z., Lombardo, K.A., Chang, E., Liu, P. and Zhang, M. (2013). Historical evaluation and future prediction of eastern North American and Western Atlantic extratropical cyclones in the CMIP5 models during the cool season. *Journal of Climate*, 26, 6882–6903.

Gastineau, G. and Soden, B.J. (2009). Model projected changes of extreme wind events in response to global warming. *Geophysics Research Letters*, 36, L10810, doi:10.1029/2009GL037500.

Grise, K.M., Son, S.-W., Correa, G.J.P. and Polvani, L.M. (2014). The response of extratropical cyclones in the Southern Hemisphere to stratospheric ozone depletion in the 20th century. *Atmosphere Science Letters*, 15, 29–36.

Kuo, Y.-H. and Reed, R.J. (1988). Numerical simulation of an explosively deepening cyclone in the Eastern Pacific. *Monthly Weather Review*, 116, 2081–2105.

Lambert, S.J. and Fyfe, J.C. (2006). Changes in winter cyclone frequencies ad strengths simulated in enhanced greenhouse warming experiments: Results from the models participating in the IPCC diagnostic exercise. *Climate Dynamics*, 26, 713–728.

Marciano, C.G., Lackmann, G.M. and Robinson, W.A. (2015). Changes in U.S. East coast cyclone dynamics with climate change. *Journal of Climate*, 28, 468–484.

Seager, R., Naik, N., Ting, M., Cane, M.A., Harnik, N. and Kushnir, Y. (2010). Adjustment of the atmospheric circulation to tropical Pacific SST anomalies: Variability of transient eddy propagation in the Pacific–North America sector. *Quarterly Journal of the Royal Meteorological Society*, 136, 277–296.

Willison, J., Robinson, W.A. and Lackmann, G.M. (2013). The importance of resolving mesoscale latent heating in the North Atlantic storm track. *Journal of Atmosphere Science*, 70, 2234–2250.

Willison, J., Robinson, W.A. and Lackmann, G.M. (2015). North Atlantic storm-track sensitivity to warming increases with model resolution. *Journal of Climate*, 28, 4513–4524.

Yin, J.H. (2005). A consistent poleward shift of the storm tracks in simulations of 21st century climate. *Geophysics Research Letters*, 32, L18701, doi:10.1029/2005GL023684.

Slide 1

How Will Storms And The Storm Track Change?

Extratropical Cyclones On A Warmer Earth

Walt Robinson: Department of Marine, Earth, and Atmospheric Sciences, North Carolina State University

Jimmy Booth: Department of Earth and Atmospheric Sciences, City College of New York

Our Warming Planet: Topics in Climate Dynamics
Lectures in Climate Change, Vol. 1
2017

Slide 2

The Cyclones That Comprise Storm Tracks

What are extratropical cyclones?

- Low-pressure systems (cyclonic rotation)

- Surface warm and cold fronts

- Produce strong surface winds, rain, and/or snow

What powers extratropical cyclones?

Primary:
Equator-to-pole temperature gradient (dT/dy)

  "baroclinic instability"

Secondary: Condensation within the storm.

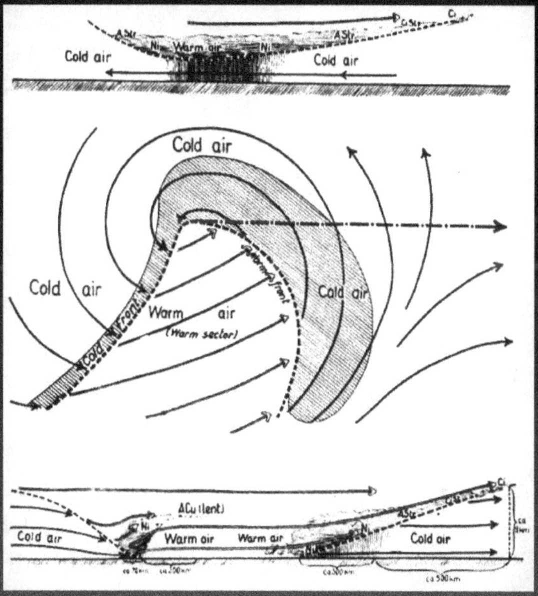

Bjerknes and Solberg (1922)

Slide 3

Slide 4

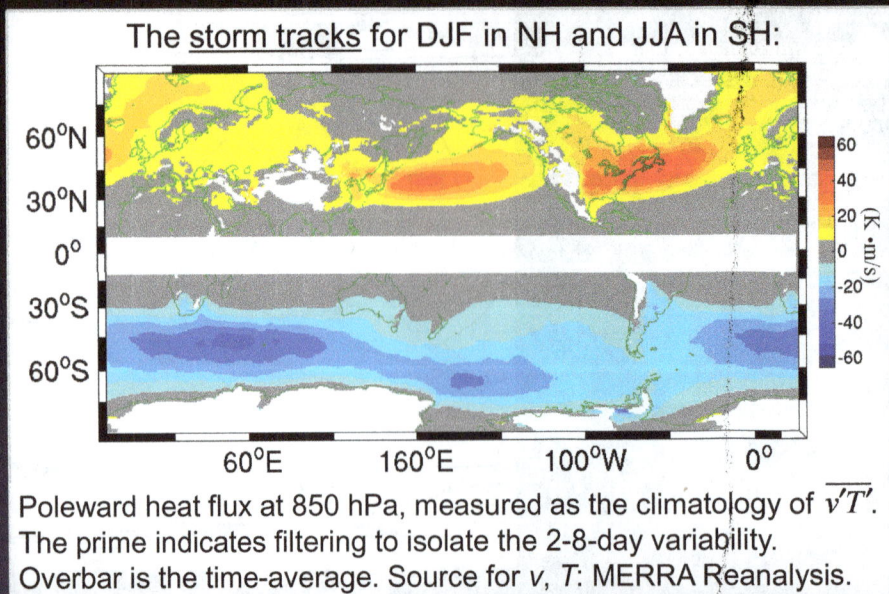

Slide 5

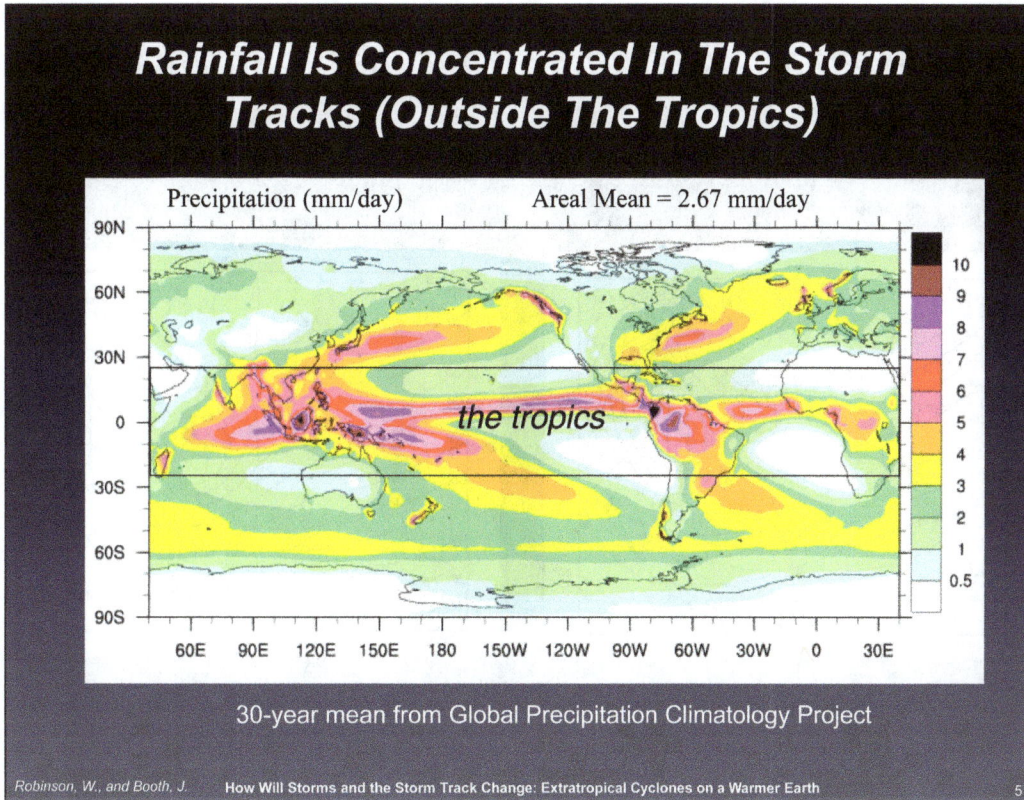

Slide 6

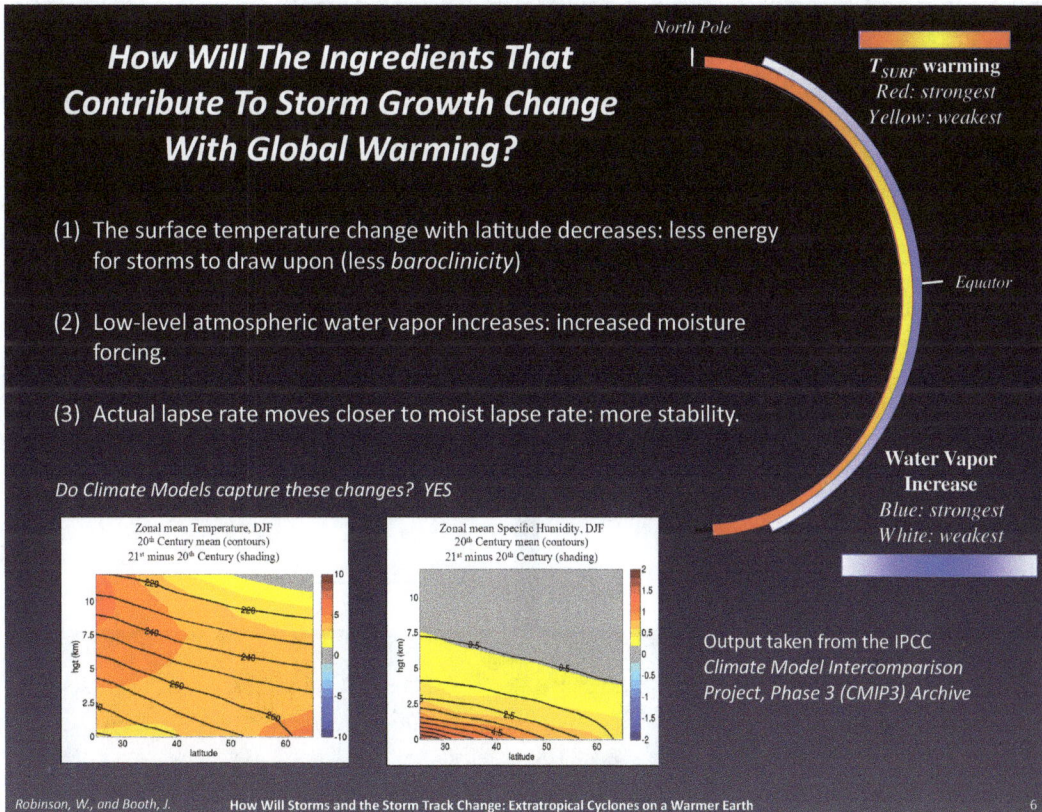

Slide 7

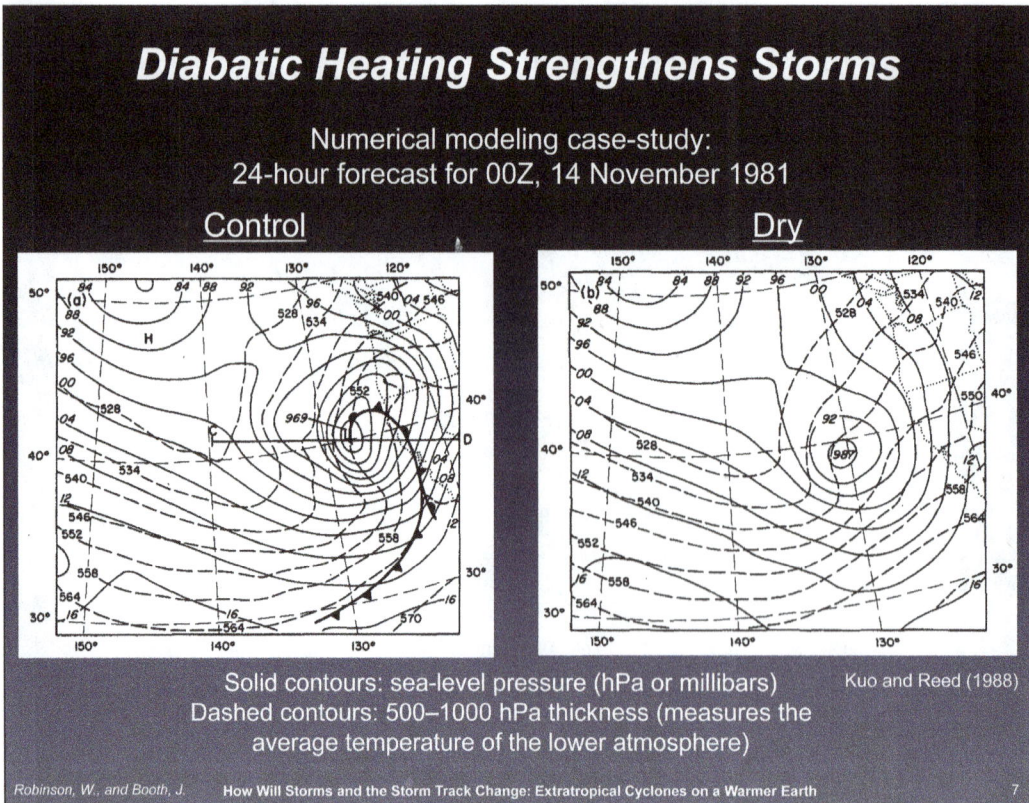

Diabatic Heating Strengthens Storms

Numerical modeling case-study:
24-hour forecast for 00Z, 14 November 1981

Control   Dry

Solid contours: sea-level pressure (hPa or millibars)   Kuo and Reed (1988)
Dashed contours: 500–1000 hPa thickness (measures the
average temperature of the lower atmosphere)

Robinson, W., and Booth, J.   How Will Storms and the Storm Track Change: Extratropical Cyclones on a Warmer Earth   7

Slide 8

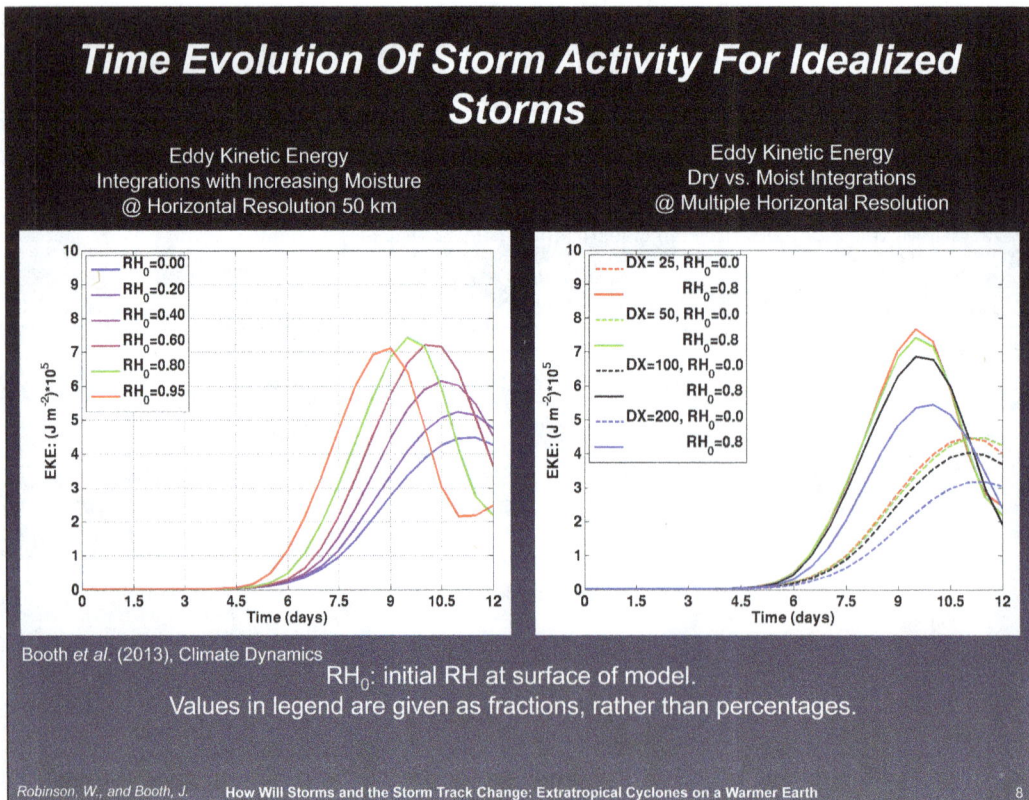

Time Evolution Of Storm Activity For Idealized Storms

Eddy Kinetic Energy
Integrations with Increasing Moisture
@ Horizontal Resolution 50 km

Eddy Kinetic Energy
Dry vs. Moist Integrations
@ Multiple Horizontal Resolution

Booth *et al.* (2013), Climate Dynamics

RH_0: initial RH at surface of model.
Values in legend are given as fractions, rather than percentages.

Robinson, W., and Booth, J.   How Will Storms and the Storm Track Change: Extratropical Cyclones on a Warmer Earth   8

Slide 9

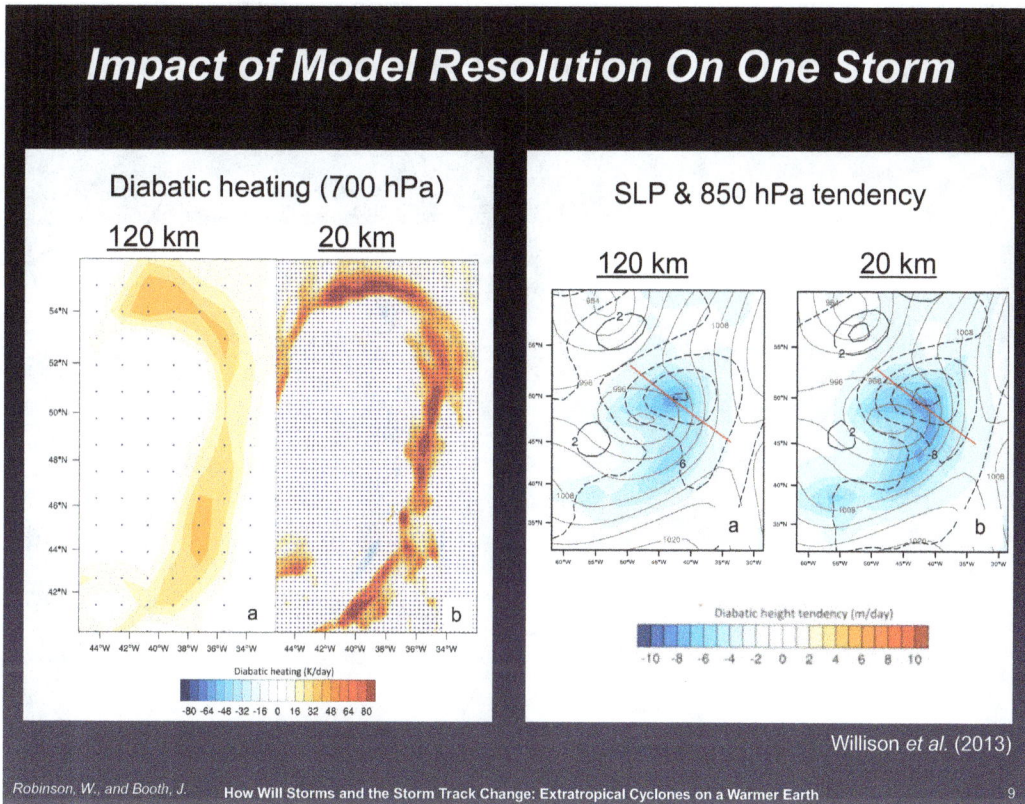

Slide 10

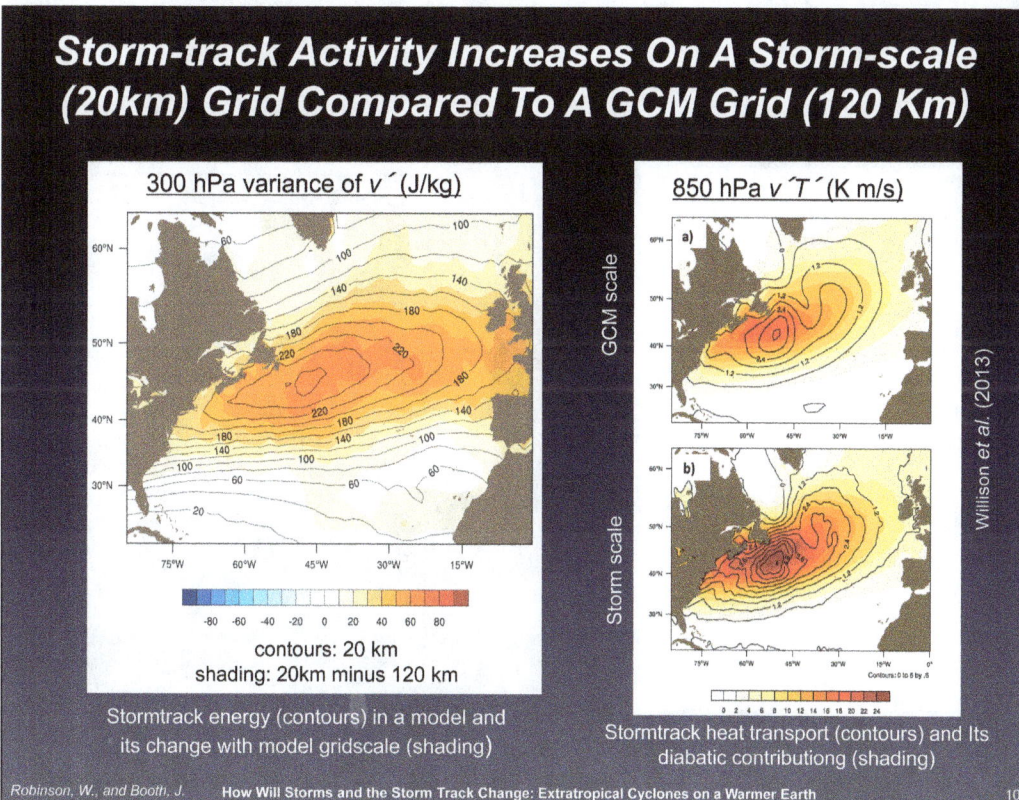

Slide 11

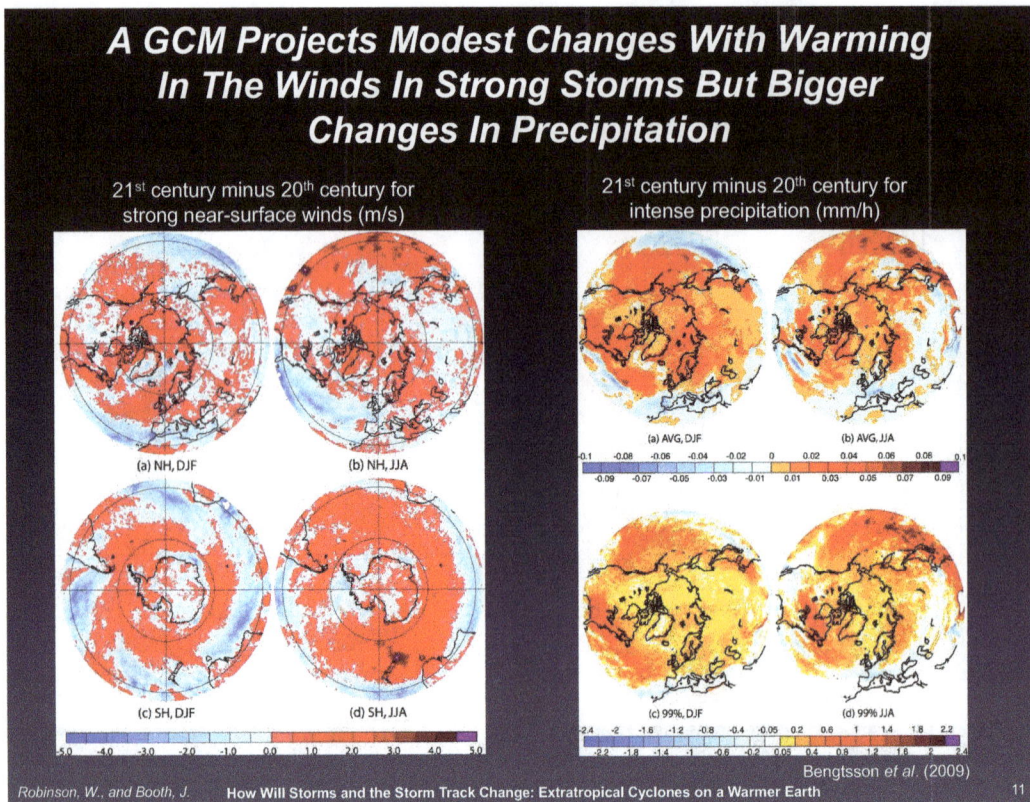

Slide 12

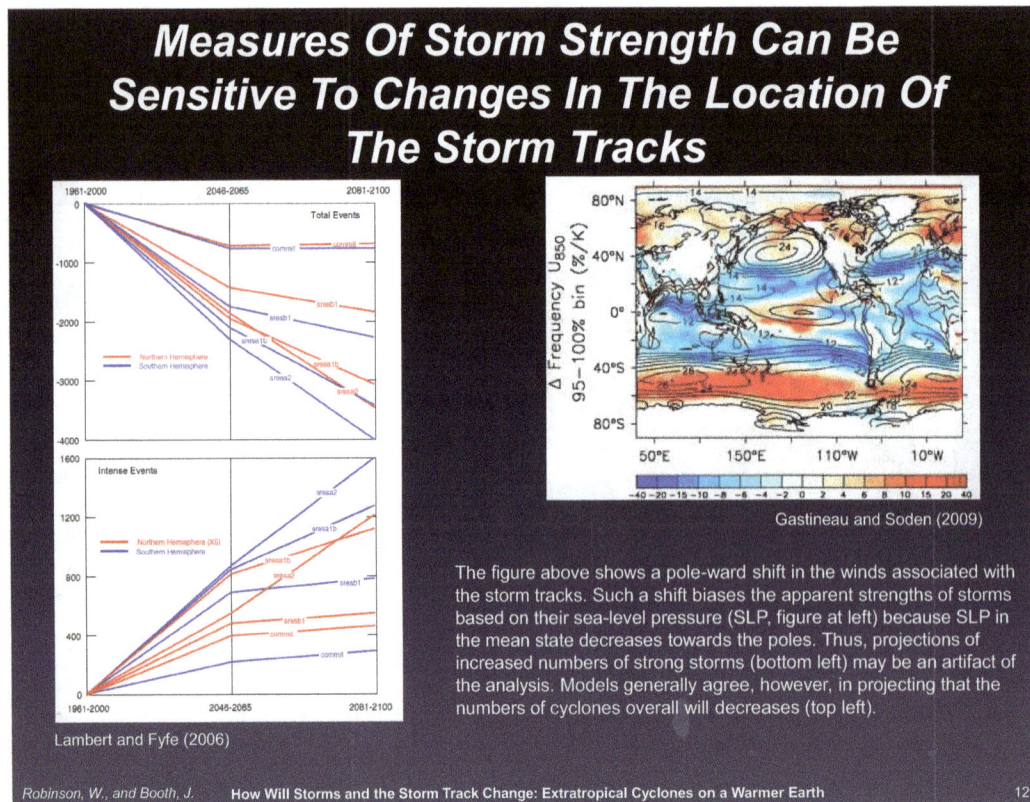

Slide 13

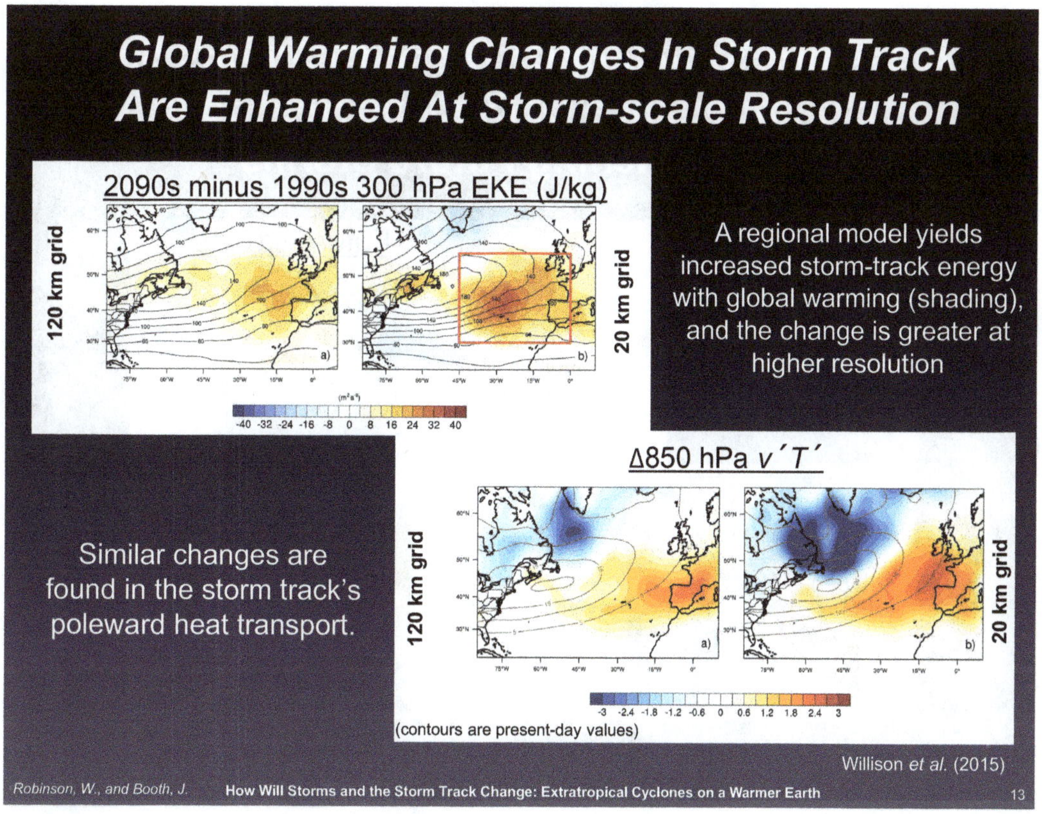
Global Warming Changes In Storm Track
Are Enhanced At Storm-scale Resolution
2090s minus 1990s 300 hPa EKE (J/kg)
120 km grid
20 km grid
-40 -32 -24 -16 -8 0 8 16 24 32 40
A regional model yields increased storm-track energy with global warming (shading), and the change is greater at higher resolution
Δ850 hPa v′T′
Similar changes are found in the storm track's poleward heat transport.
120 km grid
20 km grid
-3 -2.4 -1.8 -1.2 -0.6 0 0.6 1.2 1.8 2.4 3
(contours are present-day values)
Willison et al. (2015)
Robinson, W., and Booth, J.
How Will Storms and the Storm Track Change: Extratropical Cyclones on a Warmer Earth
13

Slide 14

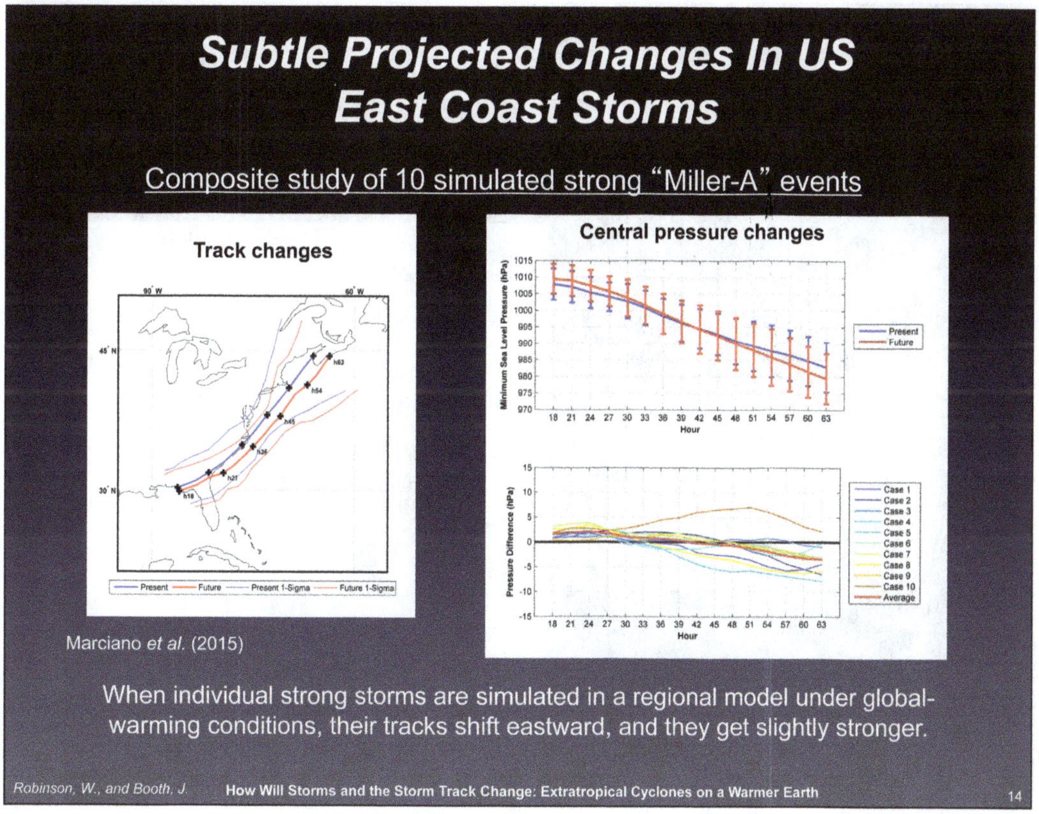
Subtle Projected Changes In US East Coast Storms
Composite study of 10 simulated strong "Miller-A" events
Track changes
Present
Future
Present 1-Sigma
Future 1-Sigma
Central pressure changes
Minimum Sea Level Pressure (hPa)
Hour
Present
Future
Pressure Difference (hPa)
Case 1
Case 2
Case 3
Case 4
Case 5
Case 6
Case 7
Case 8
Case 9
Case 10
Average
Marciano et al. (2015)
When individual strong storms are simulated in a regional model under global-warming conditions, their tracks shift eastward, and they get slightly stronger.
Robinson, W., and Booth, J.
How Will Storms and the Storm Track Change: Extratropical Cyclones on a Warmer Earth
14

Slide 15

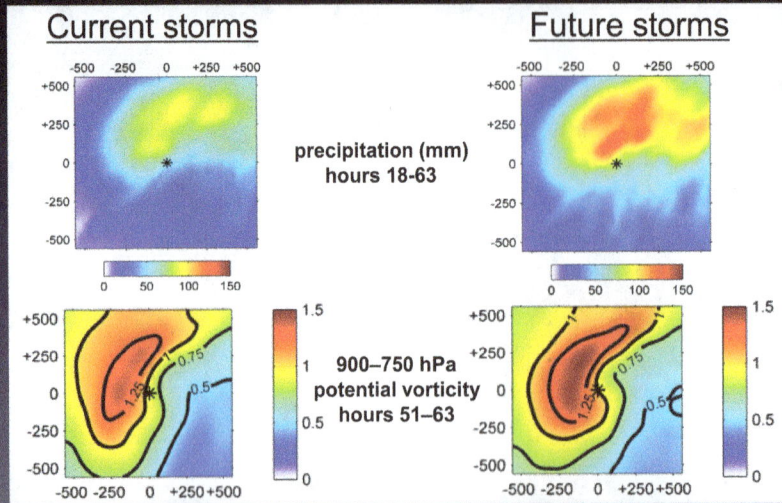

The Average Structure Of These 10 Storms Shows The Importance Of Diabatic Heating In Strengthening Future Storms

Current storms / Future storms

precipitation (mm) hours 18-63

900–750 hPa potential vorticity hours 51–63

Marciano *et al.* (2015)

Robinson, W., and Booth, J.   How Will Storms and the Storm Track Change: Extratropical Cyclones on a Warmer Earth   15

Slide 16

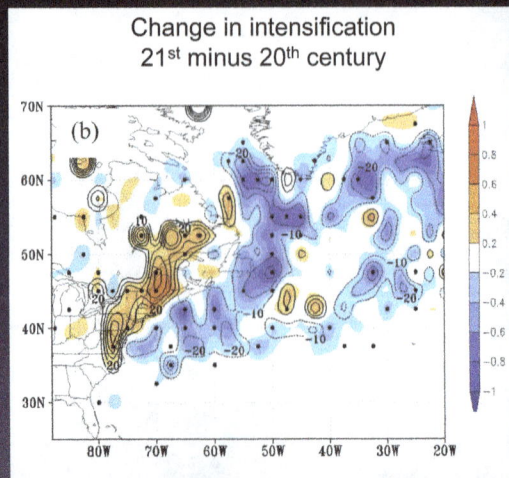

GCMs Project There Will Be More Fast-growing Storms Along The U.S. East Coast

Change in the number of 6-h cyclone deepening rates > 5 hPa (shaded as the number of cyclone tracks per 5 cool seasons) per 50,000 km^2 and the percentage change (contour every 10% with negative dashed) between the 2039–2068 future period minus the 1979–2004 historical period.

Change in intensification 21st minus 20th century

Colle *et al.* (2013)

Robinson, W., and Booth, J.   How Will Storms and the Storm Track Change: Extratropical Cyclones on a Warmer Earth   16

Slide 17

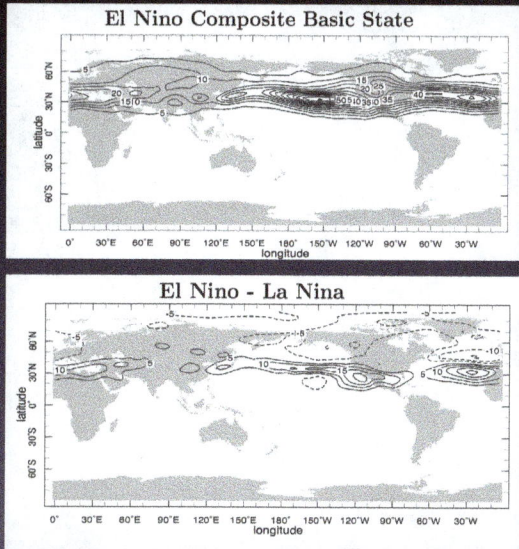

Northern Hemisphere Storm-track Changes With Global Warming: Complicated By El Nino-Southern Oscillation

The strength of the storm track (measured as the variance of the northward wind) in a model of the storm track using El Niño conditions (top) and the difference between El Niño and La Niña conditions (bottom). Units are m²/s²

El Nino Composite Basic State

El Nino - La Nina

Seager et al. (2010)

Slide 18

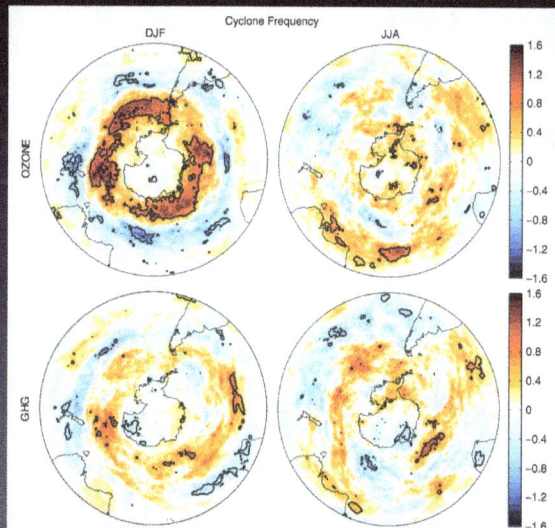

Southern Hemisphere Storm-track Changes With Global Warming: Complicated By Changes In Ozone.

Late 20th century (1960 to 2000) changes in cyclone frequency in model simulations driven by changes in ozone (top) and greenhouse gases (bottom). Units are cyclones per month within a 500-km radius.

Cyclone Frequency

Grise et al. (2014)

Slide 19

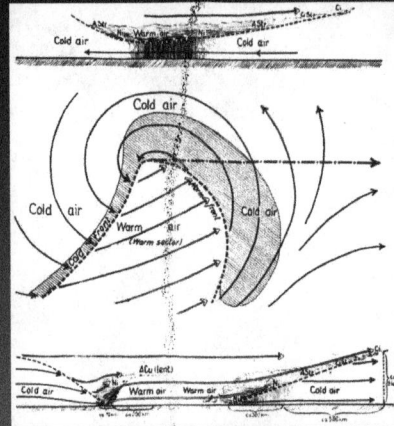

Summary

- Global scale influences—changes in mean temperatures, winds, moisture—are mostly understood

- Uncertainty is driven by their competing effects on storms & by strong variability in the climate system

- Actionable place-based answers to *"how will storms and the storm track change?"* are not yet in hand

Robinson, W., and Booth, J.   How Will Storms and the Storm Track Change: Extratropical Cyclones on a Warmer Earth   19

Slide Notes

Slide 2   Extratropical cyclones are also called mid-latitude wintertime storms. Weather forecasting and atmospheric dynamics developed from the study of these storms in Norway (hence the figure above).

Top: cross-section looking north into the warm frontal precipitation (location given by dashed arrow pointing east in the map view schematic). Warm sector air has moved northward and above the cold easterly air. *Middle*: storm schematic with warm and cold fronts labeled. Arrows show surface wind direction, but also approximate the sea-level pressure pattern. Shading shows location of precipitation. *Bottom*: cross-section looking into the cold front (left) and cross-section looking into the warm front (right).

Slide 3   Based on ship observations, this image of the storm track has all of the key items of the storm tracks: the storms propagate from west to east; there is circumpolar coverage by the storms within a fixed latitude range; a concentration of storms over the oceans, especially at the western boundary current regions (e.g., the Gulf Stream).

The picture also has a few things wrong: the lack of storms at higher latitudes, the large number of tropical cyclones that cross the Atlantic into Europe and continue eastward. Note also that the Southern Ocean storm track is cut off from the map, but whaling vessels were certainly aware of the storms down south.

Slide 4   This figure shows the poleward heat flux by synoptic eddies during winter for both hemispheres. The values for the Southern Hemisphere are negative because of the convention of assuming northward motion is positive.

Using gridded data that is continuous in time, one can isolate the transient heat flux associated with the storm track using temporal filtering. In this figure, a 2–8-day bandpass filter has been applied. This means that variability on time scales less than 2 days has been removed, and so has the variability on time scales greater than 8 days.

Slide 5   The annual mean precipitation within 30–65 latitude belts shows a pattern similar to the synoptic eddy heat flux.

In the Northern Hemisphere, the higher precipitation at the western boundary currents relates to the anomalously warm sea-surface temperature (SST) and hence the increased moisture available in those regions.

The higher precipitation on the east side of the ocean basins corresponds to the storm-track exit regions. In these locations, storms tend to stop propagating but continue to drop precipitation. Along the west coast of North America, rainfall further enhances as air rises over steep coastal topography.

Slide 6   Recall that extratropical cyclones develop in response to baroclinic instability.

Baroclinic instability develops in environments with strong vertical shear in the zonal wind, which, via the thermal wind relation, implies a strong meridional temperature gradient. However, the storms are provided a secondary boost of energy from condensational heating, via its impact on the vertical motion within the storm.

Therefore, to consider how storms will change with global warming, we must consider model projections for changes in the meridional temperature gradient, moisture availability, and vertical stability.

The figure on the bottom left shows that the models project a decrease in the low-level meridional temperature gradient. This is mainly because of polar amplification of global warming. (A 'zonal mean' is an average over all longitudes at a given latitude and height.)

The figure on the bottom right shows models project an increase in the low-level moisture. This is mainly because the models project relative humidity will remain nearly constant, while temperatures will increase.

The schematic on the right depicts the qualitative changes shown in the bottom pictures: for the sake of the storms, there is a competition between the impact of the change in the temperature gradient and the change in the moisture.

Separately, with greenhouse warming, the vertical temperature profile will move towards a moist adiabat, and hence the troposphere will be more stable.

Slide 7   Related to the precipitation, the moisture within the storms can have a big impact on storm strength, in terms of surface winds. This is a case-study of a storm off of the west coast of the United States. This storm generated devastating winds in Oregon.

Using a numerical model, it is possible to simulate the storms with and without moisture, to test the impact of moisture, via diabatic heating (the heat released when water vapor condenses into rain), on the storm.

At left, we see the control run, which includes moisture. Notice the tight packing of the SLP contours, these translate into strong winds, via the geostrophic relationship. Notice also the central pressure minimum of 969 hPa.

At right, we see a run in which the model's moisture is 0. In this case, there is no diabatic heating to provide secondary strengthening for the storm. As a result, the central pressure minimum is 987 hPa and the storm does not have tightly packed SLP contours.

In addition to the strength of the storm changing, the location of the storm central pressure has also changed. This is because the diabatic heating provides vorticity which increases the cyclonic wrapping of the storm circulation at low levels.

Slide 8   To test the influence of changes in moisture in a controlled setting, one can use idealized models.

In this case, an extratropical cyclone is modeled in a channel as it grows from an infinitesimal disturbance to a full-sized storm.

The evolution of the storm with time can be seen by calculating the eddy kinetic energy (EKE) for each time step. EKE is a measure the strength of storm activity (here the eddies are with respect to the zonal mean of the circulation).

At left, we examine the change in the EKE in response to changes in the initial relative humidity (RH) in the model. This model has water as the surface boundary condition, so as the model runs, moisture can enter the atmosphere via evaporation. Therefore, the $RH_0 = 0.0$ cases are initially dry, but moisture is added to the storm via the surface fluxes. The figure shows that the storm maximum EKE increases as the RH_0 increases from 0% to 80%. However, for the change from $RH_0 = 80\%$ to $RH_0 = 95\%$, the last boost in RH, it appears that the storms' vorticity gain from the increased moisture decreases the size of the storm. We also see that the time when the storm reaches its maximum EKE becomes earlier as the RH increases. This signifies that the intensification rate of the storm is increased due to the presence of more moisture.

At right, the EKE for the control run with $RH_0 = 80\%$ is compared with the run with $RH_0 = 0\%$, for model integrations using a different horizontal resolution. The coarsest resolution used is 200 km, which is equivalent to the resolution of the models used for global climate models (GCMs) in CMIP3 and CMIP5. Comparing the dashed and the solid line for a fixed color, we can see that the moisture leads to a stronger EKE maximum occurring at an earlier time, signifying that the boost in storm strength via diabatic heating occurs at all of the model resolutions. Comparing the solid lines only, we can also see that changing the model resolution from coarse to fine, in the presence of moisture, leads to an increase in the strength of the storms.

Slide 9   Modeling the same storm with a typical climate model grid scale (120 km) and much finer scale (20 km), atmospheric heating at the finer scale is stronger and more localized (left panels).

The dots indicate the positions of grid points, where the model keeps track of the atmospheric variables (winds, temperature, pressure, and humidity).

This leads to a stronger storm (right panels) on the finer grid. The solid contours show the sea-level pressure, and the dashed contours indicate the rate at which the storm is deepening in the lower troposphere.

Through the technique of *potential vorticity inversion*, it is possible to calculate how much of the deepening is caused by the heat released by condensation — the diabatic contribution to the storm's deepening. This contribution, shown by shading, is significantly stronger on the finer grid (far right panel). The strongest diabatic contribution is east of the low and south of where pressures are falling most rapidly. Here, diabatic effects help strengthen the *low-level jet*, which brings heat and moisture into the strengthening storm, helping it grow.

Slide 10   Here again we compare results from a regional model of the storm track, with simulations on global climate model (GCM) and storm-scale grids.

The variance of the north–south wind (left panel) is a measure of the energy in the storm track. This is increased by about 20% in the higher resolution model.

Recall that the storm tracks play a key role in the climate system by transporting heat pole-ward across the middle latitudes. This heat flux is much stronger on the storm-scale grid (right panel, bottom) than on the GCM grid (right panel, top), as is the contribution to that can be attributed to diabatic effects. Here v' is the northward wind and T' is the variation of the temperature. Their product, denoted as the 'eddy heat flux', is a measure of how effectively storms move heat northward.

Slide 11   Enhanced diabatic effects in storms on a warmer Earth might be expected to have the strongest influence on the strongest storms. A global climate model (GCM) study, however, shows only modest increases in the winds associated with the strongest 1/2 per cent of storms (left panel). The regions of decreasing winds are presumably associated with the overall pole-ward shift of the jets and storm tracks (here NH and SH are Northern and Southern Hemispheres; DJF is December–January–February and JJA is June–July–August).

Specific humidity is greater in a warmer climate, and this leads to more robust and widespread projected increases in precipitation, for both the average precipitation (right panel, top) and the strongest 1% of precipitation events (right panel, bottom)

These results are typical of those obtained with global models.

Slide 12   Each line in the left panel shows projected changes in numbers of storms and numbers of strong storms in models driven by different projections for the increases in greenhouse gases, and thus different projections in the strength of overall global warming. The general result is that a decrease in the number of extratropical cyclones is expected, but an increase in the number of the most intense cyclones. Here, strong storms are measured by the strength of their central low pressure.

Over a decade ago, it was noted that most climate models project a pole-ward shift in the storm tracks with global warming. This produces a poleward shift in the occurrence of strong winds, as shown in the right panel.

Because sea-level pressure generally decreases towards the poles, storms appear stronger if they take a more poleward track. A poleward shift in the storm track will, therefore, appear as a strengthening of storms, if storm strength is measured by depth of the central low pressure.

Slide 13   Unlike most GCMs, in a regional model of the wintertime North Atlantic storm track, the energy in the storm track and its pole-ward heat transport increase significantly with global warming, and these changes are stronger when the storm track is simulated at storm scale, as opposed to GCM scale, resolution.

With climate change, it is anticipated that the wintertime land–sea thermal contrast will be reduced. This contrast, along with the general pole-ward decrease in near surface temperatures, is the dominant source of energy for storms in the western Atlantic basin. Here, projected changes in storm-track energy are small and even negative (weaker storm track in the future).

Increases in storm-track energy (top panel) and heat transport (lower panel) are most pronounced in the eastern half of the Atlantic basin, where diabatic effects provide a greater share of storm energy. Here, v′ is the northward wind and T′ is the variation of the temperature. Their product, denoted as the 'eddy heat flux', is a measure of how effectively storms move heat northward.

Slide 14   'Miller-A' storms are single low-pressure systems that develop over or near the Gulf of Mexico and track up the US East Coast.

Each case is an observed storm, from the period 1981 to 2010, that met the criteria for being classified as a strong Miller-A event.

Consistent with findings for the storm tracks, individual strong storms along the US East Coast, simulated under present-day and global-warming conditions, increase in strength only a little with global warming. Their tracks shift eastward, presumably a result of stronger upper-troposphere westerly winds.

Slide 15   Here, we see the effects of global warming on the composite structure of these 10 storms.

The figures show their average structure, obtained by shifting the center of each storm to the same location and then averaging the quantities shown.

The role of diabatic heating increases substantially with global warming. Precipitation north of the storm center is much stronger (top panels), and this leads to a stronger dynamical role for heating, measured here using the lower tropospheric *potential vorticity* (bottom panels).

The wrapping of strong potential vorticity around to the southeast of the storm center, together with tightening of its gradient in the future storm composite, implies strengthened diabatic generation of the low-level jet, which contributes to storm growth.

Potential vorticity measures the tendency of air to rotate cyclonically (counter-clockwise).

That the storms are not substantially stronger in the future, likely results from the countervailing influences of the stronger diabatic effects, shown here, and the weakening of the temperature gradient that drives baroclinic instability.

Slide 16   This is a result from the CMIP5 collection (or ensemble) of global climate models (GCMs). The CMIP5 GCM ensemble is the set of global model simulations performed in support of the fifth assessment of climate change by the Intergovernmental Panel on Climate Change (IPCC).

Here, a subset of these models has been selected, choosing those that perform best in simulating the present-day storm track.

These models project an increase in strong storms along the US East Coast by the middle of this century. This is consistent with the results from a regional model shown on the previous slides, except that the global models do not project the eastward shift in the tracks of the storms.

More perplexing, perhaps, is that these same global models project that the changes displayed above will *weaken* by the final decades of the 21st century.

Slide 17   Intrinsic variability of the climate system also can affect the storm track, and these influences can be as great as or greater than the effects of global warming. El Niño-Southern Oscillation (ENSO) is a major mode of intrinsic variability, involving warming or cooling of ocean temperatures over large regions of the equatorial Pacific Ocean.

This slide shows the calculated effects of ENSO variability on a measure of storm-track energy, indicating that the storm tracks strengthen and shift equatorward in El Niño years.

The overall change in the storm track in a given year is a superposition of climate-change effects and those due to intrinsic variations in the climate system, although these influences need not be additive.

These results also suggest that whether the future climate becomes more 'El Niño like' or more 'La Niña like' could have significant implications for the future of the storm tracks; there is, as yet, no consensus among global models about future climate changes in the equatorial Pacific.

Slide 18   In the Southern Hemisphere, the appearance and growth of the ozone hole has, since around 1970, significantly perturbed the climate system.

The model simulations shown here indicate that, over this period, the effects of ozone depletion on the storm track have been much stronger than those of greenhouse warming.

This has implications for the future of the Southern Hemisphere storm track, since the ozone layer is expected to recover in this century, and its recovery should have effects on the storm track roughly equal and opposite to those shown in the upper panels of this figure.

Slide 19   The photo shows urban flooding in North Virginia, a city on the US East Coast, during a strong extratropical cyclone — locally called a *nor'easter* — in 2009. This event is indicative of the practical importance of possible changes in extratropical cyclones with global warming, albeit in concert with other changes in the Earth system, such as changes in sea level.

While our understanding of the processes and factors that are important for extratropical cyclones and storm tracks has grown enormously since the pioneering work of the Norwegian meteorologists a century ago, we are very far from being able to translate that understanding into useful estimates of likely changes in the risks that storms pose to places like Norfolk.

CLIMATE LECTURE 7

The Relationship Between Recent Arctic Amplification and Extreme Mid-latitude Weather

Judah Cohen

Atmospheric and Environmental Research, Inc., Lexington, MA02421, USA

Judah Cohen is Director of Seasonal Forecasting at AER [Atmospheric and Environmental Research], and a Research Affiliate at MIT's Parsons Lab. He works on the impacts of soil moisture and snow cover on other climate parameters, and provides seasonal forecasts based on these and other trends. He is also interested in decadal temperature trends and explaining those trends with large scale climate models.

Introduction

Climate change and its impacts is one of the most important challenges facing our society. Throughout his career, David Rind has been a pioneer on this topic, publishing many papers on the topic of climate change from past climates, most notably ice ages to future climates with much greater greenhouse gas (GHG) concentrations than found presently in the Earth's atmosphere. For example, David Rind's seminal paper 'The Dynamics of Warm and Cold Climates' (Rind, 1986) explored the zonal mean properties of the Earth's atmosphere under a range of past, present, and future cold and warm climates. Here, I present a short lecture on climate change, specifically how Arctic amplification (AA) might influence extreme weather in the mid-latitudes. David Rind did publish on polar amplification (e.g., 'The Consequences of not Knowing Low- and High-Latitude Climate Sensitivity', Rind, 2008) but not as it pertains to extreme weather; a topic that is still emerging. Still, I know that David is sensitive to our limitations to understand the climate and appreciates unexpected and mysterious twists and turns in the climate system.

Arctic Amplification

Global surface temperatures are projected to warm due to rapid increases in GHGs and over the entire length of the instrumental record, temperatures have undergone a warming trend (IPCC, 2013). The most significant and strongest warming globally has occurred in the most recent 40 years, with Arctic temperatures warming at nearly double the global rate (Screen and Simmonds, 2010). Coupled climate models attribute much of this warming to rapid increases in GHGs and project the strongest warming across the extratropical

Northern Hemisphere (NH) during boreal winter due to 'winter (or Arctic) amplification' (Cohen *et al.* 2014a).

Rapid Arctic warming and sea ice loss has had significant impacts locally, particularly in late summer and early fall. September sea ice has declined at a rate of 12.4% per decade since 1979 (Stroeve *et al.* 2011) so that by the summer of 2012, nearly half of the areal coverage had disappeared. This decrease in ice extent has been accompanied by an approximately 1.8 m (40%) decrease in mean ice thickness since 1980 (Kwok and Rothrock, 2009) and a 75–80% loss in volume (Overland, Wang, Walsh and Stroeve, 2014). A decrease in sea ice both reduces the surface albedo and increases latent and sensible heat fluxes into the atmosphere, resulting in warmer surface temperatures and moistening of the Arctic boundary layer.

Arctic sea ice plays an important role in modulating surface conditions at high latitudes, with ensuing feedbacks on the entire Earth climate system. With an albedo of almost 80%, sea ice moderates near-surface temperatures during the persistent sunlight in summertime. During winter, sea ice decouples the ocean surface from the overlying atmosphere, preventing moderation of Arctic air masses by latent and sensible heat fluxes from the Arctic Ocean. Furthermore, snow accumulations on sea ice mimic the cooling impacts of snow cover over land and hence amplify polar cooling during the long polar night. Given these important seasonal roles for Arctic sea ice, even small changes in sea ice extent can cause dramatic changes on Arctic climate. For example, anomalously low sea ice during the summer exposes darker (i.e. low albedo) ocean water to sunlight, producing strong Arctic warming via direct radiative impacts and anomalous latent and sensible heat fluxes. This reduction in summer sea ice extent subsequently affects fall re-growth of Arctic sea ice, allowing for warmer and moister Arctic air masses, particularly over nearby continents. The ensuing feedback contributes to amplified warming of the Arctic relative to the rest of the globe (e.g., Serreze and Francis, 2006; Screen and Simmonds, 2010).

Snow cover in spring and summer has decreased at an even greater rate than has sea ice. June snow cover alone has decreased at nearly double the rate of September sea ice (Derksen and Brown, 2012). The decrease in spring snow cover has contributed to both the rise in warm season surface temperatures over the NH extratropical landmasses and in the decrease in summer Arctic sea ice. The combined rapid loss of sea ice and snow cover in the spring and summer has contributed to the amplification of Arctic warming.

Linking AA to Extratropical NH Climate Variability

With the Arctic at record warm levels, the eight years between 2007 and 2014 have exhibited the lowest minimum sea ice extents recorded in September since satellite observations began, with an all-time record low in 2007 followed by another in 2012, when sea ice extent fell below 4 million km² for the first time in the observational record. However, while the Arctic continues to warm, NH continental winters have recently grown more extreme across the major industrialized centers. Several of these seven winters following the low sea ice minima have been unusually cold across the NH extratropical landmasses (Cohen *et al.* 2012; 2013; 2014a). The recent winters of 2013/2014 and 2014/2015 were characterized by record cold and widespread snowstorms across the eastern United States and Canada with the most intense cold-air outbreak in decades associated with the weakening of the polar vortex both winters. Cohen, Barlow and Saito (2009) argued that the occurrence of more severe NH winter weather is a two-decade long trend starting around 1988. Whether the recent colder winters are a consequence of internal variability or a response to changes in boundary forcings resulting from climate change remains an open question.

Numerous compelling hypotheses linking changes in the Arctic to recent severe weather have been presented (e.g. Liu *et al.* 2012; Francis and Vavrus, 2012). They propose that a warming Arctic results in a wavier polar vortex, leading to enhanced atmospheric blocking that often spawns extreme weather events.

The interaction between Arctic sea ice decline and the tropospheric circulation extends outside the Arctic and across multiple seasons. Several studies have examined these delayed and remote responses to Arctic sea ice decline, especially in relation to recent climate change. Francis *et al.* (2009), Jaiser *et al.* (2012), and Francis and Vavrus (2012) suggest that late summer Arctic sea-ice conditions modify static stability of the Arctic troposphere, warm the lower levels, and ultimately change the poleward thickness gradient, implying a weakening of the polar jet stream. The decrease in static stability, however, also suggests enhanced baroclinic energy at higher latitudes and a change in the mid- and high-latitude planetary wave generation and propagation (Jaiser *et al.* 2012). Such changes persist through the following winter, impacting mid-latitude weather systems and temperature patterns. Honda, Inue and Yamane (2009) and Liu *et al.* (2012) use modeling experiments to illustrate that low summer Arctic sea ice conditions promotes anomalous and persistent high pressure across the North Atlantic (i.e., the negative phase of the North Atlantic Oscillation (NAO)), allowing for anomalously cold winters across much of Eurasia and parts of North America. In an observational analysis, Hopsch, Cohen and Dethloff (2012) also found a tendency for a negative NAO in late winter following low fall sea ice.

Links between summer Arctic sea-ice variability and extratropical NH climate variability are not restricted to the following winter season, however (Tang, Zhang and Francis, 2014). Francis and Vavrus (2012) argue that AA resulting from sea-ice loss in fall and winter has imprints on the tropospheric circulation throughout the year, thus promoting weakened zonal flow and more frequent blocking episodes. Overland, Francis, Hanna and Wang (2012) argue that sea-ice melt promotes earlier snowmelt and enhanced melting of the Greenland ice sheet.

Extreme Weather, Blocking, and Links to Sea-Ice Loss

Although the term *extreme weather* is used to describe a range of phenomena, their common trait is the potential for high socioeconomic impact. Such events can be either short-lived (i.e., a strong hurricane or blizzard) or persistent for several days or weeks (i.e., droughts and floods). Deschênes and Moretti (2009) suggest that that 0.8% of annual deaths were attributed to cold temperatures, whereas Anderson and Bell (2011) found heat waves to be responsible for a 3.74% increase in US mortality rate and the August 2010 Moscow heat wave resulted in a death toll of 55,000 (Barriopedro *et al.* 2011).

Atmospheric blocking events play a key role in many extreme weather events (Coumou and Rahmstorf, 2012). Blocking patterns result from the breakdown of the background flow pattern, which makes weather systems move slower or even become stationary (Rex, 1950a, 1950b). Like boulders blocking a river, once an atmospheric block forms, its impacts are felt both upstream and downstream of the block. For example, high winter precipitation over California is oftentimes associated with persistent (or 'cutoff') low pressure in the Eastern Pacific (e.g., Carrera, Higgins and Kousky, 2004). Extreme cold temperatures during the winter months over Europe and North America are associated with blocking anticyclones over northern Eurasia and Greenland, respectively (e.g., Thompson and Wallace, 2001; Sillmann, Croci-Maspoli, Kallache and Katz, 2011). The deadly Russian heatwave of 2010 was a direct result of a stagnant upper-level ridge over western Eurasia (e.g., Dole *et al.* 2011). Blocking events have been implicated as precursors for sudden stratospheric warmings (Quiroz, 1979, 1986; Martius, Polvani and Davie, 2009), which, in turn, influence surface weather evolution for up to two months later (Baldwin and Dunkerton, 2001; Thompson, Baldwin and Wallace, 2002).

Strong evidence links the observed increase in extreme events with anthropogenic climate change (Coumou and Rahmstorf, 2012). However, the projected change in frequency and intensity of extreme weather events is considered one of the principal yet more poorly understood impacts of climate change (IPCC, 2013; Sillmann *et al.* 2013). The main culprits for poor predictions of extreme weather events are a lack of understanding of dynamics and feedback mechanisms driving variability in weather extremes and subsequent model deficiencies (Christensen, Boberg, Christensen and Lucas-Picher, 2008; Min, Hazeleger, Odenborgh and Sterl, 2013). Quantifying the contribution of reduced Arctic sea ice to anomalous circulation patterns is particularly problematic because many of the possible candidates forcing extreme weather (e.g. sea ice) exhibit significant trends in the modern satellite era. This makes it impossible to establish a causal link between sea-ice anomalies and extreme weather using statistical methods and historical data alone.

Some studies go even further arguing that the proposed evidence is sensitive to analysis parameters and that that any signal remains masked by natural variability (Barnes, 2013; Screen and Simmonds, 2013; Wallace *et al.* 2014). Specifically, Barnes, Dunn-Sigouin, Masato and Woollings (2014) find no discernable trend in the occurrence of blocking in the past 15–30 years and point to the difficulty in separating forced response from natural variability.

The Link Between Mid-latitude Extreme Weather

While climatologists debate the merits of AA and extreme mid-latitude weather, the media and public have been quick to make the connection between global, and in particular Arctic, warming and extreme weather (Hamilton and Lemcke-Stampone, 2013). While the global warming theory is consistent with record warm temperatures and more intense precipitation events, it does not directly explain cold extremes. Are the public and media leading scientists in recognizing the link between AA and extreme weather?

I argue that this unforeseen trend is likely not due to internal variability alone. Instead, evidence suggests that summer and autumn warming trends are concurrent with decreases in Arctic sea ice and increases in Eurasian snow cover, which dynamically induces large-scale wintertime cooling. Understanding this counterintuitive response to radiative warming of the climate system has the potential to improve climate predictions and our understanding of severe or extreme winter weather across the mid-latitudes.

In this lecture, I argue that the extensive winter NH extratropical cooling trend amidst a warming planet cannot be explained entirely by internal variability of the climate system. It has been shown that above normal snow cover across Eurasia in the fall leads to a negative Arctic Oscillation (AO) and cold temperatures across the Eastern United States and northern Eurasia in winter (Cohen and Entekhabi, 1999; Cohen *et al.* 2014b). Therefore, we suggest that a significant portion of the wintertime temperature trend is driven by dynamical interactions between October Eurasian snow cover, which has increased over the last two decades, and the large-scale NH extratropical circulation in the late fall and winter. I will also demonstrate that the coupling between extensive snow cover and the negative phase of the AO/weakening of the polar vortex involves both continental cooling and a warmer Arctic (Cohen *et al.* 2013; Furtado *et al.* 2015). Sea-ice loss also contributes to a warming Arctic and is likely also contributing to continental cooling.

References

Anderson, G.B. and Bell, M.L. (2011). Heat waves in the United States: Mortality risk during heat waves and effect modification by heat wave characteristics in 43 U.S. Communities. *Environmental Health Perspectives*, 119, 210–218.

Baldwin, M.P. and Dunkerton, T.J. (2001). Stratospheric harbingers of anomalous weather regimes. *Science*, 294, 581–584.

Barnes, E.A. (2013). Revisiting the evidence linking Arctic Amplification to extreme weather midlatitudes. *Geophysical Research Letters*, 40, doi:10.1002.grl.50880.

Barnes, E.A., Dunn-Sigouin, E., Masato, G. and Woollings, T. (2014). Exploring recent trends in Northern Hemisphere blocking. *Geophysics Research Letters,* 41. doi:10.1002/2013GL058745.

Barriopedro, D., Fischer, E.M., Lüterbacher, J., Trigo, R.M. and García-Gerrera, R. (2011). The hot summer of 2010: Redrawing the temperature record map of Europe. *Science*, 322, 220–224.

Carrera, M.L., Higgins, R.W. and Kousky, V.E. (2004). Downstream weather impacts associated with atmospheric blocking over the Northeast Pacific. *Journal of Climate*, 17, 4823–4839.

Christensen, J.H., Boberg, F., Christensen, O.B. and Lucas-Picher, P. (2008). On the need for bias correction of regional climate change projections of temperature and precipitation. *Geophysics Research Letters,* 35, L20709.

Cohen, J. and Entekhabi, D. (1999). Eurasian snow cover variability and Northern Hemisphere climate predictability. *Geophysics Research Letters*, 26, 345–348.

Cohen, J., Barlow, M. and Saito, K. (2009). Decadal fluctuations in planetary wave forcing modulate global warming in late boreal winter. *Journal of Climate*, 22, 4418–4426.

Cohen, J., Furtado, J., Barlow, M., Alexeev, V. and Cherry, J. (2012a). Arctic warming, increasing fall snow cover and widespread boreal winter cooling. *Environmental Research Letters*, 7, 014007. doi:10.1088/1748-9326/7/1/014007.

Cohen, J., Furtado, J. Barlow, M. Alexeev V. and Cherry, J. (2012b). Asymmetric seasonal temperature trends, *Geophysical Research Letters*, 014007 doi: 10.1029/2011GL050582.

Cohen, J., Jones, J., Furtado, J.C. and Tziperman, E. (2013). Warm Arctic, cold continents: A common pattern related to Arctic sea ice melt, snow advance, and extreme winter weather. *Oceanography*, 26(4). doi: 10.5670/oceanog.2013.70.

Cohen, J., Screen, J.A., Furtado, J.C., Barlow, M., Whittleston, D., Coumou, D., . . . Justin Jones. (2014a). Recent Arctic amplification and extreme mid-latitude weather. *Nature Geosciences*, 7, 627–637. doi:10.1038/ngeo2234.

Cohen, J., Furtado, J., Jones, J., Barlow, M., Whittleston, D. and Entekhabi, D. (2014b). Linking Siberian snow cover to Precursors of stratospheric variability. *Journal of Climate*, 27, 5422–5432.

Coumou, D. and Rahmstorf, S. (2012). A decade of weather extremes. *Nature Climate Change*, 2, 491–496.

Derksen, C. and Brown, R. (2012). Spring snow cover extent reductions in the 2008–2012 period exceeding climate model projections. *Geophysics Research Letters,* 39, L19504.

Deschênes, O. and Moretti, E. (2009). Extreme weather events, mortality and migration. *The Review of Economics and Statistics*, 9, 659–681.

Dole, R., Hoerling, M., Perlwitz, J., Eischeid, J., Pegion, P., Zhang, T., . . . Murray, D. (2011). Was there a basis for anticipating the 2010 Russian heat wave? *Geophysics Research Letters*, 38, L06702. doi:10.1029/2010GL046582.

Francis, J.A. and Vavrus, S.J. (2012). Evidence linking Arctic amplification to extreme weather in mid-latitudes. *Geophysics Research Letters*, 39, L06801. doi:10.1029/2012GL051000.

Francis, J.A., Chan, W., Leathers, D.J., Miller, J.R. and Veron, D.E. (2009). Winter Northern Hemisphere weather patterns remember summer Arctic sea-ice extent. *Geophysical Research Letters*, 36, L07503. doi:10.1029/2009GL037274.

Furtado, J.C., Cohen, J.L., Butler, A.H., Riddle, E.E. and Kumar, A. (2015). Eurasian snow cover variability, winter climate, and stratosphere-troposphere coupling in the CMIP5 models. *Climate Dynamics*, doi:10.1007/s00382-015-2494-4

Hamilton, L.C. and Lemcke-Stampone, M. (2013). Arctic warming and your weather: Public belief in the connection. *International Journal of Climatology*, 34, 1723–1728.

Honda, M., Inue, J. and Yamane, S. (2009). Influence of low Arctic sea-ice minima on anomalously cold Eurasian winters. *Geophysical Research Letters*, 36, L08707. doi:10.1029/2008GL037079.

Hopsch, S., Cohen, J. and Dethloff, K. (2012). Impact of anomalous fall Arctic sea ice concentration on atmospheric patterns in the following winter. *Tellus A*, 64, 18624.

IPCC. (2013). Summary for Policymakers. In: Stocker, T.F., Qin, D. Plattner, G.-K. Tignor, M. Allen, S. K. Boschung, J. Nauels, A. Xia, Y. Bex V. and Midgley, P. M. (Eds.), *Climate Change 2013*. The Physical Science Basis. Contribution of Working Group I to the Fifth Assessment Report of the Intergovernmental Panel on Climate Change. Cambridge and New York: Cambridge University Press.

Jaiser, R., Dethloff, K., Handorf, D., Rinke, A. and Cohen, J. (2012). Planetary- and synoptic-scale feedbacks between tropospheric and sea ice cover changes in the Arctic. *Tellus*, 64, 11595. doi:10.3402/tellusa.v64i0.11595.

Kwok, R. and Rothrock, D.A. (2009). Decline in Arctic sea ice thickness from submarine and ICES at records: 1958–2008. *Geophysical Research Letters*, 36, L15501.

Liu, J., Curry, J.A., Wang, H., Song, M. and Horton, R. M. (2012). Impact of declining Arctic sea ice on winter snowfall. *Proceedings of the National Academic Sciences*, 109, 4074–4079.

Martius, O., Polvani, L.M. and Davie, H.C. (2009). Blocking precursors to stratospheric sudden warming events. *Geophysical Research Letters*, 36, L14806.

Min, E., Hazeleger, W., Odenborgh, G.J. and Sterl, A. (2013). Evaluation of trends in high temperature extremes in north-western Europe in regional climate models. *Environmental Research Letters*, 8, 014011. doi:10.1088/1748-9326/8/1/014011.

Overland, J.E., Francis, J.A., Hanna, E. and Wang, M. (2012). The recent shift in early summer Arctic atmospheric circulation. *Tellus*, 62A, 1–9. doi:10.1111/j.1600-0870.2009.00421.

Overland, J.E., Wang, M., Walsh, J.E. and Stroeve, J.C. (2014). Future Arctic climate changes: Adaptation and mitigation timescales. *Earth's Future*, 2, 68–74.

Quiroz, R.S. (1979). Tropospheric–stratospheric interaction in the major warming event of January–February 1979. *Geophysical Research Letters*, 6, 645–648.

Quiroz, R.S. (1986). The association of stratospheric warmings with tropospheric blocking. *Journal of Geophysical Research*, 91, 5277–5285.

Rex, D.F. (1950a). Blocking action in the middle troposphere and its effect upon regional climate I. An aerological study of blocking action. *Tellus*, 2, 196–211.

Rex, D.P. (1950b). Blocking action in the middle troposphere and its effect upon regional climate (II). The climatology of blocking actions. *Tellus*, 2, 275–301.

Rind, D. (1986). The dynamics of warm and cold climates. *Journal of Atmospheric Science*, 43, 3–43.

Rind, D. (2008). The consequences of not knowing low- and high-latitude climate sensitivity. *Bulletin of the American Meteorological Society,* 89, 855–864. doi:10.1175/2007BAMS2520.1.

Screen, J.A. and Simmonds, I. (2010). The central role of diminishing sea ice in recent Arctic temperature amplification. *Nature*, 464, 1334–1337. doi:10.1038/nature09051.

Screen, J.A. and Simmonds, I. (2013). Exploring links between Arctic amplification and mid-latitude weather. *Geophysical Research Letters*, 40, 959–964.

Serreze, M.C. and Francis, J.A. (2006). The arctic amplification debate. *Climatic Change*, 76(3–4), 241–264, doi:1007/s10584-005-9017-y.

Sillmann, J., Croci-Maspoli, M., Kallache, M. and Katz, R.W. (2011). Extreme cold winter temperatures in Europe under the influence of North Atlantic atmospheric blocking. *Journal of Climate,* 24, 5899–5913.

Sillmann, J., Kharin, V.V., Zhang, X., Zwiers, F.W. and Bronaugh, D. (2013). Climate extremes indices in the CMIP5 multimodel ensemble: Part 1. Model evaluation in the present climate. *Journal of Geophysical Research Atmosphere*, 118, 1716–1733.

Stroeve, J.C., Maslanik, J., Serreze, M.C., Rigor, I., Meier, W. and Fowler, C. (2011). Sea ice response to an extreme negative phase of the Arctic Oscillation during winter 2009/2010. *Geophysical Research Letters*, 38, L02502.

Tang, Q., Zhang, X. and Francis, J. (2014). Extreme summer weather in northern mid-latitudes linked to a vanishing cryosphere. *Nature Climate Change*, 4, 45–50.

Thompson, D.W.J. and Wallace, J.M. (2001). Regional climate impacts of the Northern Hemisphere annular mode. *Science*, 293(527), 85–89. doi:10.1126/science.1058958.

Thompson, D.W.J., Baldwin, M.P. and Wallace, J.M. (2002). Stratospheric connection to Northern Hemispheric wintertime weather: Implication for prediction. *Journal of Climate*, 15, 1421–1428.

Wallace, J.M., Held, I.M., Thompson, D.W.J., Trenberth, K.E. and Walsh, J.E. (2014). Global warming and winter weather. *Science*, 343, 729–730.

Slide 1

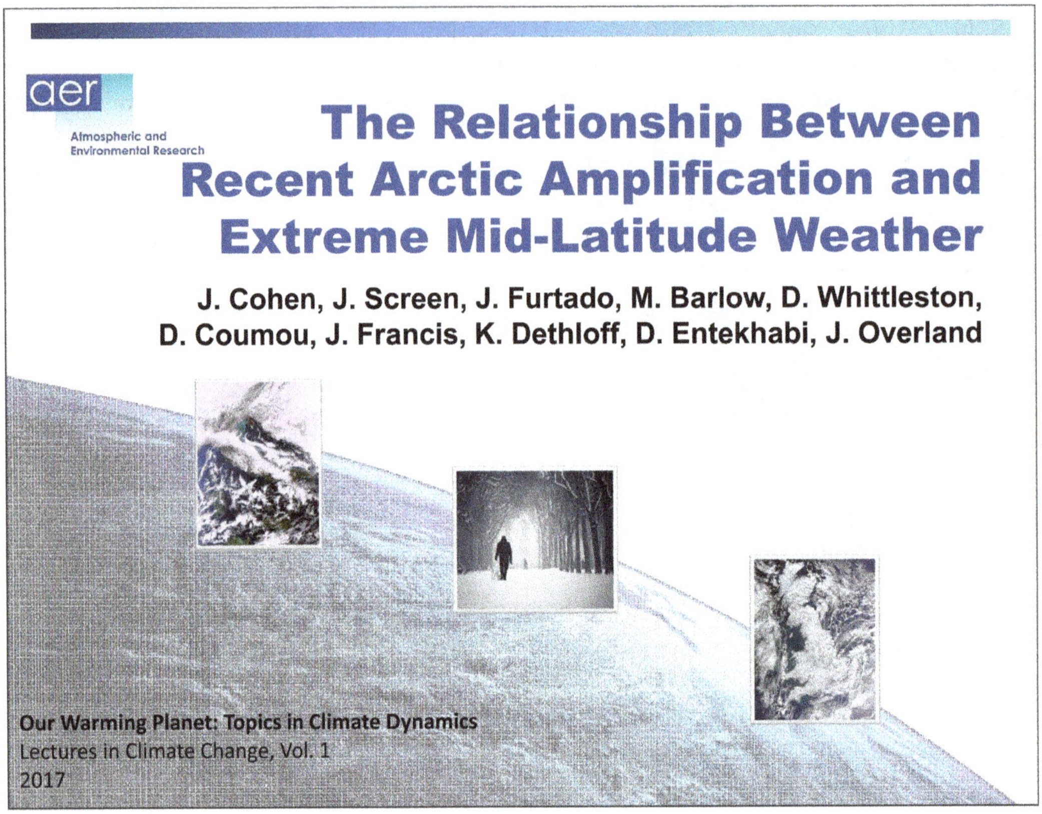

Slide 2

Outline

- Over the past two decades or so, robust warming has been observed for all spring, summer, and fall, but not winter.
- During this period the Arctic has been warming, Arctic sea ice has been melting and Eurasian fall snow cover has been increasing.
- More fall snow cover and likely less sea ice favors the negative phase of the winter Arctic Oscillation (AO).
- During negative AO winters, high latitude blocking is more common resulting in more severe weather including cold air outbreaks and snowfalls across Europe, the Eastern US, and East Asia.

aer Atmospheric and Environmental Research
Cohen, J. The Relationship Between Recent Arctic Amplification and Mid-Latitude Weather 2

Slide 3

Slide 4

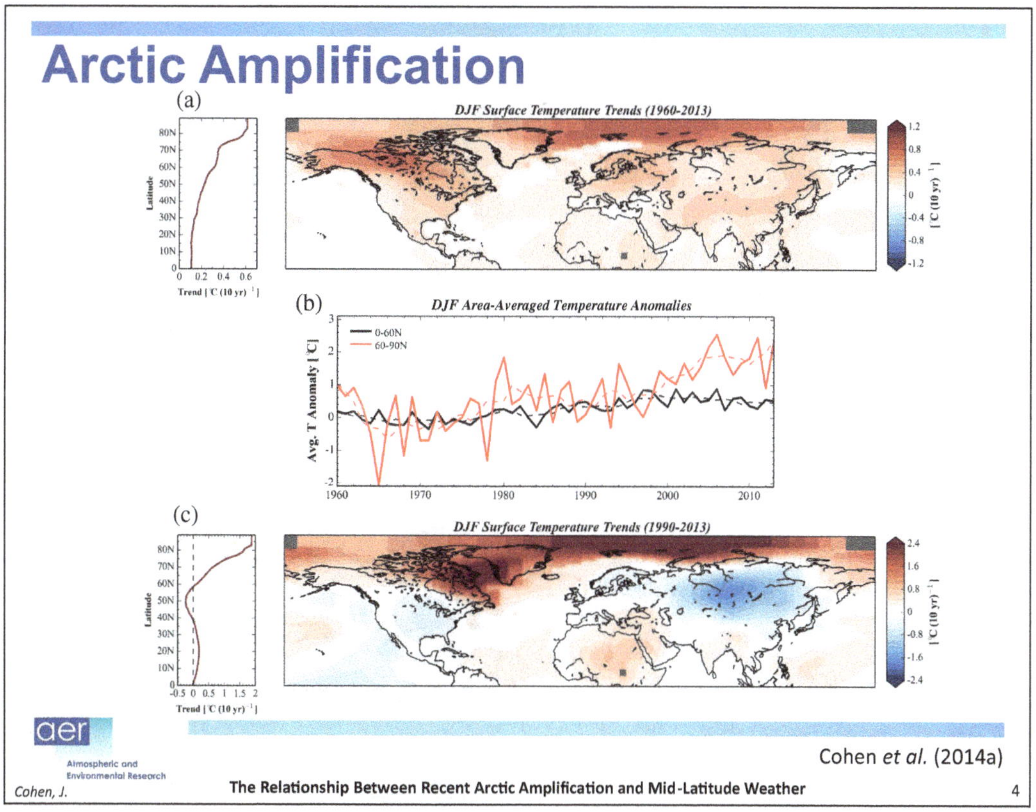

Slide 5

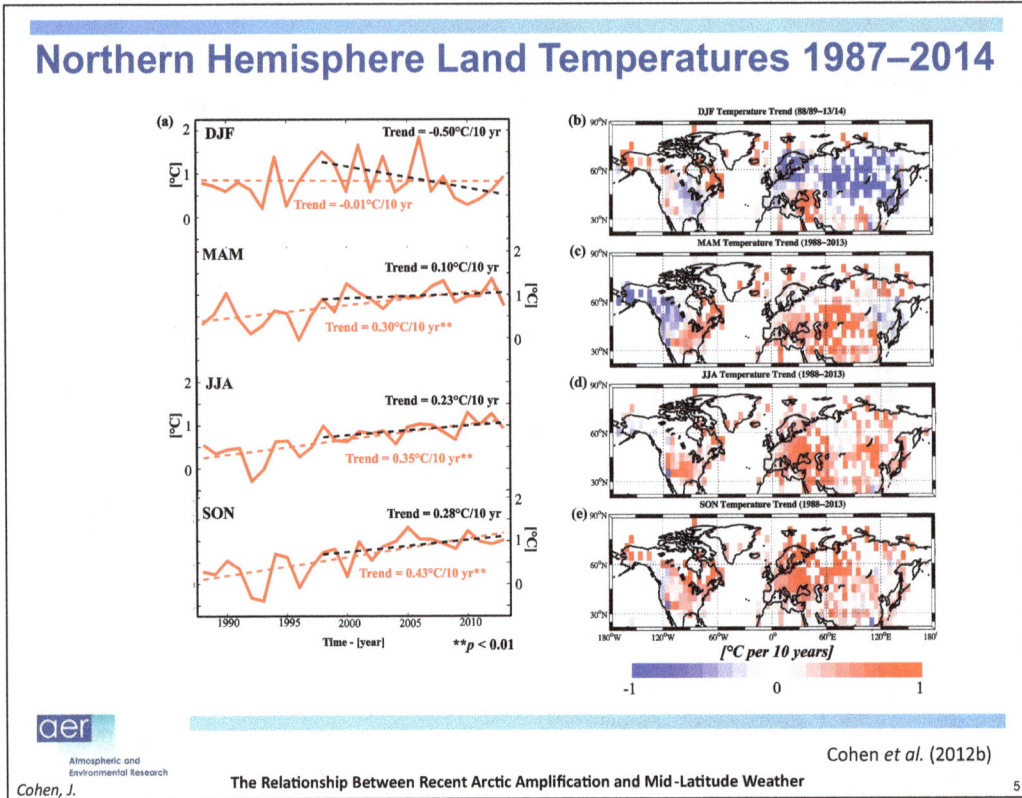

Slide 6

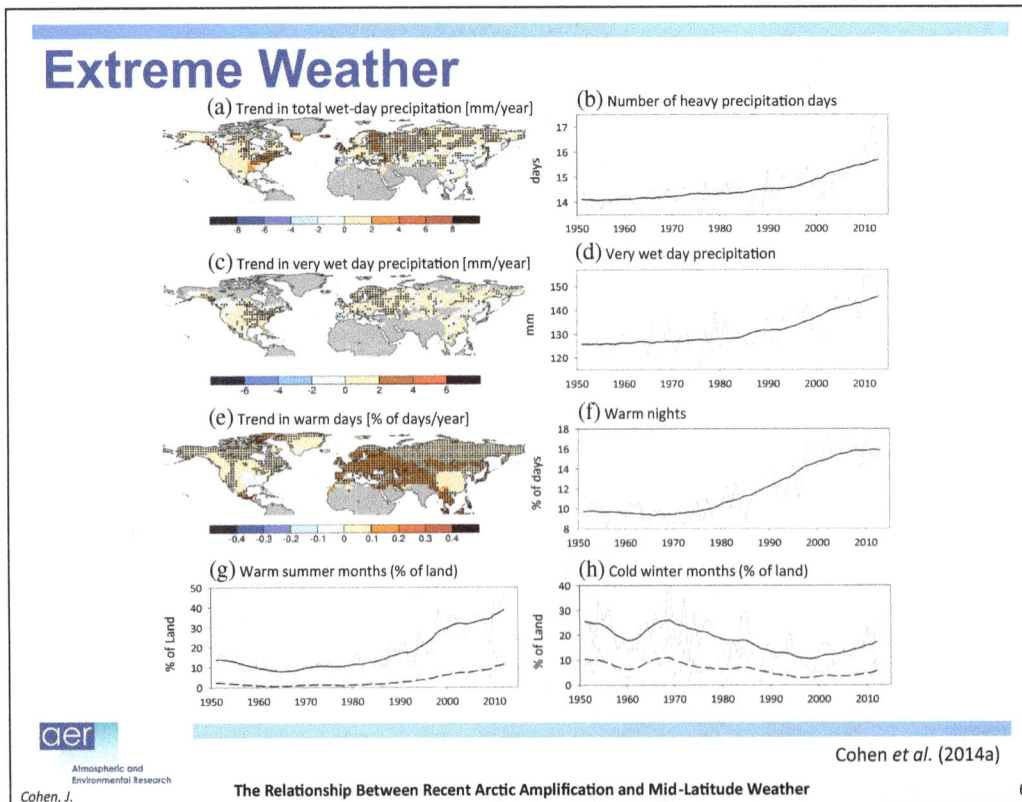

Slide 7

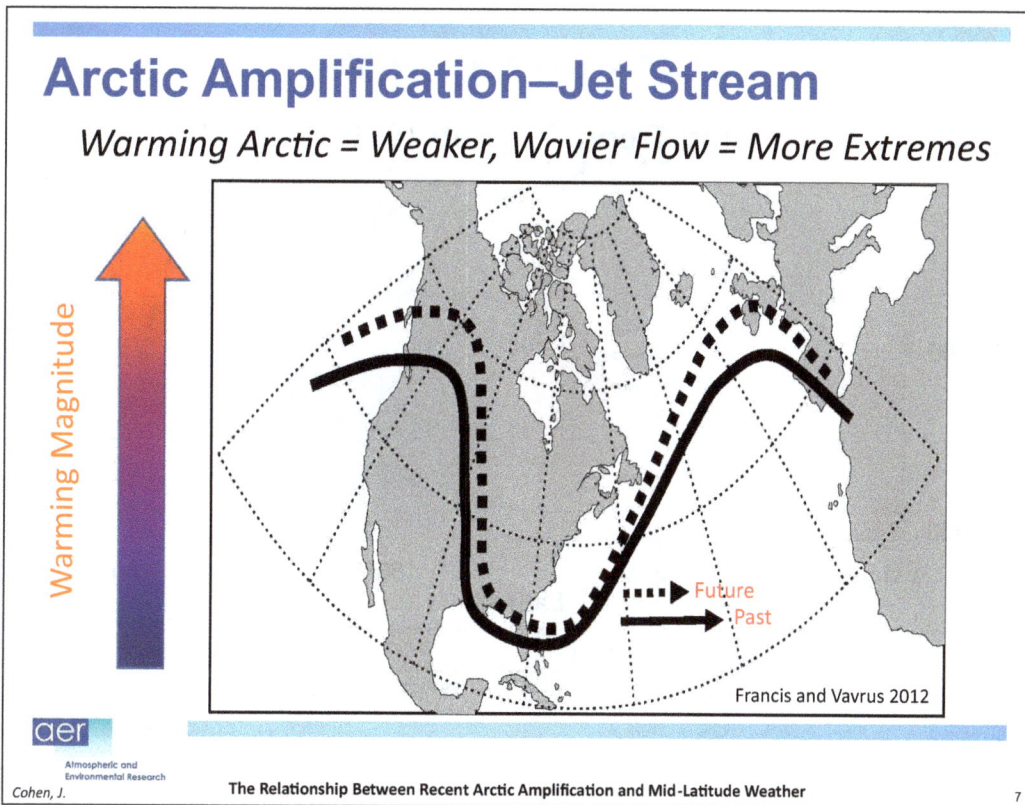

Slide 8

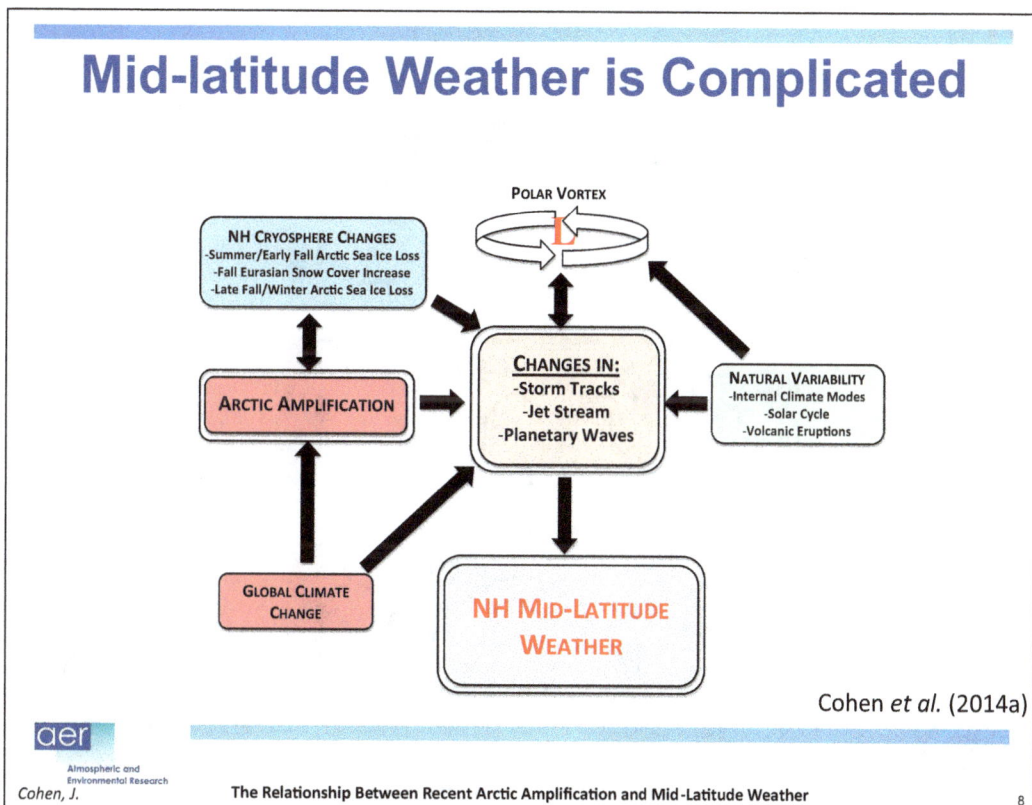

Slide 9

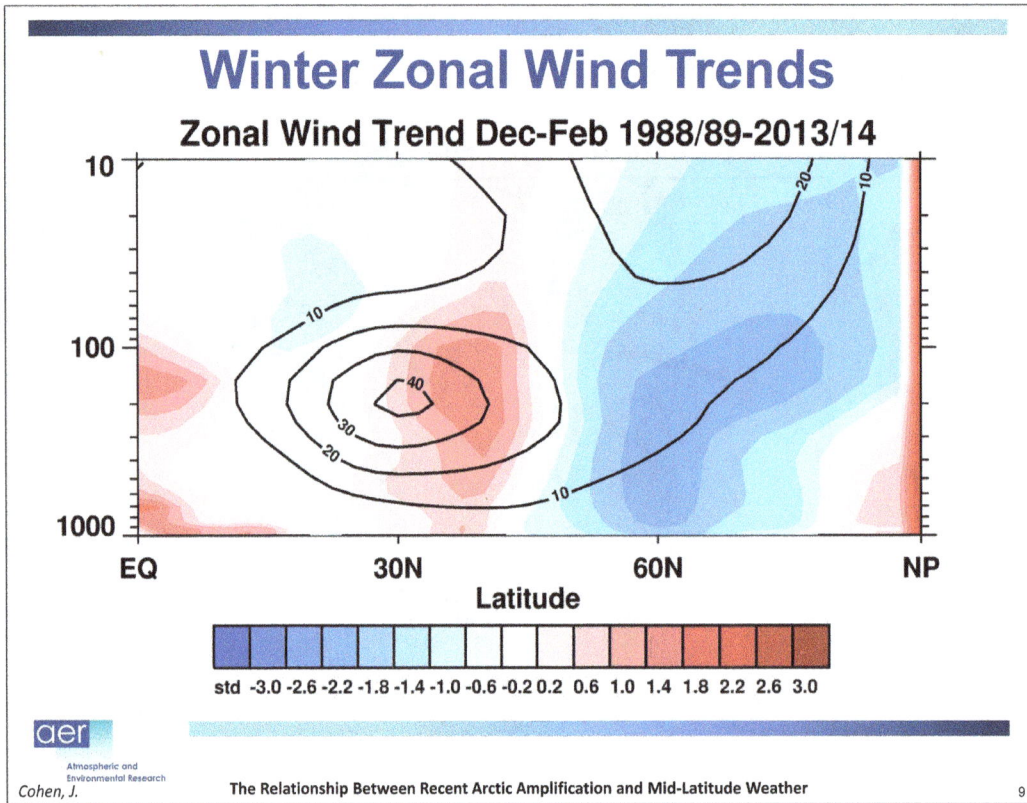

Slide 10

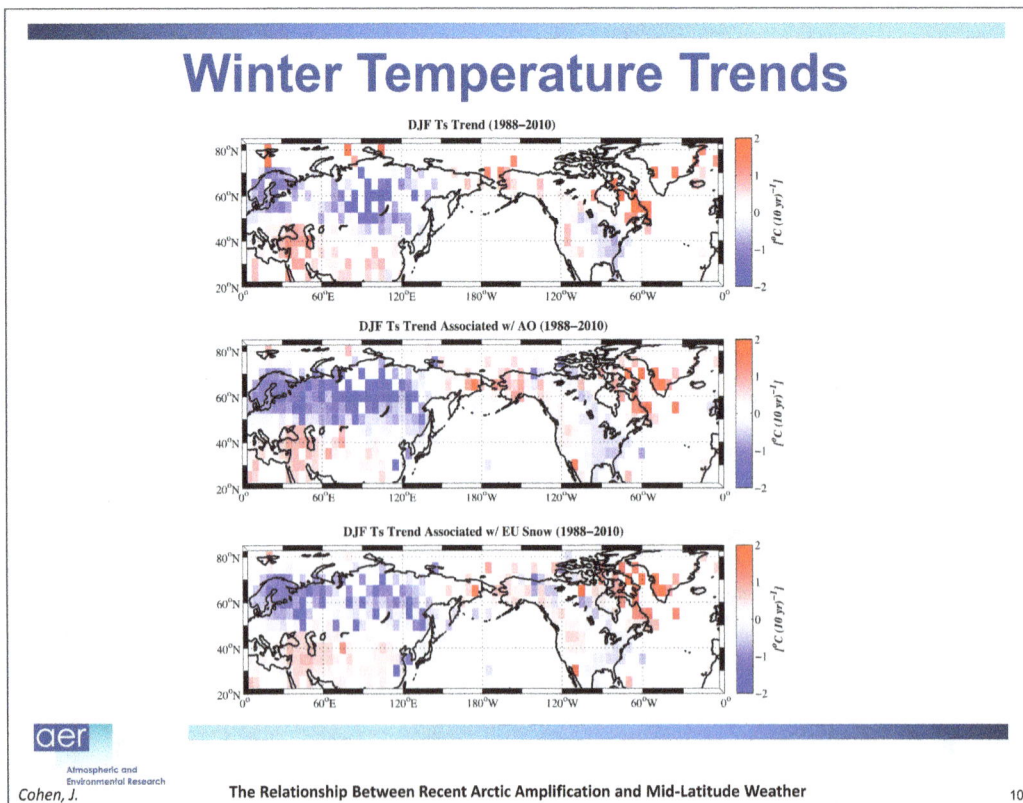

Slide 11

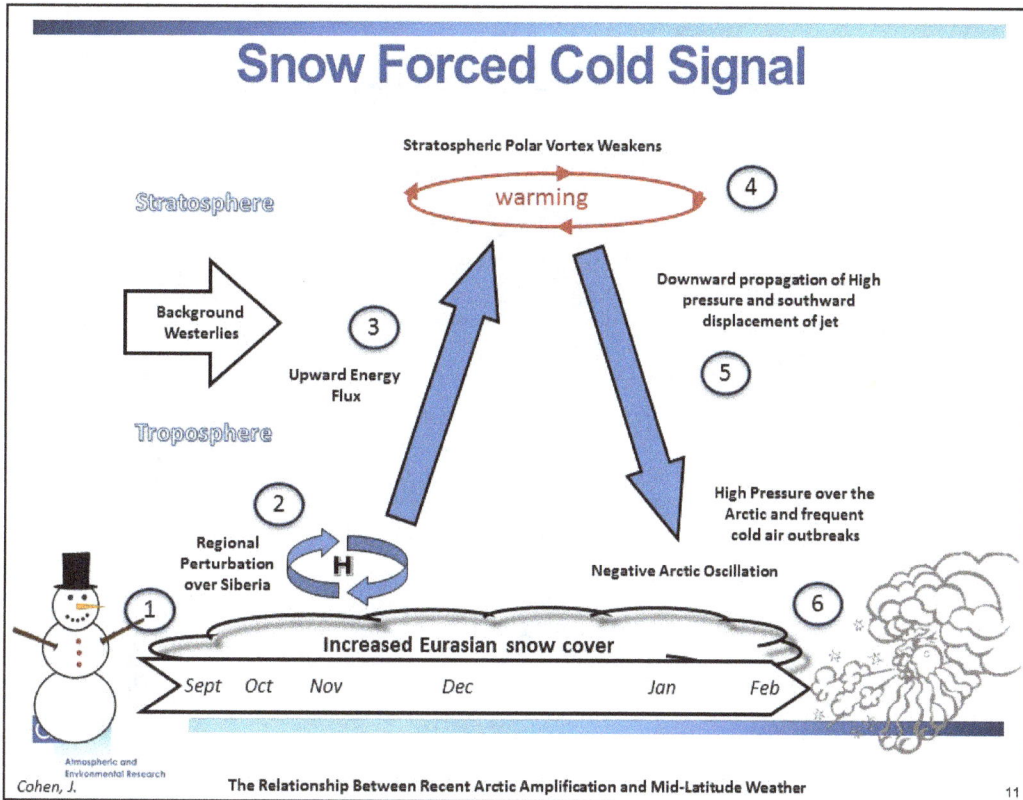

Slide 12

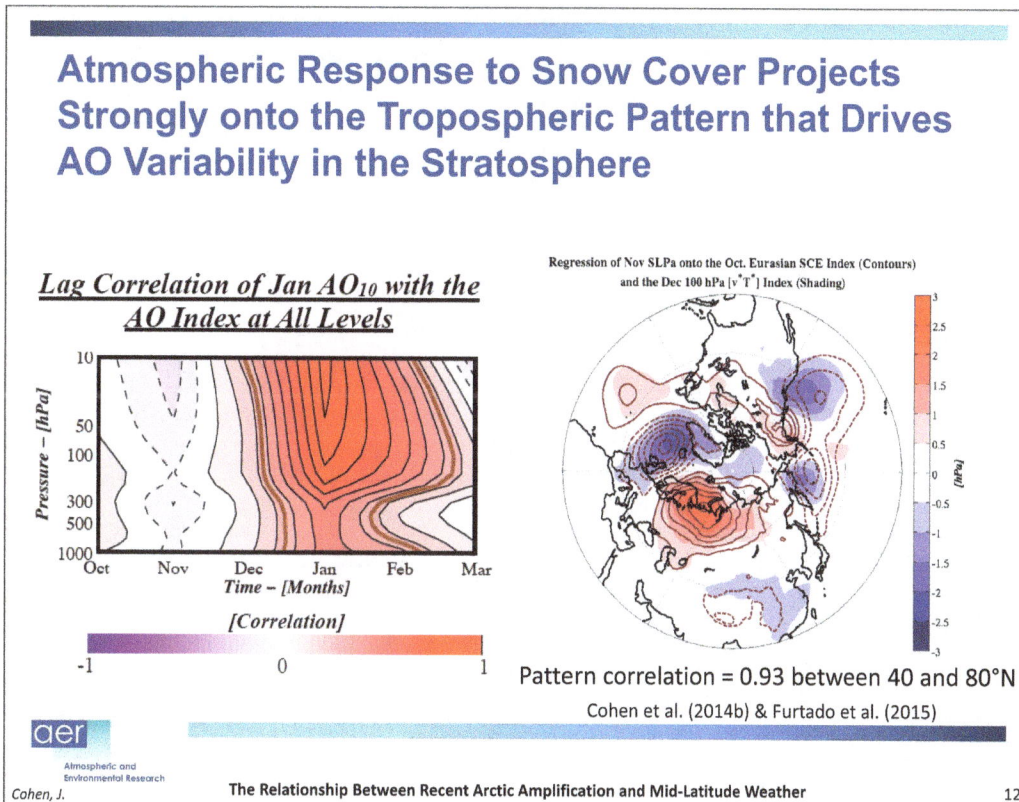

Slide 13

Warm Arctic-Cold Continents

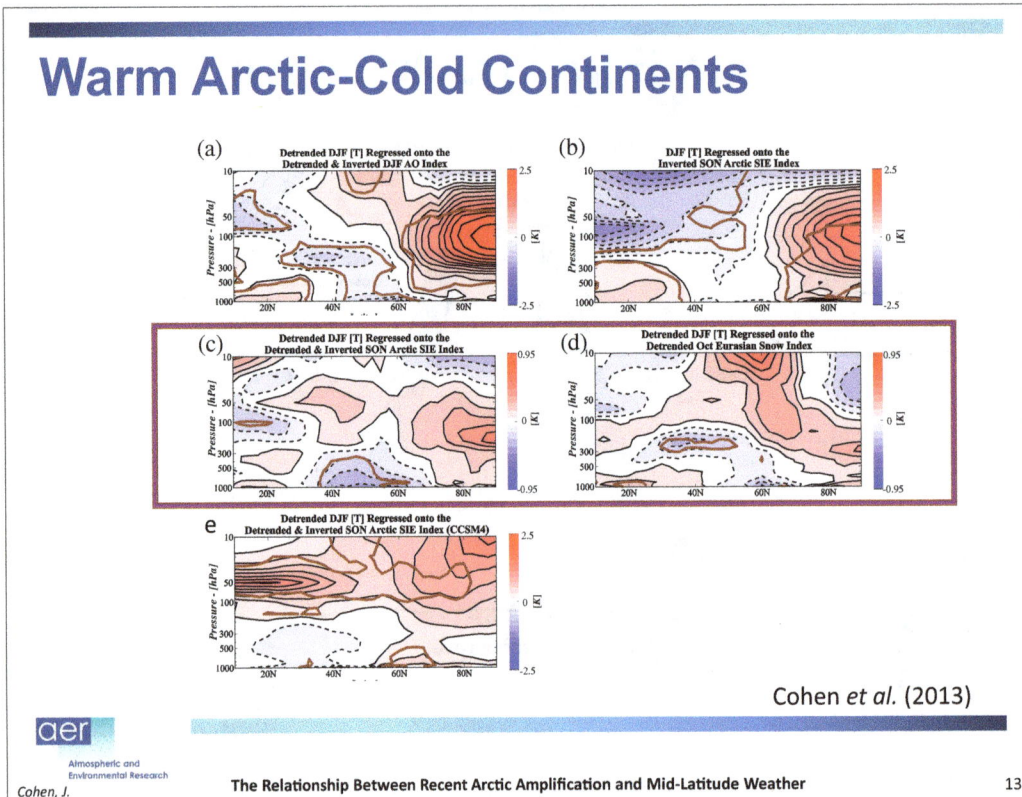

Cohen *et al.* (2013)

Slide 14

Trend in Polar Cap Height 1988/89–2013/14

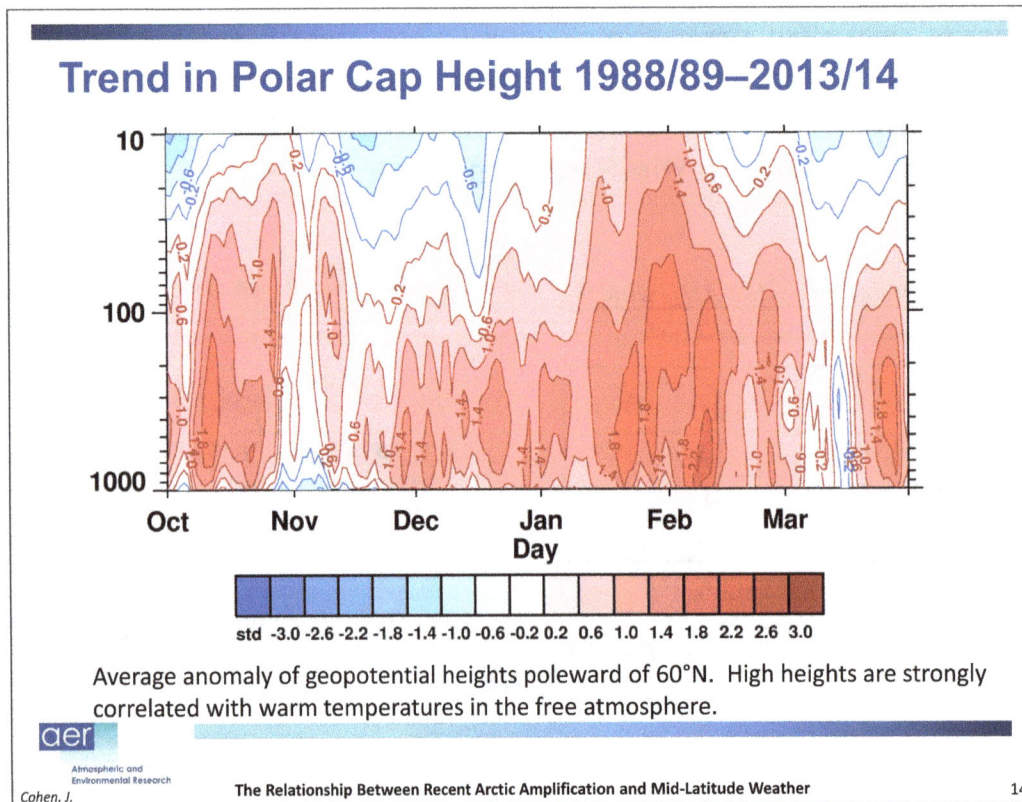

Average anomaly of geopotential heights poleward of 60°N. High heights are strongly correlated with warm temperatures in the free atmosphere.

Slide 15

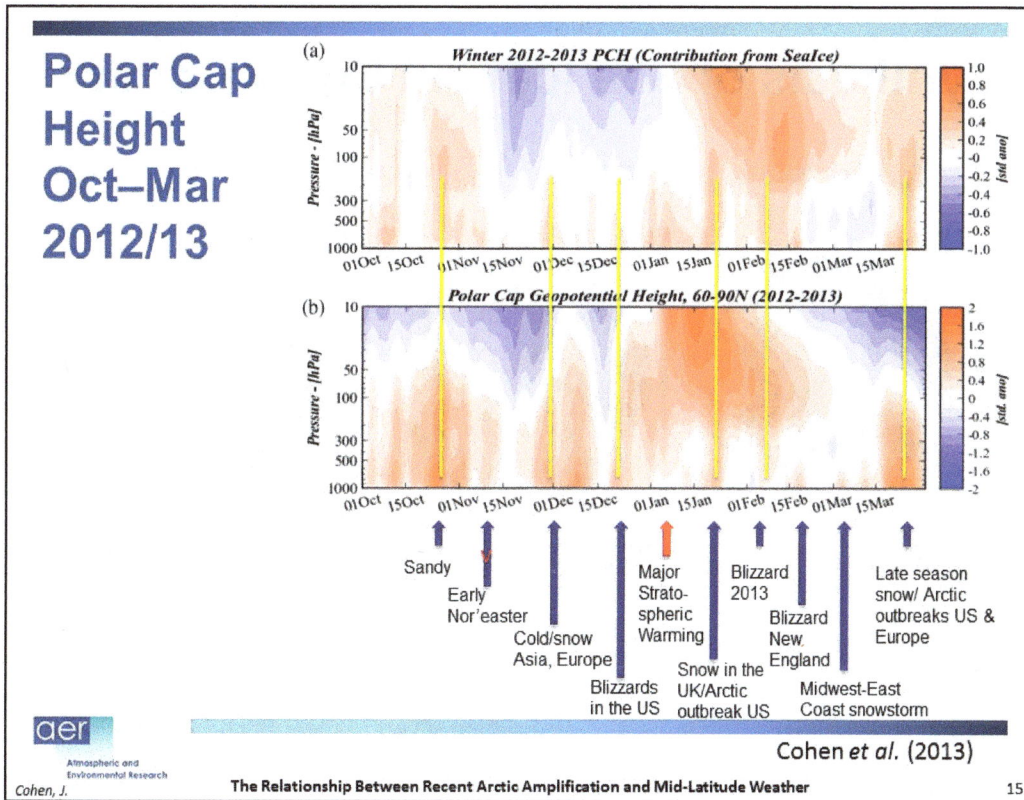

Slide 16

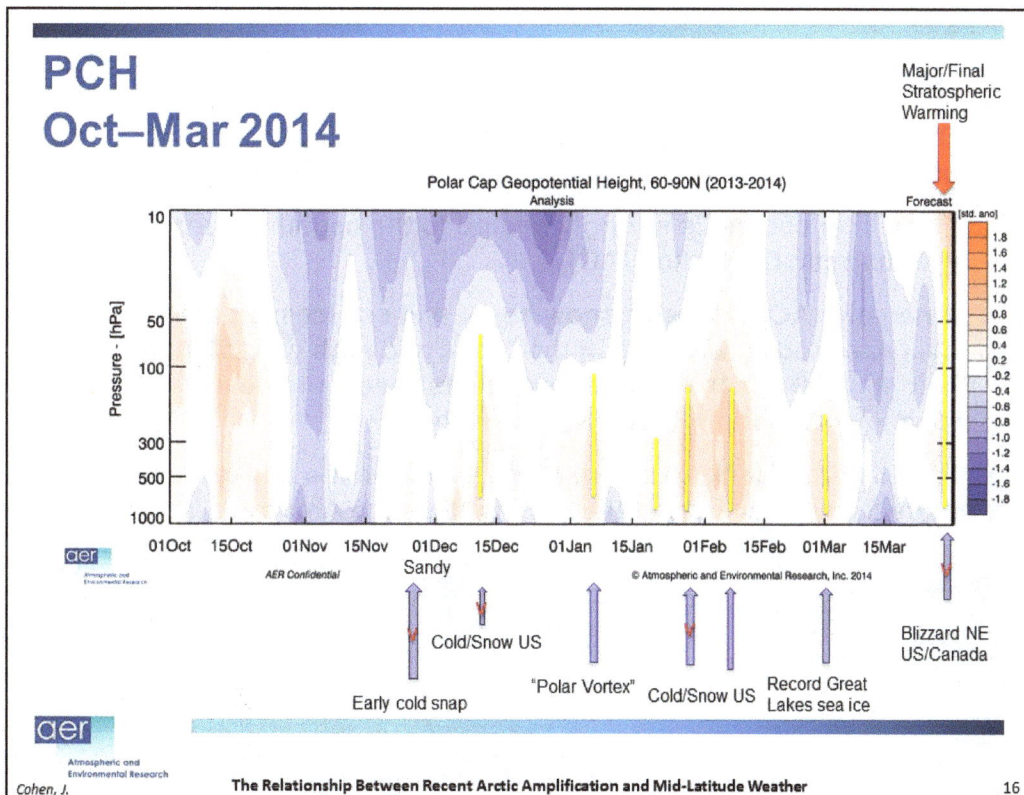

Slide 17

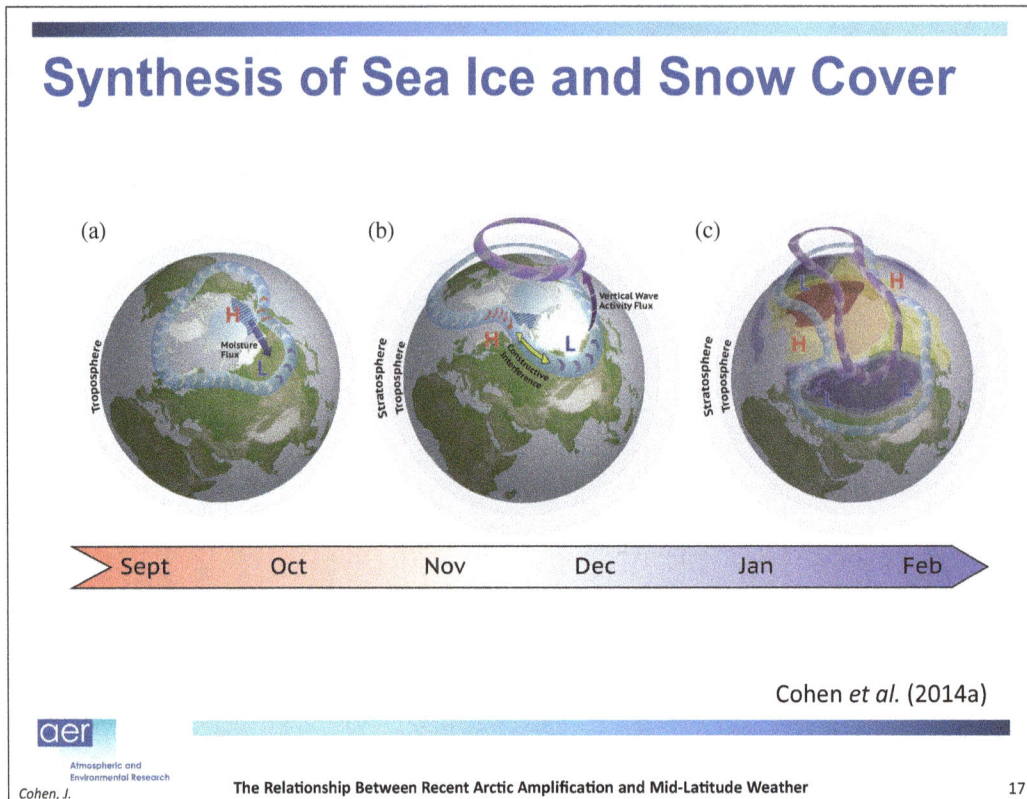

Synthesis of Sea Ice and Snow Cover

(a)　　　　　　(b)　　　　　　(c)

Sept　　Oct　　Nov　　Dec　　Jan　　Feb

Cohen *et al.* (2014a)

aer
Atmospheric and
Environmental Research
Cohen, J.　　　　The Relationship Between Recent Arctic Amplification and Mid-Latitude Weather　　　17

Slide 18

Summary

- Over the past two decades or so, robust warming has been observed for spring, summer, and fall, but not winter.

- The winter cooling is strongly related to the negative trend in the winter Arctic Oscillation (AO).

- A negative AO, extensive snow cover and sea ice melt are all characterized by warming in the Arctic and high geopotential heights. "Warm Arctic/Cold Continents" seems to be a fundamental pattern of the NH circulation.

- High latitude blocking and extreme events in the mid-latitudes are more likely when temperatures/heights in the Arctic are increased.

aer
Atmospheric and
Environmental Research
Cohen, J.

Thank you!
The Relationship Between Recent Arctic Amplification and Mid-Latitude Weather　　　18

Slide Notes

Slide 3   US winter 2015 characterized by extreme weather events!

Slide 4   The trend of Northern Hemisphere (NH) winter temperatures since 1960 is plotted in the top figure. It shows the expected almost universal warming with the greatest warming in the high latitudes or Arctic region. The greater warming in the Arctic is referred to as Arctic amplification. In the middle figure, we plot the temperature time series for the high latitudes, 60–90°N and for the remainder of the NH, 0–60°N for the same time period. For most of the period the trend of the two time series is the same, though the high latitudes shows greater variability. However, after about 1990, the time series diverge when the high latitudes warm at a much greater rate than the rest of the NH. And if we plot the NH temperature trend for the period when Arctic amplification emerged, since 1990, the trend looks different from the trend plot since 1960. The warming at high latitudes is much more pronounced and now the mid-latitudes exhibit widespread cooling. Is Arctic amplification related to mid-latitude cooling?

Slide 5   Here, we plot the NH land temperature trend for all four seasons since 1988. Three seasons show robust warming, spring summer, and especially fall. And the warming is consistent over the whole period and only the most recent half. In contrast, winter shows no warming over the period and during the most recent half shows cooling. Therefore, the winter temperature trend is divergent or asymmetric to the temperature trend of the other three seasons. On the right side, we also show the latitude-longitude spatial trend. For three of the seasons, warming is widespread through the NH. However, during winter widespread cooling has been observed across the NH continents.

Slide 6   Global climate change has been associated with an increase in heat and precipitation trends but a decrease in cold extremes across the mid-latitudes since 1950. Here, we plot different measures of temperature and precipitation extremes. Heat and precipitation extremes have been increasing, while cold extremes have been decreasing as expected. However, since the period of Arctic amplification, cold temperature extremes have reversed their trend and have been increasing as well. Those regions that are stippled for heat and precipitation extremes the trends are significant at 95% confidence. As far as the recent trend in cold extremes, given the shortness of record (~two decades) it is too early to know if the reversal in cold extremes is significant.

Slide 7   The theory that has received the most attention linking extreme weather and Arctic amplification is that of Francis and Vavrus (2012). They argue that Arctic amplification decrease the meridional temperature gradient. Through the thermal wind relationship, a weakened equator to pole temperature gradient leads to a weaker zonal Jet Stream. This further results in stronger atmospheric waves or greater wave amplitude, stronger meridional winds and slower moving atmospheric waves. Because waves move more slowly weather regimes, hot, cold wet, and dry are more persistent leading to more extreme weather.

Slide 8   Here, we show all the different phenomena and coupling that influence mid-latitude weather. As shown what determines mid-latitude weather is very complex and it is hard to isolate once single factor or cause in any observed changes in mid-latitude weather.

Slide 9   Here, I have plotted the zonal mean wind in contours and the trend in zonal mean winds since 1988. The mid-latitude tropospheric Jet Stream can be seen centered around 30°N and the stratospheric polar jet centered around 60°N. The trend in zonal winds does not necessarily show a weakening of the Jet Stream but more accurately an equator-ward shift in the Jet Stream, with a strengthening on the equator-ward flank and a weakening on the pole-ward flank. This jet shift is most commonly associated with the negative phase of the Arctic or North Atlantic Oscillation (AO/NAO).

Slide 10   To bring further support that the recent observed trends in mid-latitude weather including extremes is strongly related to variability associated with the AO and preferentially in its negative phase, we have plotted the raw NH temperature trend in the top figure and the temperature pattern associated with the trend in the phase of the AO for the same period in the middle figure. The two temperature patterns are strikingly similar. In the bottom figure, we plot the temperature pattern associated with the positive or increasing trend in October Eurasian snow cover extent (SCE) for the same period. Both the top figures match agreeably and the pattern correlation for three figures is close to 0.9, confirming the strong similarity in all three figures.

Slide 11   We have developed a six-step model to explain the physical or dynamical reasoning for why more expansive SCE in October can favor or force the negative phase of the AO and therefore colder temperatures across the NH mid-latitudes. October is the month when SCE makes its greatest advance across Eurasia. If the SCE advances more rapidly than normal, this supports colder and denser air masses in the region. The Siberian high-pressure system across the Eurasian continent is the atmospheric response to snow cover advance. If the SCE is more expansive, then the Siberian high is strengthened and more expansive than normal. A strengthened and more expansive Siberian high is more favorable for the vertical energy transfer of large atmospheric waves from the troposphere to the stratosphere. Increased energy transfer absorbed in the polar stratosphere leads to a weakened polar vortex and even possible an SSW. The surface AO is favored in its negative phase following the SSW for up to two months.

Slide 12   Extreme weather is often associated with more persistent weather regimes. It has been shown previously that extreme weather across the NH mid-latitudes when the Arctic Oscillation is in its negative phase or alternatively when the polar vortex is weak. On the plot on the left, we correlate the Arctic Oscillation at all vertical levels with the January AO at 10 hPa. A negative AO in the mid-stratosphere, which is associated with a sudden stratospheric warming (SSW) or a weakened stratospheric polar vortex favors a negative surface AO for the following two months.

On the plot on the right, we regress November sea-level pressure (SLP) anomalies both with Eurasian October SCE anomalies (shading) and with the December pole-ward heat flux anomalies in the lower stratosphere (contours). The patterns are nearly identical and are strongly suggestive that Eurasian October SCE forces a tropospheric pattern in November that drives heat into the polar stratosphere that leads to an SSW and weakened polar vortex in January. A weakening of the polar vortex in January favors more severe winter weather mid-to-late winter across the mid-latitudes. Together, these two plots demonstrate that extensive snow cover favors an SSW in the late December through mid-January time frame. An SSW at this time further increases the probability of extreme weather across the NH mid-latitudes in mid-to-late winter.

Slide 13   The negative AO state is imitated and characterized by warming in the polar or Arctic region followed by cooling in the mid-latitudes. When we regress the AO with zonal mean temperatures, as shown in the top left figure, the negative AO is characterized by warm temperatures in the Arctic and cold temperatures in the mid-latitudes. Regressing snow cover with zonal mean temperatures yields the same pattern with extensive snow cover associate with warming in the Arctic and cooling in the mid-latitudes (panel (d)). Similarly, less sea ice is associated with both warming in the Arctic and cooling in the mid-latitudes (panels (c) and (d)). Warming in the Arctic with concurrent cooling in the mid-latitudes seems to be a fundamental pattern of variability in the atmosphere and has been recently labeled as the warm Arctic-cold continents pattern.

Slide 14   In this slide, I show the trend from 1988 to 2014 in the daily polar cap geopotential heights (PCH); the averaged geopotential height anomaly from 60–90°N from the surface through 10 hPa for the months October through March. Throughout the late fall and winter, the PCH have been increasing throughout the atmospheric column. Also, geopotential heights are a good proxy for temperatures in the free atmosphere so this plot also demonstrates that the polar cap has been warming. And as shown in the previous slide, if the Arctic is warming it is likely to be accompanied by cooling and more severe winter weather in the mid-latitudes, this trend was already demonstrated on slides 4 and 10.

Slide 15   On this slide the PCH anomalies for the fall and winter of 2012 and 2013 is plotted. During this winter, the PCH anomalies were dominated by above normal heights and warm temperatures, including an SSW in early January. Above, the regression of arctic sea ice extent anomaly from September 2012 with the PCH from October 2012 through March 2013 is plotted. This figure suggests that the record low sea ice from the early fall that year contributed to the warm polar cap observed that late fall and winter. Also, extreme events from that season along the timeline are shown. Many of the mid-latitude extreme events of that fall and winter occurred during periods when the PCH warmed or pulsed. It also appears that the extremely low sea ice that fall contribute to the pulsing of the PCH that further contributed to extreme weather in the mid-latitudes.

Slide 16   The PCH anomalies are shown for the fall and winter of 2013/2014. Again warm PCH were dominant in the troposphere throughout the season and extreme mid-latitude weather events occurred at times of pulsing in the PCH.

Slide 17   In this final slide, we present our hypothesis on how low sea ice, extensive snow cover and a weaker polar vortex are all coupled to deliver severe winter weather to the mid-latitudes. In September and October, sea-ice loss has been most pronounced in the Chukchi and East Siberian seas. Warming of the atmosphere due to increased heating from newly ice-free ocean causes geopotential heights to increase in the mid-troposphere, which suppresses the jet stream southward over east Siberia. A southward shift in the storm tracks over East Asia allows for a more rapid advance of Eurasian snow cover in October. Enlarged areas of open water north of Siberia also provide increased moisture flux to the atmosphere, which precipitates as snow as the air mass is advected southward over Siberia (left globe).

In October, a more extensive snow cover cools the surface leading to lower heights and a trough in the mid-troposphere. Increased troughing over East Asia favors upstream ridging near the Barents-Kara seas (BK) and the Ural mountains. Concurrently, the large sea ice deficits and the associated strong surface heating anomalies migrate from the Chukchi and East Siberian seas in September and October to the BK in November and December. This favors mid-tropospheric ridging in the BK region with downstream troughing over East Asia. Therefore, the extensive snow cover over Siberia in October and November and the sea-ice loss over the BK in November and December produce same-signed mid-tropospheric geopotential height patterns over Eurasia. This planetary wave configuration is favorable for increased vertical propagation of Rossby waves from the troposphere into the stratosphere (middle globe).

Increased vertical propagation of Rossby wave energy from the troposphere to the stratosphere weakens the polar vortex and resulting in an SSW event. Circulation anomalies associated with the warming event appear first in the stratosphere and subsequently appear in the troposphere in January and February. These circulation anomalies resemble those associated with the negative phase of the AO and a weaker, equatorward-shifted polar jet stream. As a result, warmer conditions prevail in the Arctic regions, but colder and more severe winter weather occurs across the mid-latitude continents with a greater likelihood of snowstorms in the population centers of the NH mid-latitudes (right globe).

We propose a chain of events where less sea ice and increased open water in the Arctic and more snow cover both force the same pattern, which results in a weakened polar vortex. Because the heating anomalies are displaced longitudinally, extensive Eurasian snow cover and reduced Arctic sea ice can constructively interfere to weaken the polar vortex and hence deliver more extreme winter weather to the mid-latitudes.

It is important to keep in mind that proving the presented hypothesis is currently challenging. Natural variability is the atmosphere that is large and it is difficult to separate a signal from the noise. Also, the observational record is relatively short, so attributed changes are often not significant. Finally, the models are flawed and are especially poor at simulating cryosphere–atmosphere coupling at high latitudes. Therefore, much more research is required to study the proposed coupling in the figure.

CLIMATE LECTURE 8

The Role of Global Warming in Altering the Frequency and Intensity of Tropical and Non-Tropical Cyclones

Timothy Eichler

Saint Louis University, School of Education, MO, USA

Timothy Eichler is an Assistant Professor at the School of Education of Saint Louis University. His research interests include climate/weather connections, evaluating the potential impacts of climate change on mid-latitude cyclones, with an emphasis on Missouri agriculture. He is also interested in investigating changes in cyclone structure due to climate change in the northeastern US

Introduction

Over the last several decades, we have seen a rapid evolution in both global reanalysis datasets and climate models. In the 1970s and 1980s, models existed only with coarse spatial and temporal evolution. For example, when I first came to NASA GISS to work as a programmer, the GISS model was run on an $8° \times 10°$ (Lat/Lon) grid, with output produced monthly. By the time I received my Ph.D. in 2000, the GISS GCM was being run on a $2° \times 2.5°$ (Lat/Lon) grid, with daily output available. During my time at GISS (December 1987 to June 2001), the separation between climate and weather was quite clear. There were weather models (e.g. Nested Grid Model (NGM) and Global Forecast System (GFS)) and there were climate models. While weather models could be run at a higher resolution, the amount of computer power needed to run for more than a week real-time prevented their adaptation to climate problems. Appropriate climate model applications included only the most general aspects of the Atmospheric Circulation. Another limitation was that the older GISS model (and GCMs in general) was run with climatological sea-surface temperatures (SSTs) and later, with observed SSTs; the latter allowed for some rudimentary assessment of impacts of El Niño Southern Oscillation on the general circulation.

My training in climate I received with David as my advisor has allowed me to explore the relatively new area of climate/weather connections. With a spatial resolution of $0.5° \times 0.5°$ (Lat/Lon) grid and 3-hourly output available, I can now explore the impacts of global warming on extreme weather events such as storm tracks. Not only does this have important ramifications on where the science is going, it can now serve to communicate effectively climate science to the general public. We are now in a position as a scientific community to discuss regional climate variability (e.g. droughts, heat waves, floods, storms, etc.) as opposed to diagnostics such a globally averaged surface air temperature, which has little meaning to the public.

Global Warming and Tropical Cyclones

Hurricanes are 'warm-core' entities that derive their energy from heat from the ocean's surface. The minimum ocean condition for the maintenance of a tropical cyclone is 26.5°C through a 50-meter depth. If this were the only requirement, assessing the role of global warming in tropical cyclone development would be straight-forward: The warming of the ocean via global warming would provide additional energy for the production of more intense/frequent tropical cyclones. However, static stability and wind shear also play roles. If the atmospheric column above a tropical cyclone is too stable/dry and/or there is adverse vertical wind shear, the cyclone will fail to develop. Adding to this is the fact that hurricanes develop via different physical mechanisms. For example, tropical North Atlantic cyclones are often the offspring of African Easterly waves, while Western Pacific cyclones are often the result of the Monsoonal Trough (Barry and Carleton, 2001). Knowledge of how these features of the general circulation respond to global warming is needed to assess how tropical cyclone intensity/frequency may change.

The ability to assess the role of tropical cyclones in present and future climate has evolved. Since the coarse-grid GCMs of the 1980s and 1990s often had difficulty in simulating tropical cyclones, indices such as Gray's index (Gray, 1979) were utilized. Gray's index was designed to assess the environmental conditions conducive to tropical cyclone development instead of measuring actual tropical cyclones. It was composed of two main components: a thermodynamic portion (e.g. role of SSTs) and a shear component designed to investigate atmospheric conditions. Although this offered a rudimentary way to assess tropical cyclone environments in climate models, a drawback was that it was not tuned for future climate.

A ground-breaking study was done by Emanuel (1988). He developed an analytical equation linking the upper-limit of hurricane intensity to thermodynamic conditions. In his work, he proposed the possibility of 'hypercanes'. In this scenario, there are parameters governing this equation such that no solution exists. Under that condition, central pressure of hurricanes decrease to values beyond current hypothetical minima. Emanuel (1988) stated that SSTs would have to be at least 6–10°C warmer than current climate (while keeping the lower stratospheric temperature constant) to support such an entity.

Although Emanuel (1988) gave some perspective on the linkage of SSTs to tropical cyclones (and hence changes due to global warming), the complexity of the problem was illustrated by Vecchi and Soden (2007). They investigated wind shear in the IPCC-AR4 model suite (scenario A1B) and found an increase in vertical wind shear in the tropical Atlantic and eastern Pacific, which would be a mitigating factor for more frequent/intense tropical cyclones despite increased SSTs.

As models have become more sophisticated, it has become feasible to investigate tropical cyclones in climate models, instead of merely the general environmental conditions supporting them. Of course, a key issue is that climate models lack sufficient spatial resolution to represent the inner core of a tropical cyclone. This is especially important when evaluating how global warming may impact intense hurricanes (i.e. hurricanes of at least category 3 intensity). Emanuel, Sundararajan, and Williams (2008) mitigated this issue by randomly seeding of all ocean basins with weak, warm-core vortices, whose motion was driven by a beta-and-advection model. Intensity was derived using a coupled ocean–atmosphere numerical model. Global model data from IPCC-AR4 and reanalysis data (e.g. winds and SST) were used to drive the track and intensity models. Emanuel *et al.* (2008) found that there was a tendency toward a decrease in frequency and an increase in intensity of tropical cyclones in the Southern Hemisphere. Bender *et al.* (2010) also used a downscaling technique to assess the impacts of global warming on North Atlantic tropical cyclones. They downscaled to a Geophysical Fluid Dynamics Lab (GFDL) operational hurricane model using 18 models from the World Climate Research Program coupled model intercomparison project (CMIP3) using the IPCC A1B emissions scenario. Bender *et al.* (2010) concluded that global warming resulted in an increase in frequency of hurricanes of at least category 4 intensity.

As reported in a review article by Knutson *et al.* (2010), the general consensus of the modeling community is for a decrease in frequency and an increase in intensity of tropical cyclones with respect to global warming. As discussed by Bengtsson *et al.* (2007), the physical processes responsible for these seemingly contradictory results are an increase in static stability and an increase in SSTs. The former is associated with a decrease in frequency, while the latter is associated with an increase in intensity and precipitation. Of particular interest, high-resolution models (e.g. downscaling) show an increase in frequency of intense tropical cyclones. As model resolution continues to improve, confidence in this scenario may rise, although Chen *et al.* (2007) cautions that 1-km resolution may be needed to represent the inner-core of a tropical cyclone.

Non-tropical Cyclones

The availability of high-resolution models and datasets allows for the assessment of the role of extratropical cyclones in the current and future climate. Unlike tropical cyclones, non-tropical cyclones derive their energy from baroclinic processes (horizontal temperature gradients), as opposed to warm SSTs. The physical reasoning behind a suggested pole-ward shift in non-tropical cyclone tracks is clear: High-latitude amplification of global warming reduces the meridional baroclinicity, which reduces the frequency and intensity of cyclones in mid-latitudes. The effects over oceans are less clear, as warmer SSTs could result in enhanced latent heat available to intensify non-tropical cyclones.

Observational studies suggest that mid-latitude storm frequency has decreased, while intensity has increased. For example, McCabe, Clarke, and Serreze (2001) studied storm track trends in the National Center for Climate Prediction (NCEP) reanalysis data from 1979 to 1997 and concluded that storm track frequency has decreased in mid-latitudes but increased at high latitudes, while storm track intensity has increased at all latitudes. McCabe *et al.* (2001) suggested that the trends may be at least partly due to a positive shift in the Atlantic Oscillation since 1989.

Analyses of global warming scenarios in climate models generally support the trends seen by McCabe *et al.* (2001). For example, Jiang and Perrie (2007) used the Canadian Climate Center model and found that global warming caused storms to increase slightly in size and intensity during northern hemisphere autumn. Ulbrich *et al.* (2008), Bengtsson, Hodges, and Keenlyside (2009), and Yin (2005) all concluded that storms tend to shift their track pole-ward in response to a warmer climate. Finally, Eichler, Gaggini, and Pan (2013) investigated storm tracks in the IPCC AR5 model suite, and found a pole-ward shift in storm track frequency in response to global warming. For intensity, Eichler *et al.* (2013) found an increase pole-ward toward the Aleutian Islands, due to the maintenance of baroclinicity combined with higher SSTs. However, a weakening of the Icelandic low was also noted due to cooler SSTs southeast of Greenland.

Although Meehl *et al.* (2007) summarized the general consensus of the climate community that mid-latitude storms are decreasing in frequency and increasing in intensity due to global warming, some studies disagree with this assessment. For example, Catto, Shaffrey, and Hodges (2011) analyzed storm tracks in the High-Resolution Global Environmental Model version 1.1 in doubled and quadrupled CO_2 scenarios and found that mid-latitude storms decreased in frequency and decreased in intensity in response to global warming. Similar results were found by Bengtsson *et al.* (2009), Watterson (2006), and Geng and Sugi (2003). Catto *et al.* (2011) concluded that the differences may be attributed to either differences in model or the various methodologies used to evaluate storm tracks (e.g. SLP, band-passed geopotential height, 850 hPa vorticity, etc.)

An important aspect of studying the impacts of global warming on non-tropical cyclones is to assess their precipitation efficiency. For example, Salathe (2006) studied storm tracks in the IPCC AR4 models and found a northward shift and an increase in intensity in storm tracks without a corresponding increase in precipitation. In contrast, Champion *et al.* (2011) found an increase in extreme precipitation linked to storms in all seasons in a higher resolution version of the ECHAM5 model. Understanding how global warming alters precipitation intensity in storms will continue to be a major focus of research, which will enable the climate science community to provide improved assessments on regional climate change.

References

Barry, R.G. and Carleton, A.M. (2001). *Synoptic and Dynamic Climatology* (pp. 520–521). Routledge, New York. ISBN 978-0-415-03115-8. Retrieved 2011-03-06.

Bender, M.A., Knutson, T.R., Tuleya, R.E., Sirutis, J.J., Vecchi, G.A., Garner, S.T. and Held, I.M. (2010). Modeled impact of anthropogenic warming of the frequency of intense Atlantic hurricanes. *Science*, 327, 454–458.

Bengtsson, L., Hodges, K.L., Esch, M., Keenlyside, N., Kornblueh, L., Luo, J.-L. and Yamagata, T. (2007). How may tropical cyclones change in a warmer climate. *Tellus*, 59A, 539–561.

Bengtsson, L., Hodges, K.L. and Keenlyside, N. (2009). Will extratropical storms intensify in a warmer climate? *Journal of Climate*, 22, 2276–2301.

Catto, J.L., Shaffrey, L.C. and Hodges, K.L. (2011). Can climate models capture the structure of extratropical cyclones? *Journal of Climate*, 23, 1621–1635.

Champion, A.J., Hodges, K.I., Bengtsson, L.O., Keenlyside, N.S. and Esch, M. (2011). Impact of increasing resolution and a warmer climate on extreme weather from Northern Hemisphere extratropical cyclones. *Tellus*, 63A, 893–906.

Chen, S.S., Price, J.F., Zhao, W., Donelan, M.A. and Walsh, E. J. (2007). The CBLAST Hurricane program and the next-generation fully coupled atmosphere–wave–ocean models for hurricane research and prediction. *Bulletin of the American Meteorological Society*, 88, 311–317.

Eichler, T.P., Gaggini, N. and Pan, Z. (2013). Impacts of global warming on Northern Hemisphere winter storm tracks in the CMIP5 model suite. *Journal of Geophysical Research Atmosphere*, 118. doi:10.1002/jgrd.50286

Emanuel, K. (1988). The maximum potential intensity of hurricanes. *Journal of Atmospheric Sciences*, 45, 1143–1155.

Emanuel, K., Sundararajan, R. and Williams, J. (2008). Hurricanes and global warming: Results from downscaling IPCC AR4 simulations. *Bulletin of the American Meteorological Society*, 89, 347–367.

Geng, Q.Z. and Sugi, M. (2003). Possible change of extratropical cyclone activity due to enhanced greenhouse gases and sulfate aerosols — Study with a high-resolution AGCM. *Journal of Climate*, 16, 2262–2274.

Gray, W.M. (1979). Hurricanes: Their formation, structure, and likely role in the tropical circulation. In D.B. Shaw (Ed.), *Meteorology Over the Tropical Oceans* (pp. 155–218). Royal Meteorological Society.

Knutson, T.R., McBride, J.L., Chan, J., Emanuel, K., Holland, G., Landsea, C., . . . Sugi, M. (2010). Tropical cyclones and climate change. *Nature Geoscience*, 3, 157–163. doi:10.1038/ngeo779

McCabe, G.J., Clarke, M.P. and Serreze, M.C. (2001). Trends in Northern Hemisphere surface cyclone frequency and intensity. *Journal of Climate*, 14, 2763–2768.

Meehl, G.A., Covey, C., Delworth, T., Latif, M., McAvaney, B., Mitchell, J.F.B., . . . Taylor, K.E. (2007). The WCRP CMIP3 multimodel dataset: A new era in climate change research. *Bulletin of the American Meteorological Society*, 88, 1383–1394.

Salathe, E.P. (2006). Influences of a shift in North Pacific storm tracks on western North American precipitation under global warming. *Geophysical Research Letters*, 33, L19820. doi:10.1029/2006GL026882

Ulbrich, U., Pinto, J., Kupfer, H., Leckebusch, G., Spangehl, T. and Reyers, M. (2008). Changing Northern Hemisphere storm tracks in an ensemble of IPCC climate change simulations. *Journal of Climate*, 21, 1669–1679.

Vecchi, G.A. and Soden, B.J. (2007). Increased tropical Atlantic wind shear in model projections of global warming. *Geophysical Research Letters,* 34, L08702.

Yin, J.H. (2005). A consistent poleward shift of the storm tracks in simulations of 21st century climate. *Geophysical Research Letters*, 32, L18701. doi:10.1029/2005GL023684

Watterson, I.G. (2006). The intensity of precipitation during extratropical cyclones in global warming simulations: A link to cyclone intensity? *Tellus A*, 58, 82–97. doi:10.1111/j.1600-0870.2006.00147.x

Slide 1

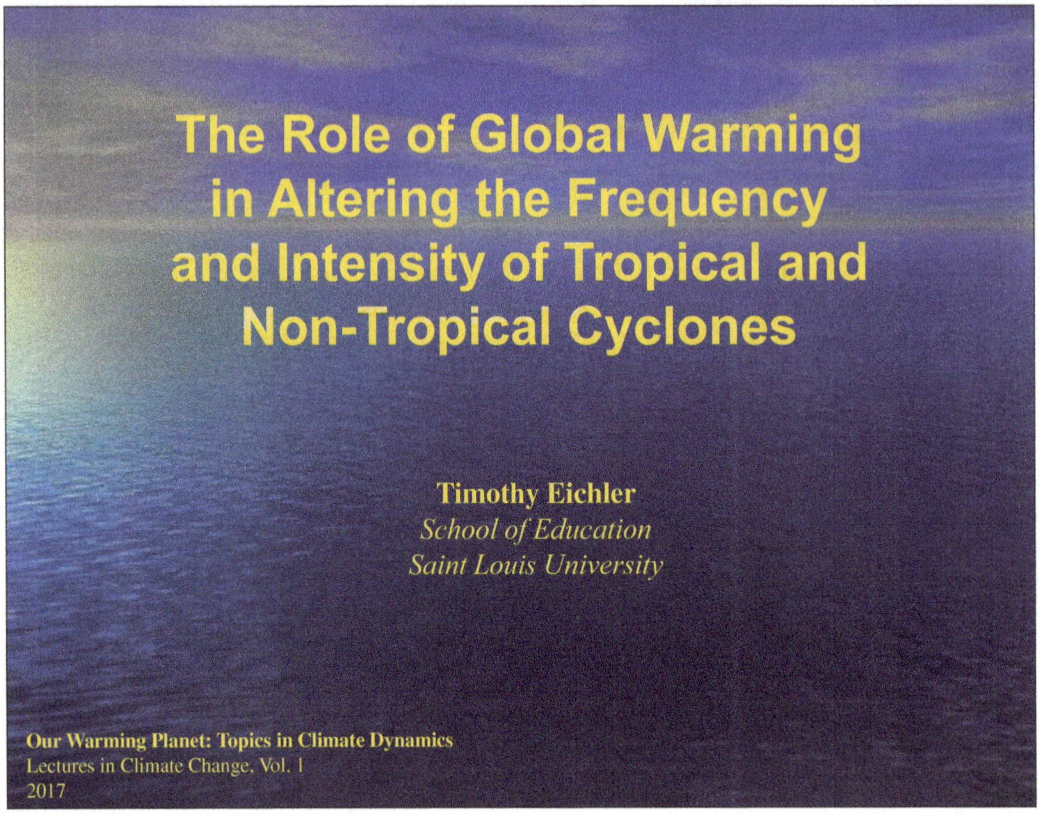
The Role of Global Warming in Altering the Frequency and Intensity of Tropical and Non-Tropical Cyclones
Timothy Eichler
School of Education
Saint Louis University
Our Warming Planet: Topics in Climate Dynamics
Lectures in Climate Change, Vol. 1
2017

Slide 2

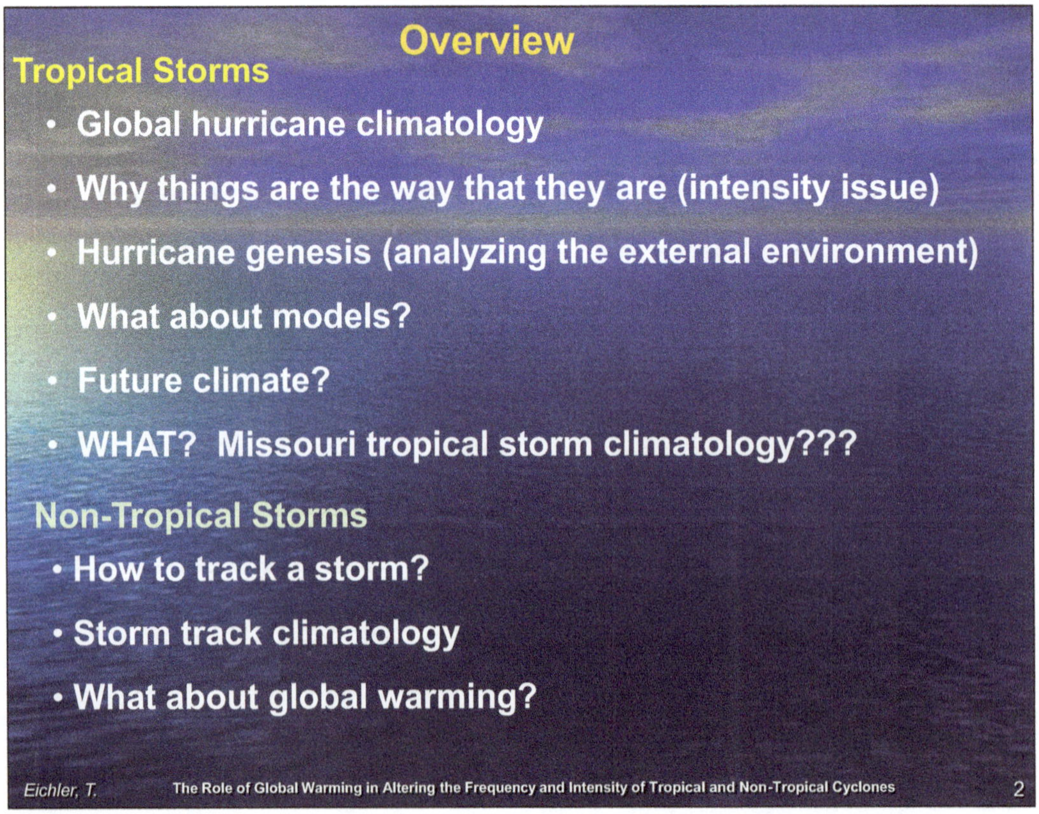
Overview
Tropical Storms
• Global hurricane climatology
• Why things are the way that they are (intensity issue)
• Hurricane genesis (analyzing the external environment)
• What about models?
• Future climate?
• WHAT? Missouri tropical storm climatology???
Non-Tropical Storms
• How to track a storm?
• Storm track climatology
• What about global warming?
Eichler, T.
The Role of Global Warming in Altering the Frequency and Intensity of Tropical and Non-Tropical Cyclones
2

Slide 3

Slide 4

Slide 5

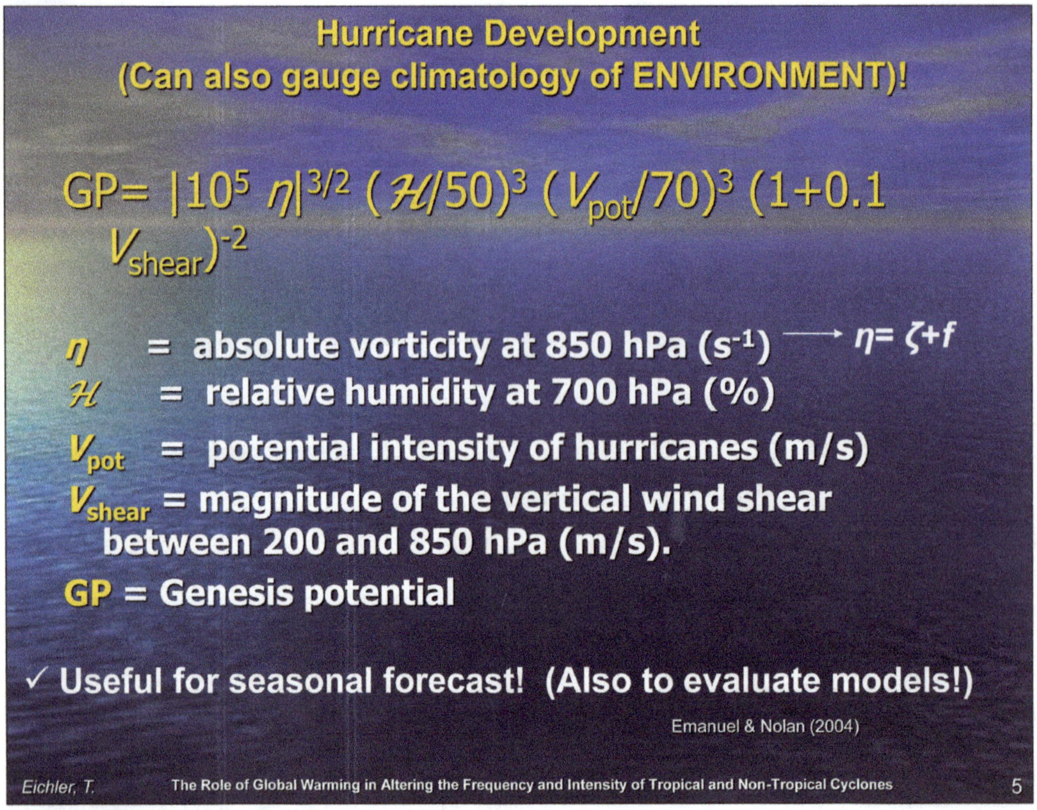

Hurricane Development
(Can also gauge climatology of ENVIRONMENT)!
$GP = |10^5\,\eta|^{3/2}\,(\mathcal{H}/50)^3\,(V_{pot}/70)^3\,(1+0.1\,V_{shear})^{-2}$
η = absolute vorticity at 850 hPa (s^{-1}) → $\eta = \zeta + f$
$\mathcal{H}$ = relative humidity at 700 hPa (%)
V_{pot} = potential intensity of hurricanes (m/s)
V_{shear} = magnitude of the vertical wind shear between 200 and 850 hPa (m/s).
GP = Genesis potential
✓ Useful for seasonal forecast! (Also to evaluate models!)
Emanuel & Nolan (2004)
Eichler, T.
The Role of Global Warming in Altering the Frequency and Intensity of Tropical and Non-Tropical Cyclones
5

Slide 6

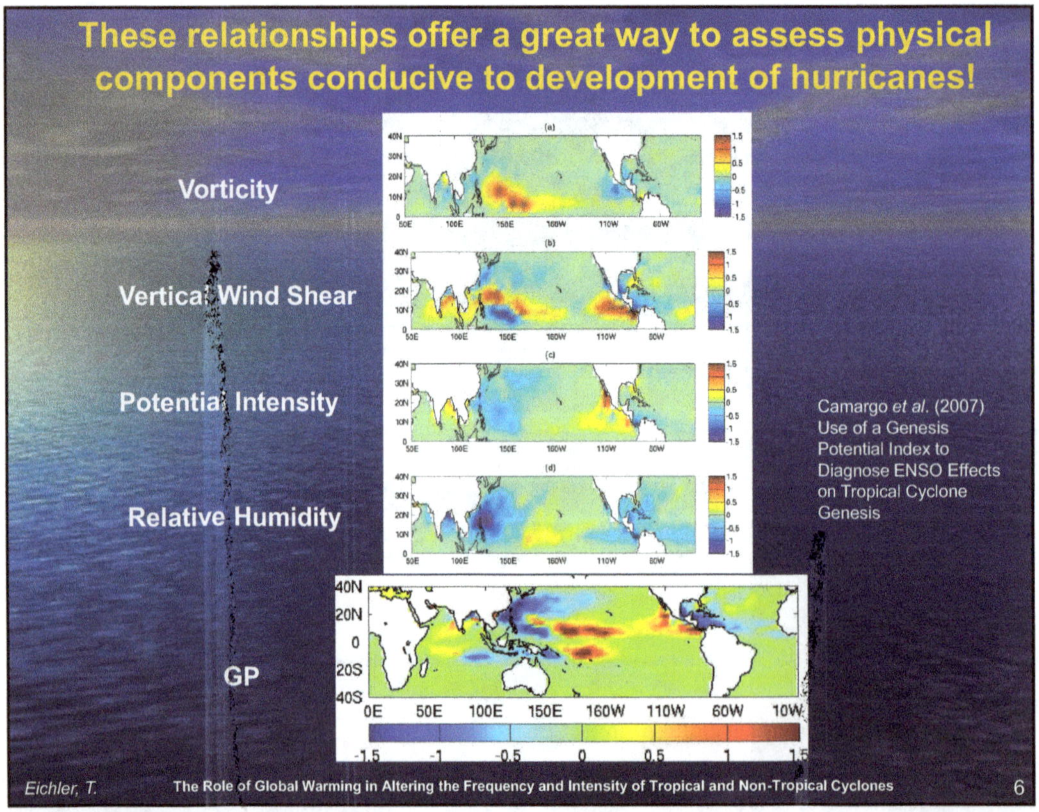

These relationships offer a great way to assess physical components conducive to development of hurricanes!
Vorticity
Vertical Wind Shear
Potential Intensity
Relative Humidity
GP
40N
20N
0
20S
40S
0E
50E
100E
150E
160W
110W
60W
10W
-1.5
-1
-0.5
0
0.5
1
1.5
Camargo et al. (2007)
Use of a Genesis Potential Index to Diagnose ENSO Effects on Tropical Cyclone Genesis
Eichler, T.
The Role of Global Warming in Altering the Frequency and Intensity of Tropical and Non-Tropical Cyclones
6

Slide 7

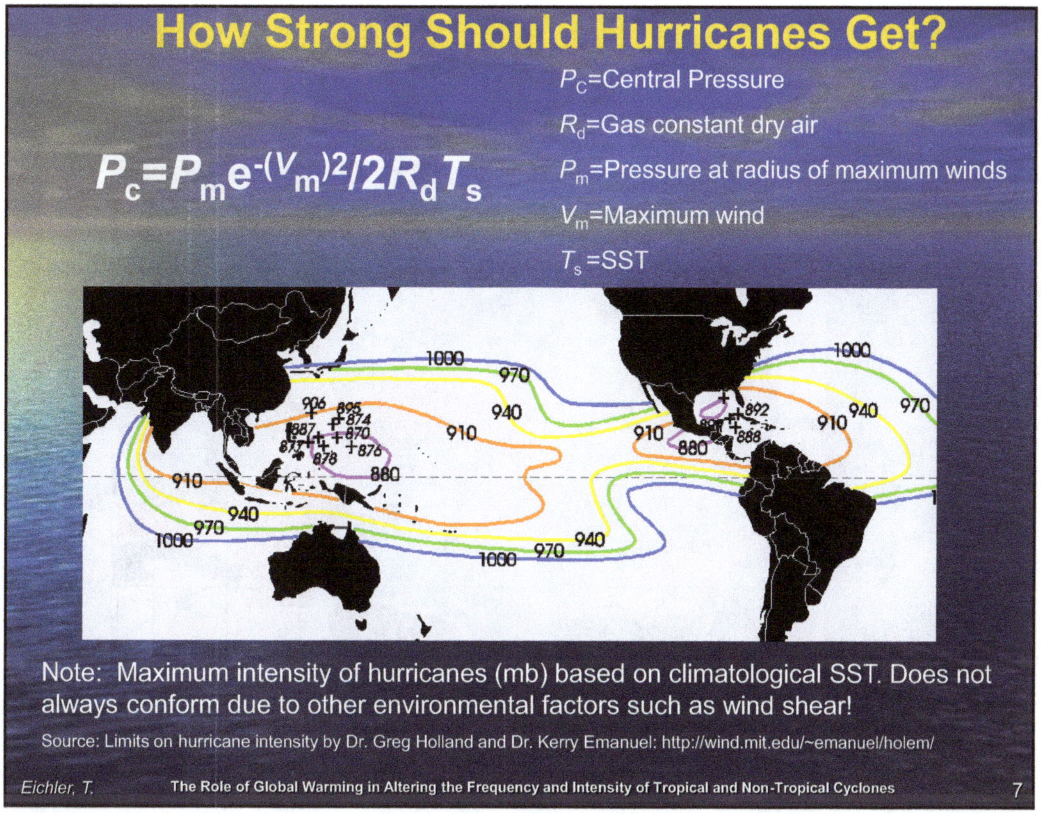
How Strong Should Hurricanes Get?
$P_c = P_m e^{-(V_m)^2/2R_d T_s}$
P_C=Central Pressure
R_d=Gas constant dry air
P_m=Pressure at radius of maximum winds
V_m=Maximum wind
T_s=SST
1000
970
940
910
880
Note: Maximum intensity of hurricanes (mb) based on climatological SST. Does not always conform due to other environmental factors such as wind shear!
Source: Limits on hurricane intensity by Dr. Greg Holland and Dr. Kerry Emanuel: http://wind.mit.edu/~emanuel/holem/
Eichler, T.
The Role of Global Warming in Altering the Frequency and Intensity of Tropical and Non-Tropical Cyclones
7

Slide 8

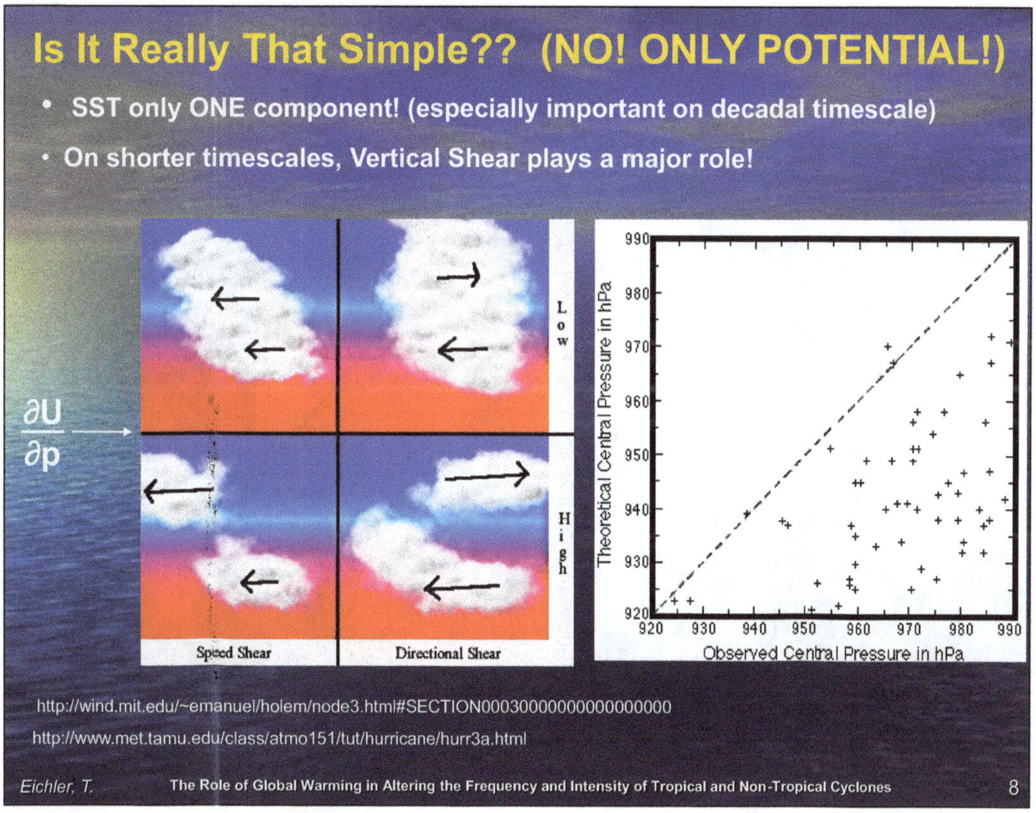
Is It Really That Simple?? (NO! ONLY POTENTIAL!)
• SST only ONE component! (especially important on decadal timescale)
• On shorter timescales, Vertical Shear plays a major role!
$\frac{\partial U}{\partial p}$
Low
High
Speed Shear
Directional Shear
Theoretical Central Pressure in hPa
Observed Central Pressure in hPa
920 930 940 950 960 970 980 990
http://wind.mit.edu/~emanuel/holem/node3.html#SECTION00030000000000000000
http://www.met.tamu.edu/class/atmo151/tut/hurricane/hurr3a.html
Eichler, T.
The Role of Global Warming in Altering the Frequency and Intensity of Tropical and Non-Tropical Cyclones
8

Slide 9

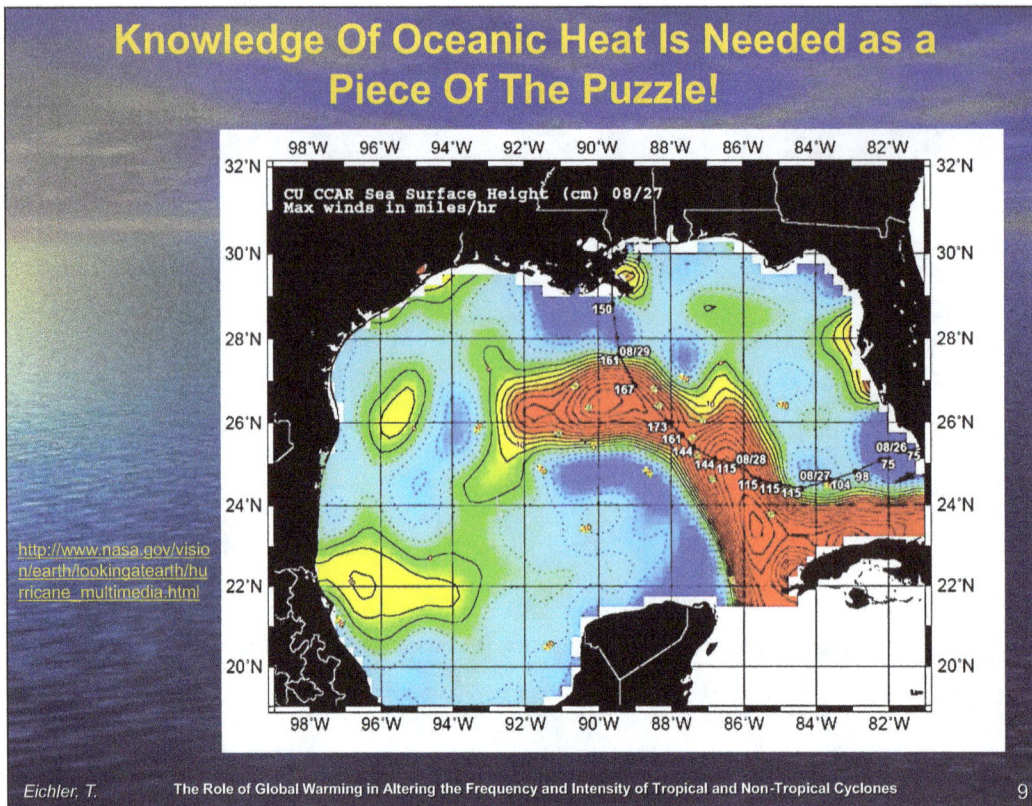

Slide 10

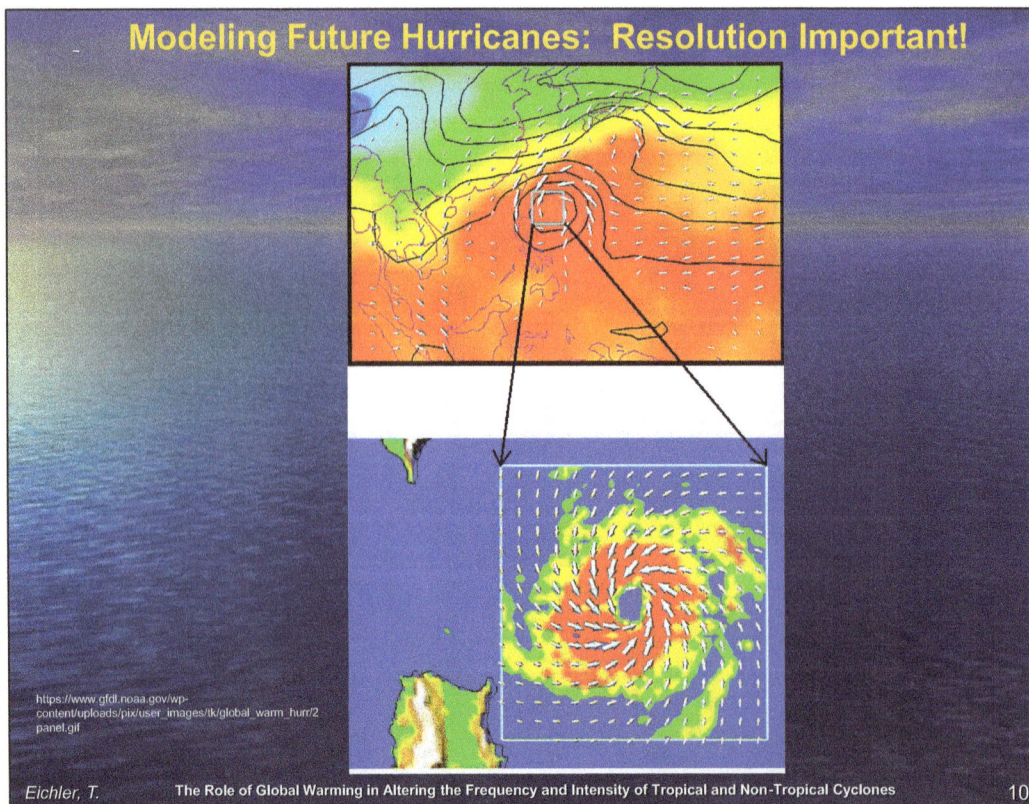

Slide 11

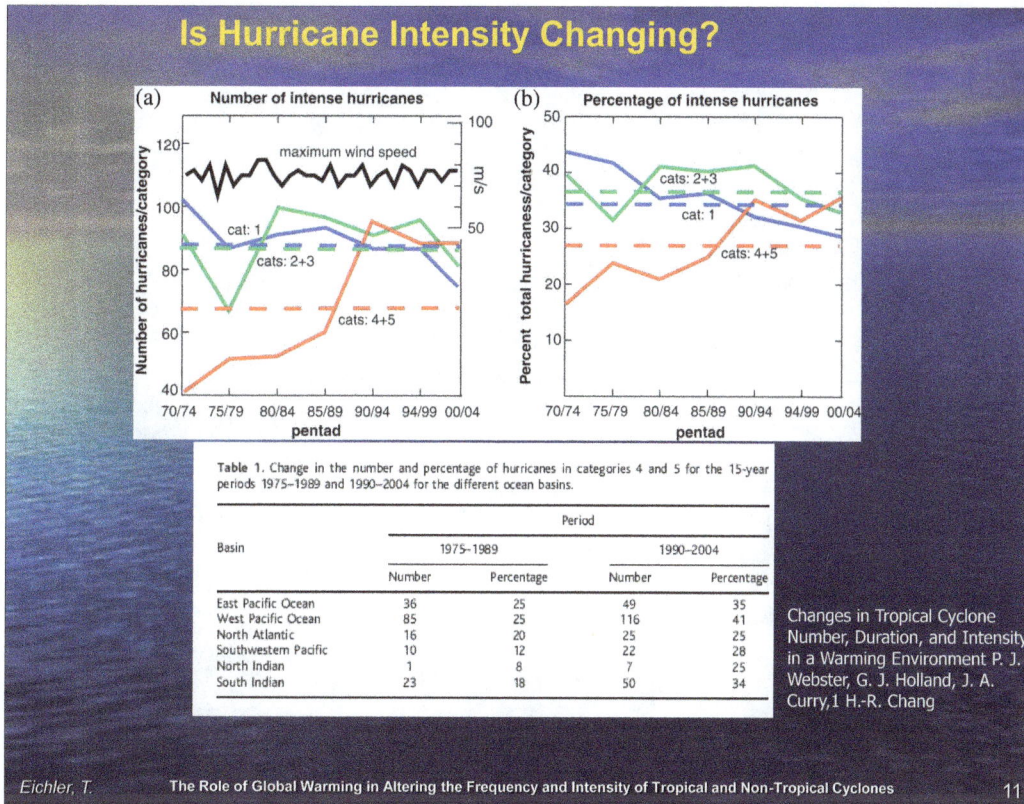

Is Hurricane Intensity Changing?

Table 1. Change in the number and percentage of hurricanes in categories 4 and 5 for the 15-year periods 1975–1989 and 1990–2004 for the different ocean basins.

Basin	1975–1989 Number	1975–1989 Percentage	1990–2004 Number	1990–2004 Percentage
East Pacific Ocean	36	25	49	35
West Pacific Ocean	85	25	116	41
North Atlantic	16	20	25	25
Southwestern Pacific	10	12	22	28
North Indian	1	8	7	25
South Indian	23	18	50	34

Changes in Tropical Cyclone Number, Duration, and Intensity in a Warming Environment P. J. Webster, G. J. Holland, J. A. Curry,1 H.-R. Chang

Eichler, T.   The Role of Global Warming in Altering the Frequency and Intensity of Tropical and Non-Tropical Cyclones   11

Slide 12

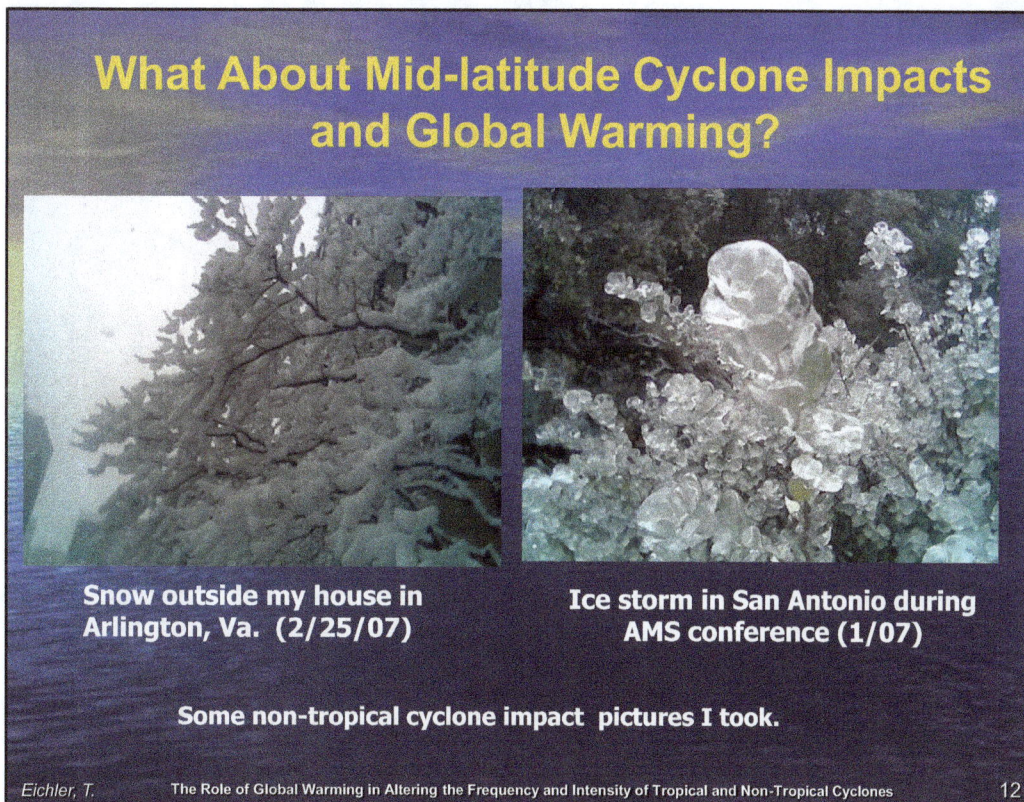

What About Mid-latitude Cyclone Impacts and Global Warming?

Snow outside my house in Arlington, Va. (2/25/07)

Ice storm in San Antonio during AMS conference (1/07)

Some non-tropical cyclone impact pictures I took.

Eichler, T.   The Role of Global Warming in Altering the Frequency and Intensity of Tropical and Non-Tropical Cyclones   12

Slide 13

Slide 14

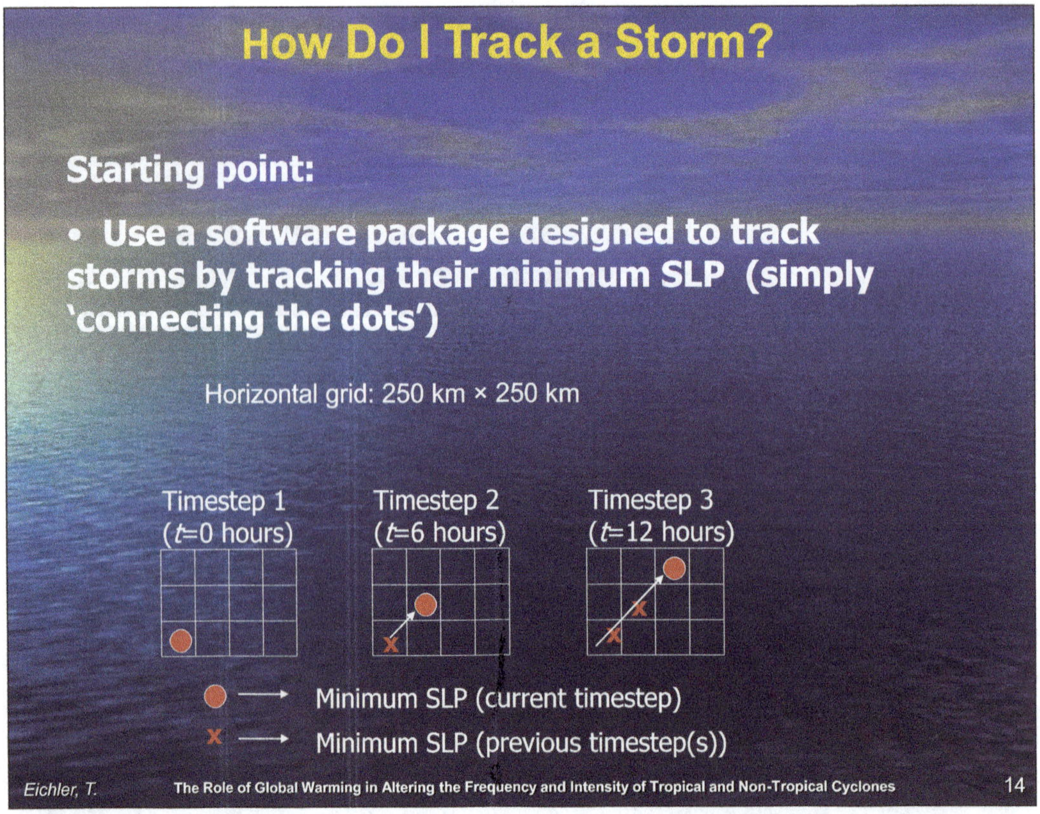

Slide 15

Slide 16

Slide 17

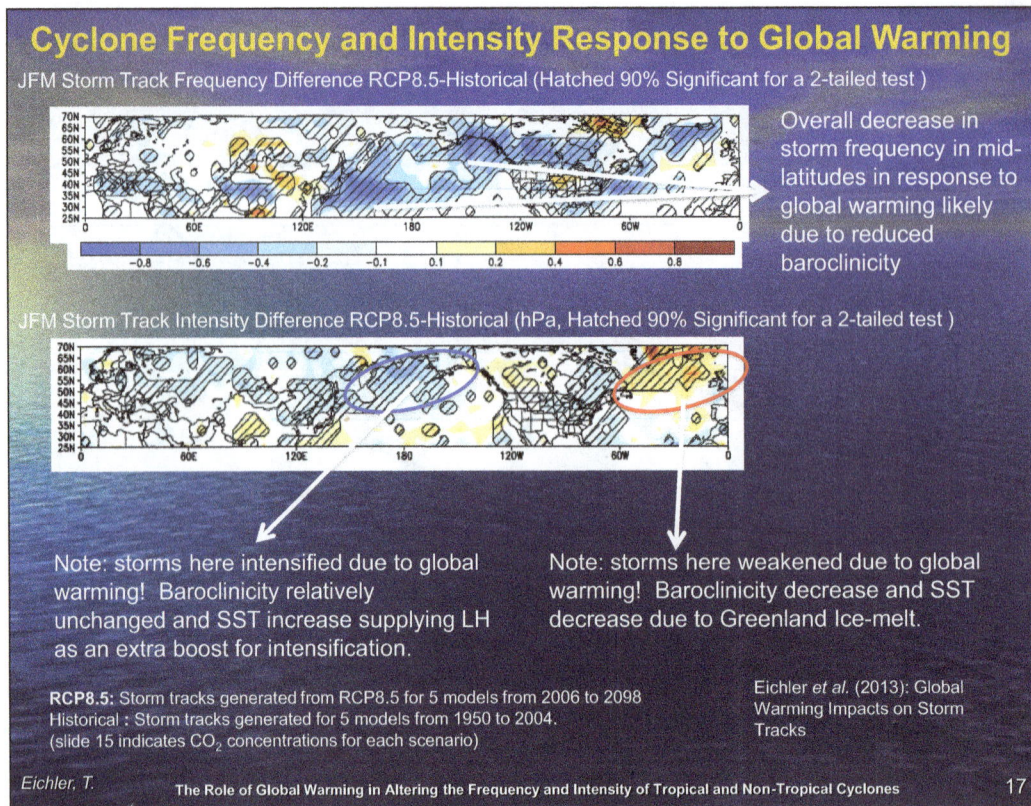

Slide 18

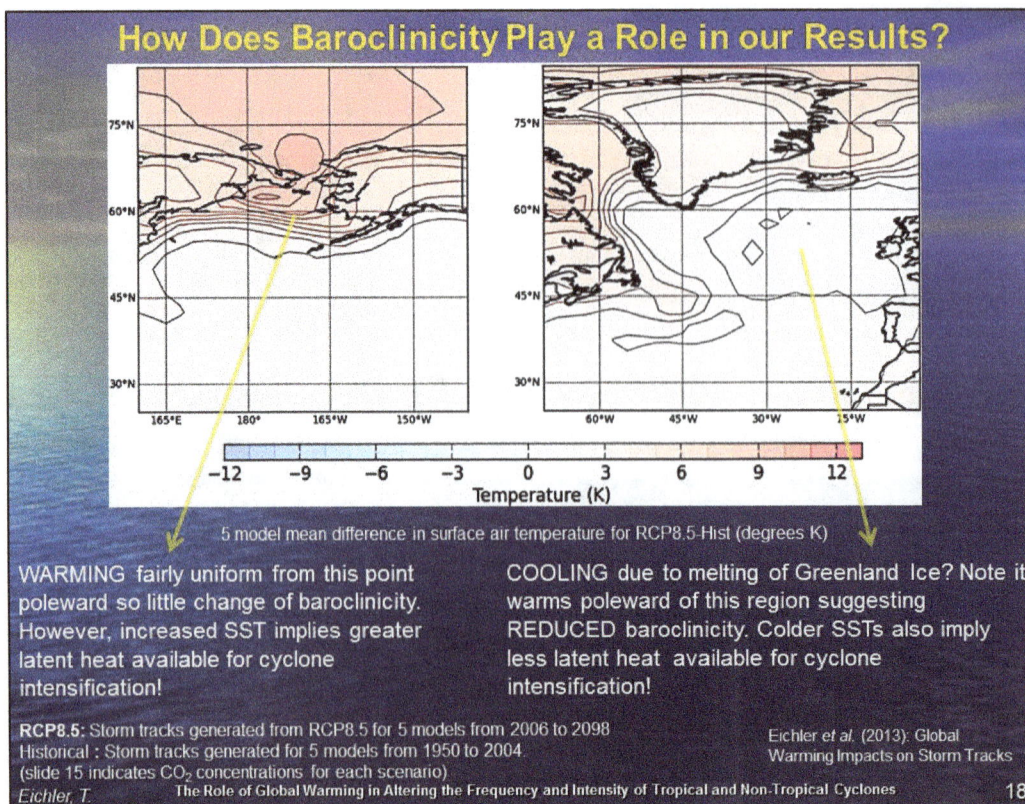

Slide 19

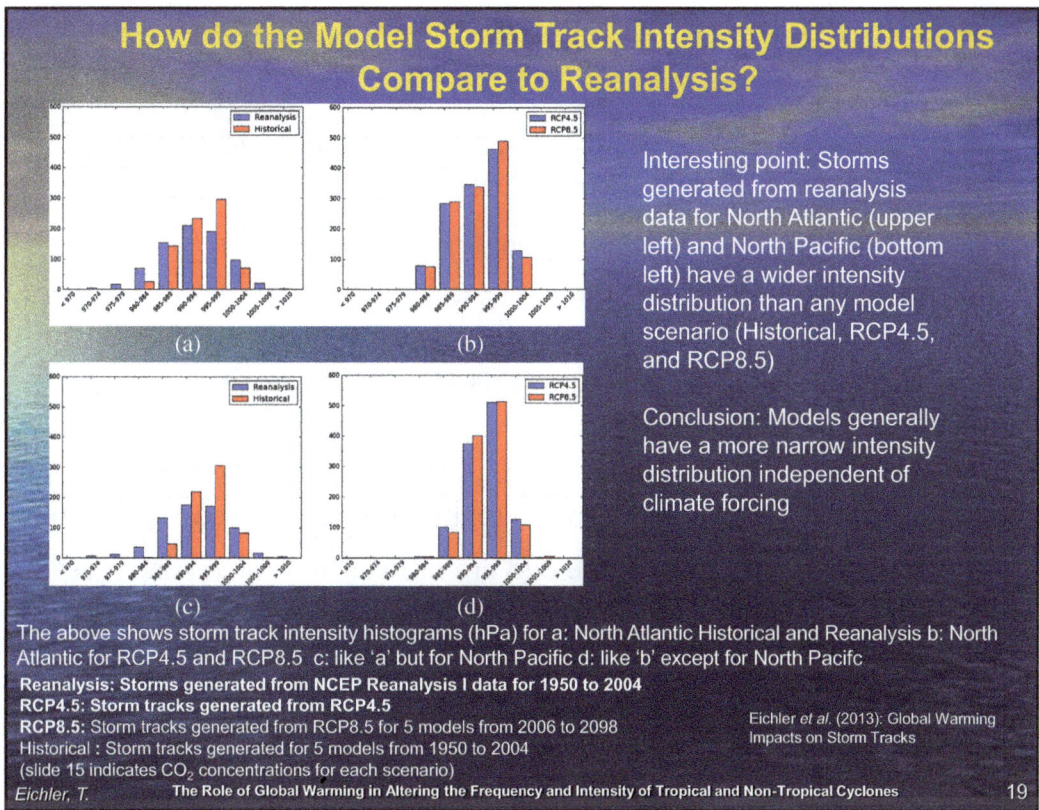

Slide 20

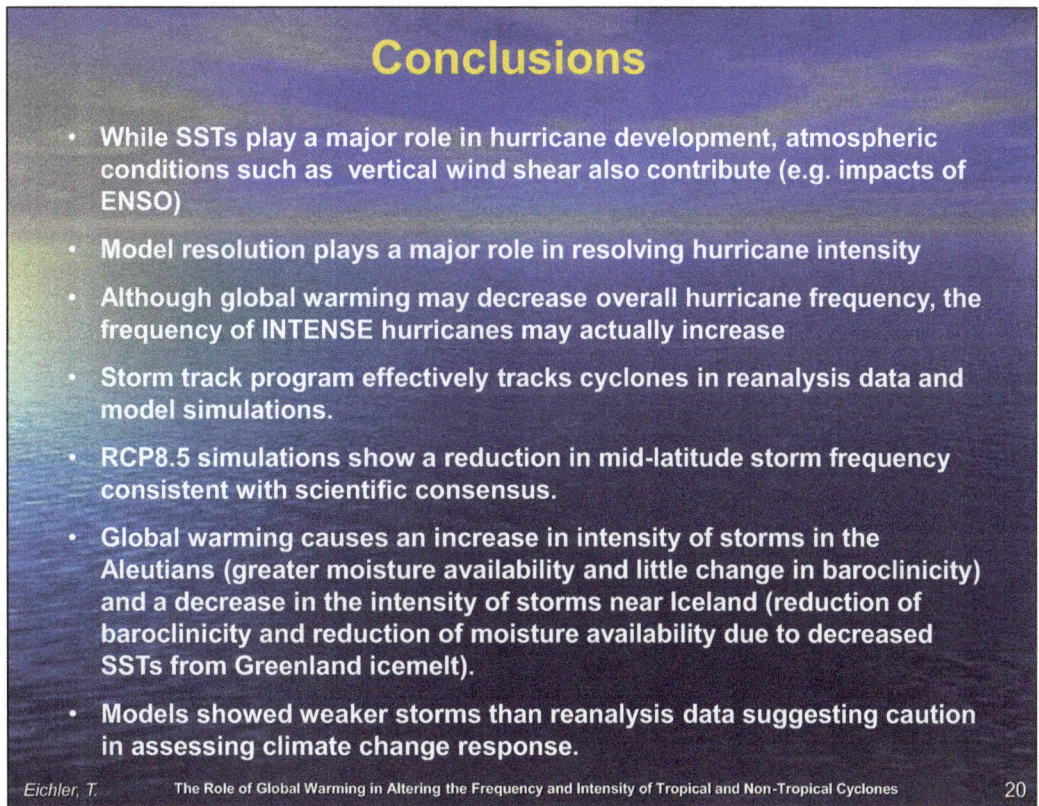

Slide Notes

Slide 3   Just a view of some great storms! (clock-wise from top left) 2006 Snowstorm, 1888 Blizzard, 1969 'Lindsay' snowstorm, hurricane Floyd (1999), hurricane Donna (1960), 1992 Nor'easter.

Slide 4   Over 150 years of data! Hurricane track and intensity (top). Bottom shows SST annual mean climatology from CPC to illustrate SST/hurricane genesis dependency.

Slide 5   One method to evaluate environment linked to hurricanes.

Slide 6   Terms from GP and GP for El Niño — La Niña for ASO. RH (drier) and vertical wind shear play roles in reducing North Atlantic hurricanes during El Niño.

Slide 7   Potential intensity of hurricanes linked to thermodynamics (does not include mitigating factors such as wind shear).

Slide 8   Wind shear plays a major role in hurricanes irrespective of SST. Plot on right shows that observed storms do not conform to theoretical SLP based on thermodynamics.

Slide 9   Ocean dynamics also play a role and are difficult to assess with regard to future climate. It is not as simple as SST. Since upwelling normally occurs due to circulatory winds of a hurricane, deeper oceanic heat content will reduce upwelling and support greater intensity. Note that Katrina intensified rapidly when crossing a warm eddy and then weakened somewhat once north of it.

Slide 10   Key to assessing hurricane intensity is spatial resolution of models. Need extremely high resolution to resolve inner core.

Slide 11   Note increase in intense hurricane frequency. Interesting that high-res. Models support this result with regard to global warming.

Slide 12   Picture on right is just a weed but looks like an ice rose!

Slide 13   NEVER thought I would see a storm like Sandy! Part of my motivation for my research.

Slide 14   Program adapted from Mark Serreze

   Interpolate SLP data onto a (Lat,Lon) = (250 km,250 km) grid.

      Find SLP minima (>1 mb below surrounding gridpoints).

      Assume a maximum distance threshold of 800 km to track storms.

   Storm track program run at CDC

      Tracks storms on a 6-hour time scale.

      Data include Lat/Lon, SLP, and a unique tag.

Slide 15   Storm tracks for JFM 1983. Climatology is acquired by binning into $5° \times 5°$ grid boxes. Storm tracks generated from NCEP 6-hourly SLP reanalysis data.

Slide 16   Slide shows models and scenarios used in my paper on global warming and storm tracks. We generated storms for NCEP reanalysis, historical, and global warming scenarios. Results are shown for ensemble mean of models at bottom.

Slide 17   Frequency shows expected (significant) decrease of storms in mid-latitudes which is most distinct for RCP8.5 differences with respect to historical (top left). Intensity results on right show more intense storms in North Pacific and weaker storms in North Atlantic for RCP8.5 relative to historical. The former is likely due to increased SSTs with little change in baroclinicity, while the latter is due to reduced baroclinicity and colder SSTs possibly due to Greenland Ice-sheet melt. Also, note bottom plots in frequency (left) and intensity (right) showing historical run ensemble relative to reanalysis. This shows that models have some bias in their climatologies.

Slide 18   Surface air temperature for RCP8.5 ensemble — historical supporting conclusions from previous slide.

Slide 19   This slide shows distribution of storm intensities for North Atlantic (top) and North Pacific (bottom) active storm track regions. Note that model simulations have a more narrow distribution of storm intensities regardless of climate scenario relative to NCEP reanalysis.

Slide 20   That's about it!

SECTION 4

Hydrologic Responses

CLIMATE LECTURE 9

Wisdom, Climate, and Water Resources

Robert Webb

NOAA/OAR/ESRL Physical Sciences Division, 325 Broadway, Boulder, CO, USA

Robert Webb is Director of the Physical Sciences Division of the NOAA Earth System Research Laboratory. He is the NOAA lead on the federal interagency Climate Change and Water Working Group and the NOAA representative on the Department of Defense Strategic Environmental Research and Development Program (SERDP) Science Advisory Board. His research has recently focused on the attribution and predictability of weather and climate extremes, and on improving the use and usability of weather and climate information to support policy, planning and decision making.

Introduction

David Rind's legacy is not only that he did great science, that he is an intellectual mentor for an impressive number of very successful scientists, but that he is an inspirational role model of how to be scientist. He grew up in the shadow of Yankee Stadium, and being a devoted Yankees fan, I believe that he first observed an atmospheric breaking wave over those hallow grounds. I always wondered if this vision over Yankee Stadium inspired him to be such a preeminent atmospheric scientist. Before spending some time discussing some of his research examining how changes in climate could impact water resources, I first want to present what I will call *David Rind's Words of Wisdom (at least as I remember):* (1) 'Hit it with a sledge hammer', (2) 'Nobody complains when they get an "A"', (3) 'No one looks very smart when you argue with an idiot', and (4) 'I would pay someone to clean out my desk'.

Words of Wisdom

'Hit it with a sledge hammer' when designing an experiment to test the model sensitivity to a hypothesized climate forcing. To design the perfect climate modeling experiment was not necessarily to design the most realistic experiment. While today we might use super ensembles running on ever-increasingly high-performance computers to assess climate system response to a forcing, David used this 'sledge hammer' approach to break through the signal-to-noise challenge in a highly non-deterministic system. If you did not see much of a modeled climate system response to an unrealistically large change in forcing, David's logic was that there was little to no chance of seeing a response when prescribing a realistic forcing. David frequently used this approach when designing climate model experiments to understand the cause and effect of paleoclimate sensitivity to various external forcings. This type of modeling approach was presented in a series of papers with Peteet and Kukla examining the initiation and growth of the continental ice sheets in response to Milankovitch forcing (e.g. Rind *et al.* 1989; Peteet, Rind and Kukla, 1992). In these studies, sledge hammer forcings of the climate system to initiate and grow continental ice sheets were (1) the prescription of a fivefold increase over the calculated reduction in summer and fall insolation, (2) the additional prescription of a 10-m-thick slab of ice in locations where continental ice sheets existed during the last glacial maximum, and (3) the prescription of full glacial sea-surface temperatures (SSTs) and atmospheric CO_2 concentrations. Model simulations revealed that even with a reduction in summer and fall insolation by a factor of 5, there was no suggestion of ice-sheet growth and there was not a year-to-year accumulation of snow. Furthermore, examination of the energy and mass budgets simulated in the model experiment revealed an imbalance where the prescribed ice slab was added, it neither grew nor was sustained, and, in fact, would have been completely removed in about five years. Only with the additional forcings of SSTs and atmospheric CO_2 concentrations prescribed at full glacial levels, was the climate model able to simulate the maintenance of the slab ice sheet in limited regions. The inability of the climate model to simulate changes observed in the geologic record in response to sledge hammer magnitude prescribed forcings raised concerns with the climate model sensitivity to known past, and perhaps future forcings, suggesting a much more complex response of mechanisms and feedbacks in the integrated earth system.

'Nobody complains when they get an "A"' when explaining why it was important to challenge the climate model results to make sure we were getting the right answer for the right reason rather than just thoroughly investigating when we got an unexpected answer. The unexpected answer to a hypothesized cause and effect was more intellectually interesting to David because it required a new rigorous explanation rather than confirmation of existing preconceived notion. I got the feeling that David was amused that the climate modeling community, and science community as a whole, spent a lot of time trying to understand what was going on when they did not get answer they expected, but were quick to accept an answer if it was what they expected. Not accepting results at face value is the reason David was continuously embarking on energy and mass balance budget analyses with numerous printouts being compared with the 822D printout on his desk, output from a few other experiments balanced on his lap, and the rest on laps of who ever happened to be in his office at the time. I think David more enjoyed when the explanation ended up being the opposite of what was expected, and required an elegant but simple explanation.

One example that comes to mind was the Overpeck, Rind, Lacis and Healy (1996) paper that looked at the role of high glacial aerosol levels in the atmosphere as a mechanism for abrupt climate change. If I remember correctly, climate model experiments were designed to test energy balance modeling results that suggested a cooling response to high glacial dust loadings. These elevated dust loadings were hypothesized to be a mechanism to trigger the millennial-scale global climate change observed in ocean and terrestrial paleoclimate record. In contrast with the energy balance modeling results, the climate modeling simulations showed an opposite response, with higher dust loadings resulting in a downstream regional warming of 5°C. Once the obvious questions of problems in the implementation of climate model experimental design were ruled out, the scientific fun began.

Examination of the global and regional energy balances revealed a different possible response to increased atmospheric dust forcing in a glacial climate than in our modern climate. Instead of cooling due primarily to backscattering of shortwave radiation, when over high albedo snow and ice covered surfaces the impact of an increase in atmospheric dust on planetary albedo is minimal. However, the greenhouse warming through the absorption of long-wave radiation by atmospheric dust is relatively speaking, significant, resulting in regional warming. While not the hypothesized cause and effect, careful analysis provided an elegant but simple explanation while reminding the climate science community that climate sensitivity and responses are, and will be, more complicated in environments dominated by snow and ice.

'No one looks very smart when you argue with an idiot' when explaining why he would not go argue the science of global warming in a Congressional hearing pitting scientists arguing for and against the science of anthropogenic climate change. It was his version of George Bernard Shaw's insight 'Never wrestle with pigs. You both get dirty and the pig likes it.' David had an insightful understanding that there is a limit to the influence of being scientifically correct. As far as I understand, rather than arguing the science in a public forum, he invested his time and energy in doing more science to advance the knowledge base. He tended to leave it to others to go do battle in the policy arena that involved the comingling of a political circus and the scientific process. While David did not expend much time and energy in the public debate on climate change and global warming, he was either the lead or co-lead author on hundreds of publications examining many characteristics, sensitivity and responses of the climate system, and authored close to 40 papers focused on various aspects of global warming and climate change. By engaging in the time-tested deliberative scientific process, David has been highly effective in moving the debate on the risk and potential dangers of anthropogenic climate change forward through his extensive volume of peer-reviewed papers while leaving the public advocacy to others. An example I recall was the Rind *et al.* (1991) analysis of SAGE II water vapor data illustrating that climate models were not overestimating the water vapor feedback. Analysis of the SAGE II was conclusive in disproving a proposed mechanism that increased convection in a warmer climate would not increase the amount of moisture but instead dry the middle and upper troposphere by means of associated compensatory subsidence. However, while politely calling out the chief proponent of this mechanism in the peer review literature, David refrained from engaging in a public jousting match to challenge the belief that the increased convection associated with projected warming would act to dry the middle and upper troposphere. Multiple times I have reminded myself of David's advice, and likewise counseled others, to take a step back and reflect on whether it will be a productive endeavor to engage in a discussion of climate science issues with people arguing for positions based on opinions and not grounded in science.

'I would pay someone to clean out my desk' when describing his career plans if he won the lottery. David told the truth. He reminded the group of us who were sitting in his office, going on about how we would use hypothetical lottery winnings to promote our idealistic goals to save the world and educate humanity, that being a climate scientist was still a job even though many of us still cannot understand how we managed to find such a fun vocation and get paid for it. While one of the hardest working and dedicated scientist I have had the honor of collaborating with, David had a balanced appreciation of science and did not allow his scientific work to be an all-consuming part of his life. If he had actually played and had won the lottery, I always imagined that David would have walked out of his office, gotten on the subway heading downtown to Times Square, and off to catch Broadway Shows, one after another. I always hoped that when David retired from NASA there would be a dramatic shift in his life, and he would spend more time enjoying the arts.

Climate and Water Resources

While primarily an atmospheric scientist, David had an interest in understanding how the surface hydrology budget may respond to changes in climate. While his interest may have started from a Bowen ratio perspective on the partitioning of the surface energy balance into sensible and latent heating, over the years his interest evolved to focus on how changes in climate could impact the surface water balance and associated changes in water-related risks.

The earliest of his studies was a Rind (1982) paper that investigated the potential of using ground moisture anomalies as a source of predictabilities in seasonal climate forecasts. Hitting it with a sledge hammer, three model simulations were run with June 1 soil moisture reduced by 25% over the continental U.S. These model simulations revealed a significant increase in surface air temperatures for June, July, and August, reductions in precipitation concentrated in June and July, a decrease in evaporation (latent heat flux), and an increase in sensible heat flux. The results of this seasonal predictability modeling study identified soil moisture as a useful antecedent condition for developing improved summer temperature forecasts.

Expanding on the role of soil moisture, Rind (1984) examined the importance of vegetation, surface albedo, water holding capacity of the soil-vegetation system, transport of water from deeper levels within the ground, the roughness of the surface, and influence of snow cover on albedo. Once again hitting it with a sledge hammer, model experiments were run changing the rate and process of exchanging water between ground levels, specifying water holding capacities for desert to all vegetation types, and assigning desert ground albedo values for all vegetation types. Analysis of these experiments revealed that the change in surface albedo and water holding capacity resulted in similar impacts with regional variability. The albedo impact was greater in wet regions and during seasons when large-scale dynamics or convection dominates precipitation, whereas impact of a change in water holding capability was more important in regions where local evaporation processes played an important role in influencing precipitation. Altering the rate and process of water exchange between ground layers predominantly resulted in shifts in seasonality within the annual cycle. Overall, the conclusions of this study underscored the importance of accurately representing the role of vegetation and more generally the land surface in modeling the global climate system.

Looking to the future, the Rind (1988) study used four double-CO_2 experiments to examine hydrologic impacts over North America in response to changes in model resolution, in specified SSTs versus ocean-heat transport calculated SSTs, and in SST gradients due to enhanced high-latitude climate sensitivity. Analyses showed that factors impacting the hydrologic cycle included: (1) changes in large-scale dynamics and atmospheric moisture convergence effects at a variety of spatial scales, (2) accurate simulation of current climate conditions was important with a bias to either too warm or too wet exaggerating responses, (3) increased atmospheric moisture transport of excess energy from the tropics to higher latitudes, and (4) the extent of polar amplification of temperature change. Although sensitive to model resolution, results for the western USA were projected to get wetter, whereas the central US response showed a response to SST gradients and the amount of mid-latitude SST warming. Results in the eastern USA were for a projected decrease in soil moisture but dependent on overall climate model sensitivity. Overall, the conclusions of this study underscored the need to better represent ground hydrology and ocean circulation in high-resolution climate models to provide more realistic estimates of projected risks to future water availability.

The Rind *et al.* (1990) study investigated how potential evaporation and drought risk might change with global warming. The likelihood of future drought was assessed using two drought indices calculated for climate model simulated projections of anthropogenic change — the Palmer drought severity index (PDSI) and the supply–demand index (SDDI). Based on these indices, drought conditions were projected to increase in the United States during the next century, with effects apparent in the 1990s. The projected increase in drought conditions resulted from large increases in potential evapotranspiration, with the largest increases in low-to-mid-latitude regions. Higher latitude regions were projected to have increases in precipitation in response to a warmer atmosphere being able to hold and deliver more moisture. Model results also suggested that global precipitation would be unable to compensate for increased terrestrial evaporative demands as surface warming over land was projected to be greater than over the ocean. PDSI-based severe drought (5% frequency under current climate) was projected to occur about 50% of the time by the mid-21st century under current greenhouse gas emission scenarios, and thus drought conditions will increase dramatically.

The Rind, Rosenzweig and Goldberg (1992) study assessed how global warming, induced changes in the hydrologic cycle could impact water availability for ecosystems, communities, and economies. An evaluation of the sensitivity of different impact models to a prescribed 4.5°C warming revealed a spectrum of responses in land-surface conditions: much larger decreases in soil moisture and runoff simulated in PSDI and forest models using Thornthwaite-based estimates for potential evapotranspiration, and less dramatic change in GCM and more physically based hydrologic process models using aerodynamic and Priestley-Taylor to estimate potential evapotranspiration. The primary conclusion from this study was that inconsistent hydrologic responses among various models used to simulate land-surface conditions and fluxes were problematic in assessing future risks and impacts. Outstanding challenges identified by Rind *et al.* (1992) included (1) a basic need to better understand hydrologic processes, (2) climate model projections of future temperature and precipitation being highly dependent on the ability of ground hydrology scheme to correctly simulate current soil moisture and runoff, (3) calculating runoff and soil moisture infiltration at the course spatial scales of GCMs, (4) resolving the proper way to calculate potential evapotranspiration, and (5) the efficacy of using offline impact models to estimate surface hydrology responses.

Two decades after the Rind *et al.* (1992) study that identified potential problems in using PDSI and Thornthwaite-based estimates for potential evapotranspiration, a number of colleagues and I revisited this issue in a study addressing the question of whether a transition to semipermanent drought conditions in the north central USA was forthcoming (Hoerling *et al.*, 2012). Our study conducted parallel diagnoses of projected changes in drought in the north central USA as inferred from PDSI and climate model calculated soil moisture and surface fluxes in order to understand different estimates of projected surface-moisture balances and associated uncertainties. Confirming Rind *et al.*'s (1992) results, our analyses also demonstrated that the suitability of the PDSI breaks down when applied in climate model projections of 21st century. While PDSI remains an excellent proxy indicator for soil moisture variations in the 20th century, with the simplified formulation of potential evapotranspiration, PDSI severely overstates future surface water imbalances and implied vegetation stresses. Our study illustrated that PSDI was excessively sensitive to the rise in surface air temperatures, inflated the implied evaporation of moisture from soils, and completely changed the physics of the surface-moisture balance relative to the 20th century. A similar dramatic change was not revealed in our analysis of land-surface fluxes and soil moisture response calculated directly by the climate model. Thus, consistent with the Rind *et al.* (1992) findings, we concluded that PDSI overestimates the future drought severity risk over the north central USA because of an unrealistic sensitivity to surface warming. While the PDSI works well as long as climate remains stationary and no warming trends are involved, the PDSI ceases to be a reliable indicator of drought and implied stress on water resources in a warming climate of the 21st century.

Conclusion

In addition to words of wisdom and an evolutionary research approach to understanding how changes in climate could impact water resources, David Rind had this strange white sculpture of a hand in his office off to the left side of his desk next to the window that would light up every so often to remind him to ask himself if he was doing something productive. While we all may not have similar sculptures in our lives, remembering to take the time to episodically assess whether we are doing something productive is a good practice to consider both in one's science and more generally in living life.

References

Hoerling, M.P., Eischeid, J.K., Quan, X.W., Diaz, H.F., Webb, R.S., Dole, R.M. and Easterling, D. (2012). Is a transition to semi-permanent drought conditions imminent in the US great plains?. *Journal of Climate*. doi:10.1175/JCLI-D-12-00449.1.

Overpeck, J., Rind, D., Lacis, A. and Healy, R. (1996). Possible role of dust-induced regional warming in abrupt climate change during the last glacial period. *Nature*, 384, 447–449. doi:10.1038/384447a0.

Peteet, D., Rind, D. and Kukla, G. (1992). Wisconsinan ice-sheet initiation: Milankovitch forcing, paleoclimatic data, and global climate modelling. In P.U. Clark and P.D. Lea (Eds.), *The Last Interglacial-Glacial Transition in North America, GSA SP-270* (pp. 53–69). Geological Society of America.

Rind, D. (1982). The influence of ground moisture conditions in North America on summer climate as modeled in the GISS GCM. *Monthly Weather Review*, 110, 1487–1494. doi:10.1175/1520-0493(1982)110<1487:TIOGMC>2.0.CO;2.

Rind, D. (1984). The influence of vegetation on the hydrologic cycle in a global climate model. In J.E. Hansen and T. Takahashi (Eds.), *Climate Processes and Climate Sensitivity, AGU Geophysical Monograph 29, Maurice Ewing* (Vol. 5, pp. 73–91). Washington, DC: American Geophysical Union.

Rind, D. (1988). The doubled CO2 climate and the sensitivity of the modeled hydrologic cycle. *Journal of Geophysical Research*, 93, 5385–5412. doi:10.1029/JD093iD05p05385.

Rind, D., Peteet, D. and Kukla, G. (1989). Can Milankovitch orbital variations initiate the growth of ice sheets in a general circulation model? *Journal of Geophysical Research*, 94, 12851–12871. doi:10.1029/JD094iD10p12851.

Rind, D., Goldberg, R., Hansen, J., Rosenzweig, C. and Ruedy, R. (1990). Potential evapotranspiration and the likelihood of future drought. *Journal of Geophysical Research*, 95(D7), 9983–10004.

Rind, D., Chiou, E.-W., Chu, W., Larsen, J., Oltmans, S., Lerner, J., McCormick, M.P. and McMaster, L. (1991). Positive water vapour feedback in climate model confirmed by satellite data. *Nature*, 349, 500–503. doi:10.1038/349500a0.

Rind, D., Rosenzweig, C. and Goldberg, R. (1992). Modelling the hydrological cycle in assessments of climate change. *Nature*, 358, 119–123. doi:10.1038/358119a0.

Slide 1

Slide 2

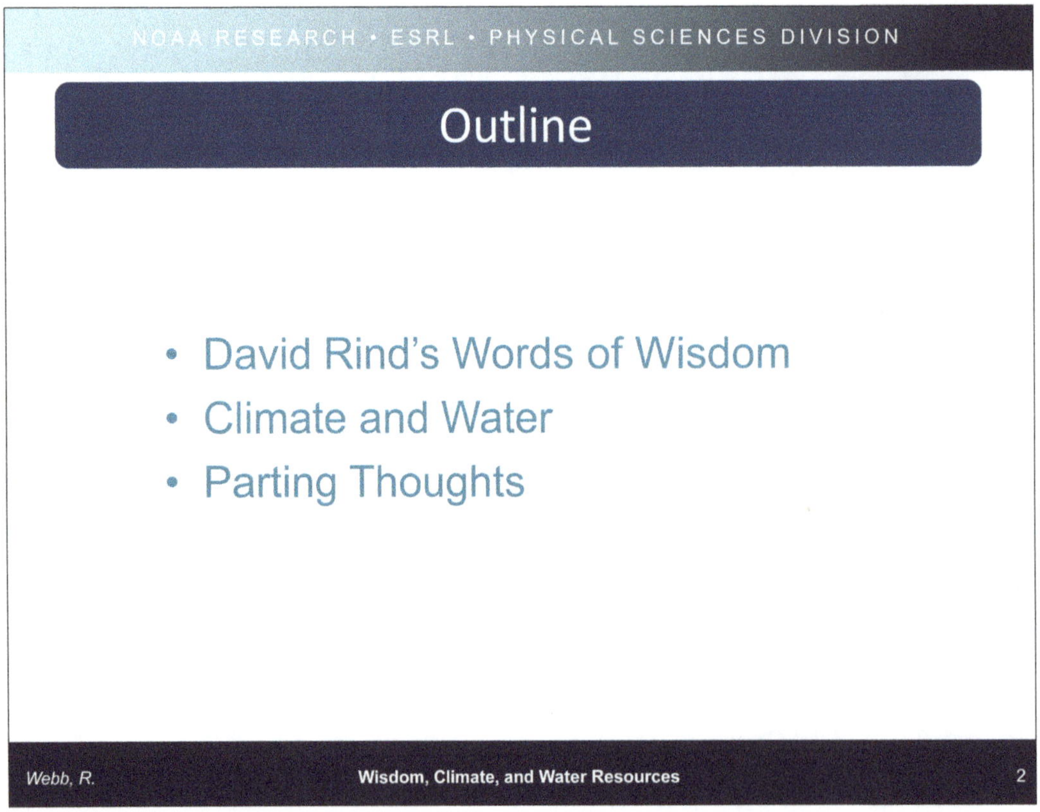

Slide 3

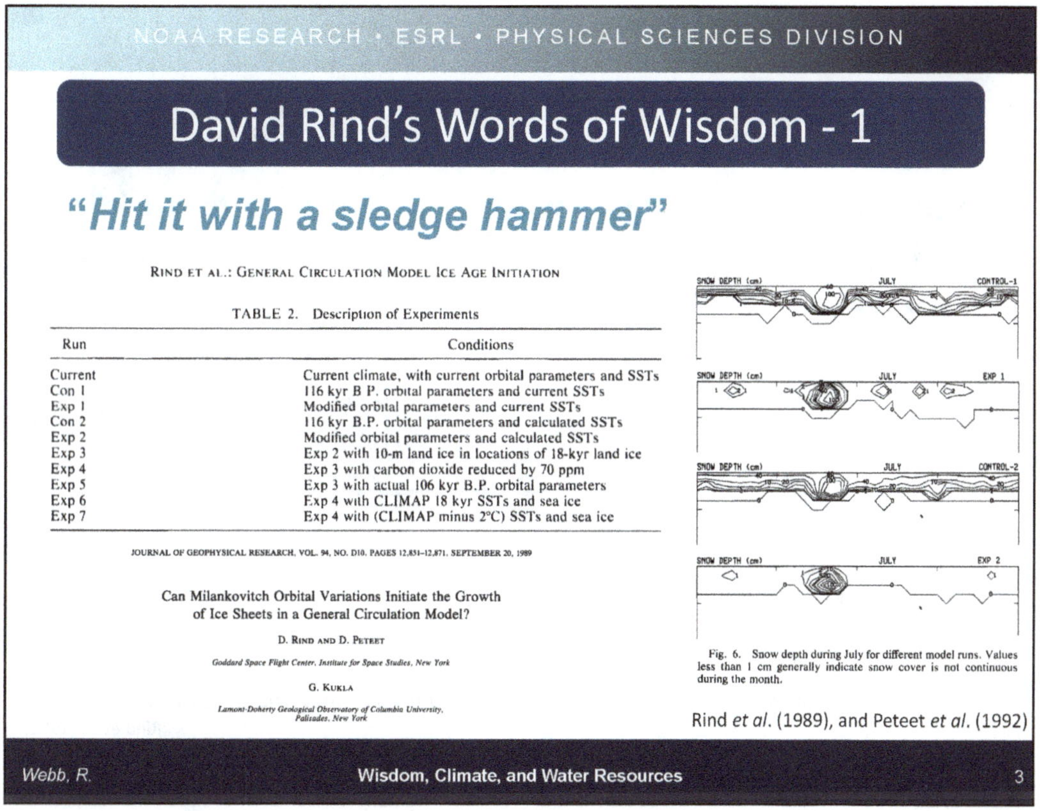

Slide 4

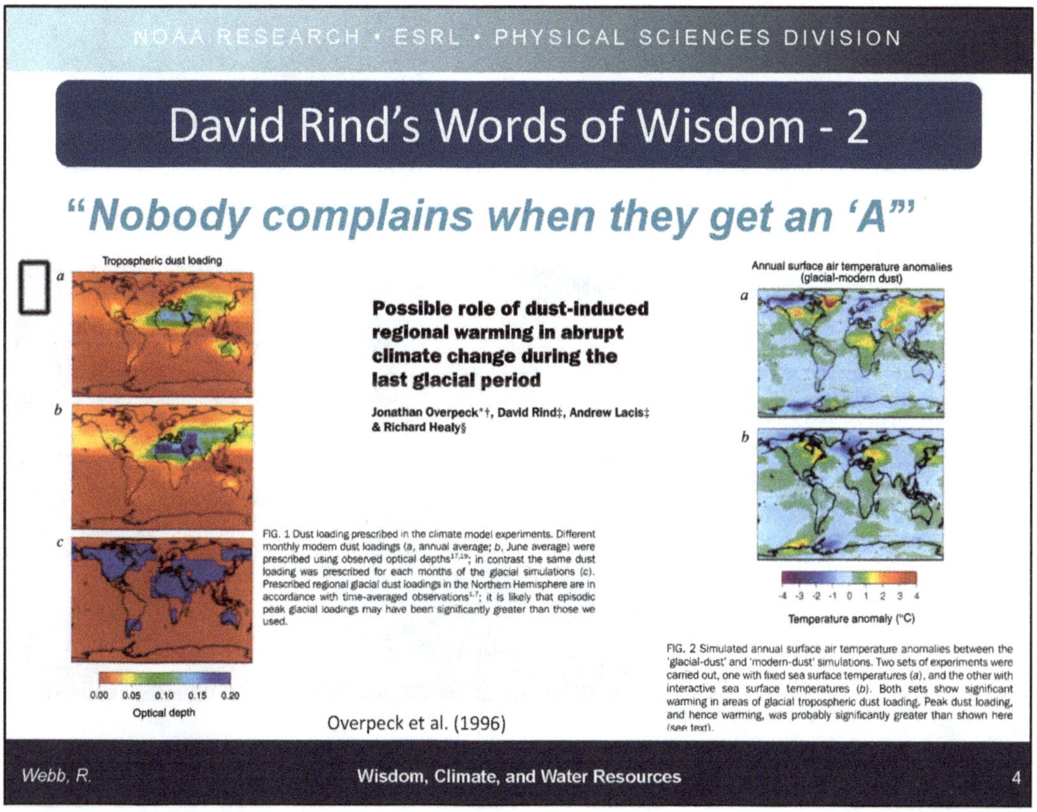

Slide 5

NOAA RESEARCH • ESRL • PHYSICAL SCIENCES DIVISION

David Rind's Words of Wisdom - 3

"No one looks very smart when you argue with an idiot"

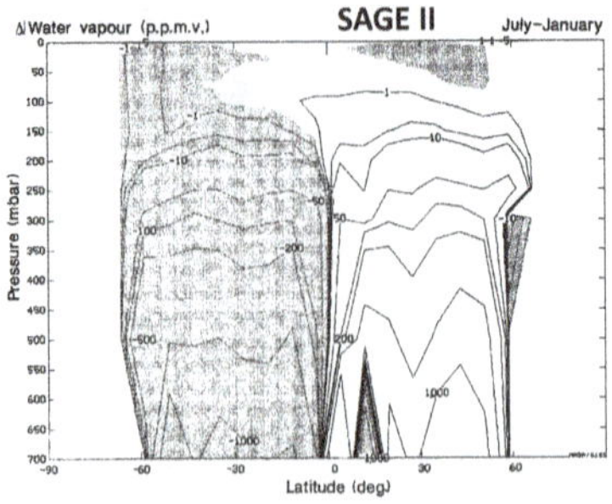

FIG. 2 Change of relative humidity *a*, between the doubled-CO_2 climate and the current climate in the GISS model; *b*, between July and January, SAGE II observations, January 1985 to July 1989; *c*, between July and January in the GISS GCM, for which three July and three January months were used from different years of a five-year simulation of the present climate. For comparison between model and observations, the levels above 700 mbar (solid line) should be considered.

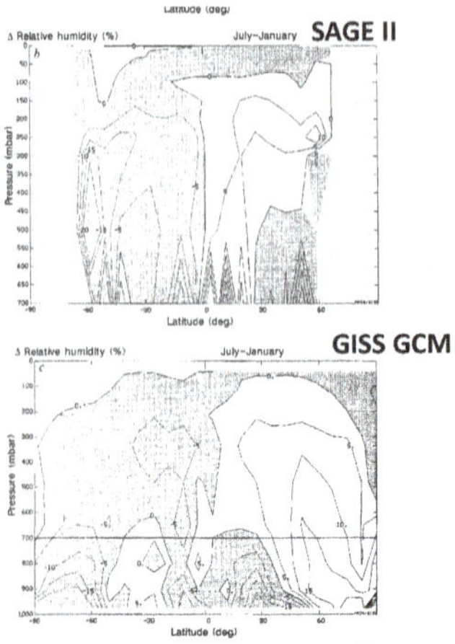

Positive water vapour feedback in climate models confirmed by satellite data

D. Rind*, E.-W. Chiou†, W. Chu†, J. Larsen†, S. Oltmans‡, J. Lerner*, M. P. McCormick† & L. McMaster†

George Bernard Shaw *"Never wrestle with pigs. You both get dirty and the pig likes it."*

Rind et al. (1991)

Webb, R. Wisdom, Climate, and Water Resources 5

Slide 6

NOAA RESEARCH • ESRL • PHYSICAL SCIENCES DIVISION

David Rind's Words of Wisdom - 4

"I would pay someone to clean out my desk"

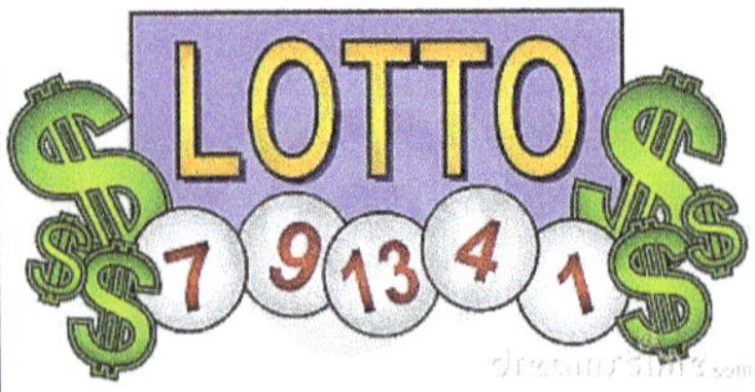

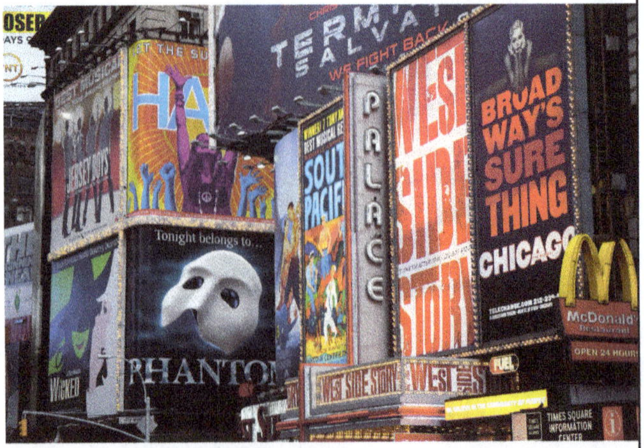

Webb, R. Wisdom, Climate, and Water Resources 6

Slide 7

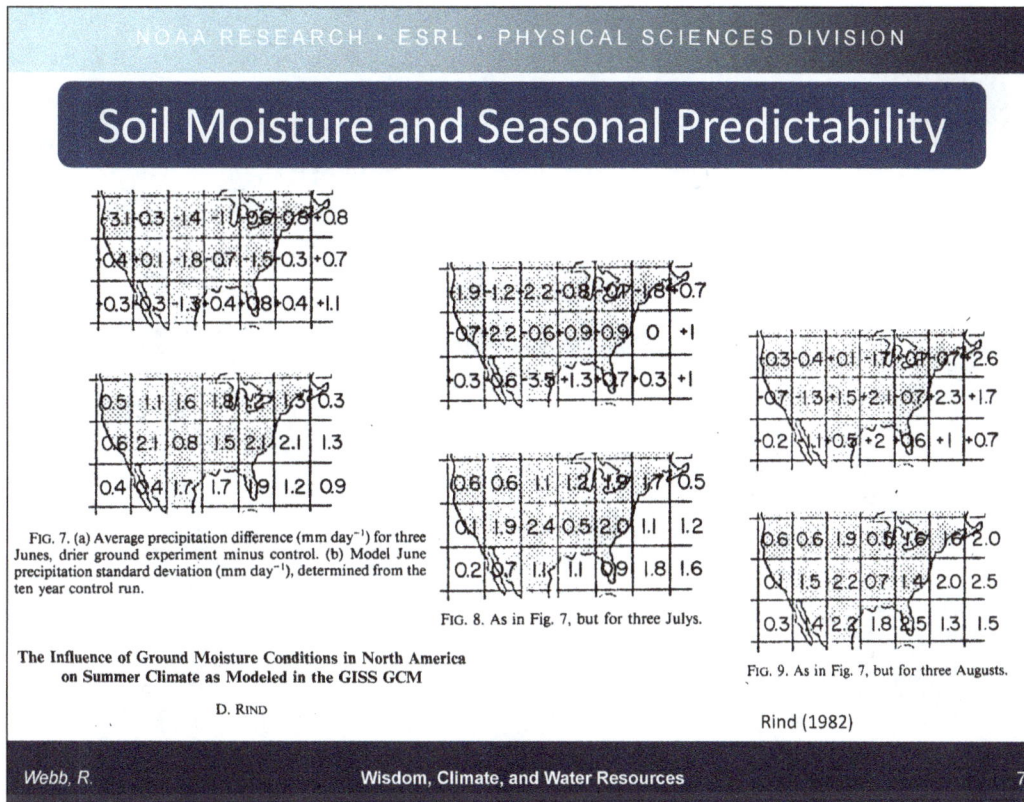

Slide 8

Land Surface Characteristics and Hydrology

Table 3.   Change in annual average precipitation (mm day^{-1}) between the experiments and the control run for different geographic regions. Also shown is the standard deviation of the control run.

REGION	DIFFUSION	CAPACITY	ALBEDO	STAND DEV
WESTERN U.S.	.06	-.72	.27	.33
MID U.S.	-.16	-.75	-.72	.11
EASTERN U.S.	-.12	.09	-.71	.63

THE INFLUENCE OF VEGETATION ON THE HYDROLOGIC CYCLE IN A GLOBAL CLIMATE MODEL

D. Rind

Rind (1984)

208   R. Webb

Slide 9

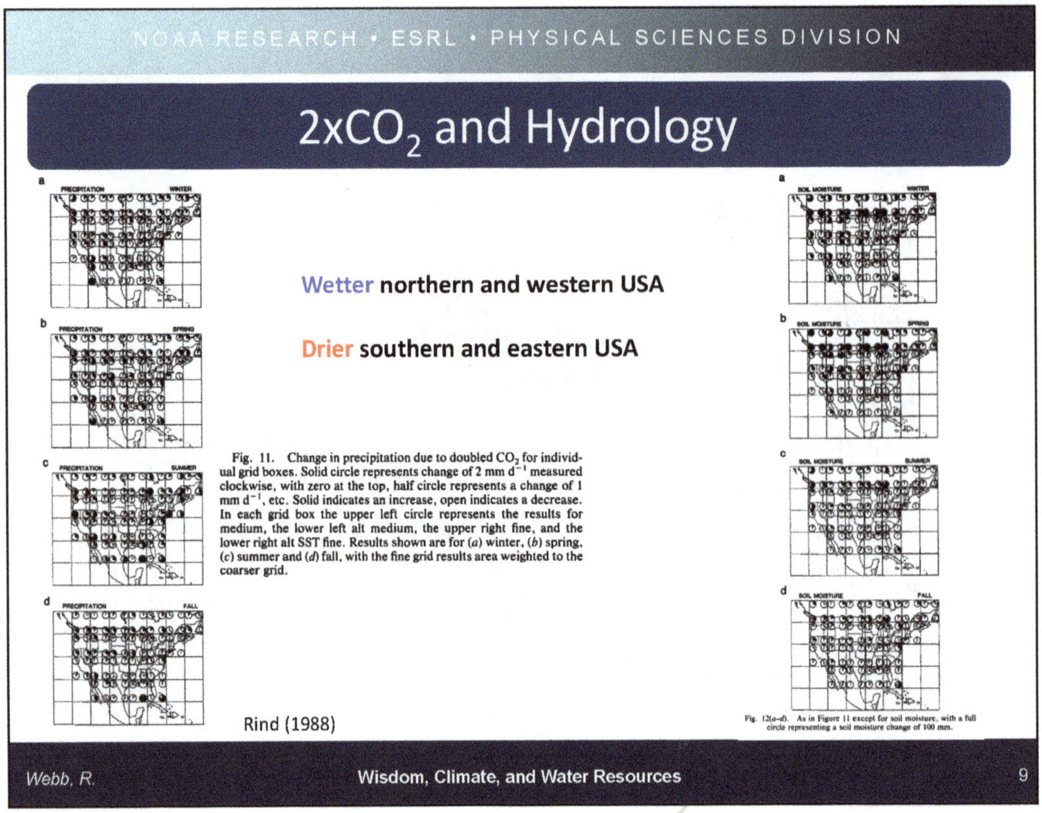

Slide 10

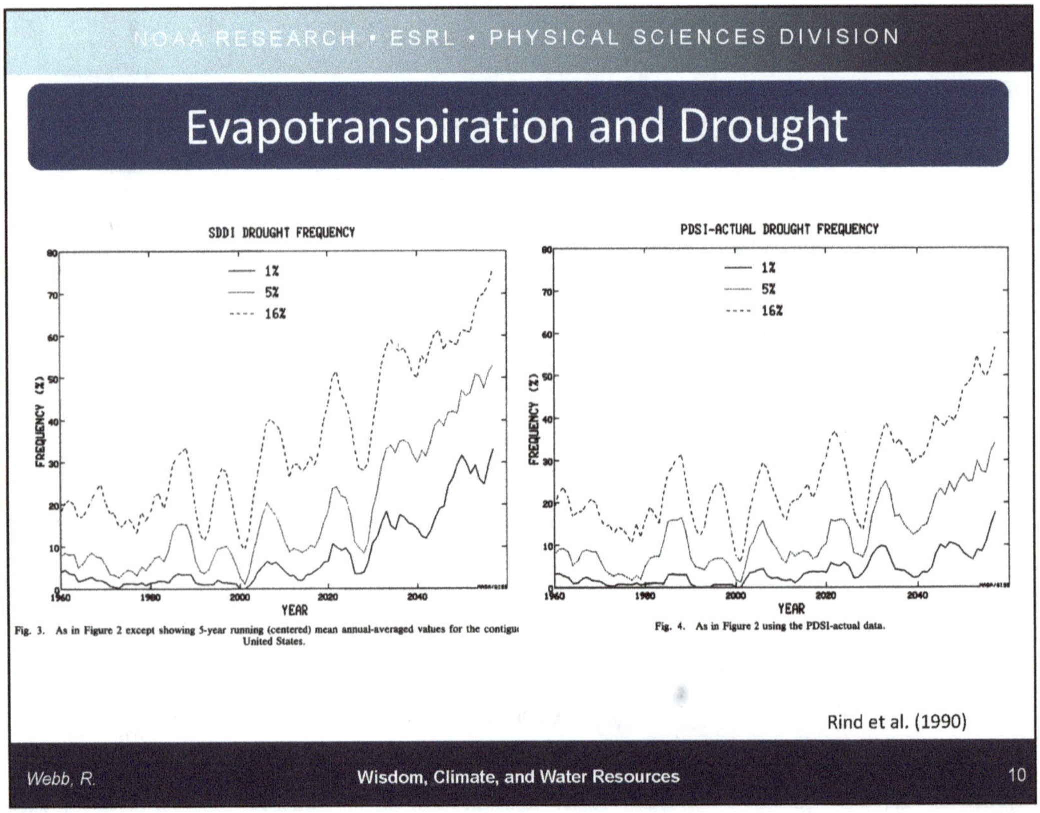

Slide 11

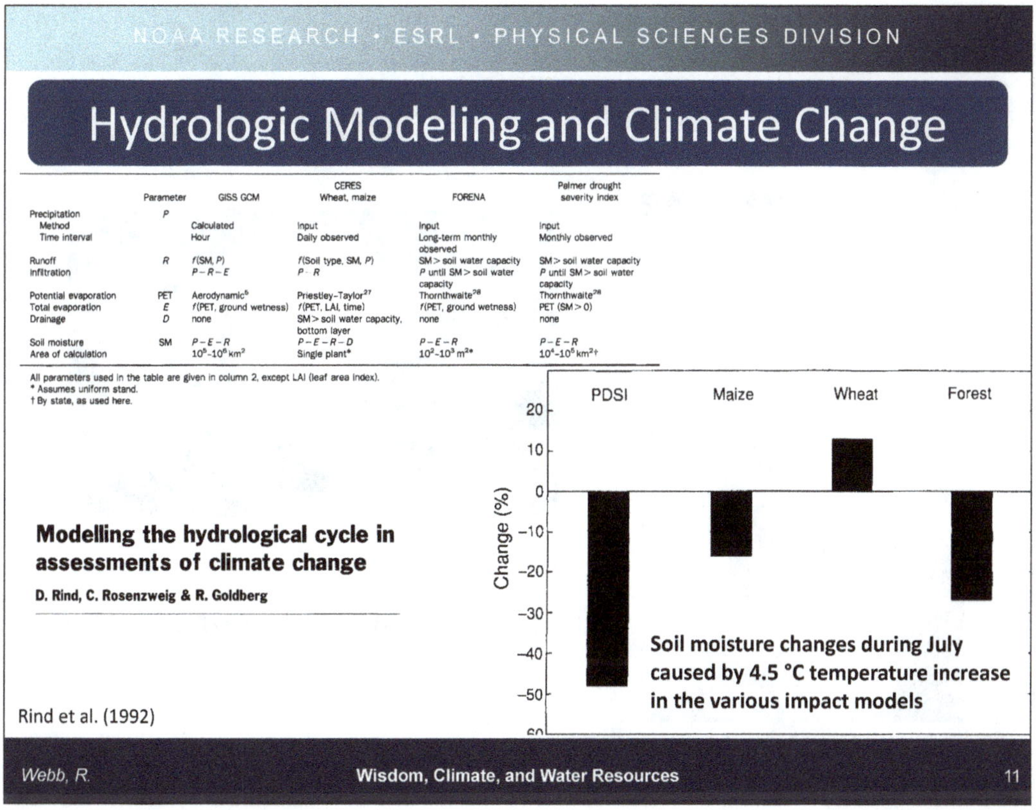

Slide 12

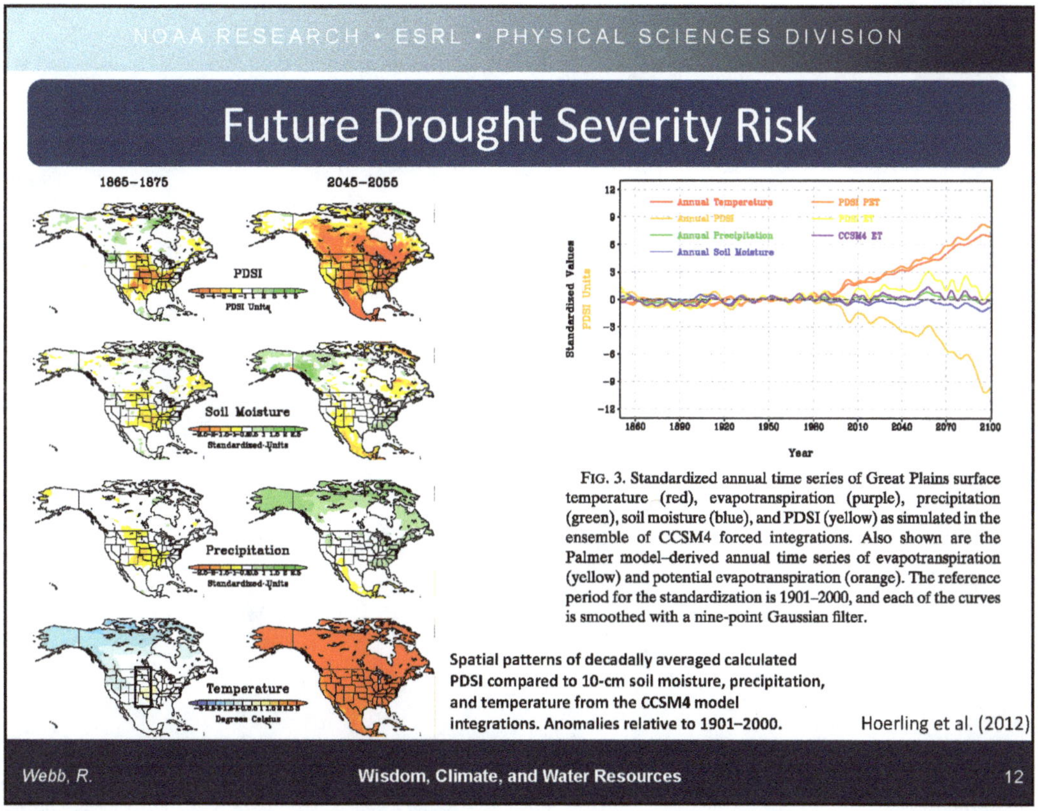

Slide 13

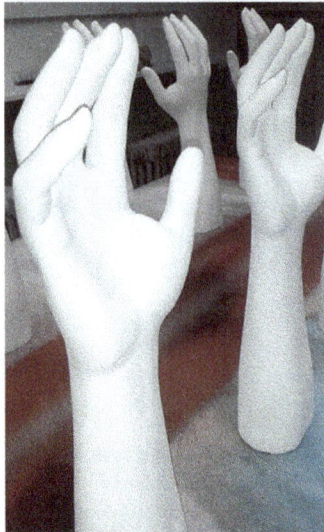

Slide Notes

Slide 3 Model simulations revealed that even with a reduction in summer and fall insolation by a factor of 5, there was no suggestion of ice-sheet growth and there was not a year-to-year accumulation of snow. Furthermore, examination of the energy and mass budgets simulated in the model experiment revealed an imbalance where the prescribed ice slab was added, it neither grew nor was sustained, and, in fact, would have been completely removed in about five years. Only with the additional forcings of sea surface temperatures and atmospheric CO_2 concentrations prescribed at full glacial levels, was the climate model able to simulate the maintenance of the slab ice sheet in limited regions. The inability of the climate model to simulate changes observed in the geologic record in response to sledge hammer magnitude prescribed forcings raised concerns with the climate model sensitivity to know past, and perhaps future forcings, suggesting a much more complex response of mechanisms and feedbacks in the integrated earth system.

Slide 4 Examination of the global and regional energy balances revealed a different possible response to increase atmospheric dust forcing in a glacial climate than in our modern climate. Instead of cooling due primarily backscattering of shortwave radiation, when over high albedo snow and ice covered surfaces the impact of the increase in dust on planetary albedo is minimal, while the greenhouse warming through the absorption of longwave radiation is relatively speaking, significant, resulting in regional warming. While not the hypothesized cause and effect, careful analysis provided an elegant but simple explanation while reminding the climate science community that climate sensitivity and responses are, and will be, more complicated in environments dominated by snow and ice.

Slide 5 Rind *et al.* (1991) analysis of SAGE II water vapor data illustrating climate models were not overestimating the water vapor feedback. Analysis of the SAGE II was conclusive in disproving a proposed mechanism that increased convection in a warmer climate would not increase the amount of moisture but instead dry the middle and upper troposphere by means of associated compensatory subsidence.

Slide 6 David had a balanced appreciation of science and did not allow his scientific work to be an all-consuming part of his life.

Slide 7 These model simulations revealed a significant increase in surface air temperatures for June, July, and August, reductions in precipitation concentrated in June and July, a decrease in evaporation (latent heat flux), and an increase in sensible heat flux. The results of this seasonal predictability modeling study identified soil moisture as a useful antecedent condition for developing improved summer temperature forecasts.

Slide 8 Model experiments were run changing the rate and process of exchanging water between ground levels, specifying water holding capacities for desert to all vegetation types, and assigning desert ground albedo values for all vegetation types. Analysis of these experiments reveals that the change in surface albedo and water holding capacity results in similar impacts with regional variability. The albedo impact was greater in wet regions and during seasons when large-scale dynamics or convection dominates precipitation, whereas impact of a change in water holding capability was more important in regions where local evaporation processes played an important role in influencing precipitation. Altering the rate and process of water exchange between ground layers predominantly resulted in shifts in seasonality within the annual cycle. Overall, the conclusions of this study underscored the importance of accurately representing the role of vegetation and more generally the land surface in modeling the global climate system.

Slide 9   Analyses showed that factors impacting the hydrologic cycle included: (1) changes in large-scale dynamics and atmospheric moisture convergence effects at a variety of spatial scales, (2) accurate simulation of current climate conditions was important with a bias to either too warm or too wet exaggerating responses, (3) increased atmospheric moisture transport of excess energy from the tropics to higher latitudes, and (4) the extent of polar amplification of temperature change. Although sensitive to model resolution, results for the western USA were projected to get wetter, whereas the central USA response showed a response to SST gradients and the amount of mid-latitude SST warming. Results in the eastern USA were for a projected decrease in soil moisture but dependent on overall climate model sensitivity. Overall, the conclusions of this study underscored the need to better represent ground hydrology and ocean circulation in high-resolution climate models to provide more realistic estimates of projected risks to future water availability.

Slide 10   The projected increase in drought conditions resulted from large increases in potential evapotranspiration with the largest increases in low-to-mid-latitude regions. Higher latitude regions were projected to have increases in precipitation in response to a warmer atmosphere being able to hold and deliver more moisture. Model results also suggested that global precipitation would be unable compensate for increased terrestrial evaporative demands as surface warming over land was projected to be greater than over the ocean. PDSI-based severe drought (5% frequency under current climate) was projected to occur about 50% of the time by the mid-21st century under current greenhouse gas emission scenarios and thus drought conditions will increase dramatically.

Slide 11   The primary conclusion from this study was that inconsistent hydrologic responses among various models used to simulate land-surface conditions and fluxes were problematic in assessing future risks and impacts. Outstanding challenges identified by Rind *et al.* (1992) included (1) a basic need to better understand hydrologic processes, (2) climate model projections of future temperature and precipitation being highly dependent on the ability of ground hydrology scheme to correctly simulate current soil moisture and runoff, (3) calculating runoff and soil moisture infiltration at the course spatial scales of GCMs, (4) resolving the proper way to calculate potential evapotranspiration, and (5) the efficacy of using offline impact models to estimate surface hydrology responses.

Slide 12   While PDSI remains an excellent proxy indicator for soil moisture variations in the 20th century, with the simplified formulation of potential evapotranspiration, PDSI severely overstates future surface water imbalances and implied vegetation stresses. Our study illustrated that PSDI was excessively sensitive to the rise in surface air temperatures, inflated the implied evaporation of moisture from soils, and completely changed the physics of the surface-moisture balance relative to the 20th century. A similar dramatic change was not revealed in our analysis of land-surface fluxes and soil moisture response calculated directly by the climate model. Thus, consistent with the Rind *et al.* (1992) findings, we concluded that PDSI overstates the future drought severity risk over the north central USA because of an unrealistic sensitivity to surface warming. While the PDSI works well as long as climate remains stationary and no warming trends are involved, the PDSI ceases to be a reliable indicator of drought and implied stress on water resources in a warming climate of the 21st century.

Slide 13   Remembering to take the time to episodically assess whether we are doing something productive is a good practice to consider both in one's science and more generally in living life.

CLIMATE LECTURE 10

Soil Moisture in the Climate System

Randal Koster

Global Modeling and Assimilation Office, NASA/GSFC, Greenbelt, MD, USA

Randal Koster is a Senior Research Scientist at NASA/GSFC. His research has focused on the development of improved treatments of land surface physics for Earth system models and on the analysis of interactions between the land and atmosphere, using these models. He coordinates the land surface modeling activity for the GSFC Global Modeling and Assimilation Office.

Soil Moisture — Atmosphere Feedback

The moisture held within the top meter or two of soil is a very tiny fraction (less than 0.01%) of the Earth's total water (Eagleson, 1970). Nevertheless, its presence at the interface of the land and atmosphere gives it inordinate importance in the context of climate variability. Simply put, soil moisture variations can help determine meteorological variations. Consider, for example, an anomalously high evapotranspiration rate induced by a high soil moisture content. The high evapotranspiration can lead to an anomalously cooled land surface and thus cooler air temperatures (Seneviratne *et al.*, 2010), and it can also lead to modifications in the evolution of the boundary layer, with concomitant impacts on the generation of convective rainfall (Betts *et al.*, 1994).

Given this potential for feedback on the atmosphere, soil moisture is particularly important in the context of prediction. Atmospheric physics and dynamics are 'fast' and contribute little to the lifetime of an atmospheric anomaly, as reflected in the well-known time scale of typical weather forecasts — these rely heavily on atmospheric initialization and are valid for only a week or so. Soil moisture processes, in contrast, are relatively 'slow'. A soil moisture anomaly may persist for a month or more and is thus predictable at such time scales. As a result, and because (as noted above) the atmosphere may respond in a predictable way to a given soil moisture anomaly, aspects of the atmosphere may also be predictable at the monthly time scale (NRC, 2010).

The degree to which soil moisture influences the atmosphere has been studied extensively within atmospheric general circulation models. The last several decades has indeed seen a variety of approaches to quantifying the feedback. Various studies have examined: (i) the ability of 'perfectly forecasted' soil moisture to help reproduce observed extreme events, such as the US drought of 1988 (e.g., Atlas *et al.*, 1993); (ii) the impact of 'perfectly forecasted' soil moisture on the statistics of interannual rainfall and air temperature variations (e.g., Delworth and Manabe, 1989); (iii) the impact of large and idealized initial soil moisture conditions on subsequent model precipitation (e.g., Rind, 1982); (iv) idealized predictability studies using non-extreme initial conditions that are not tied to observations (e.g., Schlosser and Milly, 2002); and (v) perhaps most relevant to forecast applications, the ability of realistically initialized soil moisture fields to contribute to the skill of air temperature and precipitation forecasts (e.g., Koster *et al.*, 2011). Taken together, these studies illustrate clearly that soil moisture–atmosphere feedback is an important facet of the simulated climate system.

Unfortunately, performing corresponding analyses with observational data — and thus demonstrating that soil moisture–atmosphere feedback is just as important in the real world — is very difficult. Quantifying directly the degree to which soil moisture affects, for example, precipitation is made almost impossible by the fact that the reverse direction of causality is so much stronger — any covariation between precipitation and soil moisture seen in the observations is mostly a simple reflection of the fact that precipitation tends to wet the soil. Investigators have tried using lead–lag relationships to get at the former, far more subtle connection (e.g., Findell and Eltahir, 1997), but with only limited success. Nevertheless, available observations do provide multiple *indirect* demonstrations that soil moisture affects atmospheric variability — not the least being the demonstrations that realistic initializations of soil moisture in a forecast system contribute to the skill (when compared against observations) of atmospheric forecasts (Koster *et al.*, 2011).

Soil Moisture in Land Surface Models

Taking advantage of soil moisture in a monthly-to-seasonal forecast system generally requires the initialization of soil moisture in the system's land surface model. This brings up the general issue of modeled soil moisture — how is it defined, and how can it be initialized with observations?

The definition of a land model's soil moisture variable is far less straightforward than many climate scientists assume. Regardless of the variable name employed or the units assigned to it in metadata, soil moisture is not a quantity that can be directly compared to an observed soil moisture. Given the spatial nonlinearities in the land system and the need to represent the spatial distribution of soil moisture with a limited number of prognostic variables, and given the competing and overwhelming impacts of crude evapotranspiration and runoff parameterizations on soil moisture magnitudes, soil moisture in a land model is best thought of as a model-dependent 'index of wetness', an index that increases as conditions get wetter. Different models utilize different evapotranspiration and runoff assumptions and thereby generate different climatologies of soil moisture variability — they effectively use different indices (Koster, 2015). Blindly translating soil moisture from one model (e.g., from a trusted reanalysis product) to another can lead to disastrous results; the translation can be done sensibly, but only through careful consideration of each model's distinct soil moisture climatology (Koster *et al.*, 2009).

It would likewise be a mistake to initialize a land model directly with observed, areally-averaged soil moisture values. (Note that these observations do not exist anyway; *in situ* measurements are highly localized, and satellite-based measurements only pertain to the top few centimeters of soil at most.) One reasonably safe way to initialize soil moisture in a forecast system is to drive the system's land model offline with observations-based meteorological forcing for a long period of time (years) up to the forecast start date (along the lines of Xia *et al.*, 2012). Through this process, the land model develops soil moisture states that are consistent with its own internal climatology and that reflect reasonably well the antecedent meteorology — a rainy April, for example, would lead to a wet (for the model) May 1 state that, in turn, could be used as initial conditions for a May forecast. This is the approach taken in several of today's operational forecasting centers.

Data assimilation builds on this approach. The land model is driven offline with observation-based forcing as before, but now satellite-based surface soil moisture data are assimilated into the modeling system during the model integration, taking into account differences in the underlying climatologies of the two data sources. In essence, data assimilation combines the model data and the satellite data in a mathematically optimal way, producing a merged dataset that is superior to either component dataset (e.g., Reichle *et al.*, 2014). New L-band satellite soil moisture sensors are providing unprecedented information on soil moisture contents around the world, effectively paving the way, through data assimilation, towards a Renaissance of soil moisture estimation (Entekhabi *et al.*, 2010; Kerr *et al.*, 2010).

Contributions of David Rind to Studies of Soil Moisture in the Climate System

In 1982, David Rind published a paper in *Monthly Weather Review* entitled 'The influence of ground moisture conditions in North America on summer climate as modeled in the GISS GCM' — the first paper I am aware of that examined the impact of soil moisture initialization on the subsequent simulation of rainfall, and one of the first to consider any role at all for soil moisture in the climate system. I have built my own career around this topic, and I want to thank David for pioneering it.

Of course, the soil moisture–atmosphere connection is only one small part of the myriad connections that define the climate system. David's contributions over the years impressively span the science. David, it seems to me, is the rare scientist who has a grasp for all of the connections that together define climate. Through his broad understanding of climate, he is able to integrate the various focused contributions provided by the rest of us into a solid, insightful whole.

References

Atlas, R., Wolfson, N. and Terry J. (1993). The effect of SST and soil moisture anomalies on GLA model simulations of the 1988 U.S. summer drought. *J. Climate*, 6, 2034–2048.

Betts, A.K., Ball, J.H., Beljaars, A.C.M., Miller, M.J. and Viterbo, P. (1994). Coupling between land-surface, boundary-layer parameterizations and rainfall on local and regional scales: Lessons from the wet summer of 1993. Preprints, Fifth Conf. on Global Change Studies, 74th Annual Meeting, Nashville, TN, Amer. Meteor. Soc, Boston, MA, 174–181.

Delworth, T.L. and Manabe, S. (1989). The influence of soil wetness on near-surface atmospheric variability. *J. Climate*, 2, 1447–1462.

Eagleson, P. (1970). Dynamic Hydrology. McGraw-Hill, New York.

Entekhabi, D., and Co-authors (2010). The Soil Moisture Active Passive (SMAP) mission. *Proc. IEEE*, 98, 704–716.

Findell, K.L. and Eltahir, E.A.B. (1997). An analysis of the soil moisture-rainfall feedback, based on direct observations from Illinois. *Water Resour. Res.*, 33, 725–735.

Kerr, Y.H. and Co-authors (2010). The SMOS mission: New tool for monitoring key elements of the global water cycle. *Proc. IEEE*, *98*, 666–687, doi:10.1109/JPROC.2010.2043032.

Koster, R.D., Guo, Z., Yang, R., Dirmeyer, P.A., Mitchell, K. and Puma, M.J. (2009). On the nature of soil moisture in land surface models. *J. Climate*, 22, 4322–4335.

Koster, R.D. and Co-authors (2011). The second phase of the Global Land-Atmosphere Coupling Experiment, Soil moisture contributions to subseasonal forecast skill. *J. Hydromet.*, 12, 805–822.

Koster, R. (2015). Efficiency space, a framework for evaluating joint evaporation and runoff behavior. *Bull. Am. Met. Soc.*, 96, 393–396.

National Research Council (2010). Assessment of intraseasonal to interannual climate prediction and predictability. The National Academies Press, Washington, D.C, 181 pp.

Reichle, R.H., De Lannoy, G.J.M., Forman, B.A., Draper, C.S. and Liu, Q. (2014). Connecting satellite observations with water cycle variables through land data assimilation: Examples using the NASA GEOS-5 LDAS. *Surv. Geophys.*, 35, 577–606, doi:10.1007/s10712-013-9220-8.

Rind, D. (1982). The influence of ground moisture conditions in North America on summer climate as modeled in the GISS GCM. *Monthly Weather Review*, 110(10), 1487–1494.

Schlosser, C.A. and Milly, P.C.D. (2002). A model-based investigation of soil moisture predictability and associated climate predictability. *J. Hydrometeor.*, 3, 483–501.

Seneviratne, S.I., Corti, T., Davin, E.L., Hirschi, M., Jaeger, E.B., Lehner, I., Orlowsky, B. and Teuling, A.J. (2010). Investigating soil moisture-climate interactions in a changes climate, A review. *Earth-Science Reviews*, 99, 125–161.

Xia, Y. and Coauthors (2012). Continental-scale water and energy flux analysis and validation for the North American Land Data Assimilation System project phase 2 (NLDAS-2): 1. Intercomparison and application of model products. *J. Geophys. Res.*, 117, D03109, doi:10.1029/2011JD016048.

Slide 1

Soil Moisture in the Climate System

Lecture slides by Randal Koster, Global Modeling and
Assimilation Office, NASA/GSFC, Greenbelt, MD, USA
randal.d.koster@nasa.gov

Our Warming Planet: Topics in Climate Dynamics
Lectures in Climate Change, Vol. 1
2017

Slide 2

Soil Moisture in the Climate System

The moisture in the top two meters or so of soil is a tiny fraction (<0.01%) of the Earth's water. However, because it lies at the *interface* between the land and atmosphere, it has an *inordinate impact* on climate and its variability.

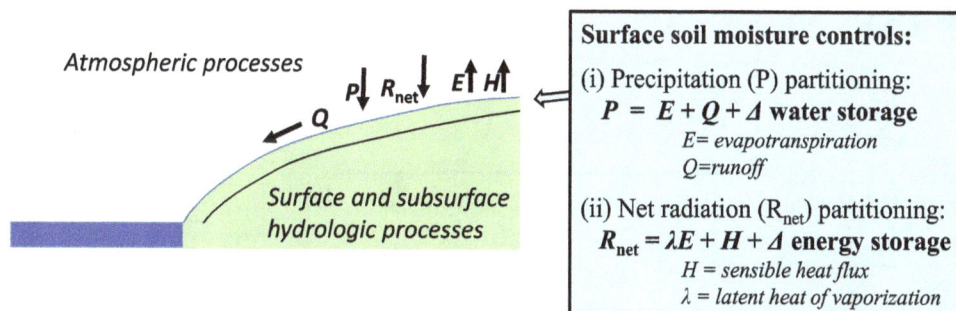

Atmospheric processes

P R_{net} E H

Q

Surface and subsurface hydrologic processes

Surface soil moisture controls:

(i) Precipitation (P) partitioning:
$P = E + Q + \Delta$ **water storage**
  E= evapotranspiration
  Q=runoff

(ii) Net radiation (R_{net}) partitioning:
$R_{net} = \lambda E + H + \Delta$ **energy storage**
  H = sensible heat flux
  λ = latent heat of vaporization

Slide 3

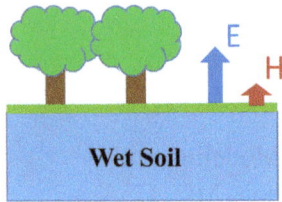

Slide 4

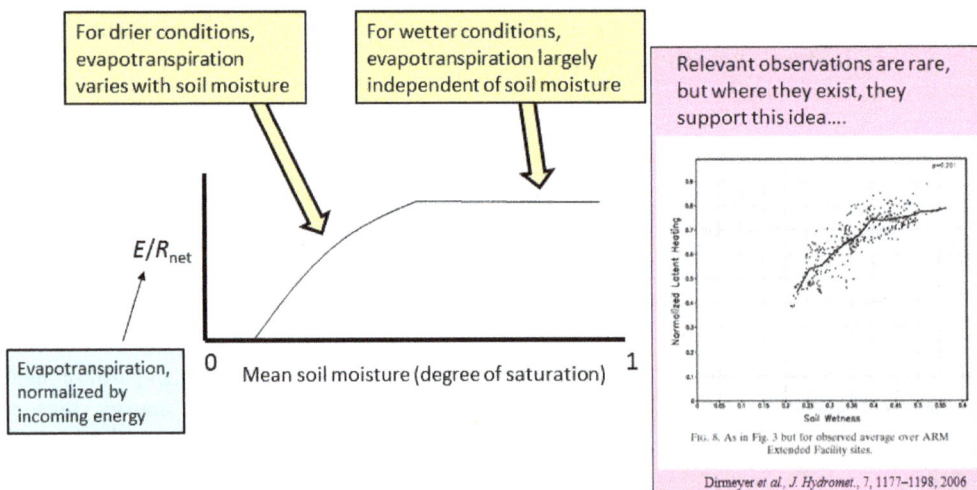

Slide 5

Because of this relationship, the connection between soil moisture and the atmosphere (through the former's effect on evapotranspiration) is strongest in the transition zones between dry and wet areas.

Shown here: results from the multi-model GLACE experiment. Indicated is where soil moisture variability helps guide short-term boreal summer rainfall variability.

Why are the transition areas important? See next slide, which focuses on North America…

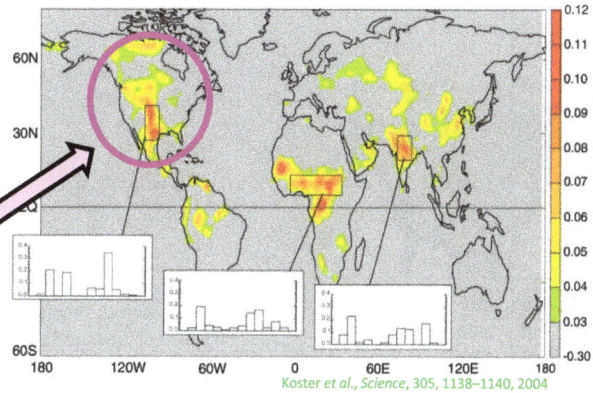

Koster et al., Science, 305, 1138–1140, 2004

Slide 6

Explanation for Why Soil Moisture Feedback on the Atmosphere is Strongest in Transition Zones

When it is really dry, E is too small (and varies too little) to have an effect.

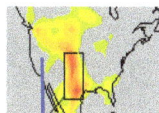

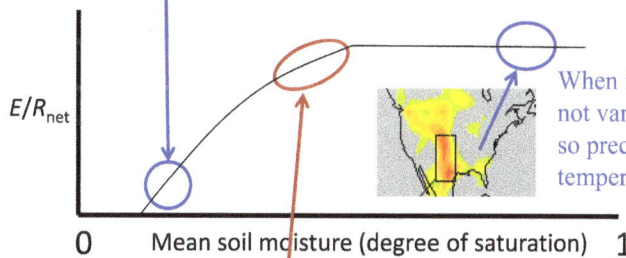

E/R_{net}

When it is really wet, E does not vary with soil moisture, so precipitation and temperature cannot, either.

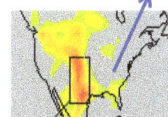

0   Mean soil moisture (degree of saturation)   1

You mainly get an impact in the "sweet spot" in between:

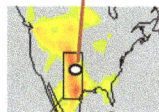

…in the transition zone, where E does vary with soil moisture and E is significantly large.

Slide 7

Direct evapotranspiration measurements are too sparse to show this pattern. However, the pattern **is** seen in measurements of E-related proxy variables.

Wanted: indication of where interannual variability of E is large. Plot instead: interannual variability of well-measured variables that vary with E. (Plot $\sigma^2_X Corr^2(X,W)$, where X is the proxy variable and W characterizes water availability.)

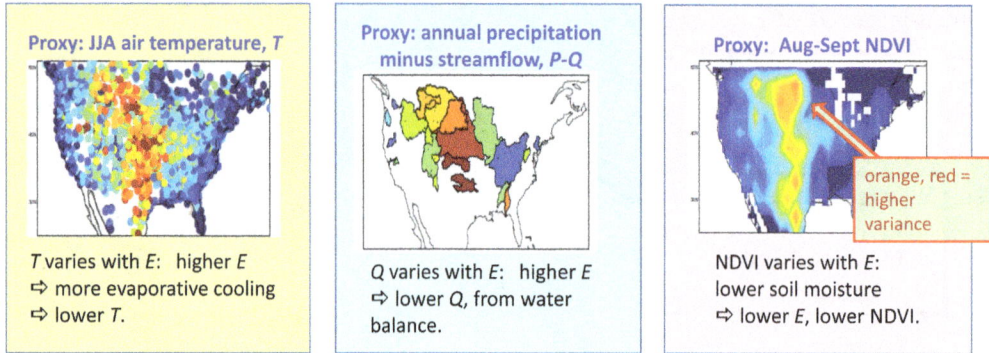

Proxy: JJA air temperature, *T*

T varies with E: higher E
⇨ more evaporative cooling
⇨ lower T.

Proxy: annual precipitation minus streamflow, *P-Q*

Q varies with E: higher E
⇨ lower Q, from water balance.

Proxy: Aug-Sept NDVI

orange, red = higher variance

NDVI varies with E:
lower soil moisture
⇨ lower E, lower NDVI.

Koster *et al., Frontiers in Earth Science*, 2015

Koster, R.   Soil Moisture in the Climate System   7

Slide 8

Soil Moisture in a Changing Climate

Climate change may lead to a shift in a region's hydroclimatic regime.

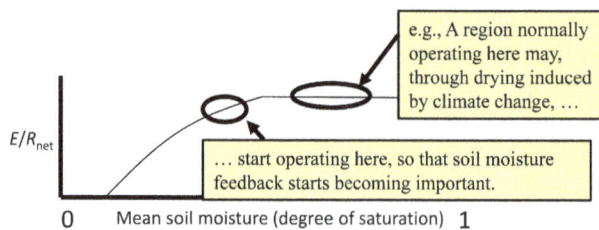

E/R_{net}

e.g., A region normally operating here may, through drying induced by climate change, …

… start operating here, so that soil moisture feedback starts becoming important.

0   Mean soil moisture (degree of saturation)   **1**

Models suggest such a shift in Europe, with soil moisture having a larger impact on climate variability during future climate.

σ_T, JJA

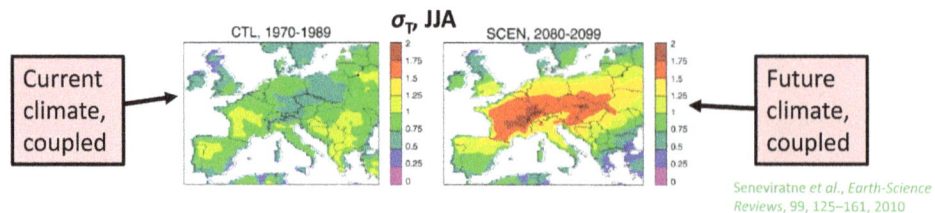

CTL, 1970-1989   SCEN, 2080-2099

Current climate, coupled

Future climate, coupled

Seneviratne *et al., Earth-Science Reviews*, 99, 125–161, 2010

Koster, R.   Soil Moisture in the Climate System   8

Slide 9

Another relevant property of soil moisture: "memory". If the soil is anomalously wet today, it will probably be anomalously wet tomorrow, next week, and maybe even next month.

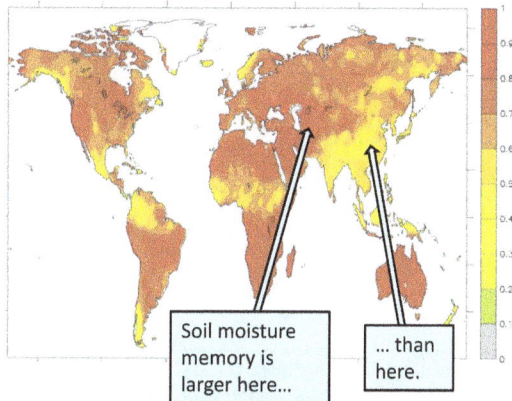

↩ Multi-model estimate of 27-day-lagged autocorrelation of soil moisture (boreal summer).

Observed estimates for individual locations are sparse, but where they exist, the models roughly agree with them.

Seneviratne *et al., J. Hydromet.,* 7, 1090–1112, 2006

Koster, R.   Soil Moisture in the Climate System   9

Slide 10

This memory, in combination with soil moisture's ability to feed back on the atmosphere, has an important implication:

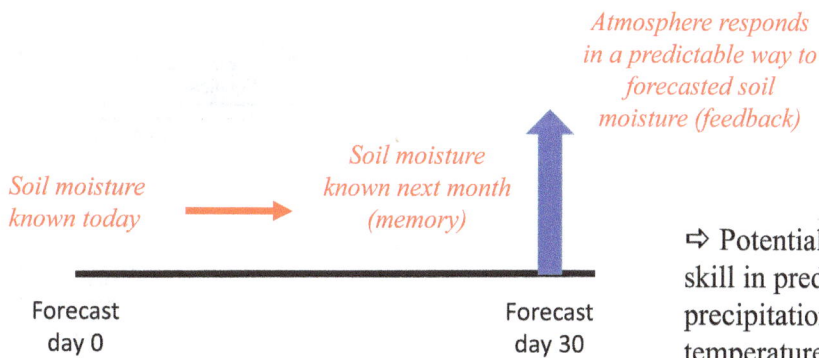

Atmosphere responds in a predictable way to forecasted soil moisture (feedback)

Soil moisture known today

Soil moisture known next month (memory)

Forecast day 0

Forecast day 30

⇨ Potential for some skill in predicting precipitation and temperature at 30-day (or more) leads!

Koster, R.   Soil Moisture in the Climate System   10

Slide 11

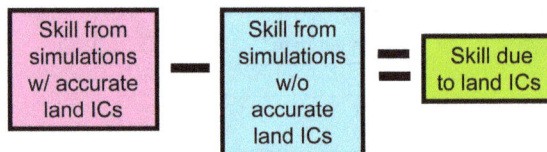

Estimations of Forecast Skill Associated with Soil Moisture Initialization

GLACE-2: The second phase of the Global Land-Atmosphere Coupling Experiment (an international, multi-institution project)
Koster *et al., J. Hydromet.*, 12, 804-822, 2011

Gist of experiment:

1. Perform two sets of forecast simulations:
   (i) <u>with</u> accurate soil moisture initial conditions (ICs)
   (ii) <u>without</u> accurate soil moisture ICs

2. Compare forecasted *P*, *T* to obs.

3. Compute soil moisture contribution to forecast skill:

| Skill from simulations w/ accurate land ICs | − | Skill from simulations w/o accurate land ICs | = | Skill due to land ICs |

Koster, R.   Soil Moisture in the Climate System   11

Slide 12

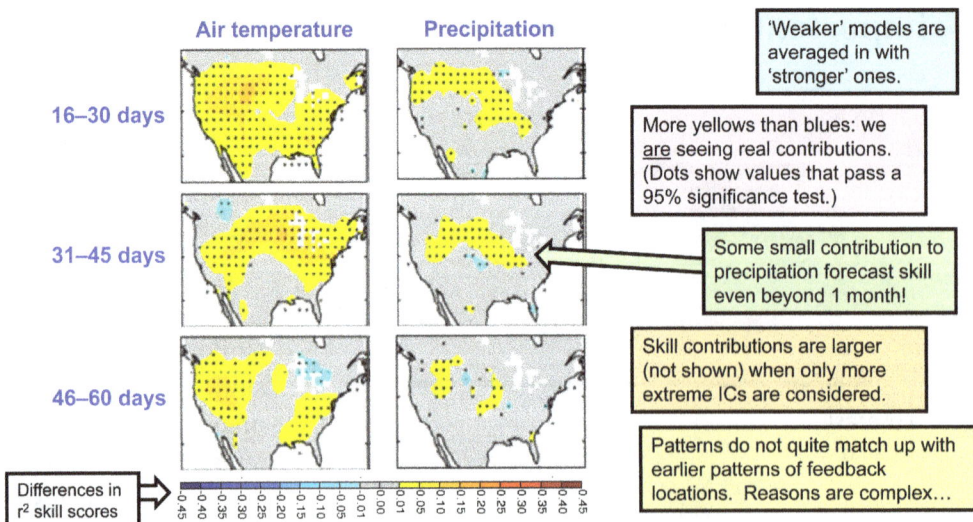

GLACE-2 Results: 'Consensus' Skill Contribution from Land Initialization (JJA)

Air temperature   Precipitation

16–30 days

31–45 days

46–60 days

Differences in r² skill scores
-0.45 -0.40 -0.35 -0.30 -0.25 -0.20 -0.15 -0.10 -0.05 -0.01 0.00 0.01 0.05 0.10 0.15 0.20 0.25 0.30 0.35 0.40 0.45

'Weaker' models are averaged in with 'stronger' ones.

More yellows than blues: we <u>are</u> seeing real contributions. (Dots show values that pass a 95% significance test.)

Some small contribution to precipitation forecast skill even beyond 1 month!

Skill contributions are larger (not shown) when only more extreme ICs are considered.

Patterns do not quite match up with earlier patterns of feedback locations. Reasons are complex…

Koster, R.   Soil Moisture in the Climate System   12

Slide 13

Soil Moisture and Streamflow Forecast Skill

Snow (or rainfall) over <u>wet</u> soil: most of the meltwater runs off into streams, reservoirs.

Snow (or rainfall) over <u>dry</u> soil: much of the meltwater infiltrates the soil and is lost to evapotranspiration ⇨ less to streams, reservoirs.

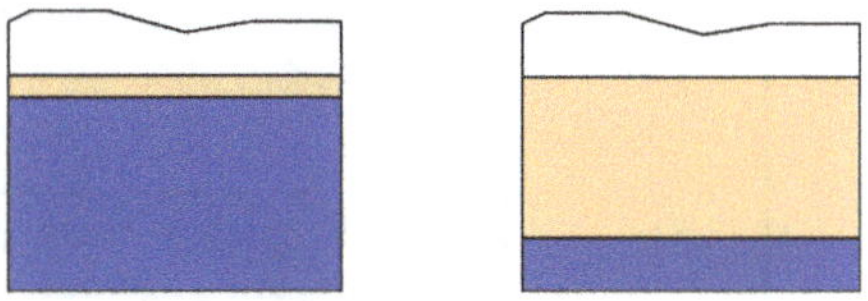

Forecast experiment: compute skill of March–July streamflow forecasts (vs. obs) associated solely with accurate soil moisture initialization:

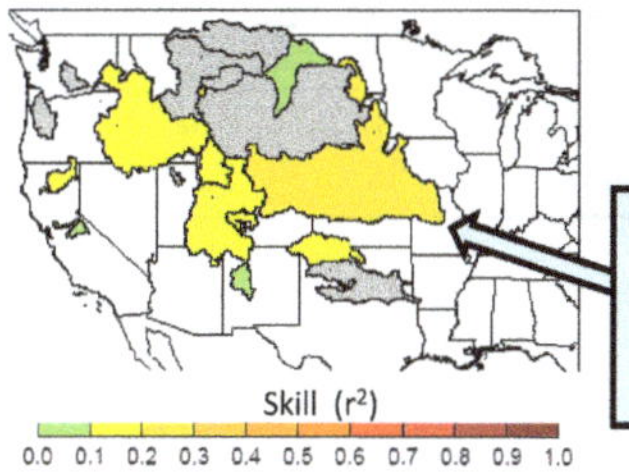

Knowledge of winter soil moisture, coupled with soil moisture memory, provides springtime streamflow forecast skill.

Real skill found in many basins!

Skill (r²)

0.0 0.1 0.2 0.3 0.4 0.5 0.6 0.7 0.8 0.9 1.0

Koster et al., Nature Geoscience, 3, 613–616, 2010.

Koster, R.   Soil Moisture in the Climate System   13

Slide 14

Soil Moisture in Earth System Models

How can we initialize soil moisture, e.g., for forecasts? From direct, in situ observations? *No. In situ observations (certainly across large areas) are sparse to non-existent.*

Modeling approach: Force a gridded array of land surface model elements with arrays of observations-based meteorological forcing ⇨ let modeled soil moistures evolve in response to the forcing.

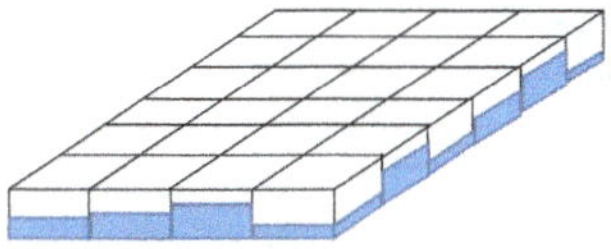

Force array of land models with arrays of P, R_{net}, etc.

Compute E, Q, etc. at each land element; let all computed fluxes update soil moisture

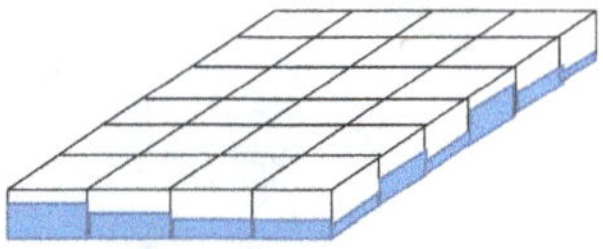

Array of soil moistures at beginning of time period (crude representation)

Array of soil moistures at end of time period: reflects antecedent meteorology

Koster, R.   Soil Moisture in the Climate System   14

Slide 15

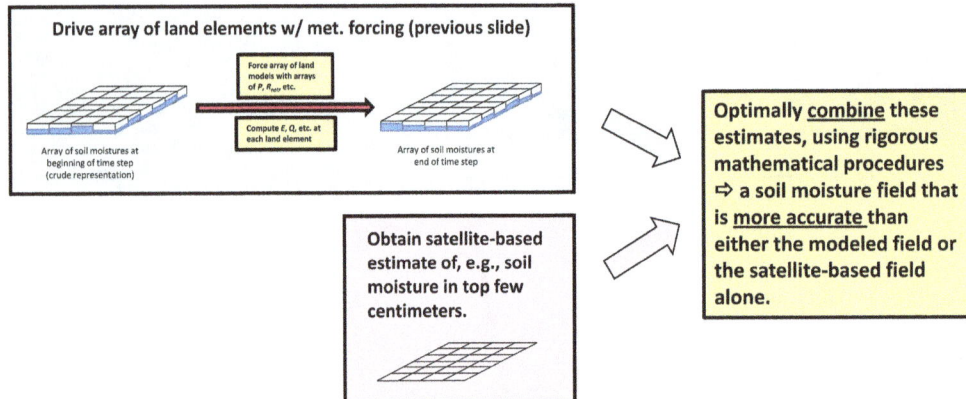

Slide 16

Slide 17

Soil Moisture in Earth System Models: Some Caveats

As discussed earlier, evapotranspiration is controlled by soil moisture.

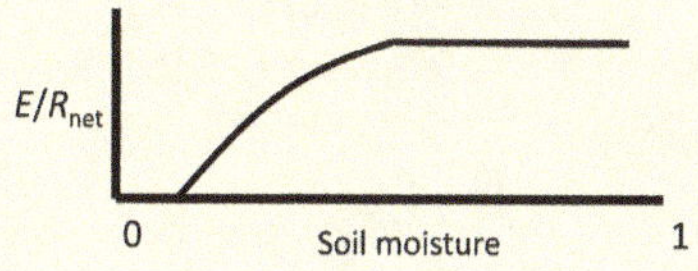

Similarly, the ratio of runoff to precipitation is controlled by soil moisture.

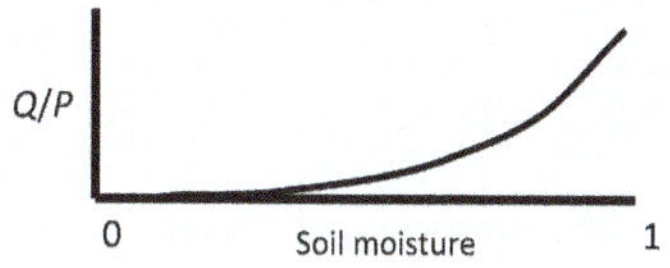

A land surface model has multiple complex formulations that interact with each other, implicitly producing some form of these two relationships (approximate–scatter will be large around them).

It is the **interaction** between these underlying relationships that determines how a land surface model behaves. ⇨

Important Modeling Take-home Lesson:
A 'perfect' evapotranspiration formulation does not imply accurate evapotranspiration rates. Accuracy in E is limited by the weaker of the evapotranspiration and runoff formulations. Model development requires suitable **balance**.

Koster and Mahanama, *J. Hydromet.*, 13, 1604–1620, 2012.
Koster, *Bull. Am. Met. Soc.*, 96, 393–396.

Koster, R. Soil Moisture in the Climate System 17

Slide 18

Also, soil moisture in a land surface model is a very model-dependent variable, more an 'index of wetness' than an actual physical quantity.

⇩

Another Important Modeling Take-home Lesson:
Taking soil moisture from one model (e.g., from a reanalysis) and using it in another can lead to serious problems. Scaling is essential.

Demonstration of problem: seasonal precipitation forecast skill

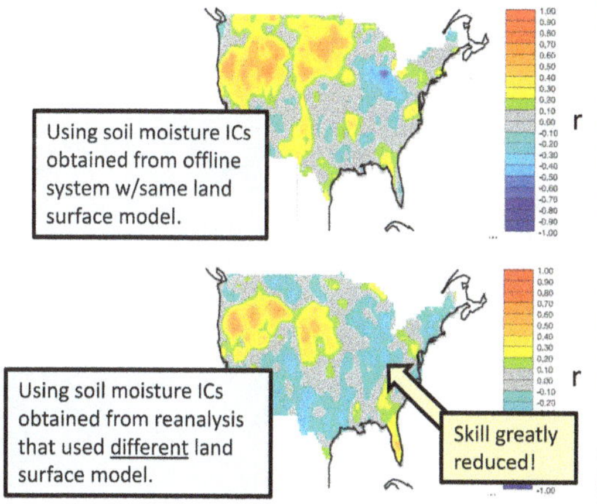

Koster *et al.*, *J. Climate*, 22, 4322–4335, 2009.

Koster, R. Soil Moisture in the Climate System 18

Slide 19

A Few Summary Statements

Variability in near-surface soil moisture can causally translate to variability in meteorological variables through the influence of soil moisture on the energy and water budgets at the land-atmosphere interface, particularly in transition zones.

This potential for feedback, combined with soil moisture's intrinsic memory, provides the basis for utilizing soil moisture information in monthly-to-seasonal forecasts.

The outlook for improved soil moisture estimation (e.g., for initializing forecasts) is optimistic given the advent of new satellite data, which can be assimilated into land surface models. Use of models, however, comes with important caveats:

- A model's overall performance is only as good as the weaker of its evapotranspiration and runoff formulations (not often appreciated!)
- Inserting soil moisture from one model (even from a trusted reanalysis) into another without performing climatology transformations can lead to significant problems (appreciated even less!)

Koster, R. Soil Moisture in the Climate System 19

Slide Notes

Slide 4   'Normalized latent heating' in the observations panel is the same as E/R_{net}. These observations represent averages over ARM Extended Facility Sites (see Dirmeyer *et al.*, 2006).

Slide 5   In GLACE, each participating model isolated soil moisture impacts on precipitation or temperature variability by prescribing time series of soil moisture across an ensemble of simulations (see Koster *et al.* reference). Shown here are the average results over a dozen participating models.

Slide 7   Temperature data are from GHCN stations. Runoff data are naturalized and collected from various sources (see Mahanama *et al.*, *Journal of Hydrometeorology*, 13, 189–203, 2012). NDVI (Normalized Difference Vegetation Index, a measure of how green and "leafy" the vegetation is) is from the GIMMS dataset. W, used to characterize water availability, is the October–September precipitation at the location plotted. The plotted quantity is, in effect, the variance of the given variable "explained" by variations in W.

Slide 8   The lower plots indicate an increase in temperature variability shown by models to be induced by a future increase in land-atmosphere feedback.

Slide 9   The autocorrelations plotted are, in effect, measures of soil moisture memory. These data were in fact derived from the GLACE experiment products.

Slide 11   Accurate land ICs come from offline simulations of the land models utilizing observations-based meteorological forcing (see the later slide).

Slide 12   To compute these skill increases, all model forecasts were considered together against observations; results for individual models vary, with larger predictive skill for some.

Slide 13   Soil moisture in fact leads to even higher skill contributions in other seasons; see Mahanama *et al.*, *Journal of Hydrometeorology*, 13, 189–203, 2012.

Slide 16   SMOS, with ~40 km resolution, was launched in November, 2009; SMAP was launched in January, 2015. The high resolution radar instrument on SMAP is no longer functioning; the radiometer on SMAP, however, is still providing valuable 36 km data.

Slide 18   These plots were computed from a GLACE-2 type study with a single forecasting system. Use of reanalysis soil moisture as the 'truth' with which to initialize an independent forecast system is a common enough mistake that leads to reduced skill. Before translating a reanalysis soil moisture into an independent land model, the climatologies of both the reanalysis and the independent land model must be established. Using these distinct climatologies, a given anomaly in the reanalysis must be scaled to a corresponding anomaly appropriate for the independent land model. (See the listed reference.)

CLIMATE LECTURE 11

Projections of Future Drought

Jennifer Aminzade

Jennifer Aminzade has a Ph.D. degree in Earth Sciences from Columbia University (2011). Her principal area of expertise is the analysis of GCM model output, in particular aspects related to water availability. Dr Rind was her thesis advisor.

The Definition of Drought

The Intergovernmental Panel on Climate Change (IPCC) reported widespread agreement among climate model projections on emerging patterns in the global hydrologic cycle in the 21st century, such as increased precipitation at high latitudes, decreased precipitation in the subtropics, and increases in the frequency of heavy precipitation events, all of which would lead to permanent changes in annual and seasonal runoff and groundwater recharge (IPCC, 2007, 2013). These changes in the hydrologic cycle could fundamentally alter the nature of drought and flooding in many regions. In areas that are currently using a large proportion of their available water resources, these changes could prove challenging.

In addition to precipitation changes in the 21st century, rising temperatures are likely to impact evaporation as well (IPCC, 2007). Simply looking at the change in precipitation does not take in to account changes in evapotranspiration or vegetation, which are sensitive to temperature changes and therefore likely to influence 21st-century climate. According to the Clausius–Clapeyron equation, every 1°C increase in temperature increases the water-holding capacity of the atmosphere (instantaneously) by approximately 7% (IPCC, Climate Change 2007: Working Group I: The Physical Science Basis, "FAQ 3.2 How is Precipitation Changing?", URL http://www.ipcc.ch/publications_and_data/ar4/wg1/en/faq-3-2.html). As temperatures increase, the atmosphere is able to hold more water vapor, and thus a global increase in atmospheric moisture demand is projected. Today's water availability deficits often result from decreased precipitation. Future droughts may occur primarily because of temperature increases, and may occur even where precipitation is increasing. This is why looking at projected changes in precipitation alone is not sufficient to characterize droughts.

There are many ways to define drought. A study by Burke and Brown (2008) found significant differences in doubled CO_2 drought projections when using different measures. This lecture continues in the same vein and focuses on two types of water availability measures: soil moisture and drought index values, specifically the Supply Demand Drought Index (SDDI) (Rind *et al.*, 1990), and the Palmer Drought Severity Index (PDSI) (Palmer, 1965).

The SDDI is calculated using monthly precipitation and temperature, which are available for most of the past century over much of the world. The PDSI requires additional inputs that are often unavailable globally. Although efforts have been made to create a PDSI that could be used globally, Wells, Goddard and Hayes (2004) created a self-calibrating PDSI (SC-PDSI) that replaces the empirically derived constants in the formulation with dynamically calculated values.

Soil moisture measurements have proven to be elusive. Measuring soil moisture is a difficult task since values can vary significantly over small areas. Furthermore, many countries do not share hydrologic data. Thus, it is difficult to compile global datasets that are not sparse or infrequently sampled. This has led the agricultural industry to rely heavily on drought indices (Robock *et al.*, 1998; Henderson-Sellers *et al.*, 2002, 2003; Varis, Kajander and Lemmela, 2004). Today, when using global climate models (GCMs) to simulate future climate, we have both soil moisture projections and the variables needed to compute drought indices, for studying future drought.

Soil Moisture

In theory, soil moisture provides the most essential information needed for drought determination. However, there is a lack of long-term (i.e., >50 years) global soil moisture measurement. Not only are global soil moisture measurements lacking, meaningful local ones are also hard to come by. Measuring soil moisture is extremely difficult because of the great heterogeneity that exists on small spatial scales. Passive radiometers on satellites can observe large areas but higher spatial resolution requires lower orbits and large diameter antennas, both require more fuel to stay in space. Hence, most soil moisture sensors (e.g., European Remote Sensing satellites/Meteorological Operational satellite program (ERS/MetOp, the Advanced Microwave Scanning Radiometer for EOS, and the Soil Moisture and Ocean Salinity) provide spatial resolutions in tens of kilometers that are too coarse for catchment hydrology applications (e.g., Lakshmi, 2013) and difficult to compare with the heterogeneous *in-situ* observations. Separating the contribution from vegetation layers above the soil remains an area of research. In addition, satellite sensors tend to observe the upper levels of the soil (on the order of centimeters in depth), whereas the soil moisture of interest for climate purposes extends down to rooting depths (on the order of meters). It is for those reasons that *modeled* soil moisture is often used for climate and climate change research.

Drought Indices

Across the world, farmers, researchers, journalists, and policy-makers use drought indices to simplify hydrologic conditions and quantify drought severity using frequently limited data. There are many drought indices to choose from. This study utilizes the PDSI and a PDSI-derivative index, the SDDI.

Palmer Drought Severity Index

The most common meteorological drought index in the United States is the PDSI (Palmer, 1965), which ranges from –10 (driest) to +10 (wettest). At time (i), $PDSI(i) = 0.897PDSI(i-1) +$

$Z_{\text{PDSI}(i)}/3$, where $Z_{\text{PDSI}} = K \times d_{\text{PDSI}}$. K is a weighting factor dependent on the average local water supply and demand and $d_{\text{PDSIi}} = P - (\alpha E_p + \beta R_p + \delta RO_p - \varepsilon L_p)$, where P is the precipitation, E_p is the potential evapotranspiration, RO_p is the potential runoff, and L_p is the potential soil moisture loss. The coefficients α, β, δ, and ε are the ratios of actual to potential values for evapotranspiration, recharge, runoff, and soil moisture loss, respectively. They have been generated for the United States using observations but are not available globally, which limits large-scale studies using PDSI. Dai *et al.* (2004) created a global PDSI dataset for the years 1870–2002. However, the utilized PDSI coefficients were tuned to US soil characteristics in that work.

Supply Demand Drought Index (SDDI)

Another approach to estimating a surplus or deficit of soil moisture is by calculating the difference between precipitation (the atmospheric supply) and potential evapotranspiration (the atmospheric demand), which is the basis of the SDDI. SDDI requires monthly mean precipitation and temperature as inputs for each location without needing the regional coefficients required for the widely used PDSI. SDDI can be calculated globally and is highly correlated with PDSI in the United States (Rind *et al.*, 1990). SDDI is calculated using the change in the differences of precipitation (P) and potential evapotranspiration (E_p) between the future simulation and the average of the control for each grid box, $d_{\text{SDD1}} = P - E_p - (P - E_p)_{\text{control ave}}$, where E_p is calculated using the Thornthwaite (1948) method and Z, the 'SDDI moisture anomaly index', is defined as $Z_{\text{SDDI}} = d_{\text{SDDI}}/\sigma(P - E_p)$, where $\sigma(P - E_p)$ is the standard deviation of $(P - E_p)$ for the control run. SDDI is defined in terms of SDDI for the previous month and the SDDI moisture index anomaly index is somewhat similar to the PDSI formulation, $\text{SDDI}(i) = 0.897\,\text{SDDI}(i - 1) + \text{ZSDDI}(i)$.

Projections of Future Water Availability

Five GCMs were used: Canadian Centre for Climate Modeling and Analysis CGCM3.1 [CCCma], Geophysical Fluid Dynamics Laboratory CM2.1 [GFDL], Goddard Institute for Space Studies ER [GISS], Hadley Centre for Climate Prediction Model HadCM3 [HadCM3], and University of Tokyo Center for Climate System Research V3.2 [MIROC] (see IPCC (2007) for appropriate references) as a representative sampling of the 23 IPCC Fourth Assessment Report [AR4] models (IPCC, 2007). The basic comparisons shown here between soil moisture and the drought indices do not change substantially with the AR5 (IPCC, 2013) climate model results. The five chosen models span the spectrum of AR4 resolutions and equilibrium climate sensitivities (i.e., simulations run sufficiently long that the ocean temperatures are no longer changing noticeably, on the order of hundreds of years). In this study, GFDL represents the highest resolution at $2.0° \times 2.5°$ (latitude $\times$ longitude), while GISS ($4° \times 5°$) has the lowest in this study and among the IPCC AR4 models. The full range of equilibrium climate sensitivities among AR4 models is 2.1–4.4°C. The models used in this study show increased mean surface temperatures between 2.7°C and 4°C in the GISS and MIROC models, respectively, in response to doubled CO_2. The chosen models are thus representative in a broad sense of the range in global warming estimates. The IPCC AR4 SRES A2 (strong population growth, self-reliance, and preservation of local identities, hence 'business as usual' industrial development and trace gas release) was used for the years 2001–2100.

Over the next 85 years, global-average SDDI predictions over land are projected to decrease, indicating severe drying globally by the year 2100 [Slide 5]. In contrast, global-average soil moisture projections show almost no or little change in the 21st century in all models except the GFDL model, which shows a steady decrease [Slide 3]. IPCC (2007, 2013) found that droughts have become more common, especially in the tropics and subtropics, since the 1970s, which is in agreement with modeled and observed SDDI.

Because raw SDDI and soil moisture values give little intuitive understanding of the severity of drought, we investigated the change of water availability relative to prior conditions, as a percentage of time during which those conditions occurred in the past. Five percent drought indicates a drought of severity seen during only 5% of the control time period (1928–1978), in, respectively, the SDDI index or soil moisture, while 5% flood indicates a similar measure of the occurrence of wetter conditions. For soil moisture, there is an increase in both dry and wet extreme conditions, which results in very little change in the global-average values [Slide 7]. In the SDDI analysis, extreme drought outweighs extreme flooding conditions although both show increases, which explains why global-average SDDI shows an overall drying trend [Slide 8]. For the 5% drought, the percentage increase of this condition from soil moisture considerations are, in general, smaller than those from SDDI, especially in summer, while the 5% flood condition increases are greater than for the SDDI in both seasons.

The impact of the IPCC SRES A2 storyline for the 21st century is in most cases smaller than the differences between the GCMs. It is therefore likely that all SRES storylines (most of which assume reduced greenhouse gas emissions compared to the A2 scenario used here) would have a global drying trend in SDDI due to increased temperatures.

Soil moisture percentage changes between the end of the 20th and 21st centuries [Slide 9] show the equator getting slightly wetter over land consistent with increased atmospheric upwelling, driven by the warmer temperatures and greater atmospheric moisture. Drying is projected around 35N–40N and 35S–40S and in some models around 10N–30N and 10S–30S. The lower latitude tropical portions of this region (10N–15N and 10S–15S) are due to changes in the mean circulation and precipitation over land. The subtropical changes seem to be more associated with warming temperatures and increased evaporation. For the SDDI, in both summer and winter, there are decreases (i.e., drier conditions) in the tropics ranging from mild (GISS and GFDL) to extreme (HadCM3). There is more agreement among the GCMs that there are extreme decreases in the mid latitudes. In the Arctic, the models show increased SDDI (wetter conditions) ranging from moderate to extreme. Generally, the GCM results look similar in both seasons, which implies that the lifetimes of both SDDI and soil moisture characteristics are long enough that they impose a consistent response throughout the year. In general, then, both measures of drought show strong increases in the subtropics, but the SDDI extends these large increases through portions of both the tropics and mid-latitudes.

Bringing the Two Measures of Drought Closer Together

The fact that modeled soil moisture and drought indices project different severities of water availability change over the next century should be a cause for concern in the modeling and climate impacts communities. The discrepancies between soil moisture and SDDI are largely related to differences in the way that evaporation is calculated.

In the SDDI formulation, the evaporation represents the atmospheric demand for moisture and is therefore set to a maximum value, the potential evapotranspiration, which is the evaporation expected over a completely wet surface. SDDI uses the Thornthwaite (1948) method to calculate the potential evapotranspiration. SDDI is shown to be sensitive to the chosen method of potential evapotranspiration calculation [Slide 15]; SDDI calculated with Hargreaves E_p (Hargreaves and Sumani, 1982) shows the most similarity to soil moisture. An additional SDDI sensitivity test, using multiple regression analysis, showed that SDDI is more sensitive to temperature than is soil moisture. However, the Thornthwaite E_p used to calculate SDDI is less sensitive to temperature than the aerodynamic E_p used to determine soil moisture in GCMs. The aerodynamic E_p depends on the gradient in moisture between an assumed saturated ground (with the temperature of the actual ground) and the moisture in the air at a height of 10 m.

Evaporation in GCMs is determined by multiplying the aerodynamic potential evapotranspiration by a scaling factor that will be referred to as β. β ranges from zero in very dry conditions to one in well-watered conditions, and is physically (and numerically) related to the ratio of the vegetative or canopy surface moisture conductance (determined from measured conductivity coefficients) to that of the atmosphere (a function of wind speed and stability). The aerodynamic method for calculating potential evapotranspiration produces much larger values than the Thornthwaite method or any other potential evapotranspiration method by an order of magnitude [Slide 14]. For this reason, the β factor must be artificially small in GCMs, which means the two variables used to calculate it, atmospheric and surface conductance, must also conform; the atmospheric conductance must always be larger than the vegetative surface conductance although, in the real world, there is no such incongruence.

This scaling factor β is the link between E_p and evaporation and it has been found to play a large role in bridging the gap between soil moisture and SDDI projections. When β is added to the SDDI formula, and especially when it is allowed to change (i.e., decrease as the soil dries), the evaporation is dampened and SDDI becomes more similar to soil moisture [Slide 16]. SDDI and soil moisture also become more alike when the vegetative surface conductance is increased in the GCM calculation of evaporation from the surface [Slide 18]; soil moisture then displays global mean drying due to increased evaporation and SDDI conditions become wetter due to increased precipitation and decreased temperatures. Soil moisture in this simulation also shows increased memory, closer to SDDI than previously. While these factors might seem to alleviate the discrepancies in future drought projections somewhat, the changes to SDDI make its results *less* like those of the PDSI [Slide 17], which has been validated to some degree by comparison with observations (e.g., tree ring data, [Cook *et al.*, 2004] and its standard use by the National Weather Service to depict current drought conditions). This is despite the PDSI having a β term in its formulation.

Conclusions

At face value, it would appear that GCMs provide a more sophisticated approach for the calculation of soil moisture changes under climate change than do drought indices. For example, the calculated soil moisture in these models is able to respond to short timescale precipitation events, while SDDI could not since it uses average monthly precipitation. Two other expected changes are regional shifts in vegetation and changes in stomatal conductance due to increased CO_2. GISS as well as many other GCMs are working towards a functional fully coupled dynamic vegetation scheme but they are not there yet. The SDDI remains blind to vegetation, although it can be incorporated in subsequent formulations. Both soil moisture and drought indices reflect changes due to ongoing temperature increases, but with different sensitivities.

Nevertheless, it is possible that a simple formula like SDDI may be better than a complex land scheme used to calculate soil moisture. More complex parameterizations may differ substantially from reality, the high aerodynamic E_p and low β values in GCMs being just one prominent example. Vegetative response and the nature of its changes are extremely complicated to assess. The simple measures, as suggested by the relevance of PDSI to observed paleo-drought situations (e.g., tree rings, Cook *et al.*, 2004), might be better able to capture the 'emergent behavior' of future droughts.

Ultimately, which measure should we use, drought indices or soil moisture? We know that droughts are likely to increase in the future, but the two measures strongly disagree as to how much. The two water availability measures can be altered to become more like one another in a number of ways, although there is still no full convergence, and it is not clear that there is an advantage in doing so. What we really want to know is which one is better when assessing changes due to climate change? Since neither measure is perfect, it is important to understand the abilities of each. Since they have different strengths and weaknesses, perhaps the best approach is to use both; when they disagree, we can decide which is most appropriate to the situation, and when they agree, the strongest case can be made for a GCM's water availability projections. A more complete discussion of this topic is available in my Ph.D. thesis (Aminzade, 2011).

References

Aminzade, J. (2011). *Water Availability in a Warming World*. Ph.D. Thesis, Columbia University, pp. 248.

Burke, E.J. and Brown, (2008). Evaluating Uncertainties in the Projection of Future Drought *J. Hydrometeor*, 9, 292 299.

Cook, E.R., Woodhouse, C.A., Eakin, C.M., Meko, D.M. and Stahle, D.W. (2004). Long-term aridity change in the western United States. *Science*, 306, 1015–1018.

Dai, A., Trenberth, and Qian (2004). A global data set of Palmer Drought Severity Index for 1870–2002: Relationship with soil moisture and effects of surface warming *J Hydrometeor*, 5, 1117–1130.

Hargreaves, G.H. and Samani, Z.A. (1982). Estimating potential evapotranspiration. Tech. Note. *Journal of Irrigation and drainage Engineering*, ASCE, 108(3), 225–230.

Henderson-Sellers, A., Pitman, A.J., Irannejad, P. and McGuffie, K. (2002). Land-surface simulations improve atmospheric modeling. *Eos Transactions American Geophysical Union*, 83, 145–152.

Henderson-Sellers, A., Irannejad, P., Sharmeen, S., Phillips, T.J., McGuffie, K. and Zhang, H. (2003). Evaluating GEWEX CSEs' simulated land-surface water budget components. *GEWEX News*, 13(3), International GEWEX Project Office, Silver Spring, MD, pp. 3–6.

IPCC Climate Change (2007). The physical science basis. In S. Solomon, D. Qin, M. Manning, Z. Chen, M. Marquis, K.B. Averyt, M. Tignor and H.L. Miller (Eds.), *Contribution of Working Group I to the Fourth Assessment Report of the Intergovernmental Panel on Climate Change* (pp. 996). Cambridge, United Kingdom and New York: Cambridge University Press.

IPCC Climate Change (2013). The physical science basis. In T.F. Stocker, D. Qin, G.-K. Plattner, M. Tignor, S.K. Allen, J. Boschung, A. Nauels, Y. Xia, V. Bex and P.M. Midgley (Eds.), *Contribution of Working Group I to the Fifth Assessment Report of the Intergovern- mental Panel on Climate Change* (pp. 1535). New York: Cambridge University Press.

Lakshmi, V. (2013). Remote sensing of soil moisture. *ISRN Soil Science*, 2013, Article ID 424178, 33. doi:10.1155/2013/424178.

Palmer, W.C. (1965). *Meteorological Drought*. Research Paper No. 45, U.S. Department of Commerce Weather Bureau, Washington, DC.

Rind, D., Goldberg, R., Hansen, J., Rosenzweig, C. and Ruedy, R. (1990). Potential evapotranspiration and the likelihood of future drought. *Journal of Geophysical Research,* 95, 9983–10,005.

Robock, A., Vinnikov, K.Y., Schlosser, C.A., Speranskaya, N.A. and Xue, Y. (1998). Evaluation of the AMTP soil moisture simulations. *Global Planet Change*, 19, 181–208.

Thornthwaite, C.W. (1948). An approach toward a rational classification of climate. *Geographical Review*, 38(1), 55–94. doi:10.2307/210739.

Varis, O., Kajander, T. and Lemmela, R. (2004). Climate and water: From climate models to water resources management and vice versa. *Climate Change*, 66, 321–344.

Wells, N., Goddard, S. and Hayes, M.J. (2004). A self-calibrating palmer drought severity index. *Journal of Climate*, 17, 2335–2351.

Slide 1

Projections of Future Drought

Jennifer Aminzade

Ph.D., Earth Sciences, Columbia University (2011)

Our Warming Planet: Topics in Climate Dynamics
Lectures in Climate Change, Vol. 1
2017

1

Slide 2

Models Differ in Soil Moisture Amounts

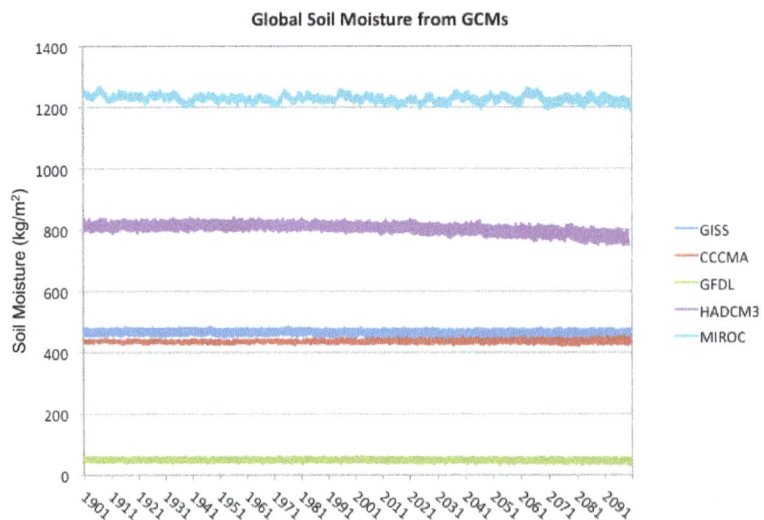

Global average soil moisture as a function of time in five GCMs (Aminzade, 2011).

Slide 3

Modeled Changes in Soil Moisture

Global GCM Soil Moisture Anomalies from the 1979-1989 Average

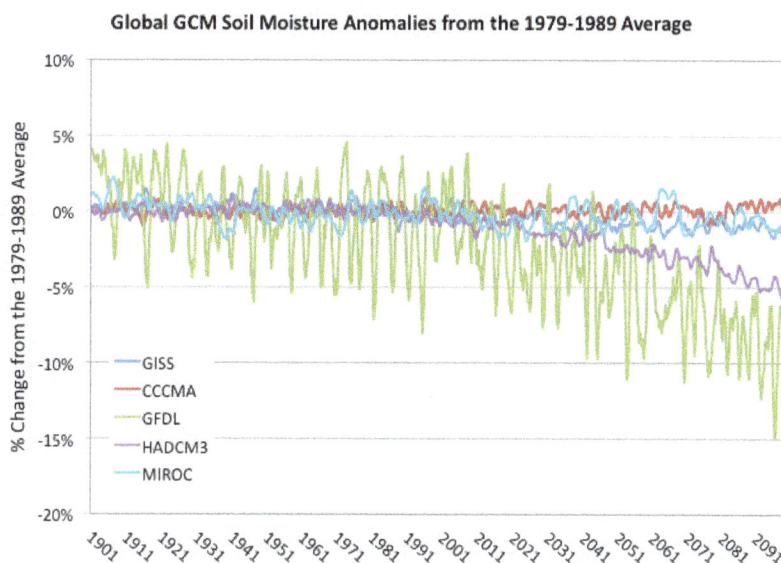

Soil moisture anomalies in the five GCMs projected relative to their 1979–1989 averages (Aminzade, 2011).

Slide 4

Drought and Flood Categories

PALMER CLASSIFICATIONS	
4.0 or more	Extremely wet
3.0 to 3.99	Very wet
2.0 to 2.99	Moderately wet
1.0 to 1.99	Slightly wet
0.5 to 0.99	Incipient wet spell
0.49 to -0.49	Near normal
-0.5 to -0.99	Incipient dry spell
-1.0 to -1.99	Mild drought
-2.0 to -2.99	Moderate drought
-3.0 to -3.99	Severe drought
-4.0 or less	Extreme drought

Palmer Drought Severity Index (PDSI) classifications, which can also be applied to the Supply/Demand Drought Index (SDDI).

Slide 5

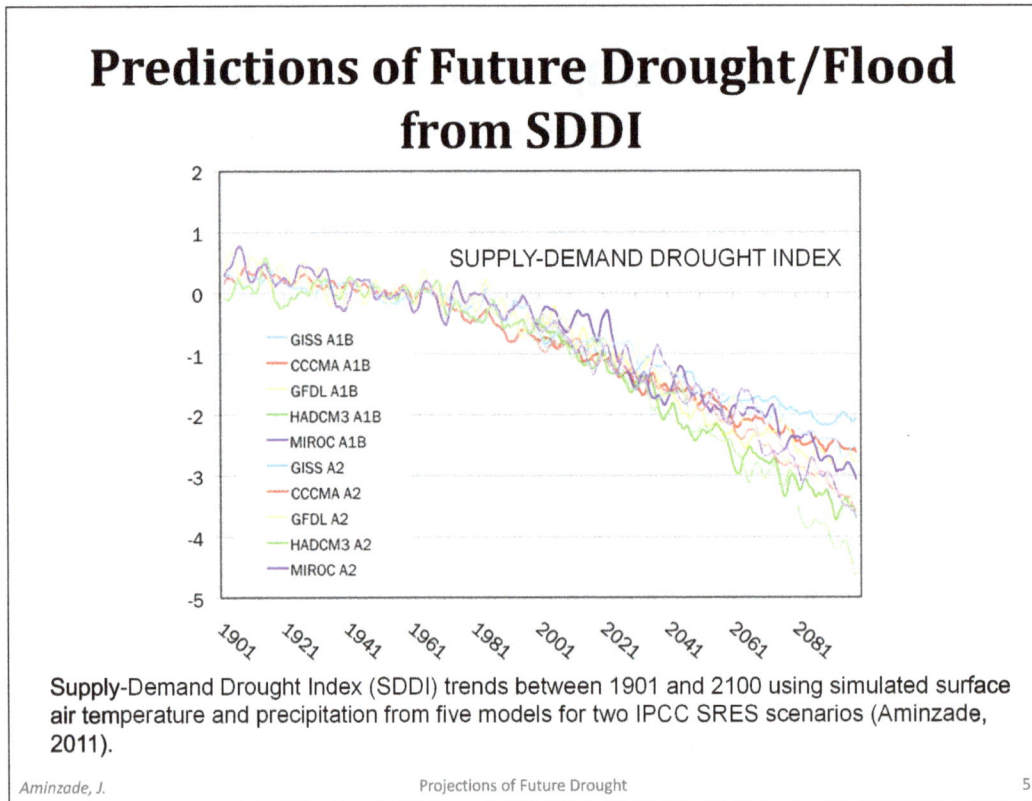

Predictions of Future Drought/Flood from SDDI

Supply-Demand Drought Index (SDDI) trends between 1901 and 2100 using simulated surface air temperature and precipitation from five models for two IPCC SRES scenarios (Aminzade, 2011).

Slide 6

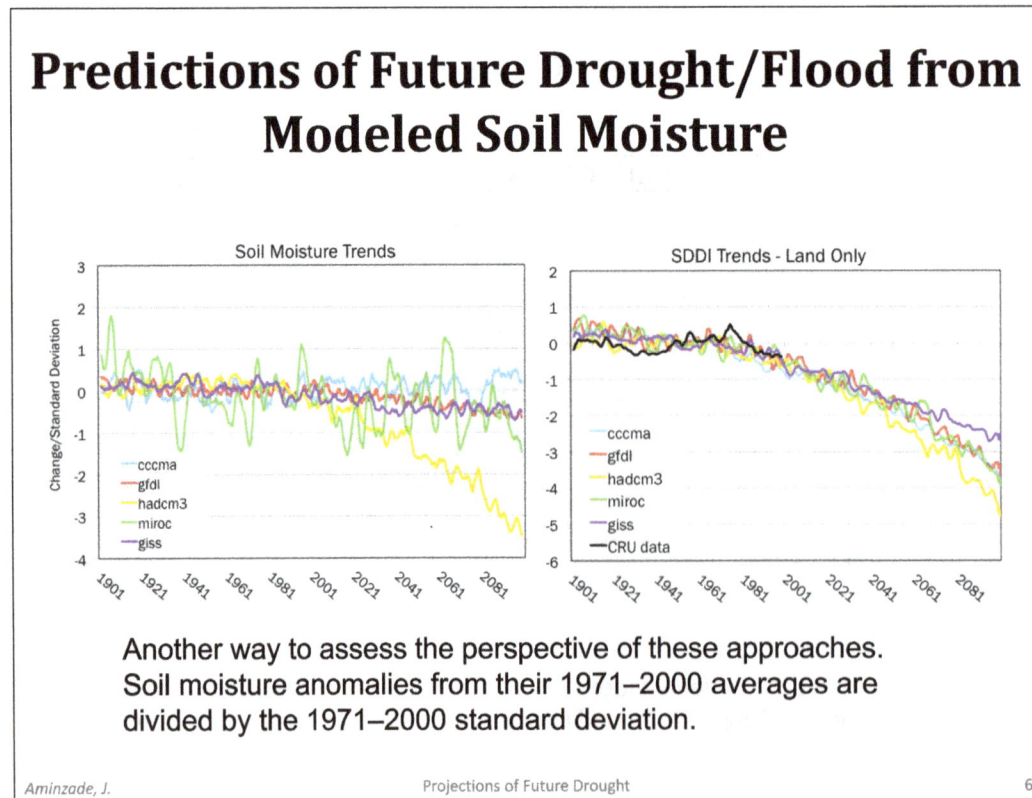

Predictions of Future Drought/Flood from Modeled Soil Moisture

Another way to assess the perspective of these approaches. Soil moisture anomalies from their 1971–2000 averages are divided by the 1971–2000 standard deviation.

Slide 7

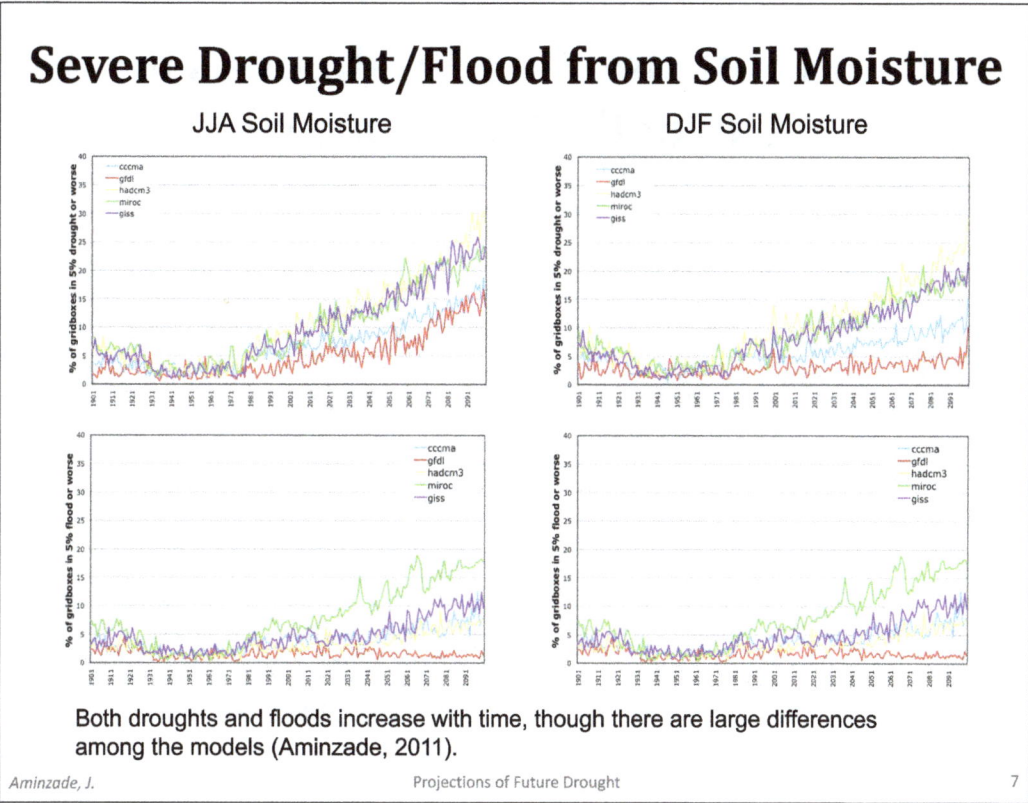

Severe Drought/Flood from Soil Moisture

JJA Soil Moisture   DJF Soil Moisture

Both droughts and floods increase with time, though there are large differences among the models (Aminzade, 2011).

Slide 8

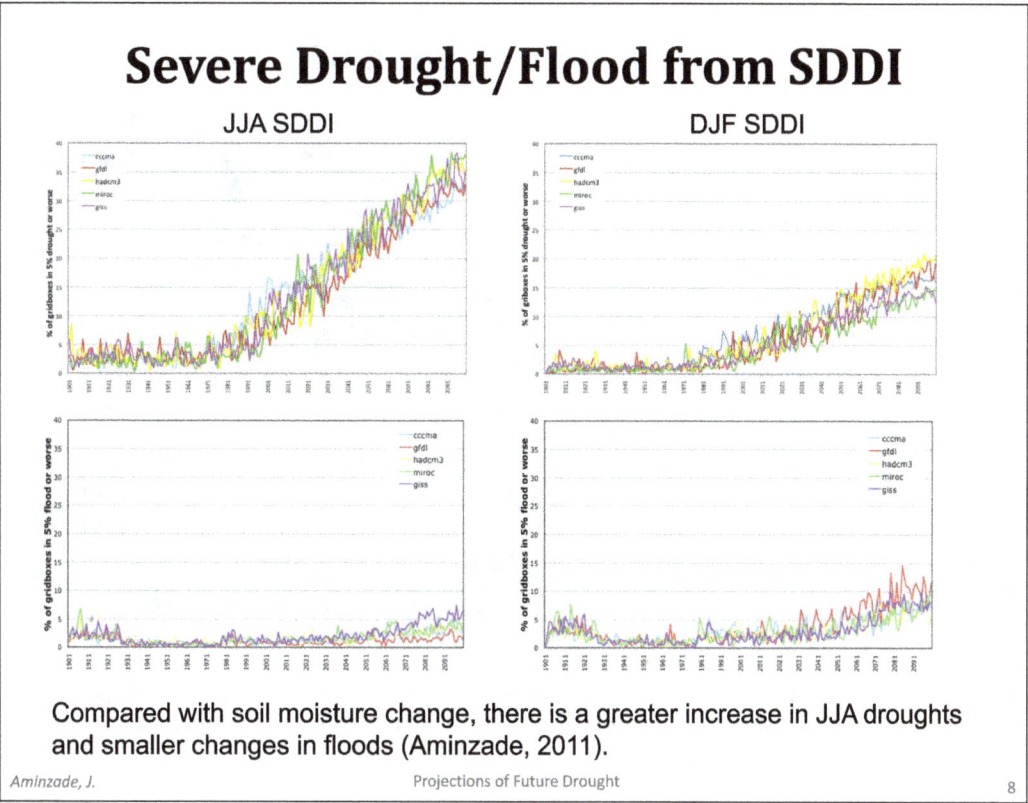

Severe Drought/Flood from SDDI

JJA SDDI   DJF SDDI

Compared with soil moisture change, there is a greater increase in JJA droughts and smaller changes in floods (Aminzade, 2011).

Slide 9

SDDI and Soil Moisture Predictions for the End of the 21st Century

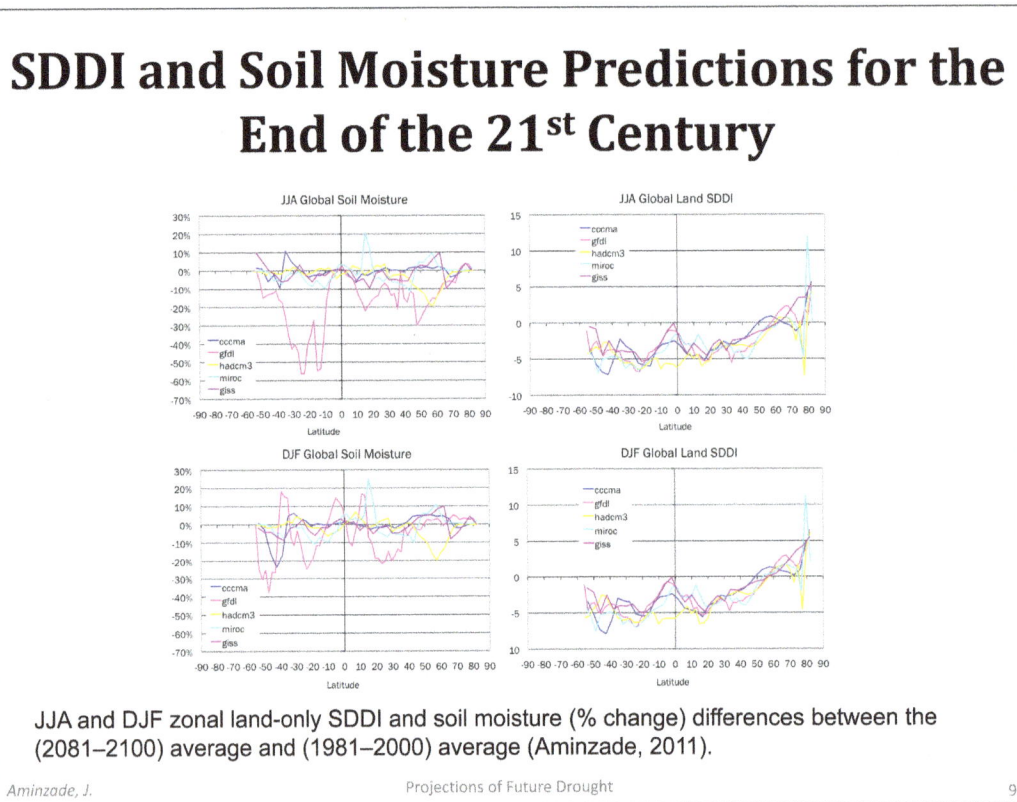

JJA and DJF zonal land-only SDDI and soil moisture (% change) differences between the (2081–2100) average and (1981–2000) average (Aminzade, 2011).

Slide 10

Soil Moisture and its Change with Climate

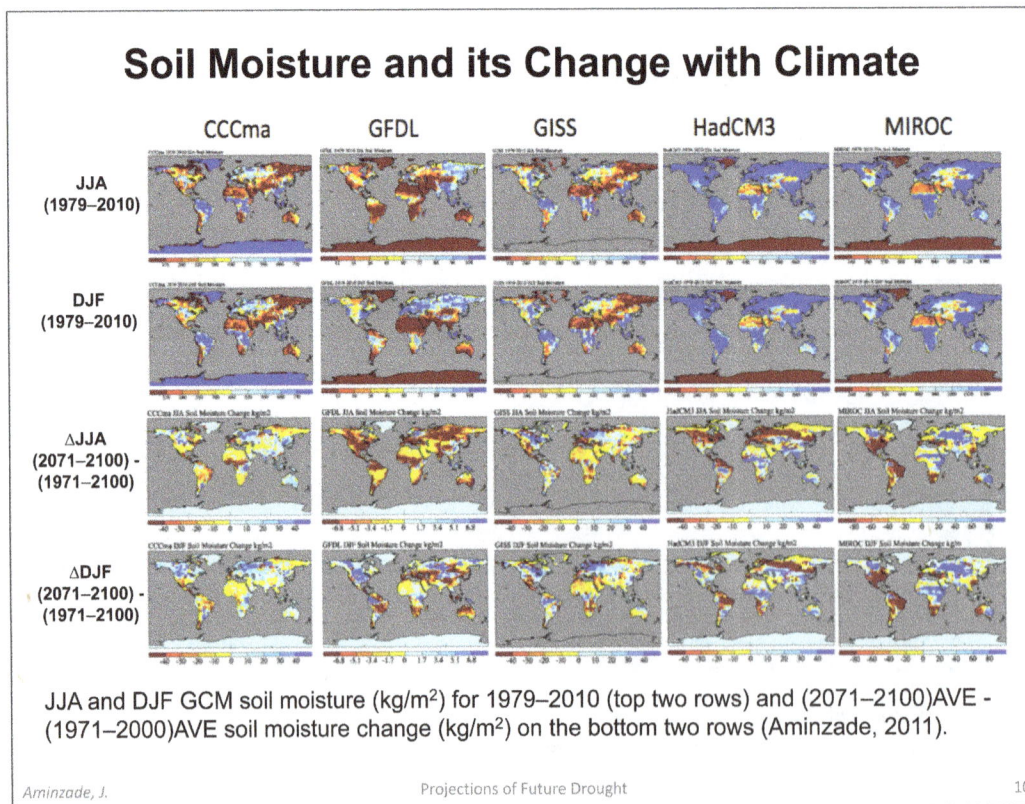

JJA and DJF GCM soil moisture (kg/m²) for 1979–2010 (top two rows) and (2071–2100)AVE - (1971–2000)AVE soil moisture change (kg/m²) on the bottom two rows (Aminzade, 2011).

Slide 11

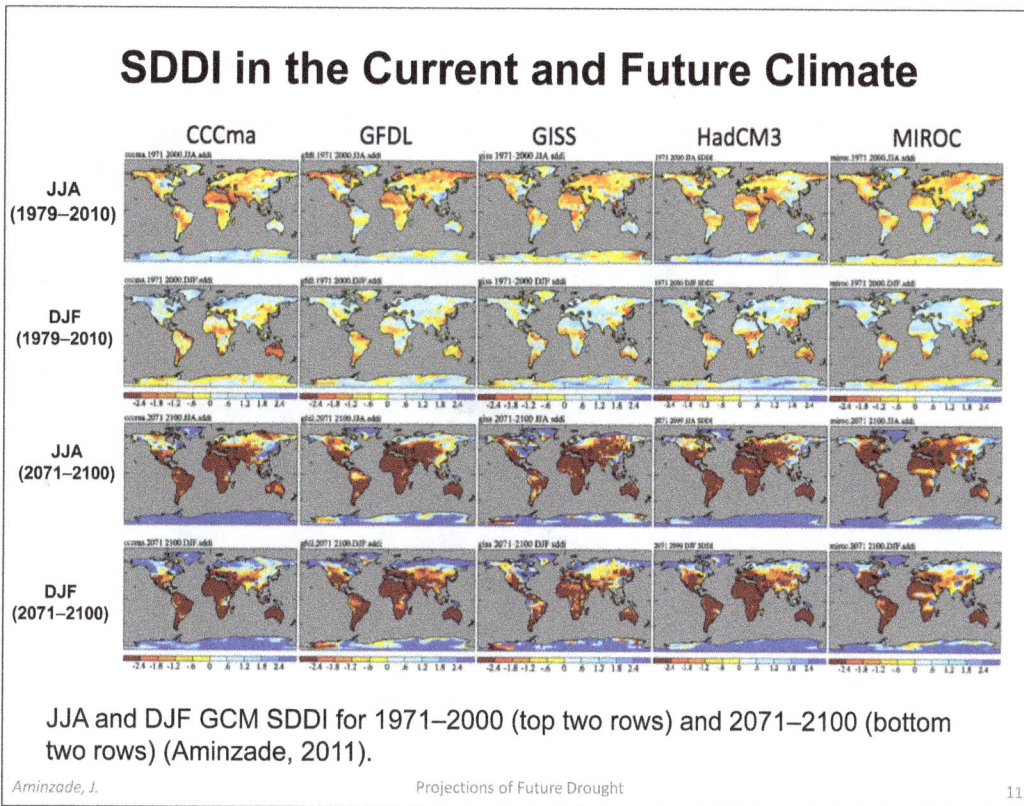

SDDI in the Current and Future Climate

JJA and DJF GCM SDDI for 1971–2000 (top two rows) and 2071–2100 (bottom two rows) (Aminzade, 2011).

Slide 12

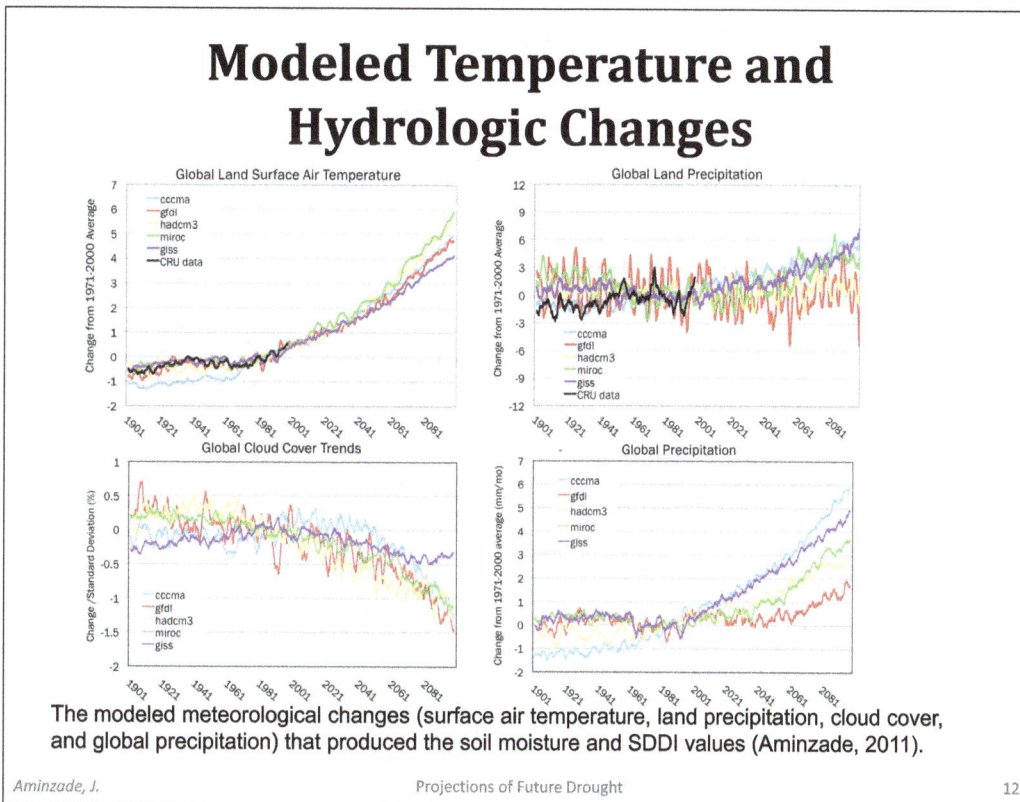

Modeled Temperature and Hydrologic Changes

The modeled meteorological changes (surface air temperature, land precipitation, cloud cover, and global precipitation) that produced the soil moisture and SDDI values (Aminzade, 2011).

Slide 13

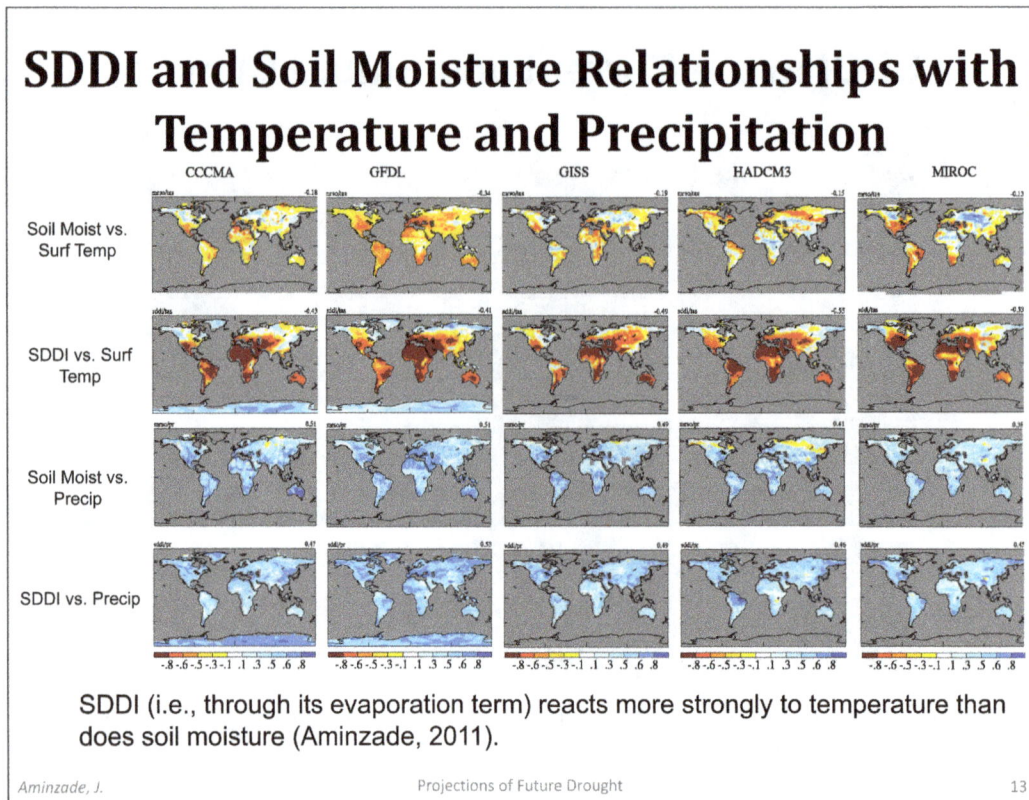

SDDI (i.e., through its evaporation term) reacts more strongly to temperature than does soil moisture (Aminzade, 2011).

Slide 14

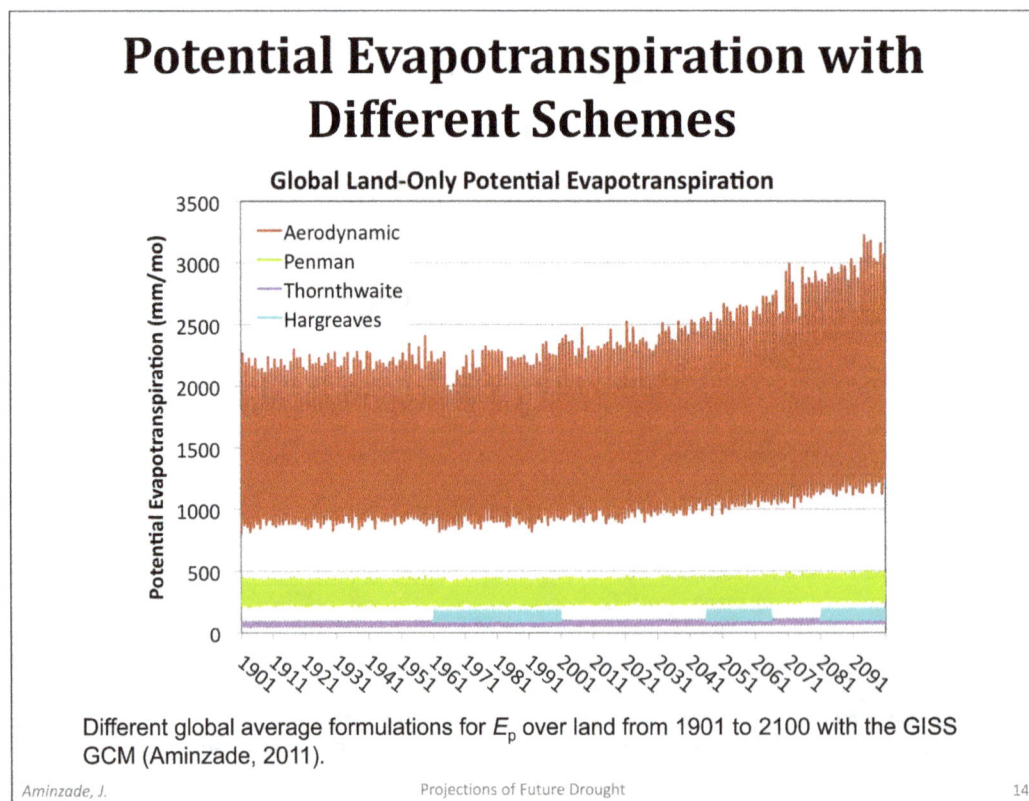

Different global average formulations for E_p over land from 1901 to 2100 with the GISS GCM (Aminzade, 2011).

Slide 15

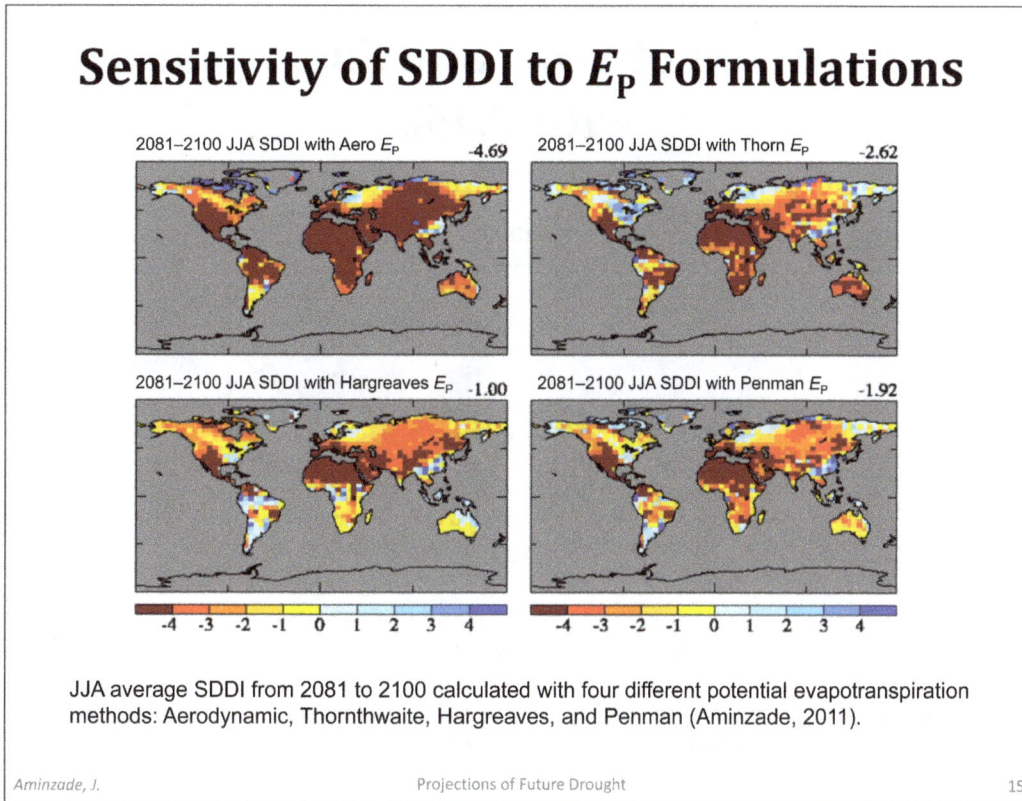

Sensitivity of SDDI to E_P Formulations

2081–2100 JJA SDDI with Aero E_P -4.69

2081–2100 JJA SDDI with Thorn E_P -2.62

2081–2100 JJA SDDI with Hargreaves E_P -1.00

2081–2100 JJA SDDI with Penman E_P -1.92

JJA average SDDI from 2081 to 2100 calculated with four different potential evapotranspiration methods: Aerodynamic, Thornthwaite, Hargreaves, and Penman (Aminzade, 2011).

Slide 16

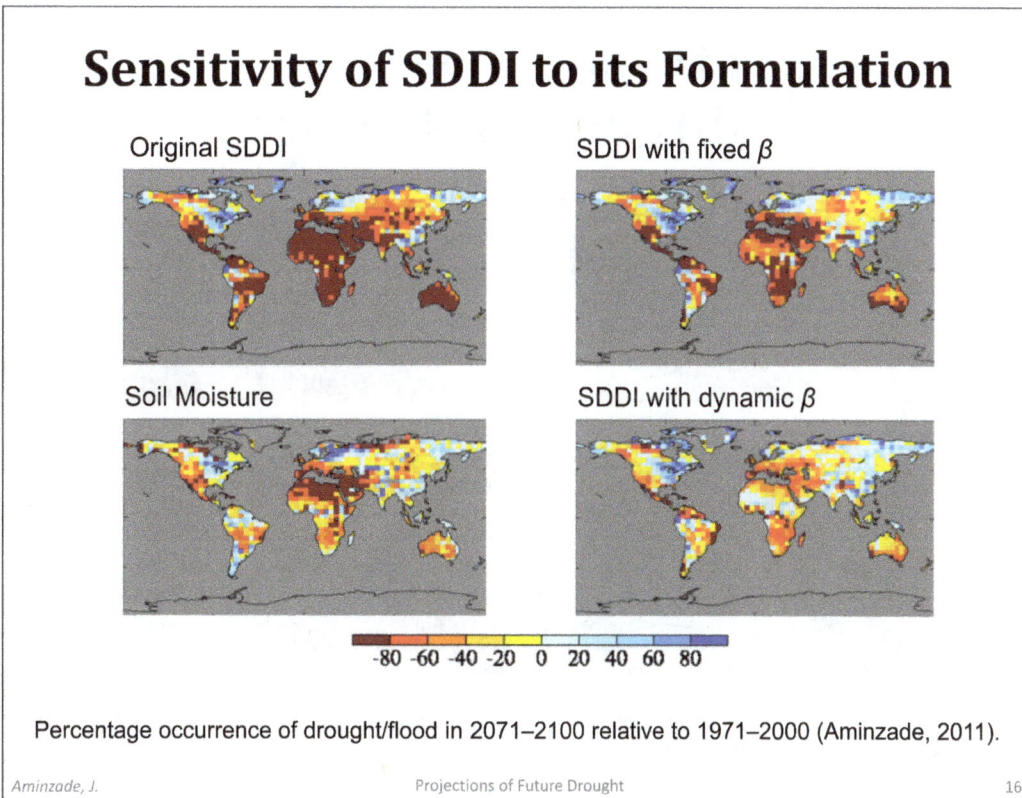

Sensitivity of SDDI to its Formulation

Original SDDI

SDDI with fixed β

Soil Moisture

SDDI with dynamic β

Percentage occurrence of drought/flood in 2071–2100 relative to 1971–2000 (Aminzade, 2011).

Slide 17

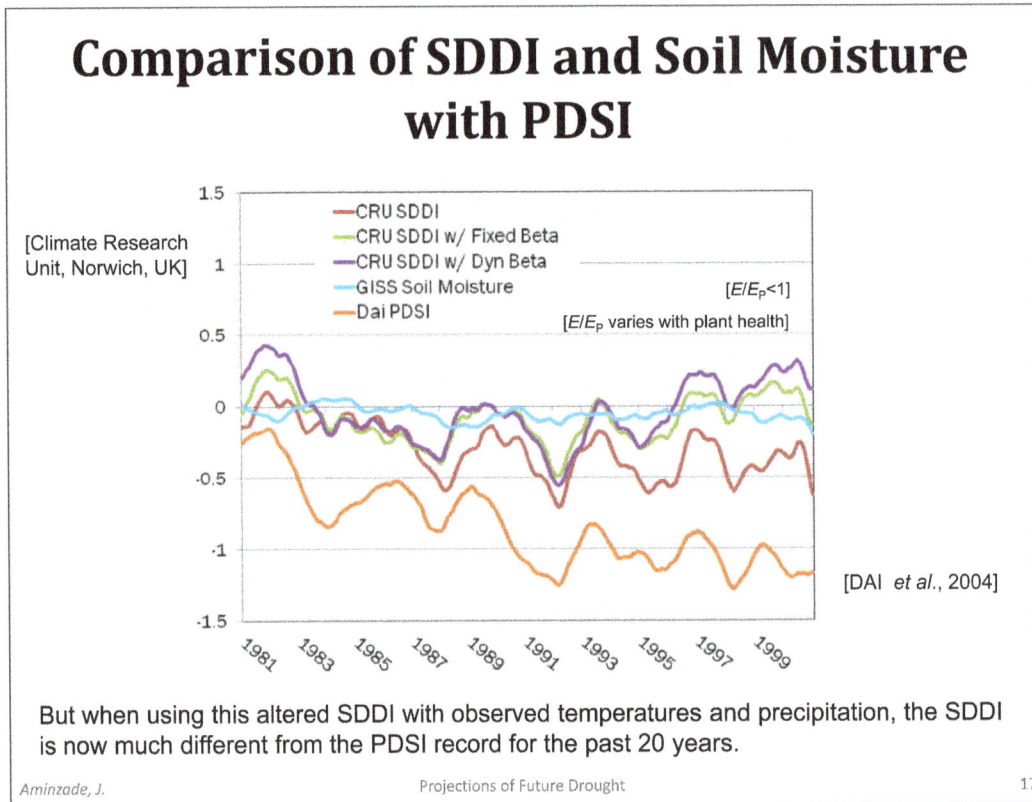

Comparison of SDDI and Soil Moisture with PDSI

[Climate Research Unit, Norwich, UK]

[DAI *et al.*, 2004]

But when using this altered SDDI with observed temperatures and precipitation, the SDDI is now much different from the PDSI record for the past 20 years.

Slide 18

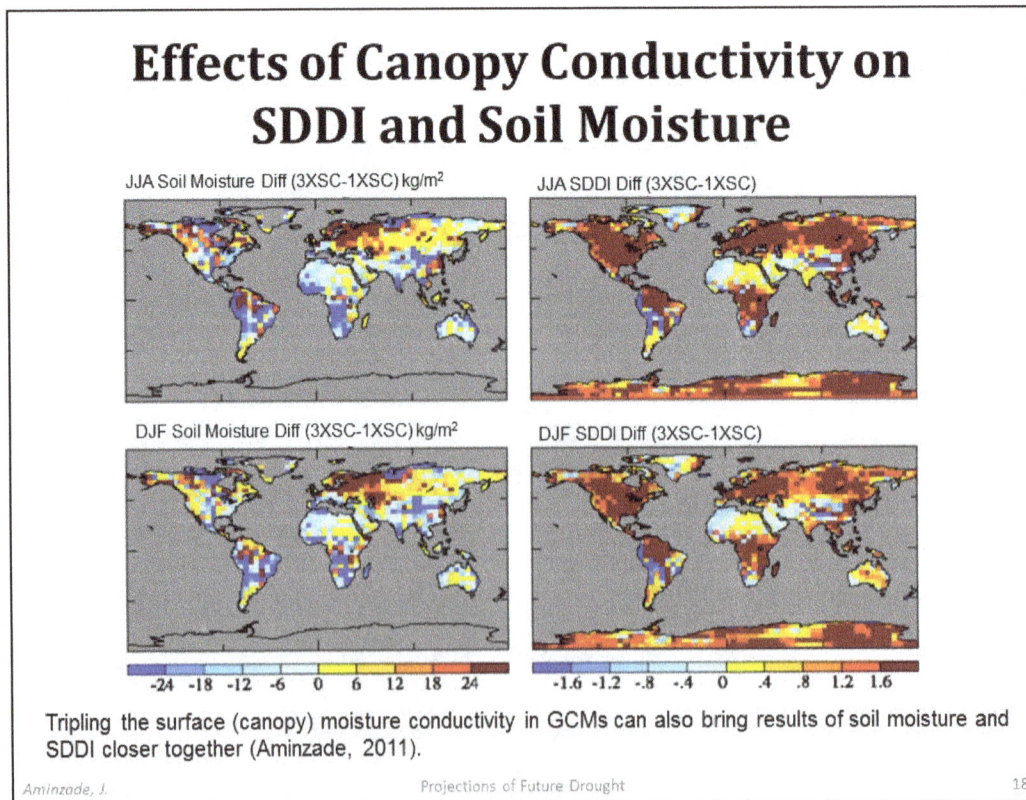

Effects of Canopy Conductivity on SDDI and Soil Moisture

JJA Soil Moisture Diff (3XSC-1XSC) kg/m²

JJA SDDI Diff (3XSC-1XSC)

DJF Soil Moisture Diff (3XSC-1XSC) kg/m²

DJF SDDI Diff (3XSC-1XSC)

Tripling the surface (canopy) moisture conductivity in GCMs can also bring results of soil moisture and SDDI closer together (Aminzade, 2011).

Slide 19

Conclusions

- Two different measures of future drought/flood obtained from GCMs—soil moisture changes and drought indices—give somewhat different results.
- Drought index changes are much more extreme, showing intensifying drought from the tropics through mid-latitudes, while soil moisture assessments show primarily subtropical drought.
- Results can be brought closer together by limiting the evaporative response in drought indices and by increasing canopy conductivity in GCMs.
- A combination of the two assessments can be used to increase our confidence in future water availability changes.

Aminzade, J. Projections of Future Drought 19

Slide Notes

Slide 2   Global average soil moisture as a function of time in five GCMs. Most show little change globally as the climate warms. Notice also how different the values are in the models. This is due primarily to lack of observations to calibrate this quantity. Satellite observations should eventually help guide the models.

Slide 3   Soil moisture anomalies in the five GCMs projected relative to their 1979–1989 averages. Only the GFDL model, which had the lowest control run value, shows strong decreases with time.

Slide 5   Supply-Demand Drought Index (SDDI) trends between 1901 and 2100 using simulated surface air temperature and precipitation from five models for two IPCC SRES scenarios, A2 (strong population growth, self-reliance, and preservation of local identities) and A1B (rapid economic growth with a balance between fossil and non-fossil energy sources). All show strong increases in global drought conditions as the 21st century progresses.

Slide 6   Another way to assess the perspective of these approaches. Soil moisture anomalies from their 1971–2000 averages are divided by the 1971–2000 standard deviation. The actual value of the SDDI is shown, calculated using modeled and observed surface air temperature and precipitation. Projections are from five GCMs for 1901–2100; negative SDDI indicates drier conditions. While SDDI is showing extensive drying in all GCMs, only HadCM3 has consistent soil moisture changes greater than one standard deviation.

Slide 7   These figures show the percentage of grid boxes in each model experiencing 5% drought (or drier) conditions in JJA (upper left) and DJF (upper right), as well as 5% flood (or wetter) conditions in JJA (bottom left) and DJF (bottom right) as measured by soil moisture. Both droughts and floods increase with time, though there are large differences among the models.

Slide 8   These figures show the percentage of gridboxes in each model experiencing 5% drought (or drier) conditions in JJA (upper left) and DJF (upper right), as well as 5% flood (or wetter) conditions in JJA (bottom left) and DJF (bottom right) as determined from SDDI. Compared with soil moisture change, there is a greater increase in JJA droughts and smaller changes in floods.

Slide 9   JJA and DJF zonal land-only SDDI and soil moisture (% change) differences between the (2081–2100) average and (1981–2000) average. Negative latitudes are in the southern hemisphere.

Slide 10   Note that the color bar scales for GFDL and MIROC are different from CCCma, GISS, and HadCM3. Both control run values and climate changes vary among the GCMs (Aminzade, 2011).

Slide 11   There is much better agreement among the models, for both control runs and climate changes.

Slide 12   The modeled meteorological changes (surface air temperature, land precipitation, cloud cover, and global precipitation) that produced the soil moisture and

SDDI values, in particular the changes over land (top row). Note that the GCM hydrologic projections disagree (Aminzade, 2011).

Slide 13   This series of maps shows the time series correlation between soil moisture and surface air temperature on the top row, the correlation between SDDI and surface air temperature on the second row, the correlation between soil moisture and precipitation on the third row, and the correlation between SDDI and precipitation on the bottom row for all GCMs.

Slide 14   Different global average formulations for E_p over land from 1901 to 2100 with the GISS GCM (which uses the aerodynamic formulation for its soil moisture calculations). The differing magnitudes can affect evaporation and its change with climate.

Slide 15   SDDI (and PDSI) normally use Penman; values are a little less extreme with Hargreaves, although the general patterns are similar.

Slide 16   By reducing the evaporation in the SDDI to below potential values (via β, the ratio between evaporation and potential evaporation) (upper right), and also by allowing β to change (e.g., when vegetation would be wilting (lower right)), SDDI projections are brought closer to the GCM soil moisture changes (lower left).

Slide 18   For the 2051–2100 period, the soil moisture values are reduced (left hand-side), while with cooler surface air temperatures, SDDI values are more positive.

CLIMATE LECTURE 12

Lightning and Climate Change

Colin Price

School of Geosciences, Tel Aviv University, Tel Aviv, Israel

Colin Price is a Professor in Atmospheric Sciences in the School of Geosciences, Tel Aviv University. His main field of research deals with global lightning activity, and the connections between lightning activity and the Earth's climate. His work involves the relationships between lightning and forest fires, atmospheric chemistry, and severe storms.

Introduction

Lightning is one of nature's most beautiful and awesome sights. Yet it can also be extremely dangerous, presenting a major natural hazard in many different environments, from power utility companies, to civil aviation, to golfers, and more. Thousands of people are killed every year by lightning bolts, while tens of thousands are injured as well (Cooray, Cooray and Andrews, 2007). Lightning impacts both our daily commercial and recreational activities. In the United States alone, damages due to lightning strikes amounts to tens of millions of dollars annually (Curran, Holle and Lopez, 2000). In recent years, with great interest in renewable energy, wind turbines have become extremely vulnerable to lightning damage (Glushakow, 2007). Furthermore, most commercial airliners are struck about once a year by lightning; however, due to the protective metal skin, generally little damage is incurred. Tens of thousands of fires are also ignited by lightning every year, generally in temperate or high latitudes (e.g. Canada, Siberia, etc.) (Stocks *et al.*, 2002). In such cases, tens of fires can be ignited locally on the same day as a storm passes through, causing major problems for fire crews and fire management. Hence, knowledge of how lightning activity may change as the Earth's temperature changes is of critical importance and interest.

David Rind was my advisor for my Ph.D. at Columbia/Goddard Institute for Space Studies in the early 1990s, and was the first to suggest trying to model global lightning activity in global climate models (GCMs). As a result, our parameterization (Price and Rind, 1992) is still used more than 25 years later in climate studies related to changes in lightning activity (Mareev and Volodin, 2014), atmospheric chemistry (Fang *et al.*, 2010), and even for forecasting severe weather (Giannaros, Kotroni and Lagouvardos, 2015).

Global Distribution of Thunderstorms

While it was thought for many decades that the global lightning frequency was ~100 flashes/sec, recent satellite observations give the best estimate of 45 flashes/sec (Christian

251

et al., 2003). The distribution of these lightning discharges around the globe is not random, following the general circulation patterns of the atmosphere, driven by solar heating (Price, 2006). Solar heating in the tropics results in the creation of warm moist air that initiates vertical convection and mixing of the atmosphere. Lightning activity in thunderstorms is an indication of the intensity of atmospheric convection. Atmospheric convection occurs under unstable atmospheric conditions, either due to the heating of the boundary layer by solar radiation during the day, or by the mixing of air masses of different densities. Depending on the level of atmospheric instability, the convection will lead to the development of thunderstorms of different intensities. However, this region of tropical convection decreases with increasing latitude due to the circulation patterns in the atmosphere that result in sinking air (subsidence) in the subtropical regions around 30N and 30S. A secondary area of convergence along the polar front results in an additional region of thunderstorms in mid-latitudes.

In addition to solar heating, water vapor, and particularly the release of latent heat during condensation and freezing play a vital role in thunderstorm development. Since the saturation water vapor concentrations increase ~7% for every one degree increase in temperature (Clausius–Clapeyron relationship), the tropical atmosphere has an order of magnitude more water vapor to condense into rain and ice than the polar atmospheres. Hence, there is much more energy available for the formation of thunderstorms in the tropical atmosphere compared with the higher latitudes.

Temperature and Lightning

It has been shown by many studies that lightning activity, thunderstorm days, or indices linked to global lightning activity (ionospheric potential, Schumann resonances, etc.) are sensitive indicators of surface temperature changes (Williams, 1992; Price, 1993; Williams, 2005; Reeve and Toumi, 1999; Markson and Price, 1999; Price and Asfur, 2006; Markson, 2007; Williams, 2009). These studies show that on different temporal and spatial scales, small increases in surface temperature result in large non-linear increases in thunderstorm and lightning activity (Williams, 2005, 2009; Price, 2014).

Future Predictions

The recent Intergovernmental Panel on Climate Change (IPPC, 2013) report predicts a global warming of 1–4°C by the end of this century, depending on the scenario we assume for future uses of energy and land use. However, one of the weaknesses of all the models used for these predictions is the simulation of convective clouds, which are a sub-grid scale process that needs to be parameterized in climate models (Del Genio, Mao-Sung and Jonas, 2007; Futyan and Del Genio, 2007). While climate models have problems accurately modeling convective clouds, they are even more problematic modeling lightning activity, although this has been attempted (Price and Rind, 1994a, 1994b; Shindell *et al.*, 2006; Grenfell, Shindell and Grewe, 2003).

For understanding lightning changes in a warmer world, we need to look at not only surface temperatures, but also the temperature profile (lapse rate) in the atmosphere as greenhouse gases increase. There are three possibilities regarding the mean vertical temperature profile in the lower troposphere as greenhouse gases increase, and surface temperatures warm: (1) If the surface warms more than the upper troposphere, the atmosphere will become more unstable (on average), and we would expect more convection and thunderstorms. (2) If the surface and the upper troposphere warm at the same rate, then no change will be seen in the lapse rate, and hence there will be no change in the mean stability (or instability) of the troposphere. (3) If the upper troposphere warms more than the surface temperatures, this will stabilize the atmosphere, with fewer thunderstorms developing in a warmer world. Climate models of all complexity and sizes show that as the climate warms at the surface, the tropical upper troposphere (exactly the location of most of the global thunderstorms) warms even more (IPCC, 2013). The reason for this is that increased convection transports additional water vapor into the upper atmosphere where it acts as a strong greenhouse gas, absorbing infrared radiation emitted from the surface of the Earth. The increase in water vapor results in a larger warming in the upper troposphere than at the surface, resulting in the average stabilization of the tropical atmosphere.

However, within the thunderstorms themselves the instability, measured by the convective available potential energy (CAPE) tends to increase in a warmer climate (Del Genio et al., 2007), especially for the most intense thunderstorms. And increases in CAPE in the present climate show clear increases in lightning activity (Williams et al., 1992; Pawar, Lal and Murugavel, 2011; Siingh et al., 2012). Therefore, when these same climate models are run under a scenario with a doubled-CO_2 atmosphere (a situation we will reach by the middle of this century), the models show increases in lightning activity, of approximately 10% for every 1C global warming (Price and Rind, 1994b; Grenfeld et al., 2003; Shindell et al., 2006). Locally that increase can be much larger. In the tropics, the model shows an approximate 40% increase for a doubled-CO_2 atmosphere.

This apparent paradox (a stabilizing atmosphere producing more lightning) was dealt with in more depth by Del Genio et al. (2007) where they showed that in a doubled-CO_2 climate the updrafts strengthen by ~1 m/s, due to a rise in the height of the freezing level in the model. They showed that in certain regions, such as the western United States, the drying in a warmer climate reduces the frequency of thunderstorms, but the strongest storms (highest CAPE values) occur 26% more often. In other words, the drier climate produces less thunderstorms overall, but those storms that do develop are more intense in a warmer climate (Price, 2009). This agrees with observations (Williams et al., 2005) showing increased electrification in the drier regions of the Great Plains of the United States, and increased lightning activity in southeast Asia during the dry warm El Niño periods (Hamid, Kawasaki and Mardiana, 2001).

Another factor that may be important for future thunderstorm development is related to atmospheric aerosols. These microscopic particles may increase in the future due to increased industrialization, increased biomass burning, dust storms, etc. Cloud microphysics and lightning activity appears to be fairly sensitive to the amount of aerosols and cloud condensation nuclei in the atmosphere. It is possible that warmer, drier climates will result in more suspended aerosols and cloud condensation nuclei, hence influencing cloud microphysics and cloud electrification (Rosenfeld et al., 2008). However, while small increases in aerosol loading may result in enhanced thunderstorm activity, for very polluted atmospheres we see a decrease in thunderstorm activity (Altaratz *et al.*, 2010). Hence, future changes in lightning activity will depend on the changes in surface temperature, changes in the atmospheric lapse rate, changes in atmospheric circulation, and changes in aerosol loading of the atmosphere.

Conclusions and Discussion

Lightning is a major natural hazard that impacts many commercial and recreational sectors, while also often causing death or severe injury due to the high temperatures and high peak currents in the lightning channel. Global lightning and thunderstorm activity is driven first and foremost by the Earth's climate, which is driven by solar insolation that varies with latitude, longitude (land/ocean), season, and hour. The climate drives circulation patterns that promote thunderstorms in the tropics and mid-latitudes, and inhibit thunderstorms in the subtropics and polar regions. Locally, thunderstorm activity depends on surface temperature, water vapor, the tropospheric lapse rate, as well as aerosol loading. These parameters can impact the intensity of lightning activity in thunderstorms.

On short time scales (hourly, daily, semi-annual, and annual), there is a robust positive correlation between tropical lightning activity and surface temperature. There is also evidence for increasing thunderstorm activity over the 20th century, although we have limited data to study long-term trends over the last century (Price, 2014). When looking into the future, climate models support the positive correlation between lightning and global temperatures, with increases of approximately 10% in lightning activity for every 1°C of global warming due to increasing greenhouse gas concentrations in the atmosphere. Since lightning itself is a major source of NOx in the atmosphere, a precursor of tropospheric ozone (Price, Penner and Prather, 1997), future changes in lightning could result in a positive temperature feedback, amplifying the initial warming signal.

References

Altaratz, O., Koren, I., Yair, Y. and Price, C. (2010). Lightning response to smoke from Amazonian fires. *Geophysical Research Letters*, 37, L07801. doi:10.1029/2010GL042679.

Christian, H.J., Blakeslee, R.J., Boccippio, D.J., Boeck, W.L., Buechler, D.E., Driscoll, K.T., . . . Stewart, M.F. (2003). Global frequency and distribution of lightning as observed from space by the Optical Transient Detector. *Journal Geophysical Research*, 108, 4005. doi:10.1029/2002JD0023,.

Cooray, V., Cooray, C. and Andrews, C.J. (2007). Lightning caused injuries in humans. *Journal of Electrostatics*, 65, 386–394.

Curran, E.B., Holle, R. L. and Lopez, R.E. (2000). Lightning casualties and damages in the United States from 1959–1994. *Journal of Climate*, 13, 3448–3464.

Del Genio, A.D., Mao-Sung, Y. and Jonas, J. (2007). Will moist convection be stronger in a warmer climate? *Geophysical Research Letters*, 34, L16703. doi:10.1029/2007GL030525.

Fang, Y., Fiore, A.M., Horowitz, L.W., Levy, II H., Hu, Y. and Russell, A.G. (2010). Sensitivity of the NOy budget over the United States to anthropogenic and lightning NOx in summer. *Journal Geophysical Research*, 115, D18312. doi:10.1029/2010JD014079.

Futyan, J.M. and Del Genio, A.D. (2007). Relationships between lightning and properties of convective cloud clusters, *Geophysical Research Letters*, 34, L15705. doi:10.1029/2007GL030227.

Giannaros. T.M., Kotroni, V. and Lagouvardos, K. (2015). Predicting lightning activity in Greece with the Weather Research and Forecasting (WRF) model, *Atmospheric Research,* 156, 1–13.

Glushakow, B. (2007). Effective lightning protection for wind turbine generators, *IEEE Transactions on Energy Conversion,* 22, 214–222.

Grenfell, J.L., Shindell, D.T. and Grewe, V. (2003). Sensitivity studies of oxidative changes in the troposphere in 2100 using the GISS GCM, *Atmospheric Chemical Physics Discussion,* 3, 1805–1842.

Hamid, E.Y., Kawasaki, Z. and Mardiana, R. (2001). Impact of the 1997–98 El Nino on lightning activity over Indonesia. *Geophysical Research Letters,* 28, 147–150.

Intergovernmental Panel on Climate Change (IPCC) (2013). *Climate Change 2013: The Physical Science Basis.* World Meteorological Organization (WMO) and UN Environment Programme (UNEP).

Mareev, E.A. and Volodin, E.M. (2014). Variation of the global electric circuit and ionospheric potential in a general circulation model. *Geophysical Research Letters,* 41, 9009–9016. doi:10:1002/2014GL062352.

Markson, R. (2007). The global circuit intensity: Its measurement and variation over the last 50 years. *Bulletin American Meteorological Society,* 88, 1–19.

Markson, R. and Price, C. (1999). Ionospheric potential as a proxy index for global temperatures, *Atmospheric Research,* 51, 309–314.

Pawar, S.D., Lal, D.M. and Murugavel, P. (2001). Lightning characteristics over central India during Indian summer monsoon. *Atmospheric Research,* 106, 44–49.

Price, C. (1993). Global surface temperatures and the atmospheric electrical circuit. *Geophysical Research Letters,* 20, 1363–1366.

Price, C. (2006). Global thunderstorm activity. In M. Fullekrug et al. (Eds.), *Sprites, Elves and Intense Lightning Discharges* (pp. 85–99). Amsterdam: Springer.

Price, C. (2009). Will a drier climate result in more lightning? *Atmospheric Research,* 91, 479–484.

Price, C. (2014). Lightning and global temperature change. In V. Cooray (Ed.), *The Lightning Flash* (pp. 861–878). London: The Institute of Engineering and Technology (IET) Press.

Price, C. and Rind, D. (1992). A simple lightning parameterization for calculating global lightning distributions. *Journal of Geophysical Research,* 97, 9919–9933.

Price, C. and Rind, D. (1994a). Modeling global lightning distributions in a General Circulation Model. *Monthly Weather Review,* 122, 1930–1939.

Price, C. and Rind, D. (1994b). Possible implications of global climate change on global lightning distributions and frequencies. *Journal of Geophysical Research,* 99, 10823–10831.

Price, C., Penner, J. and Prather, M. (1997). NOx from lightning, Part I: Global distribution based on lightning physics. *Journal of Geophysical Research,* 102, 5929–5941.

Price, C. and Asfur, M. (2006). Can lightning observations be used as an indicator of upper-tropospheric water vapor variability? *Bulletin American Meteorological Society,* 87, 291–298.

Reeve, N. and Toumi, R. (1999). Lightning activity as an indicator of climate change. *Quarterly Journal of Royal Meteorological Society,* 125, 893–903.

Rosenfeld, D., Lohmann, U., Raga, G.B., O'Dowd, C.D., Kulmala, M., Fuzzi, S., Reissell, A. and Andrea, M. (2008). Flood or drought: How do aerosols affect precipitation? *Science,* 1309–1313.

Shindell, D.T., Faluvegi, G., Unger, N., Aguilar, E., Schmidt, G.A., Koch, D.M., Bauer, S.E. and Miller, R.L. (2006). Simulations of preindustrial, present-day, and 2100 conditions in the NASA GISS composition and climate model G-PUCCINI. *Atmospheric Chemical Physics,* 6, 4427–4459.

Siingh, D., Ramesh Kumar, P., Kulkarni, M.N., Singh, R.P., Singh, A.K. (2012). Lightning, convective rain and solar activity — Over the South/Southeast Asia. *Atmospheric Research,* 120–121, 99–111.

Stocks, B.J., Mason, J.A., Todd, J.B., Bosch, E.M., Wotton, B.M., Amiro, B.D., . . . Shinner, W.R. (2002). Large forest fires in Canada, 1959–1997. *Journal of Geophysical Research,* 107, 8149. doi:10.1029/2001JD000484.

Williams, E.R. (1992). The Schumann resonance: A global tropical thermometer. *Science*, 256, 1184–1187.

Williams, E.R. (2005). Lightning and climate: A review. *Atmospheric Research*, 76, 272–287.

Williams, E.R. (2009). The global electric circuit: A review. *Atmospheric Research*, 91, 140–152.

Williams, E.R., Mushtak, V., Rosenfeld, D., Goodman, S. and Boccippio, D. (2005). Thermodynamic conditions favorable to superlative thunderstorm updraft, mixed phase microphysics and lightning flash rate. *Atmospheric Research*, 76, 288–306.

Slide 1

Slide 2

Slide 3

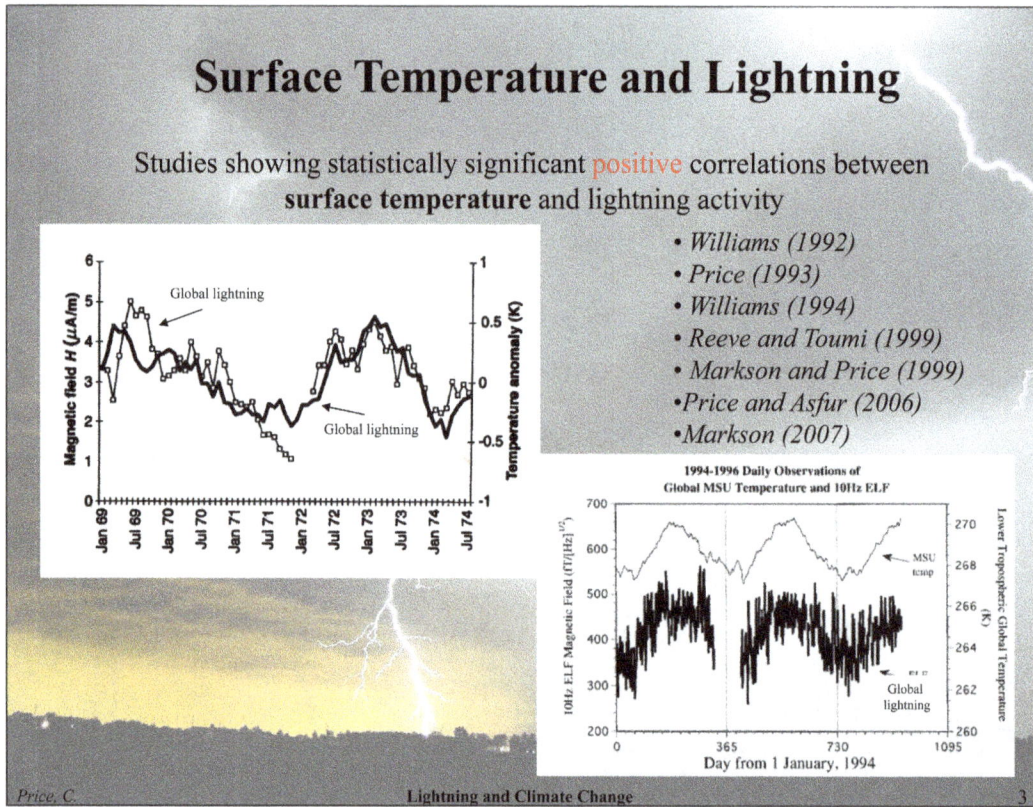

Slide 4

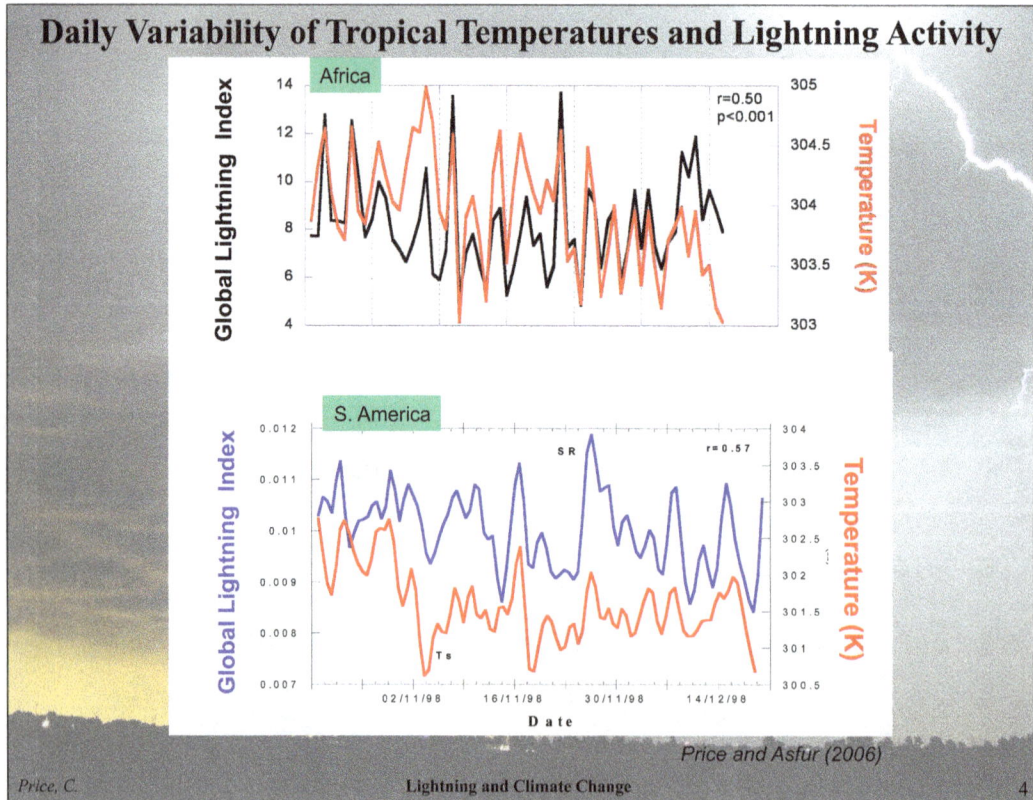

Slide 5

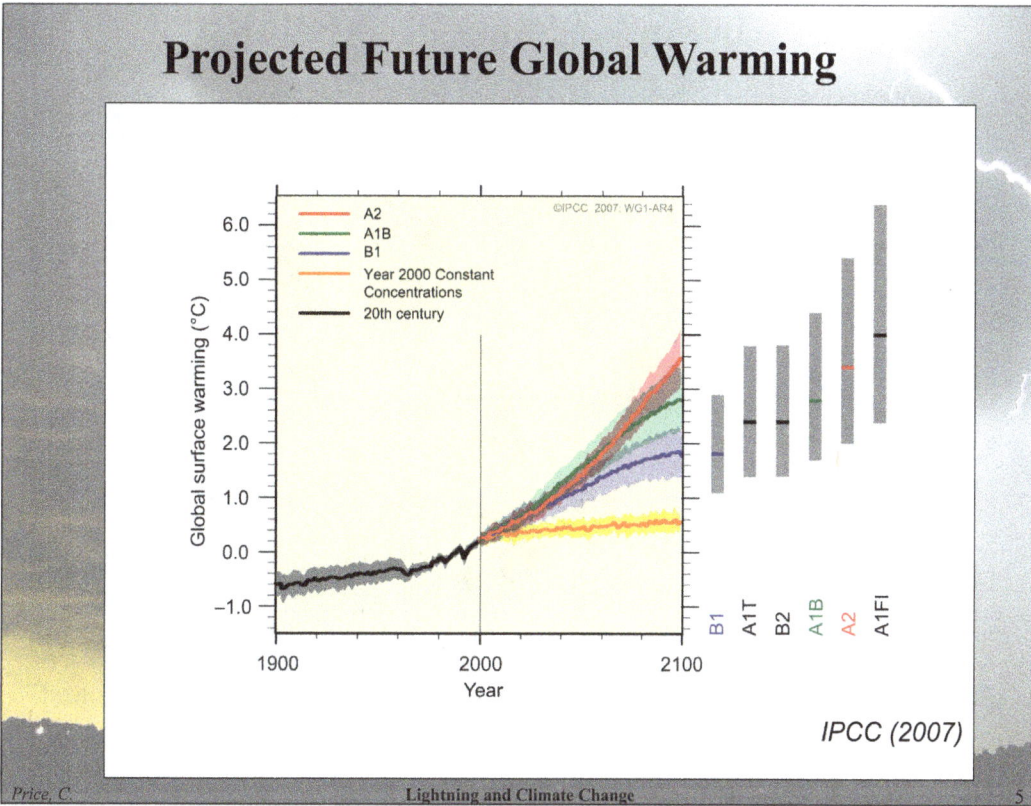

Slide 6

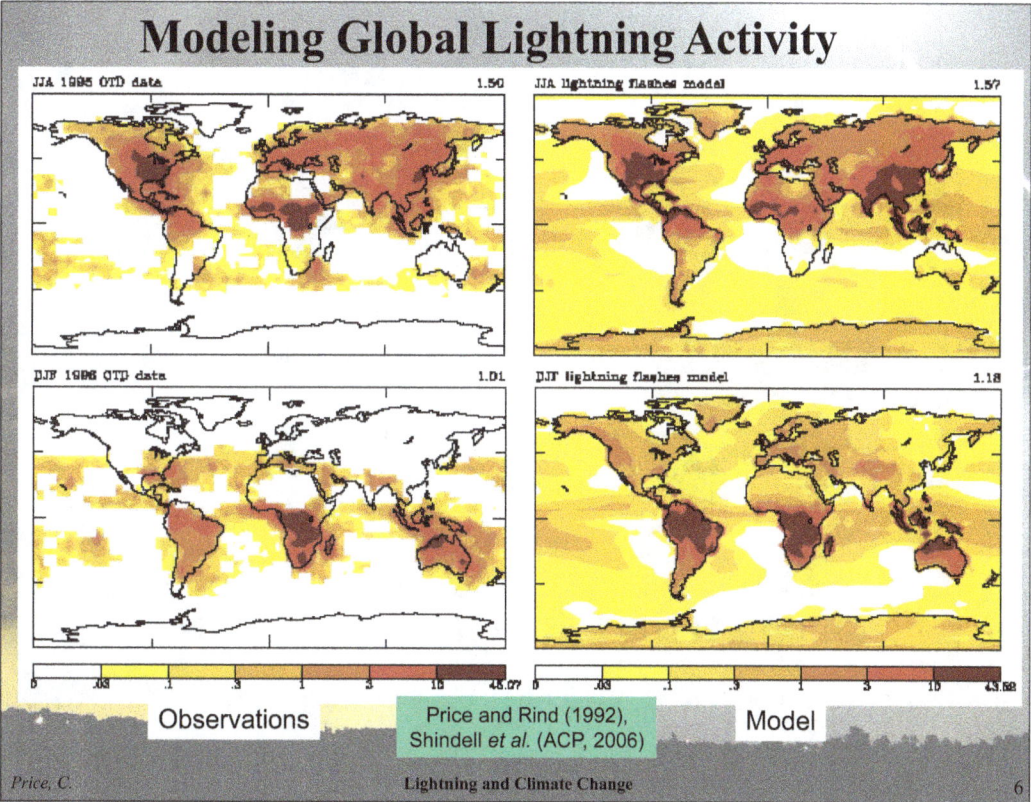

Slide 7

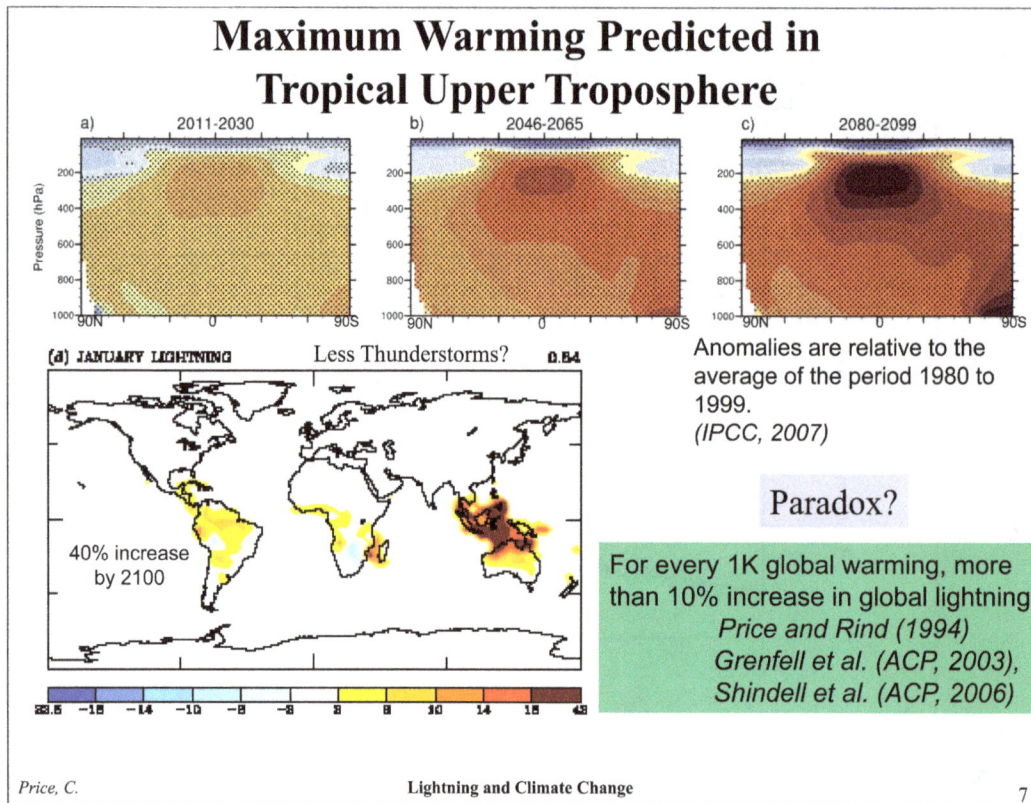

Maximum Warming Predicted in Tropical Upper Troposphere

Anomalies are relative to the average of the period 1980 to 1999.
(IPCC, 2007)

Paradox?

For every 1K global warming, more than 10% increase in global lightning
Price and Rind (1994)
Grenfell et al. (ACP, 2003),
Shindell et al. (ACP, 2006)

Price, C.   **Lightning and Climate Change**   7

Slide 8

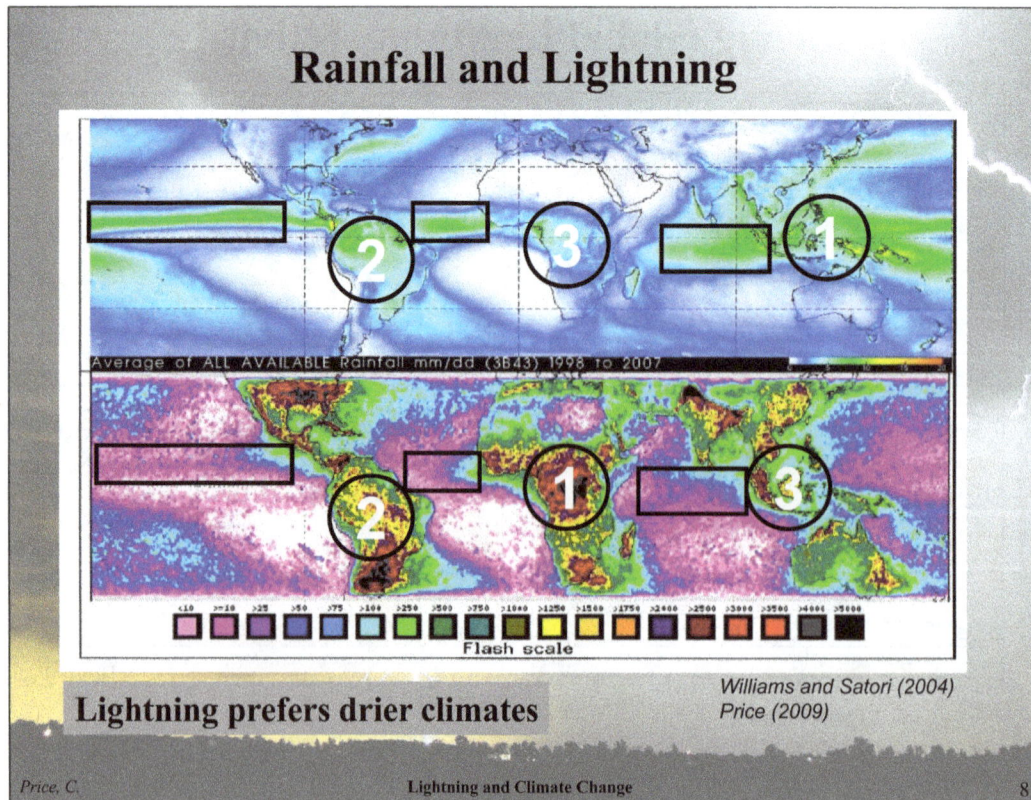

Rainfall and Lightning

Average of ALL AVAILABLE Rainfall mm/dd (3B43) 1998 to 2007

Flash scale

Lightning prefers drier climates

Williams and Satori (2004)
Price (2009)

Price, C.   **Lightning and Climate Change**   8

Slide 9

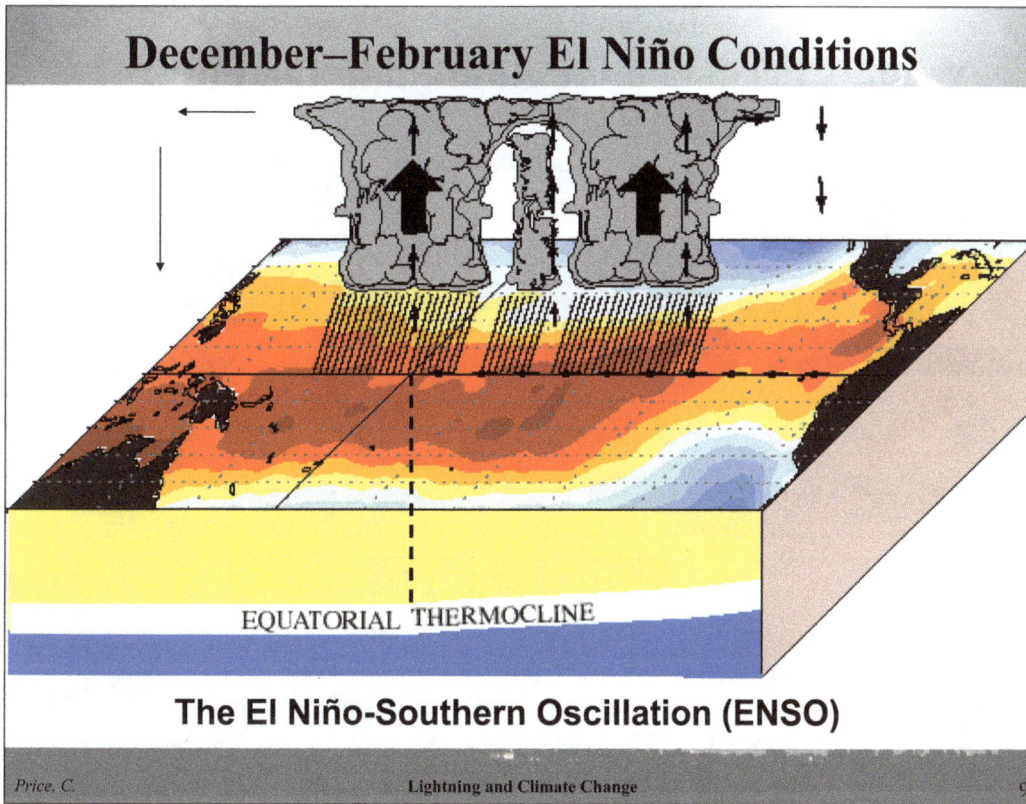

Slide 10

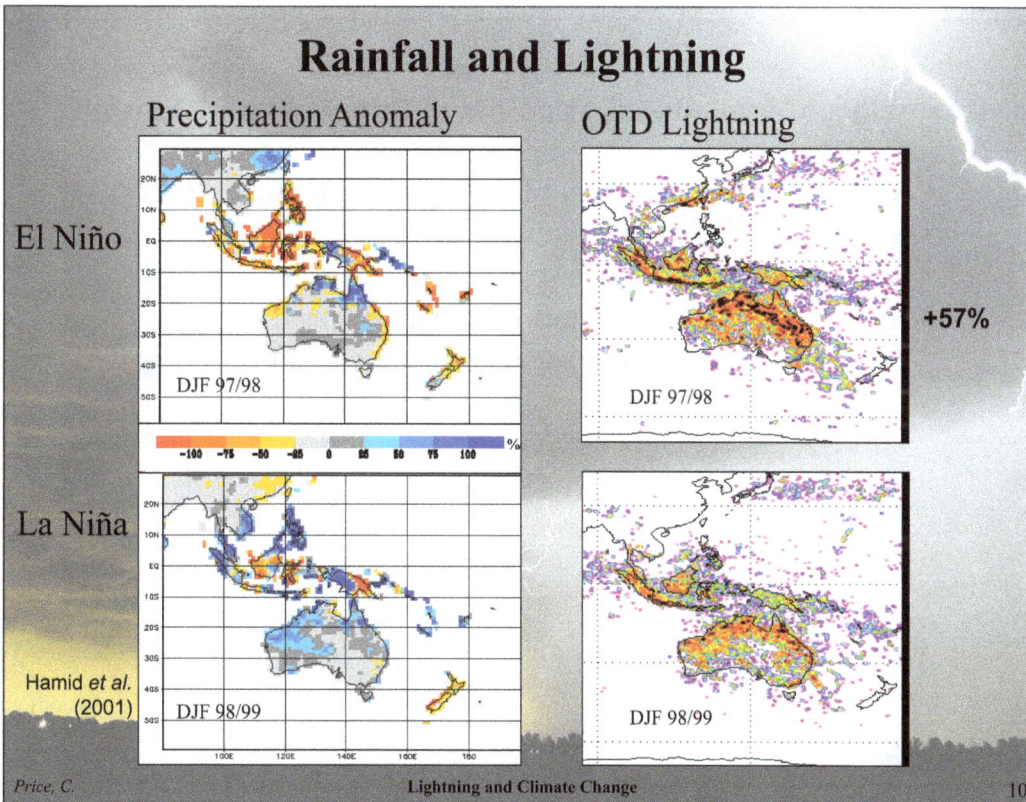

Slide 11

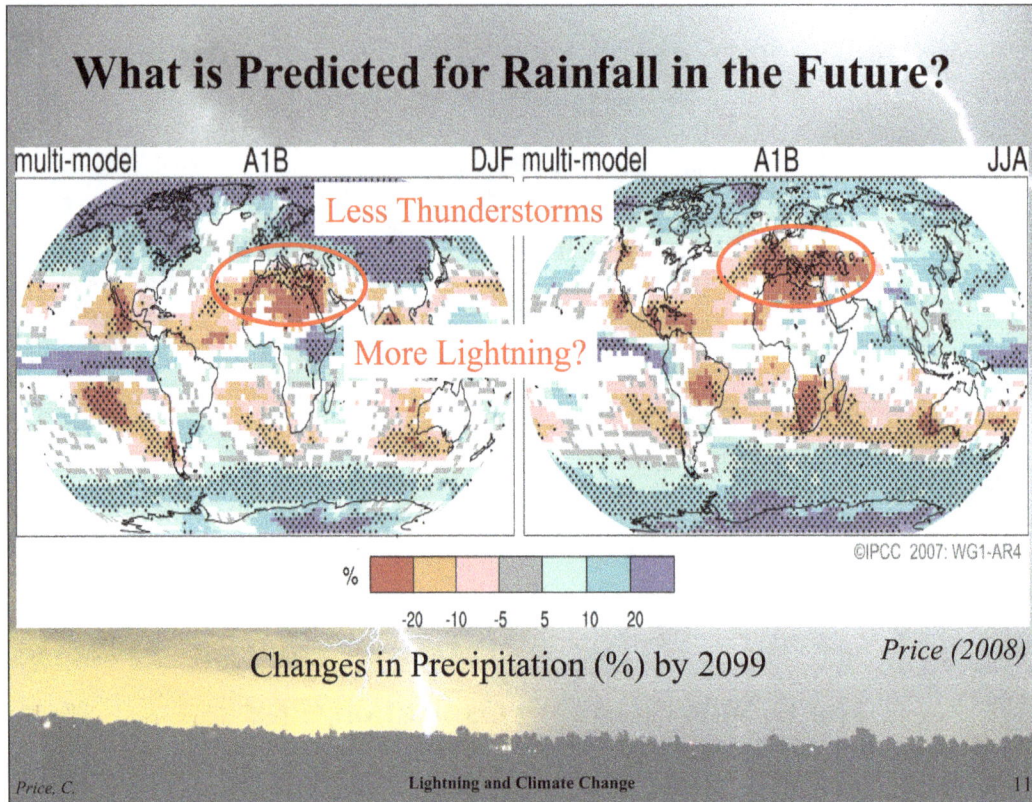

Slide 12

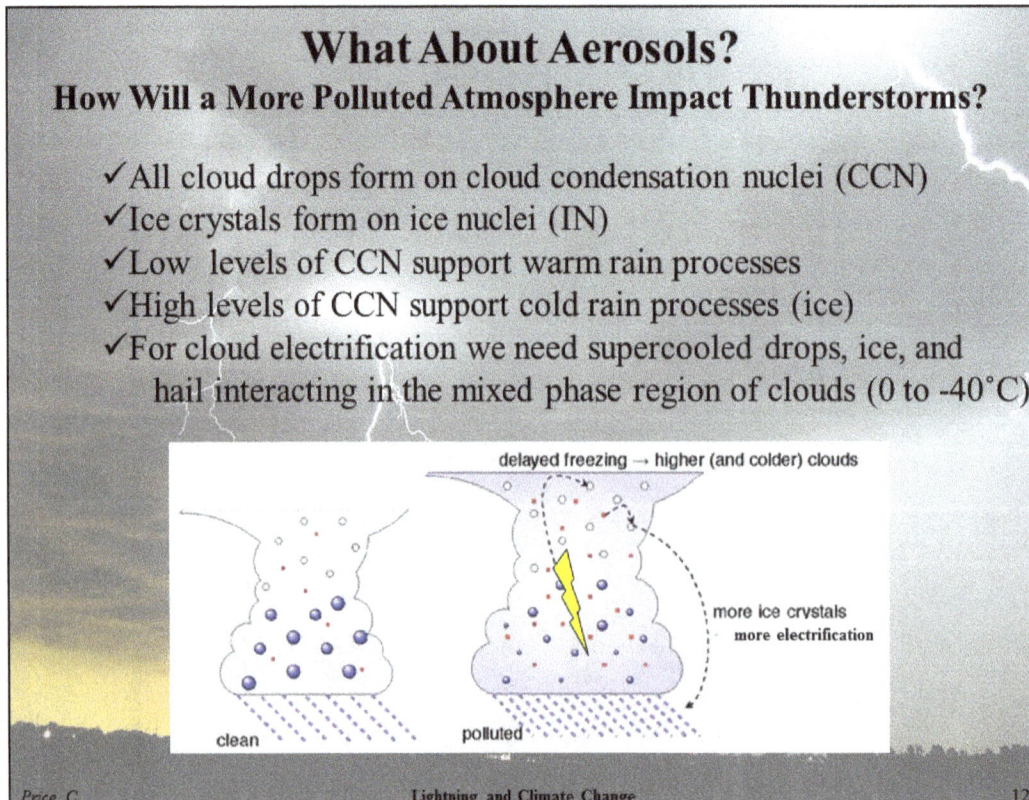

Slide 13

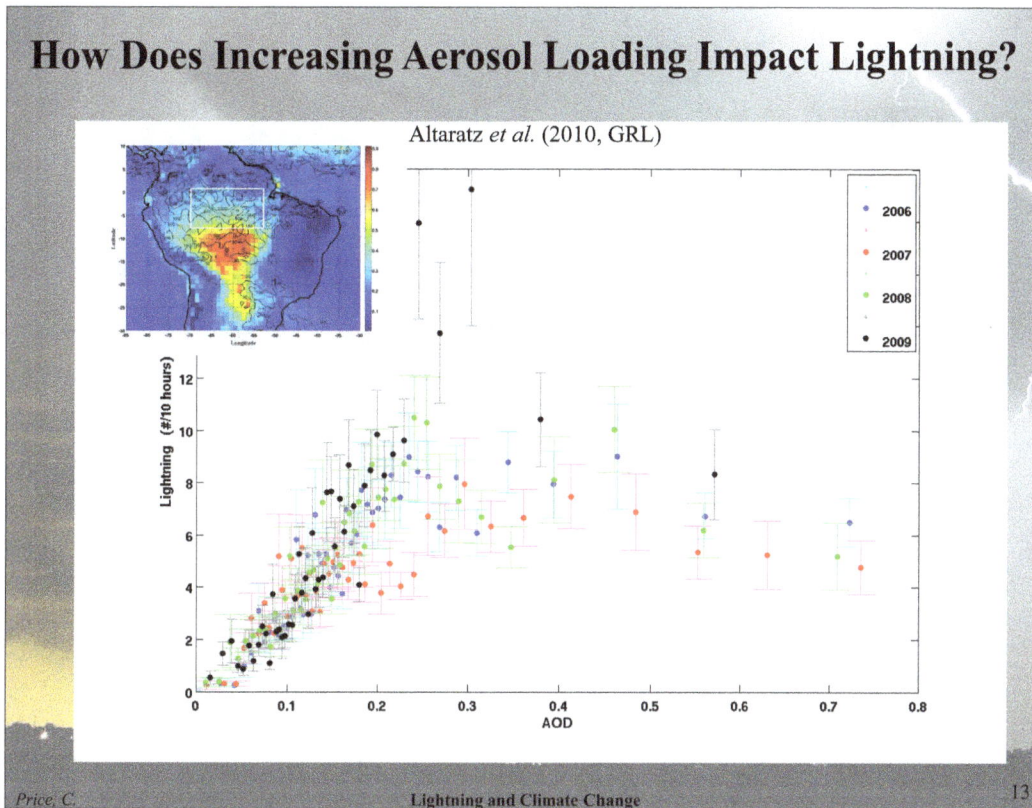

Slide 14

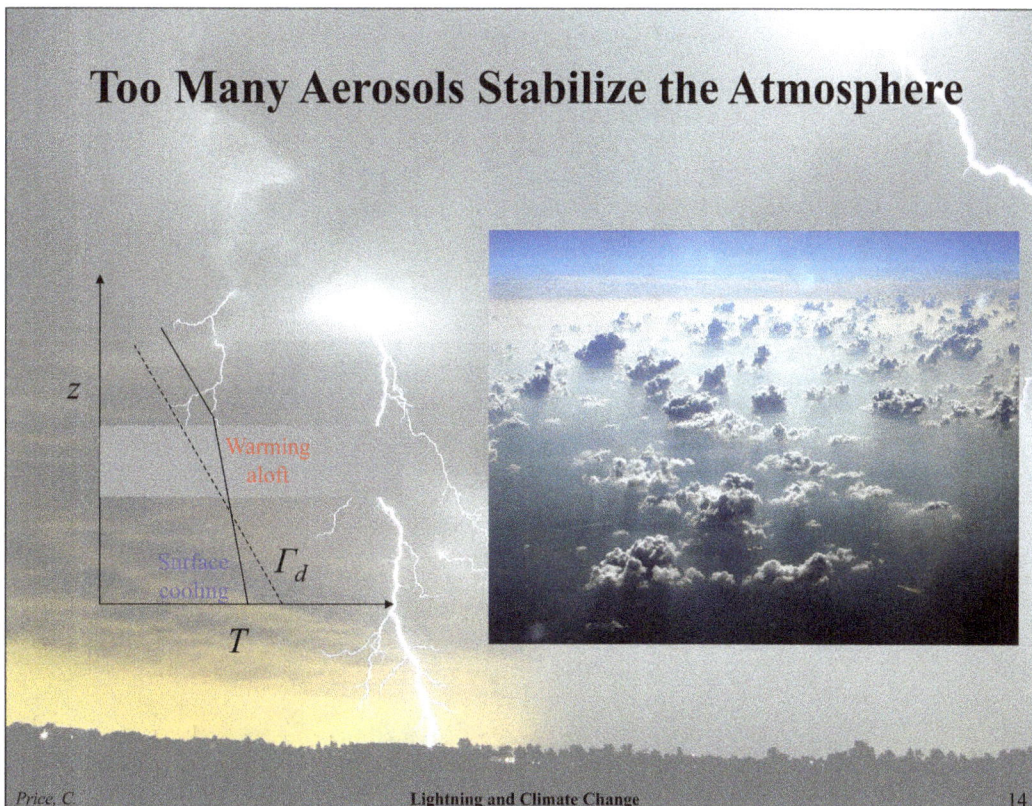

Slide 15

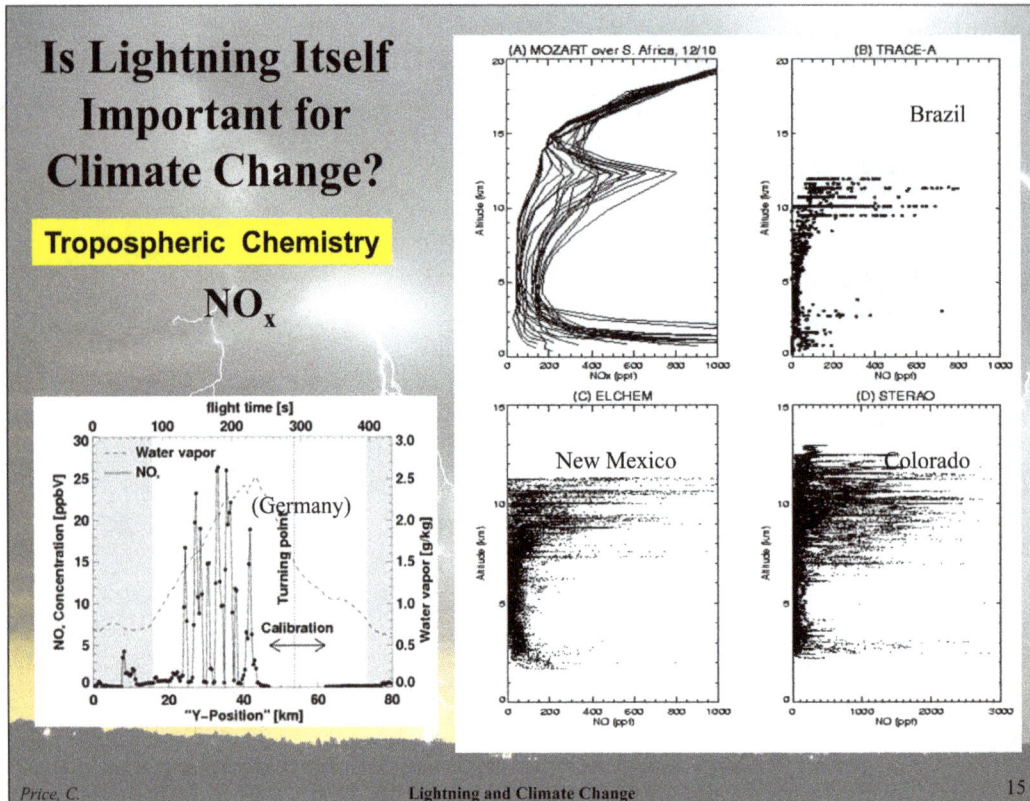

Price, C.   Lightning and Climate Change   15

Slide 16

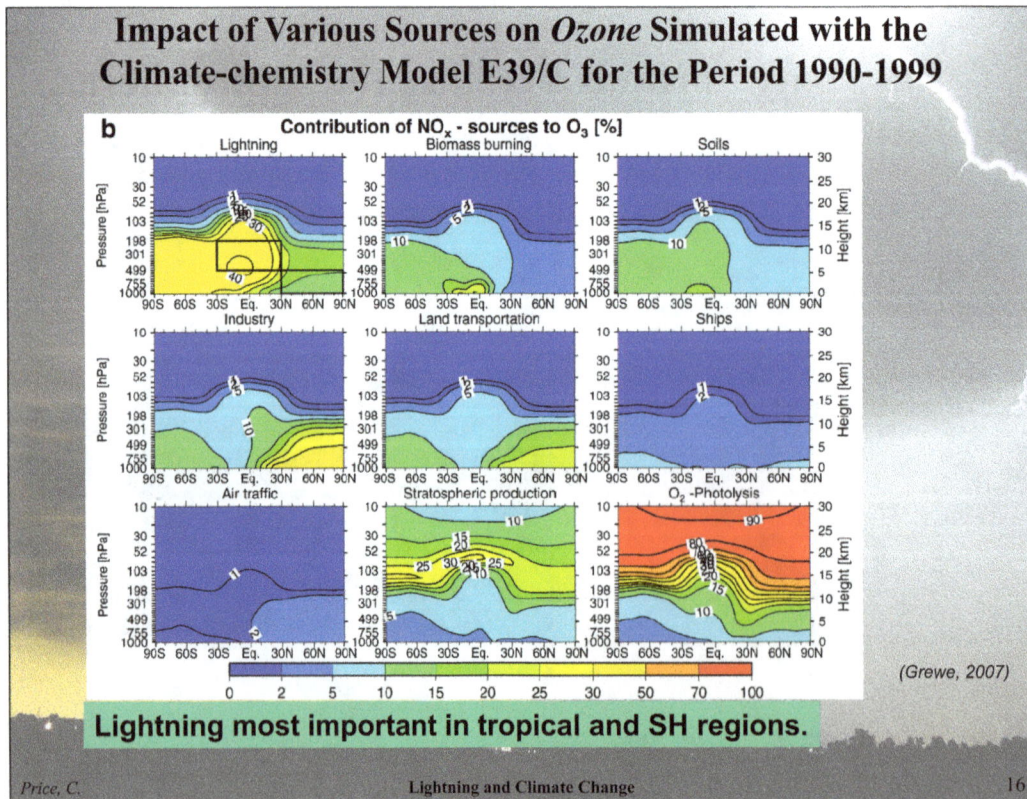

Price, C.   Lightning and Climate Change   16

Slide 17

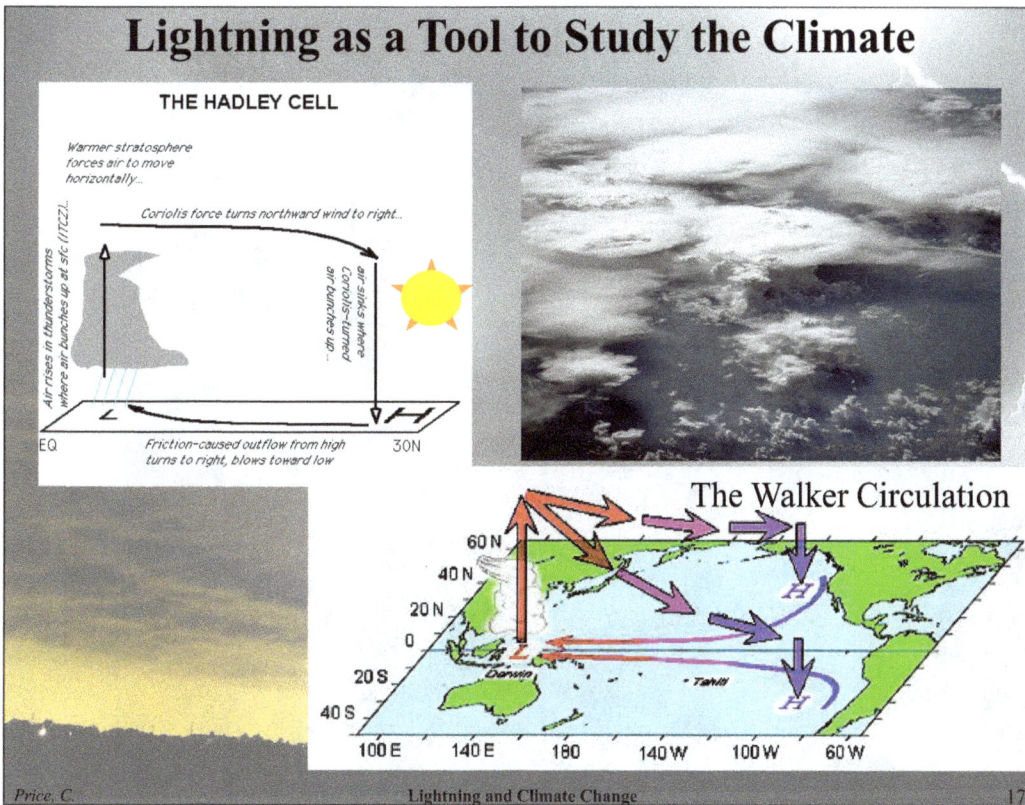

Slide 18

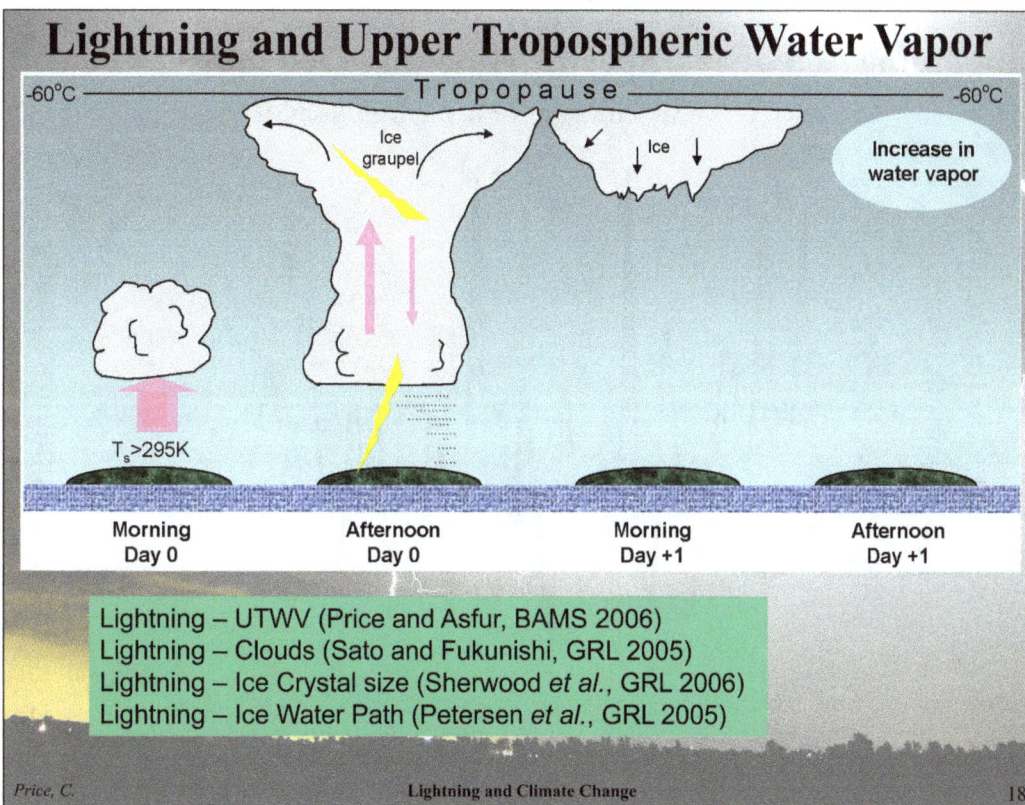

Slide 19

Slide 20

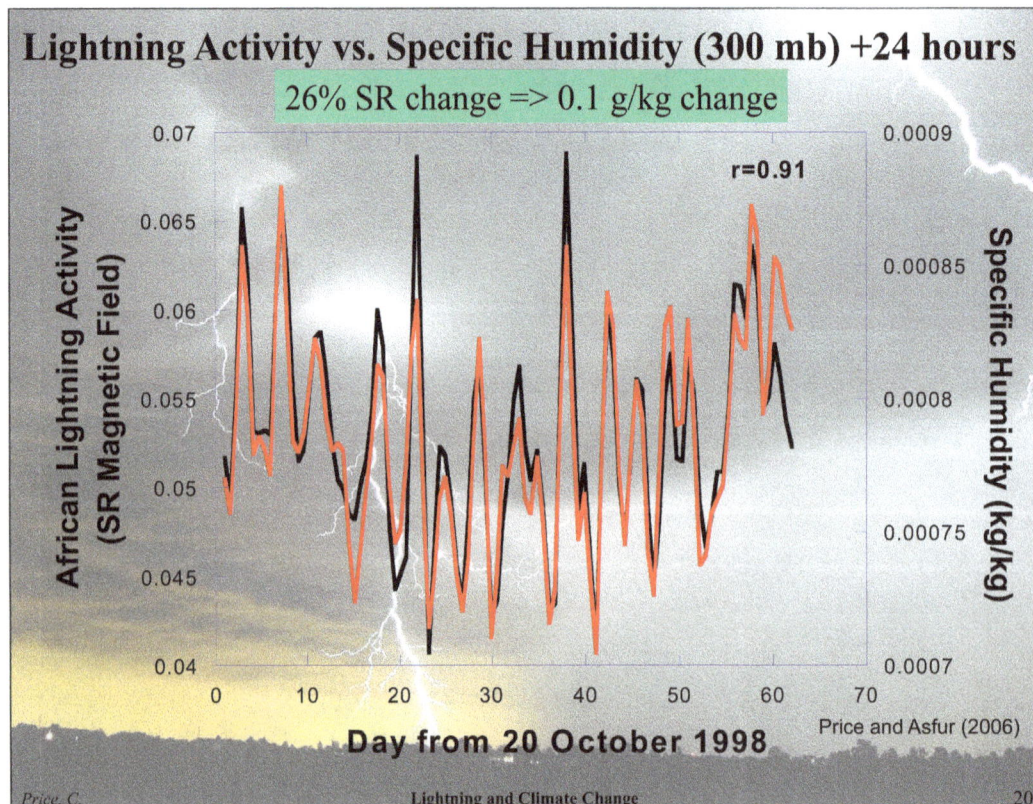

Slide 21

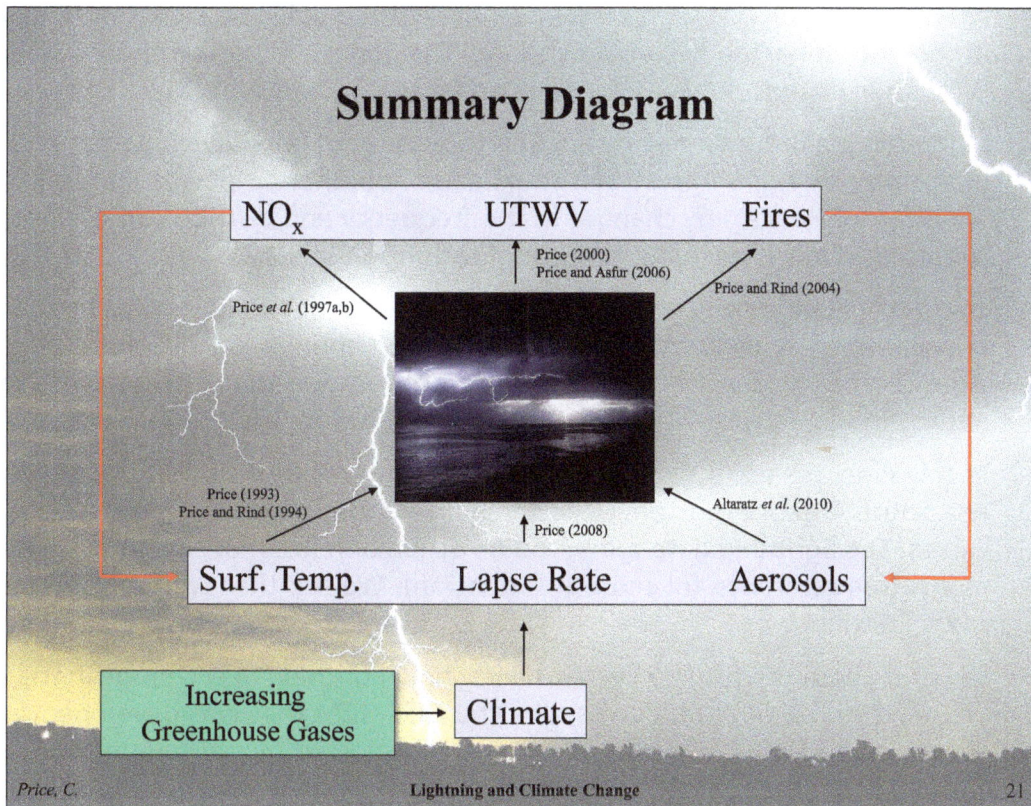

Slide Notes

Slide 1   This presentation gives a general overview of the link between lightning activity in the atmosphere, and climate change. This topic was part of the Ph.D. thesis of Prof. Colin Price (Tel Aviv University) under the guidance of David Rind.

Slide 2   Lightning is a natural hazard that kills thousands of people every year, and causes major property and infrastructure damage, whether to power lines, aircraft, or wind farms. Hence, any changes in the frequency and distribution of lightning in a warmer climate will have many implications on society.

Slide 3   There have been many studies looking at the link between surface temperatures and lightning activity. All these studies, even though on different time scales from diurnal to annual, show a positive correlation between surface warming and lightning activity. Warmer temperatures produce more lightning.

Slide 4   In this slide, we see the daily tropical temperatures (red curves) over Africa (top) and South America (bottom), and the daily lightning activity in the black and red curves. The lightning data are based on Schumann Resonance lightning measurements (see reference for more information). We see that warmer days result in more lightning.

Slide 5   So what about the future? Well, to study how lightning may change in the future, we need to use climate models, like the GISS GCM that was developed over the years by David Rind. The figure shows the projected global temperatures out to the year 2100 based on the IPCC model runs.

Slide 6   David Rind was key in developing the first lightning parameterization in climate models, and this parameterization (Price and Rind, 1992) is still used today in many climate models. Here, we see a comparison between the GISS model lightning (right) and satellite observations of lightning (left) from the Optical Transient Detector (OTD) for northern hemisphere summer (top) and winter (bottom).

Slide 7   But while the surface warms in a warmer climate, the upper atmosphere warms even more, stabilizing the atmosphere (top). In fact, in the tropics we have the largest stabilizing effect. So, we would think that we should have LESS thunderstorms. But the model still predicts MORE lightning in a warmer world (bottom). And at a rate of about 10% increase per 1 K warming. This seems to present a paradox. More stable, but more lightning?

Slide 8   If we take a look at this 10-year climatology of rainfall (top) and lightning (below) from the same TRMM satellite, we see a strange thing. While the most active centers of lightning rank (1) Africa, (2) Americas, and (2) South East Asia, the rainfall centers rank in the opposite order. (1) SE Asia most rainfall, then (2) South America, and (3) Africa. And in the boxes over the ocean with lots of rain, there is no lightning. Lightning prefers the drier climate.

Slide 9   We can also see this when we look at the El Niño Southern Oscillation (ENSO) phenomenon. During El Niño, there is less rainfall in Indonesia and SE Asia, while during La Niña, there is more rainfall.

Slide 10 But during the dry El Niño there is 57% MORE lightning. The drying out of the climate appears to increase lightning activity. And in the moist La Niña period, there is less lightning. So on climatological scales (climate change) not only is temperature important, but also whether regions will become drier or wetter.

Color bar for lightning is the same as in Slide 8.

Slide 11 When we look at the IPCC rainfall changes by 2099 we see that some areas will get wetter and others drier. Based on what we just saw for the global rainfall/lightning link, and the ENSO analogue, will we get more lightning in the areas that will become drier, like the Mediterranean?

Slide 12 Another factor we need to consider is aerosols. Aerosols are tiny particles in the atmosphere that act as cloud condensation nuclei (CCN) and influence rainfall, cloud properties, and cloud electrification. How will changes in aerosol loading (due to air pollution, biomass burning, dust storms, etc.) impact lightning activity. Having more CCN in the atmosphere may result in smaller drops that are lighter and easier to be lofted into the mixed phase (ice and supercooled drops) region of the cloud where lightning is generated. More droplets turning to ice will also release more latent heat and perhaps invigorate the clouds producing more lightning.

Slide 13 To check this, we looked at the lightning activity over the Amazon from 2006 to 2009 during the dry season in South America. We looked to see how the aerosols (measured from satellites as the aerosol optical depth [AOD]) influenced lightning. What we found was that for clean air, with AOD<0.25 increases in aerosols increased the lightning activity. But when AOD>0.25 we got a boomerang effect, where the lightning decreased.

Slide 14 What is happening here? With a little aerosol, the microphysical effect dominates, and the aerosols delay the precipitation process, and help the electrification process, increasing lightning. But if there are too many aerosols, the aerosols themselves in the air absorb solar radiation and heat the layer of the aerosols, while cooling the surface, stabilizing the lower atmosphere, choking the thunderstorms, reducing the lightning activity.

Slide 15 Now can lightning itself impact the Earth's climate? Yes, lightning is a major source of NOx gas, which is a precursor of ozone (O3) that is a strong greenhouse gas. Flights in thunderstorms show high concentrations of NOx and O3, and if lightning increases in a warmer world, the O3 increases associated with the lightning may feedback on the climate and amplify any initial warming.

Slide 16 Here, we can see a model output where the different global sources of NOx are separated to show how significant each is as a function of latitude and height (pressure). In the tropics, and southern hemisphere troposphere, lightning is the largest source of ozone in the atmosphere.

Slide 17 In addition, lightning observations themselves can be useful in monitoring changes in different climate parameters. Lightning occurs along the Intertropical Convergence Zone (ITCZ) and in the West Pacific, defining the rising branch of the Hadley circulation and the Walker circulation. More lightning is related to stronger convection.

Slide 18   Numerous studies have shown that in addition to links between lightning and surface temperature, lightning is also related to upper tropospheric water vapor, an important climate feedback, as well as clouds, ice crystal size, ice water path, etc.

Slide 19   Here, we can see how thunderstorms (white) contribute to the upper tropospheric water vapor (gray). The outflow of thunderstorms moistens the upper troposphere, and then transports this water vapor away from the tropics where it can act as an efficient greenhouse gas.

Slide 20   Here, we see a 62-day time series of the mean lightning over Africa (same as African lightning plot on Slide 4) together with the 300-mb-specific humidity (absolute humidity) above Africa. It should be noted that the red curve was shifted left by 24 hours to show the agreement. The lightning activity peaks one day before the peak in humidity in the upper atmosphere.

Slide 21   So, we have seen that the Earth's climate determines the surface temperature, lapse rate, and aerosols distribution around the planet, and these parameters are important for thunderstorm and lightning production. The thunderstorms transport water vapor to the upper atmosphere, while also generating NOx and ozone, increasing the greenhouse gas feedbacks. Although not discussed here, lightning can also generate wildfires that also produce aerosols and greenhouse gases that can feedback on the climate system. And if climate change results in additional changes in temperature, drying, and aerosols, there will be significant impacts on thunderstorms and lightning, while the impact may differ from region to region.

SECTION 5

Polar Responses

CLIMATE LECTURE 13

Polar Sea Ice Coverage, Its Changes, and Its Broader Climate Impacts

Claire L. Parkinson

NASA Goddard Space Flight Center, Greenbelt, MD, USA

Claire L. Parkinson is a Climate Change Senior Scientist at NASA's Goddard Space Flight Center. Her research emphasis is on polar sea ice and its connections to the rest of the climate system and to climate change, utilizing both observations and models. Her work on sea ice monitoring includes the development and analysis of satellite-derived sea ice data sets and the use of those data sets to quantify the changes in the Arctic and Antarctic sea ice covers over the course of the satellite record.

Introduction

Sea ice is a crucial component of the climate system in both polar regions, reflecting solar radiation back to space, hindering exchanges of heat, mass, and momentum between ocean and atmosphere, and having numerous other impacts both on climate and on polar eco-systems. In view of these impacts and the interconnectedness of the global climate system as a whole, David Rind early on recognized the potential importance of sea ice changes to latitudes far equatorward of where the sea ice is and led a study to quantify those changes (described below).

Sea ice is ice formed from the freezing of seawater. It spreads over large areas of both polar regions all year but especially during the respective polar winters. Sea ice coverage in the Arctic typically reaches a maximum in March and a minimum in September, and in the Antarctic it typically reaches a maximum in September and a minimum in February. In both polar regions, wintertime sea ice spreads over an area far larger than the entire continent of Europe, making it quite impractical to monitor the sea ice coverage routinely by any means other than through satellite observations.

Satellite passive-microwave instruments have been particularly useful in obtaining a climate record of sea ice because they allow data collection all year, irrespective of day or night conditions, and through most cloud covers. Some satellite passive-microwave data were collected in the early and mid-1970s, but the near-continuous record of satellite passive-microwave data started in late 1978, following the October 1978 launch of the Scanning Multichannel Microwave Radiometer (SMMR) on NASA's Nimbus 7 satellite. The SMMR was followed by a sequence of Special Sensor Microwave Imager (SSMI) instruments on satellites

273

of the U.S. Defense Meteorological Satellite Program (DMSP), the first launched in June 1987, and the DMSP SSMI Sounder (SSMIS) instruments, first launched in November 2006, and by passive-microwave instruments from other countries.

Averaged over 1979–2014 from the SMMR/SSMI/SSMIS record, the average annual cycle of monthly sea ice extents in the Arctic ranged from 6.3×10^6 km^2 in September to 15.2×10^6 km^2 in March and in the Antarctic ranged from 3.1×10^6 km^2 in February to 18.6×10^6 km^2 in September, showing a far larger annual range in the Antarctic, where the central polar region is covered by the Antarctic continent, limiting the area for summertime ice, and the ice is not geographically constrained in its equatorward expansion during winter. ['Sea ice extent' is the summed area of all grid elements with at least 15% sea ice concentration, where the grid elements used in the SMMR, SSMI, and SSMIS calculations are approximately 25 km × 25 km and 'sea ice concentration' is the percent areal coverage of sea ice in a given grid element or other area of interest.]

For 2014 specifically, the monthly average ice extents in the Arctic were 14.6×10^6 km^2 in March and 5.3×10^6 km^2 in September, much lower than the 1979–2014 average, and the ice extents in the Antarctic were 3.8×10^6 km^2 in February and 19.8×10^6 km^2 in September, much higher than the 1979–2014 average. For comparison, the area of the United States is 9.9×10^6 km^2 and the area of Canada is 10.0×10^6 km^2.

The large area of sea ice coverage has a variety of impacts on the rest of the polar climate system. Among the most important, the sea ice reflects 50–98% of the solar radiation that reaches it, in sharp contrast to the surrounding liquid water, which absorbs by far the majority of the solar radiation reaching it. Thus, the very presence of sea ice tends to keep the polar regions colder than they would be without it. Sea ice also restricts exchanges of heat, mass, and momentum between ocean and atmosphere, thereby affecting, for instance, the amount of precipitation (snowfall and/or rainfall) in surrounding regions. Other climate impacts include changes to ocean circulation brought on by salinity changes caused by the ice formation, aging, movement, and decay. These at times include overturning in the water column beneath the ice.

Recognizing the highly interconnected nature of the Earth's climate system and the likelihood that polar sea ice changes could affect climate far removed from the polar regions, David Rind led a study in the mid-1990s that examined in detail the specific case of how the projected warming from doubled CO_2 is affected by sea ice, doing so by calculating the effects of doubled CO_2 in a major global climate model (from the Goddard Institute for Space Studies, or GISS) both when including sea ice calculations and when not allowing the sea ice annual cycle to change. The results showed that inclusion of sea ice calculations in the model resulted in more warming at every latitude. As expected, the increased warming was greatest in the sea ice regions (where warming reduces sea ice coverage, which in turn reduces the reflection of solar radiation back to space and thereby further increases the warming), but even near the equator, the additional warming produced simply by incorporating sea ice calculations was greater than 0.5 K. Overall, the global warming simulated for a doubling of CO_2 was 4.17 K with sea ice calculations included and 2.61 K without them, meaning that 37% of the simulated 4.17 K global warming could be attributed explicitly to the inclusion of sea ice calculations (Rind, Healy, Parkinson and Martinson, 1995). This is a whopping percentage for a phenomenon that only covers about 4–5% of the Earth's surface.

Other studies have suggested that the observed greater warming in recent decades in the Arctic than in lower latitudes, resulting in part from the loss of sea ice coverage, is currently affecting and will continue to affect weather patterns far south of the sea ice cover (Walsh, 2013). For instance, Mori et al. (2014) connect the Arctic sea ice decreases to recent increased frequency of severe winters in mid-latitude Eurasia, and Francis and Vavrus (2012) suggest that the amplification of warming in the Arctic might be causing changes in atmospheric wave patterns that increase the probability of mid-latitude extreme weather events, including cold spells, heat waves, and increased drought and flooding. However, not all studies find a clear connection between the sea ice losses and the lower latitudes (e.g., Barnes, Dunn-Sigouin, Masato and Woollings, 2014), and sorting out the sea ice impacts will probably remain an area of active research for many years.

The combination of the large areal coverage of sea ice, its substantial changes over time, and the potential widespread impacts of those changes all contribute to an interest in closely monitoring the sea ice cover, and this has indeed been done since the start of the SMMR record in late 1978. By the end of the 1980s, the satellite passive-microwave data showed a slight decrease in Arctic sea ice coverage since late 1978. In light of significant interannual variability, the slight overall decrease was not yet a convincing long-term trend, but it was enough to alert sea ice researchers to keep an eye on the continuing changes. By the end of the 1990s, the downward trend was more convincing, and in the subsequent years it became even more convincing, especially with a very prominent plummeting of the sea ice cover in the summer of 2007, reaching what at that time was a record low in Arctic sea ice extent, falling to 4.1×10^6 km^2 in September 2007. The ice rebounded somewhat in subsequent summers, but reached a new record low of 3.4×10^6 km^2 in September 2012 before rebounding again in 2013 and 2014. The prominent plummeting of the sea ice cover in the summer of 2007 was strikingly large compared to the entire preceding satellite record and came as a shock to the sea ice research community, leading some to predict that an ice-free Arctic might be upon us far quicker, within a decade, than had previously been thought. Case studies have been done for both the 2007 and 2012 sea ice retreats, and in both cases weather conditions in those particular years have been found to be a factor in the ice losses.

When considering the full 1979–2014 record, the Arctic sea ice extent has decreased in every season and every month. The month with the greatest decreases is September, with a trend of –84,900 ± 8900 km^2/yr (–10.9 ± 1.1%/decade), but even the month with the smallest decreases, May, has reductions of –30,100 ± 4100 km^2/yr (–2.2 ± 0.3%/decade). On a yearly average basis, the trend in Arctic sea ice extents is –53,100 ± 3600 km^2/yr (–4.3 ± 0.3%/decade), equaling an ice extent loss each year of an area slightly exceeding the area of Costa Rica (51,100 km^2). [The Arctic sea ice trends are updates from Parkinson et al. (1999) and subsequent studies.]

Several other measures further confirm reductions in Arctic sea ice coverage. These include reductions in ice thickness (e.g., Kwok and Rothrock, 2009), reductions in the amount of multiyear ice (e.g., Comiso, 2012), shortening of the length of the sea ice season in much of the Arctic (Parkinson, 2014b), and earlier occurrence of melt onset (e.g., Bliss and Anderson, 2014). The length of the sea ice season (number of days with sea ice present) can be calculated each year for each grid element, enabling detailed maps to be made of the trends. When this is done, trends for 1979–2014 show a shortening of the sea ice season almost everywhere in the Arctic sea ice region outside the region of year-round sea ice coverage except in the Bering Sea and small portions of the Canadian Archipelago. This is a significant contrast from the corresponding trends for the first 10 years of the record, 1979–1988, when lengthened sea ice seasons dominated the season length changes in the Western hemisphere and shortened sea ice seasons dominated in the Eastern hemisphere (Parkinson, 2014b).

The Arctic sea ice decreases are complemented by a wide range of additional changes in the Arctic, including atmospheric warming, ocean warming, permafrost thawing, shorter snow seasons, loss of mass in land-based glaciers and the Greenland ice sheet, and increased coastal erosion (e.g., Jeffries, Overland, and Perovich, 2013). They are not, however, complemented by corresponding decreases in the sea ice extents in the Antarctic, at least for the period through 2014.

The same satellite passive-microwave instruments used to monitor the Arctic sea ice cover are also used to monitor the Antarctic sea ice cover, and for the Antarctic, instead of showing sea ice decreases, every month shows sea ice increases overall for the 1979–2014 period. These increases are notable but not nearly as large as the ice extent decreases in the Arctic. December has the largest increases, at 34,100 ± 9900 km^2/year (3.5 ± 1.0%/decade), and February has the smallest, at 13,200 ± 5800 km^2/year (4.6 ± 2.0%/decade). On a yearly average basis, the overall trend in Antarctic sea ice extent over the 1979–2014 period is 22,400 ± 4300 km^2/year (2.0 ± 0.4%/decade). This equals an ice extent gain each year of an area greater than the area of El Salvador (21,000 km^2). [The Antarctic sea ice trends are updates from Zwally et al. (2002) and subsequent studies.]

Several scientists have explored what might have caused the Antarctic sea ice to expand overall between 1979 and 2014, and here the pattern of changes in the ice cover have proved helpful. In particular, there were marked sea ice decreases in the Bellingshausen Sea (to the immediate west of the Antarctic Peninsula) and marked increases in the Ross Sea (farther to the west, with the Amundsen Sea in between). This pattern is consistent with more persistent low-pressure systems centered over the Amundsen Sea, sending relatively warm air from lower latitudes southward over the Bellingshausen Sea and Antarctic Peninsula and cold air from over Antarctica northward out over the Ross Sea, atmospheric changes that some have thought could be connected to the Antarctic ozone hole (e.g., Turner et al., 2009). Other suggestions as to why the sea ice has expanded around parts of Antarctica focus on changes in ocean circulation and others on melt from the Antarctic ice sheet adding near-freezing, fresh water to the Southern Ocean (Bintanja et al., 2013). Sorting out the causes of the Antarctic sea ice increases remains an area of active research and could be helped with the additional data from 2015 through early 2017, when the Antarctic experienced significant sea ice decreases.

Whatever the cause of the Antarctic sea ice increases between 1979 and 2014, they were far outweighed by the sea ice decreases in the Arctic. Global sea ice extents typically reach a minimum in February, as in the Antarctic, and a maximum in either October or November, with the 1979–2014 global trends in every month being downward. On a yearly average basis, the global sea ice extent trend over the period 1979–2014 was $-30{,}600 \pm 6100$ km^2/year ($-1.3 \pm 0.3\%$/decade) (updated from Parkinson, 2014a), equating to a global ice extent loss each year of an area slightly exceeding the area of Belgium (30,500 km^2).

The satellite results for 1979–2014 hence give a strong picture of overall decreasing Arctic sea ice coverage, increasing Antarctic sea ice coverage, and decreasing global sea ice coverage, with considerable interannual variability in both polar ice covers. Such data-centered results are extremely important for establishing the changes that have occurred in key components of the climate system, in this case sea ice. But the data alone are not sufficient for everything that scientists, policy-makers, and others want to know about sea ice and the rest of the climate system, as the data alone cannot predict the future or establish cause and effect, both of which require a model of one type or another. Climate models are far from perfect, and their results could be flawed, but they do allow researchers to simulate into the future and to establish cause-and-effect at least for the particular model. As David Rind has so well illustrated, with a model a scenario can be run twice, with only one condition changed from the first to the second run, thereby establishing what impacts that one change has on the model results. Doing so with sea ice, Rind and his colleagues established, as elaborated above, that at least in the GISS model at the time, 37% of the global warming predicted from the doubling of atmospheric CO_2 was attributable exclusively to changes in the sea ice cover (Rind *et al.*, 1995). Such model results contribute to increasing the understanding of Earth's very interconnected climate system.

References

Barnes, E.A., Dunn-Sigouin, E., Masato, G. and Woollings, T. (2014). Exploring recent trends in Northern Hemisphere blocking. *Geophysical Research Letters*, 41. doi:10.1002/2013GL058745.

Bintanja, R., van Oldenborgh, G.J., Drijfhout, S.S., Wouters, B. and Katsman, C.A. (2013). Important role for ocean warming and increased ice-shelf melt in Antarctic sea-ice expansion. *Nature Geoscience*, 6, 376–379. doi:10.1038/NGEO1767.

Bliss, A.C. and Anderson, M.R. (2014). Snowmelt onset over Arctic sea ice from passive-microwave satellite data: 1979–2012. *The Cryosphere*, 8, 2089–2100. doi:10.5194/tc-8-2089-2014.

Comiso, J.C. (2012). Large decadal decline of the Arctic multiyear ice cover. *Journal of Climate*, 25, 1176–1193. doi:10.1175/JCLI-D-11-00113.1.

Francis, J.A. and Vavrus, S.J. (2012). Evidence linking Arctic amplification to extreme weather in mid-latitudes. *Geophysical Research Letters*, 39, L06801. doi:10.1029/2012GL051000.

Jeffries, M.O., Overland, J.E. and Perovich, D.K. (2013). The Arctic shifts to a new normal. *Physics Today*, 66 (10), 35–40. doi:10.1063/PT.3.2147.

Kwok, R. and Rothrock, D.A. (2009). Decline in Arctic sea ice thickness from submarine and ICESat records: 1958–2008. *Geophysical Research Letters*, 36, L15501. doi:10.1029/2009GL039035.

Mori, M., Watanabe, M., Shiogama, H., Inoue, J. and Kimoto, M. (2014). Robust Arctic sea-ice influence on the frequent Eurasian cold winters in past decades. *Nature Geoscience*, 7, 869–873. doi:10.1038/NGEO2277.

Parkinson, C.L. (2014a). Global sea ice coverage from satellite data: Annual cycle and 35-yr trends. *Journal of Climate*, 27 (24), 9377–9382. doi:10.1175/JCLI-D-14-00605.1.

Parkinson, C.L. (2014b). Spatially mapped reductions in the length of the Arctic sea ice season. *Geophysical Research Letters*, 41, 4316–4322. doi:10.1002/2014GL060434.

Parkinson, C.L., Cavalieri, D.J., Gloersen, P., Zwally, H.J. and Comiso, J.C. (1999). Arctic sea ice extents, areas, and trends, 1978–1996. *Journal of Geophysical Research*, 104 (C9), 20837–20856. doi:10.1029/1999JC900082.

Rind, D., Healy, R., Parkinson, C. and Martinson, D. (1995). The role of sea ice in $2 \times CO_2$ climate model sensitivity. Part 1: The total influence of sea ice thickness and extent. *Journal of Climate*, 8 (3), 449–463.

Turner, J., Comiso, J.C., Marshall, G.J., Lachlan-Cope, T.A., Bracegirdle, T., Maksym, T., Meredith, M.P., Wang, Z. and Orr, A. (2009). Non-annular atmospheric circulation change induced by stratospheric ozone depletion and its role in the recent increase of Antarctic sea ice extent. *Geophysical Research Letters*, 36, L08502. doi:10.1029/2009GL037524.

Walsh, J.E. (2013). Melting ice: What is happening to Arctic sea ice, and what does it mean for us? *Oceanography*, 26 (2), 171–181. doi:10.5670/oceanog.2013.19.

Zwally, H.J., Comiso, J.C., Parkinson, C.L., Cavalieri, D.J. and Gloersen, P. (2002). Variability of Antarctic sea ice 1979–1998. *Journal of Geophysical Research*, 107 (C5), 3041. doi:10.1029/2000JC000733.

Slide 1

Polar Sea Ice Coverage, Its Changes, and Its Broader Climate Impacts

Claire L. Parkinson

NASA Goddard Space Flight Center

Greenbelt, Maryland

Our Warming Planet: Topics in Climate Dynamics
Lectures in Climate Change, Vol. 1
2017

Slide 2

What Is Sea Ice?

Sea ice is ice formed from the freezing of sea water.

Parkinson, C.   Polar Sea Ice Coverage, Its Changes, and Its Broader Climate Impacts   2

Slide 3

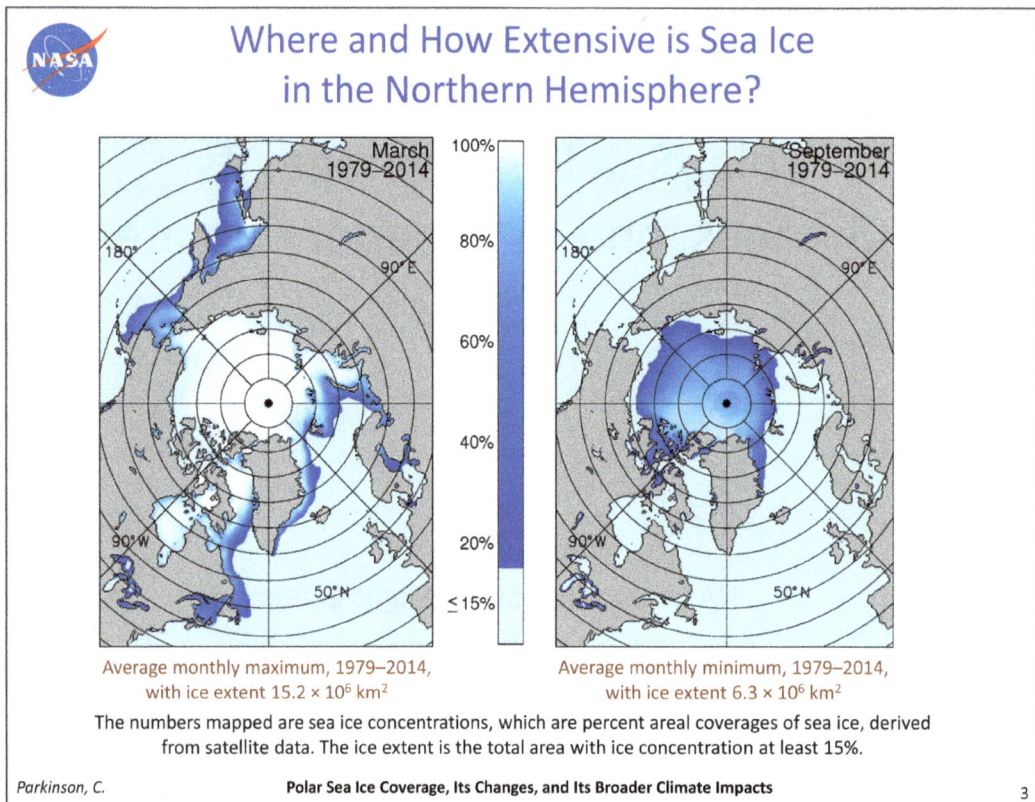

Slide 4

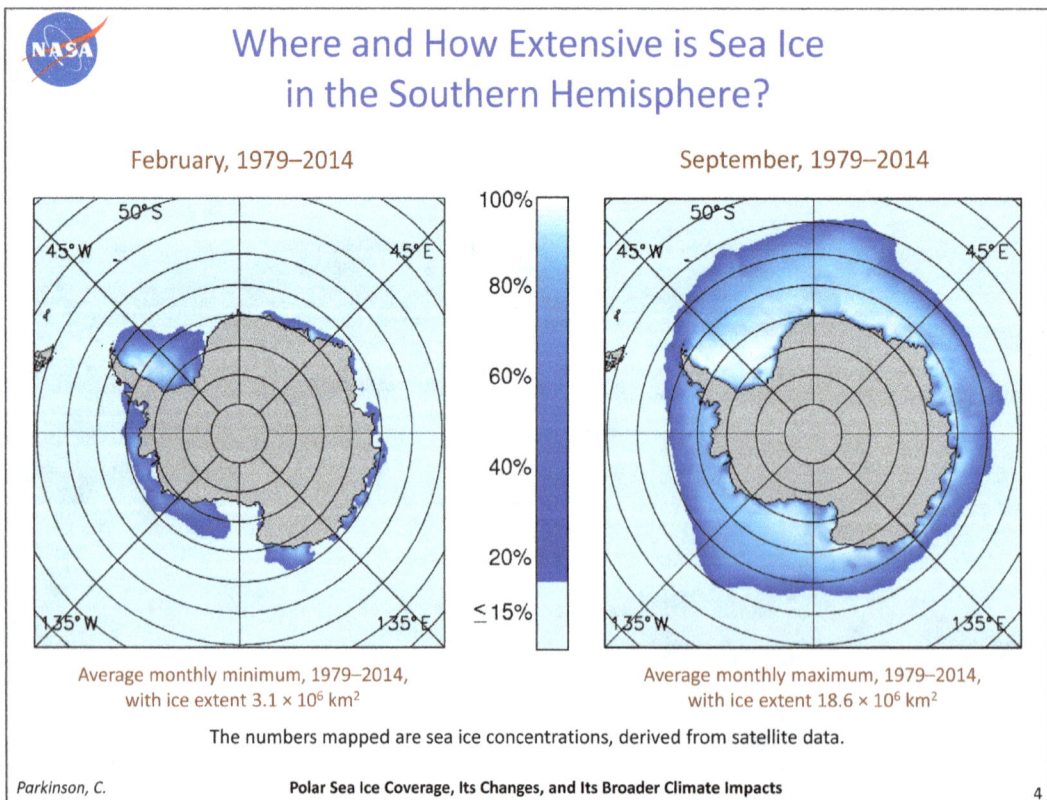

Slide 5

Through What Mechanisms Does Sea Ice Impact the Rest of the Polar Climate System?

- It reflects solar radiation back to space.
- It serves as an insulator between the ocean and atmosphere.
- As it forms and ages, it releases salt to the underlying water; as it moves, it transports cold, relatively low salinity H_2O; and as it melts, it releases this cold, relatively low salinity water. All these processes affect ocean circulation.

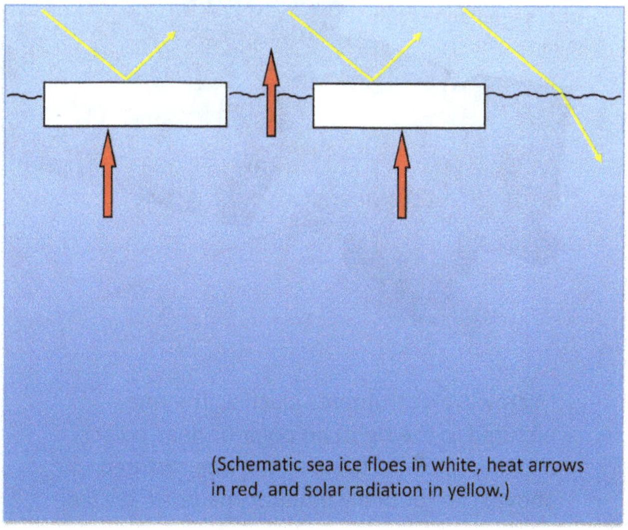

(Schematic sea ice floes in white, heat arrows in red, and solar radiation in yellow.)

Parkinson, C.   Polar Sea Ice Coverage, Its Changes, and Its Broader Climate Impacts   5

Slide 6

Do Sea Ice Impacts Extend Beyond the Polar Regions?

Yes; but the details are uncertain (this is an area of active research).

Projected global warming from doubled CO_2, with and without sea ice included in the calculations (redrawn and relabeled from Rind *et al.* 1995).

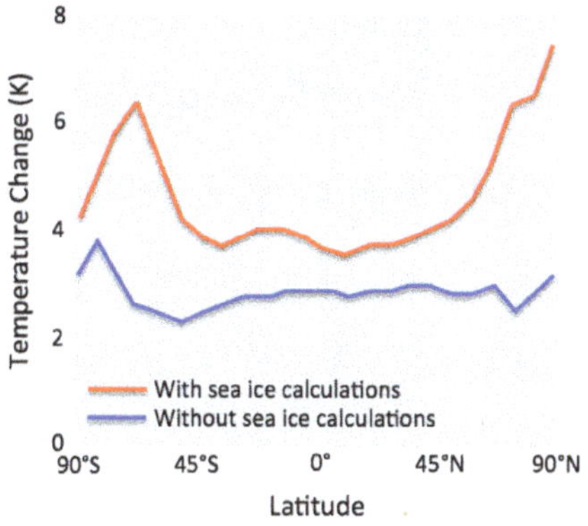

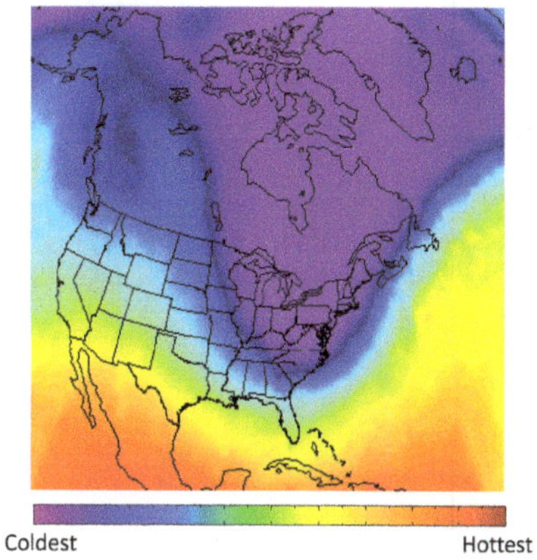

With sea ice calculations
Without sea ice calculations

Temperature pattern over North America on February 19, 2015. It is possible (but by no means certain) that Arctic sea ice reductions have increased the probability of such excursions of cold Arctic air into the midlatitudes. (Image from data from the Atmospheric Infrared Sounder on the Aqua satellite.)

Parkinson, C.   Polar Sea Ice Coverage, Its Changes, and Its Broader Climate Impacts   6

282 *C. L. Parkinson*

Slide 7

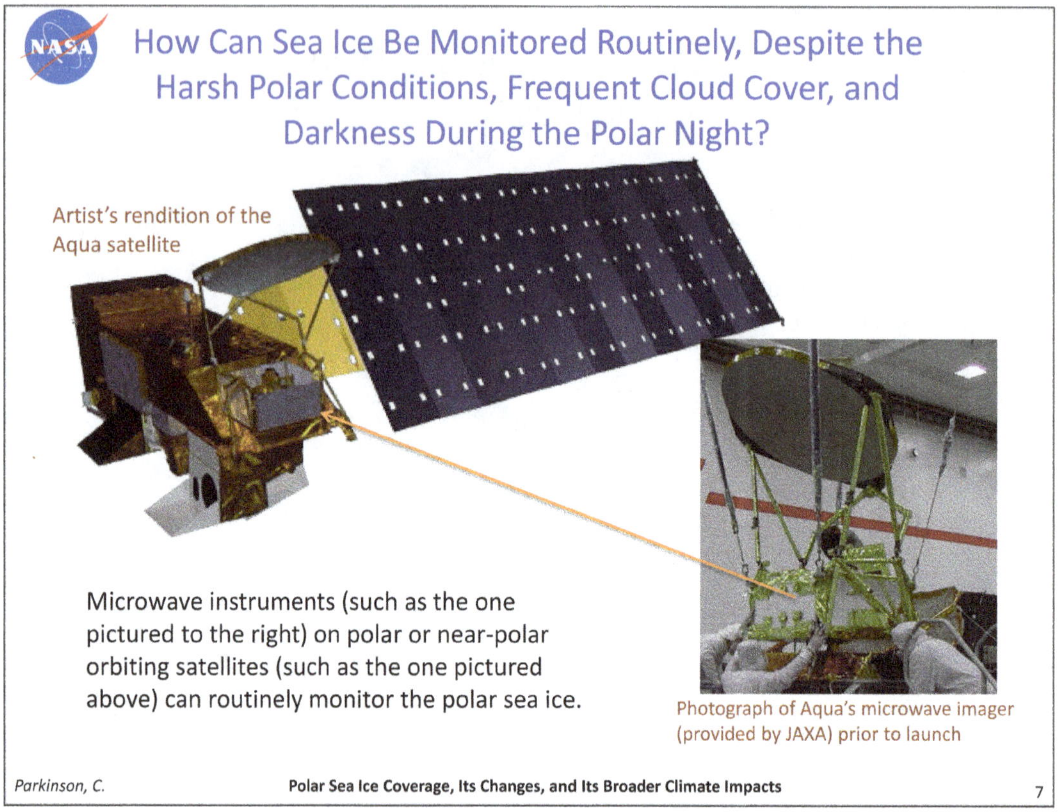

Slide 8

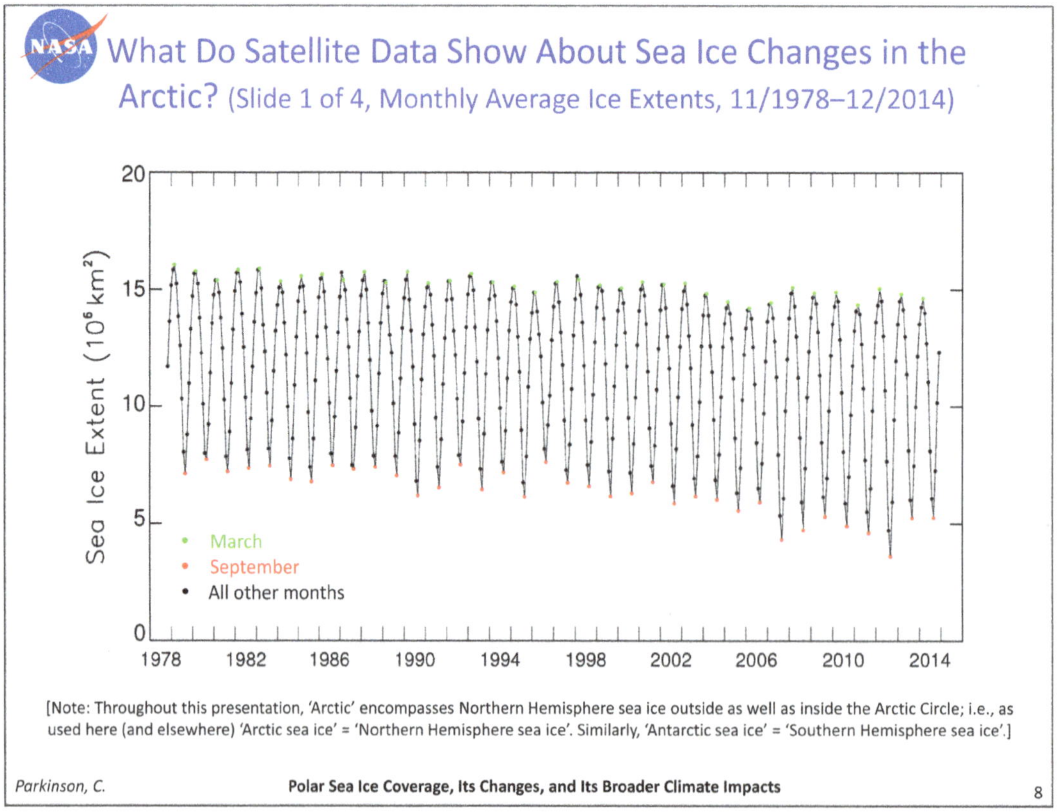

Slide 9

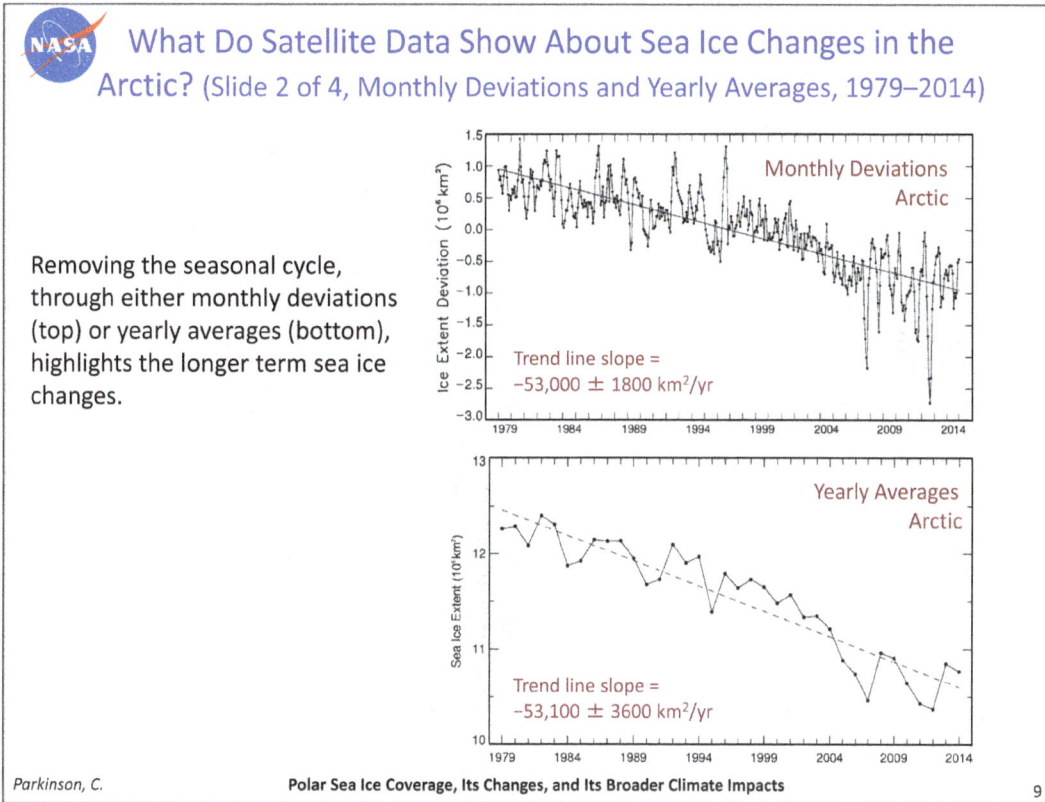

Slide 10

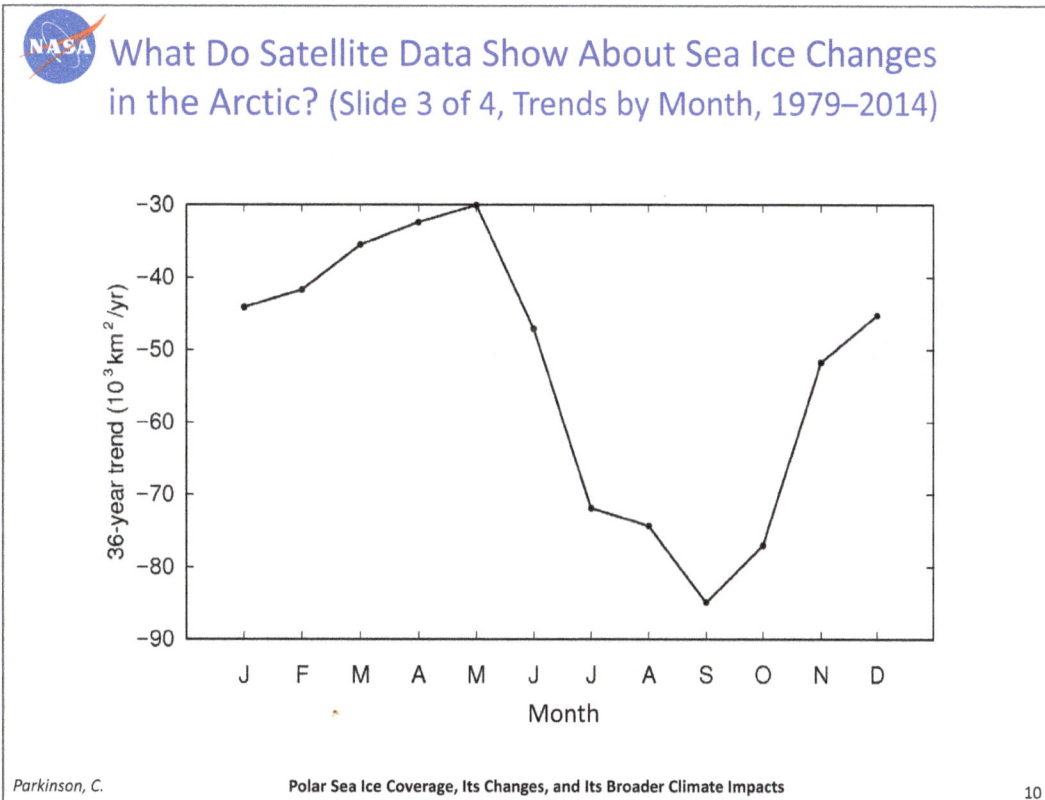

Slide 11

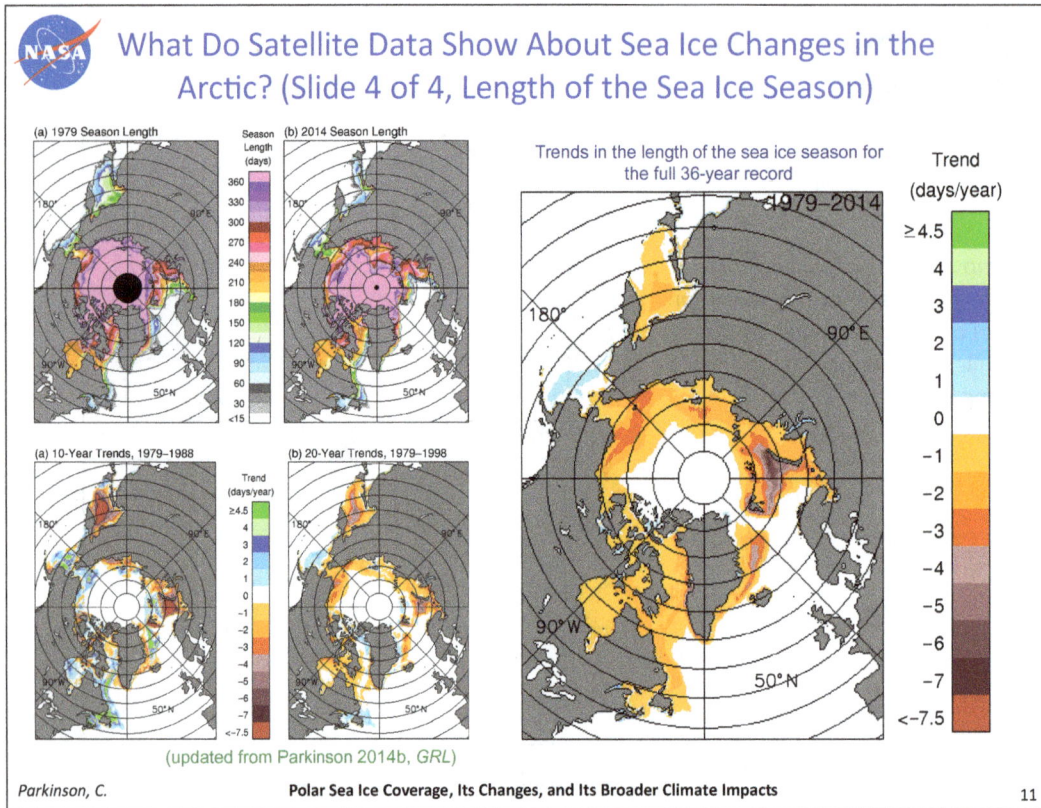

Slide 12

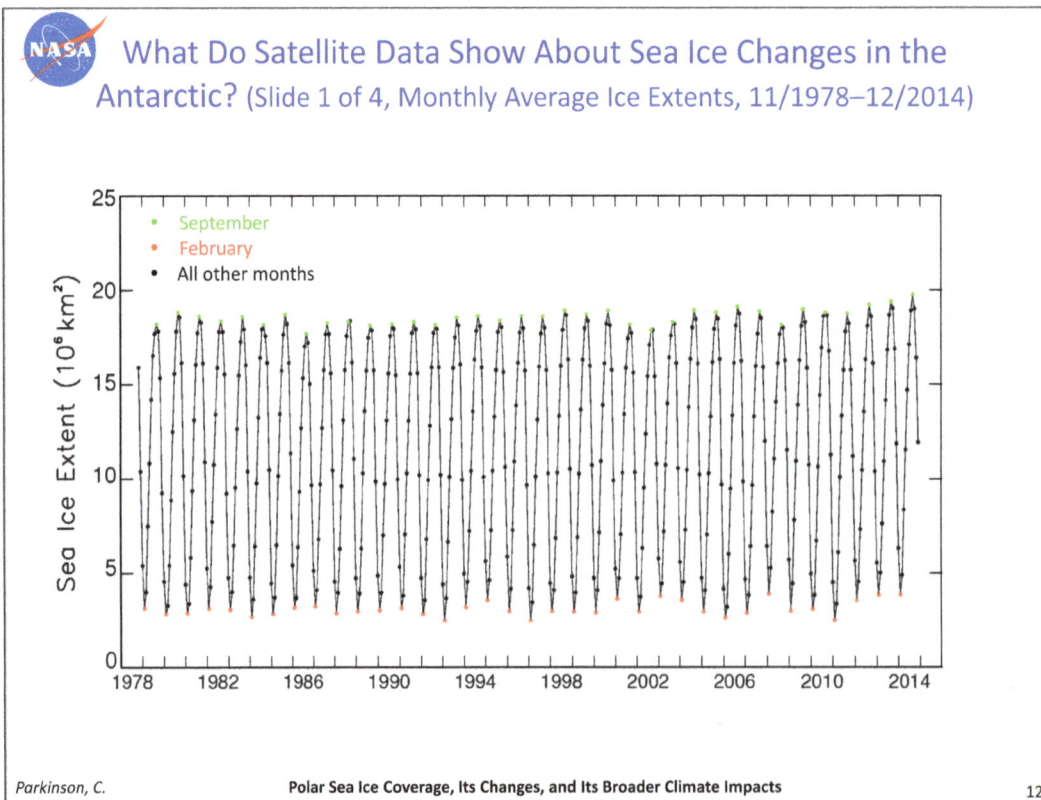

Slide 13

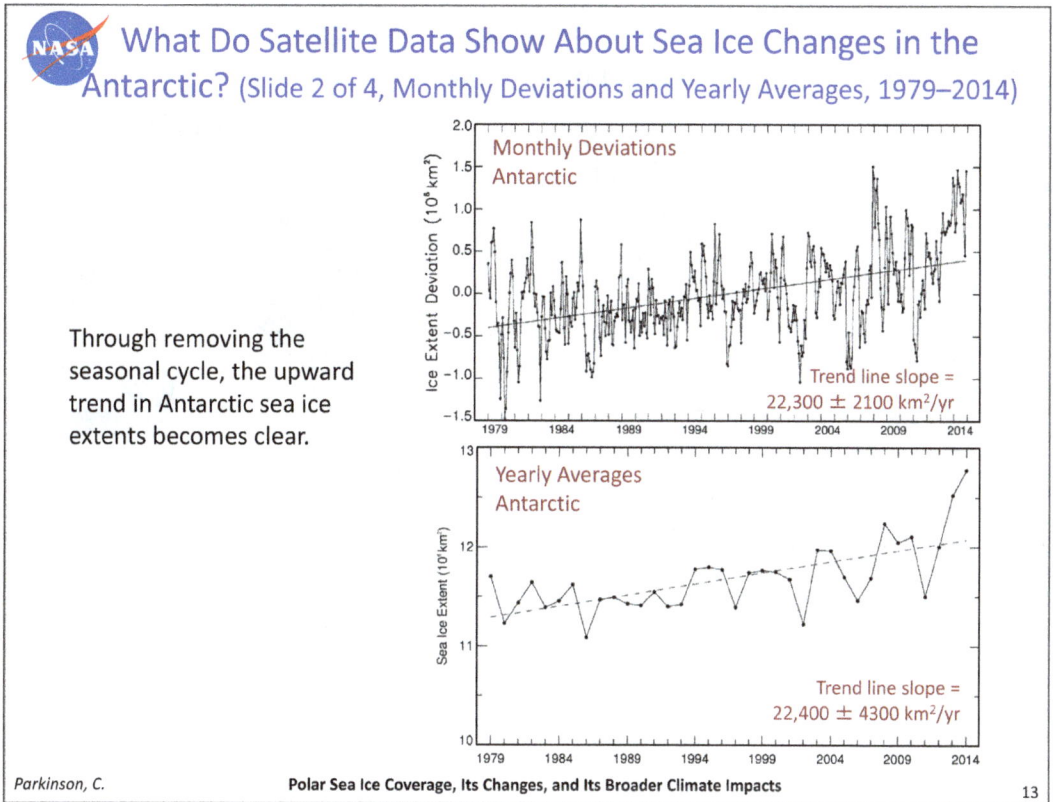

Slide 13: What Do Satellite Data Show About Sea Ice Changes in the Antarctic? (Slide 2 of 4, Monthly Deviations and Yearly Averages, 1979–2014)

Through removing the seasonal cycle, the upward trend in Antarctic sea ice extents becomes clear.

Parkinson, C.   Polar Sea Ice Coverage, Its Changes, and Its Broader Climate Impacts   13

Slide 14

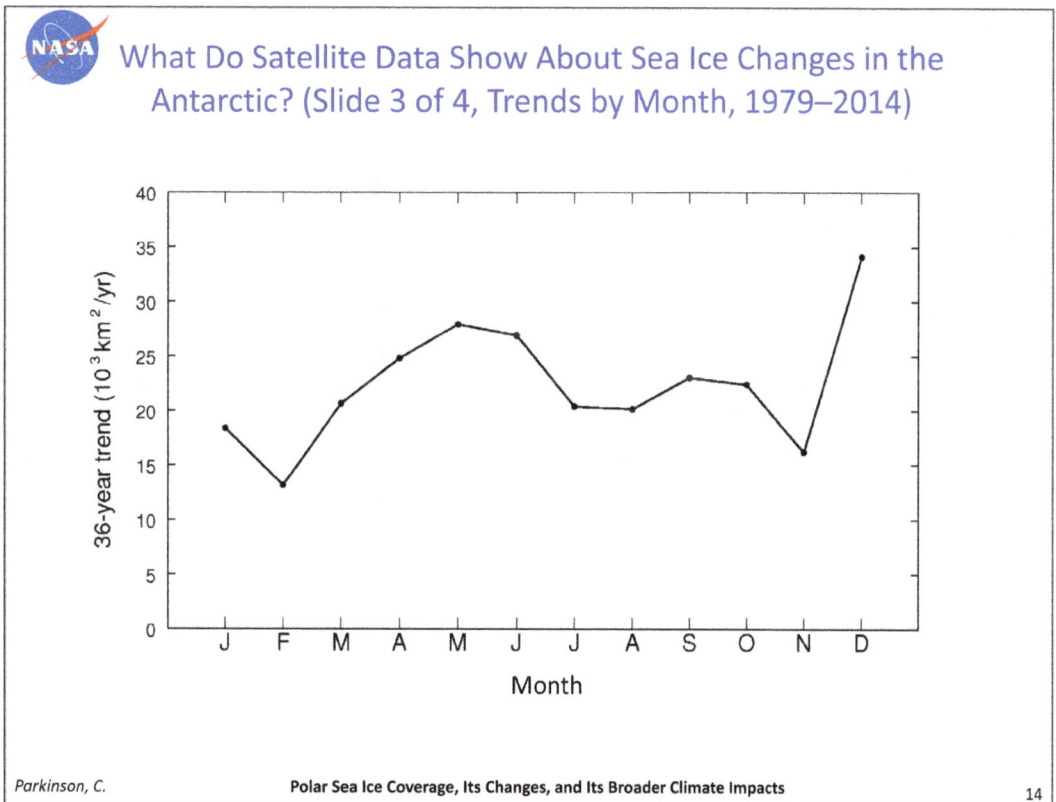

Slide 14: What Do Satellite Data Show About Sea Ice Changes in the Antarctic? (Slide 3 of 4, Trends by Month, 1979–2014)

Parkinson, C.   Polar Sea Ice Coverage, Its Changes, and Its Broader Climate Impacts   14

Slide 15

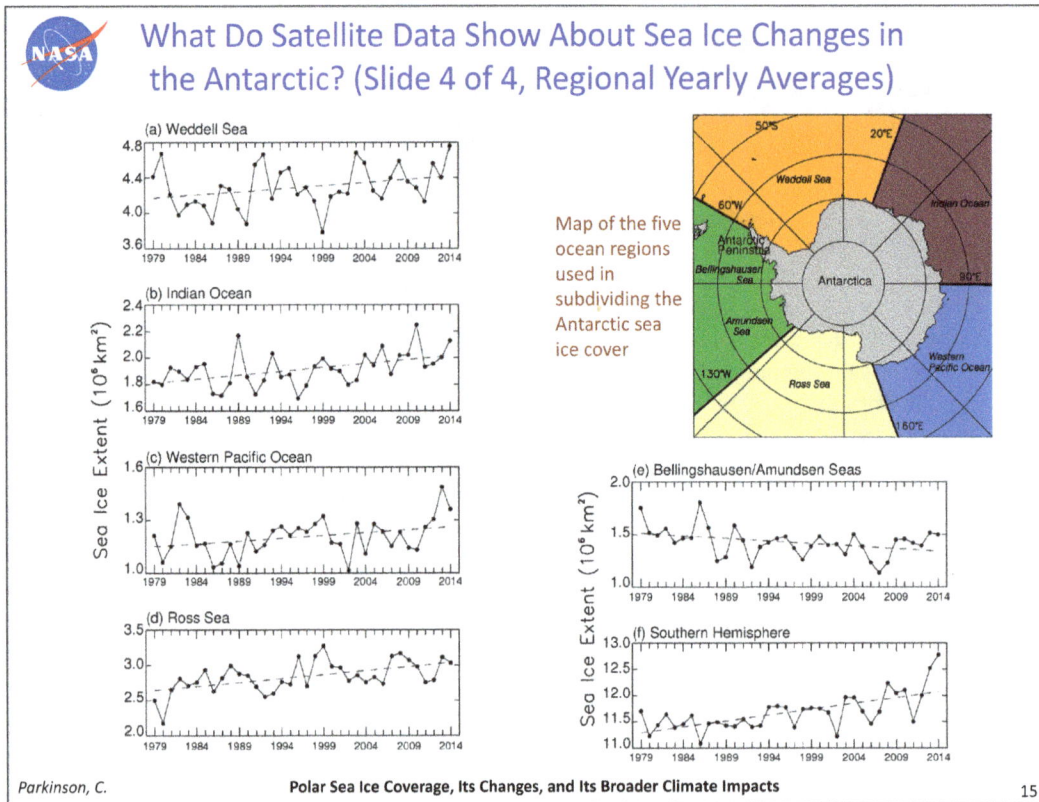

Slide 16

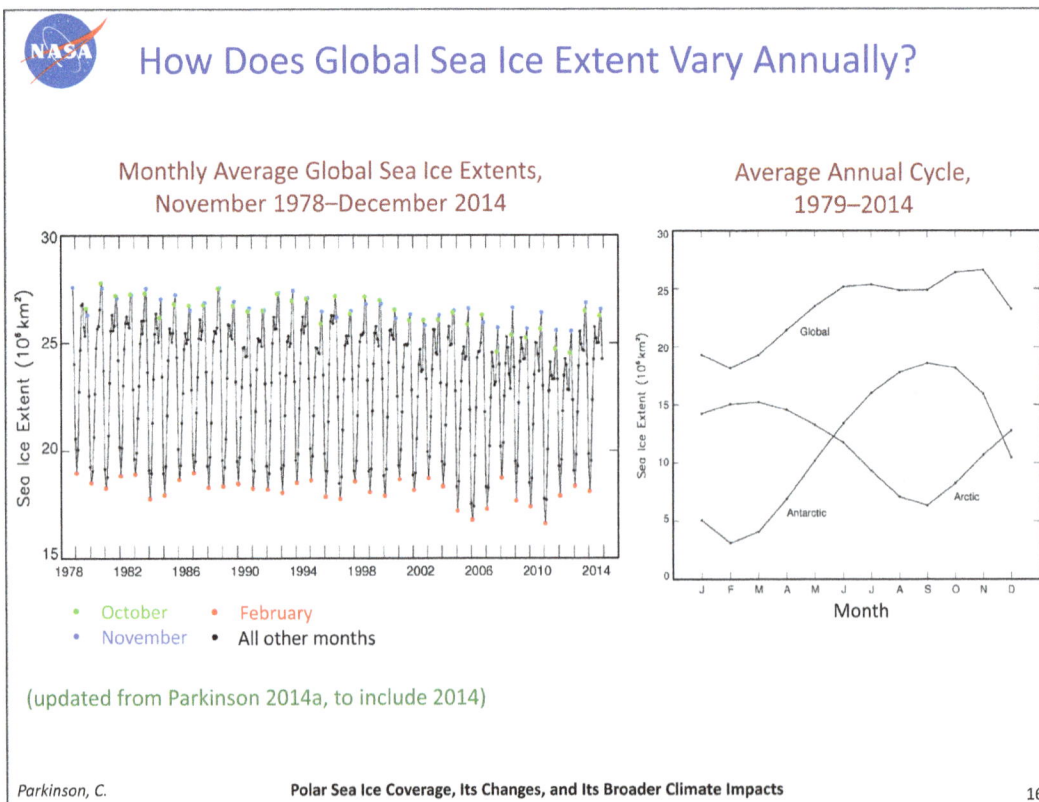

Slide 17

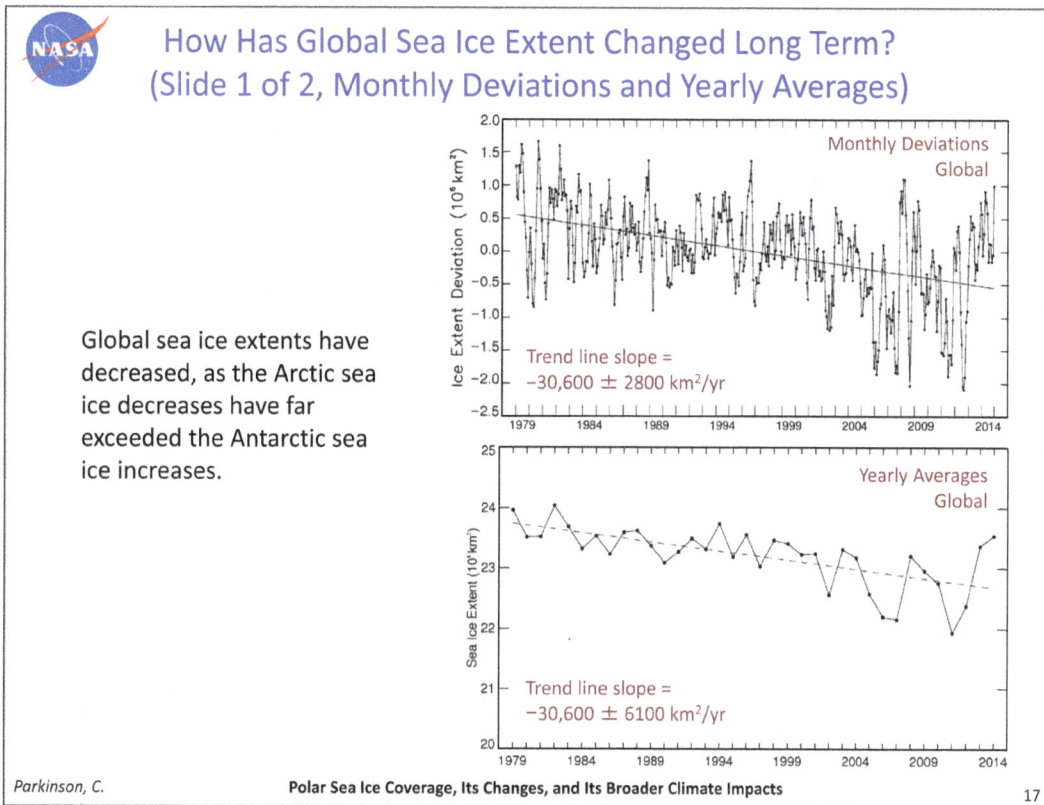

How Has Global Sea Ice Extent Changed Long Term?
(Slide 1 of 2, Monthly Deviations and Yearly Averages)

Global sea ice extents have decreased, as the Arctic sea ice decreases have far exceeded the Antarctic sea ice increases.

Monthly Deviations Global

Trend line slope = −30,600 ± 2800 km²/yr

Yearly Averages Global

Trend line slope = −30,600 ± 6100 km²/yr

Parkinson, C.   Polar Sea Ice Coverage, Its Changes, and Its Broader Climate Impacts   17

Slide 18

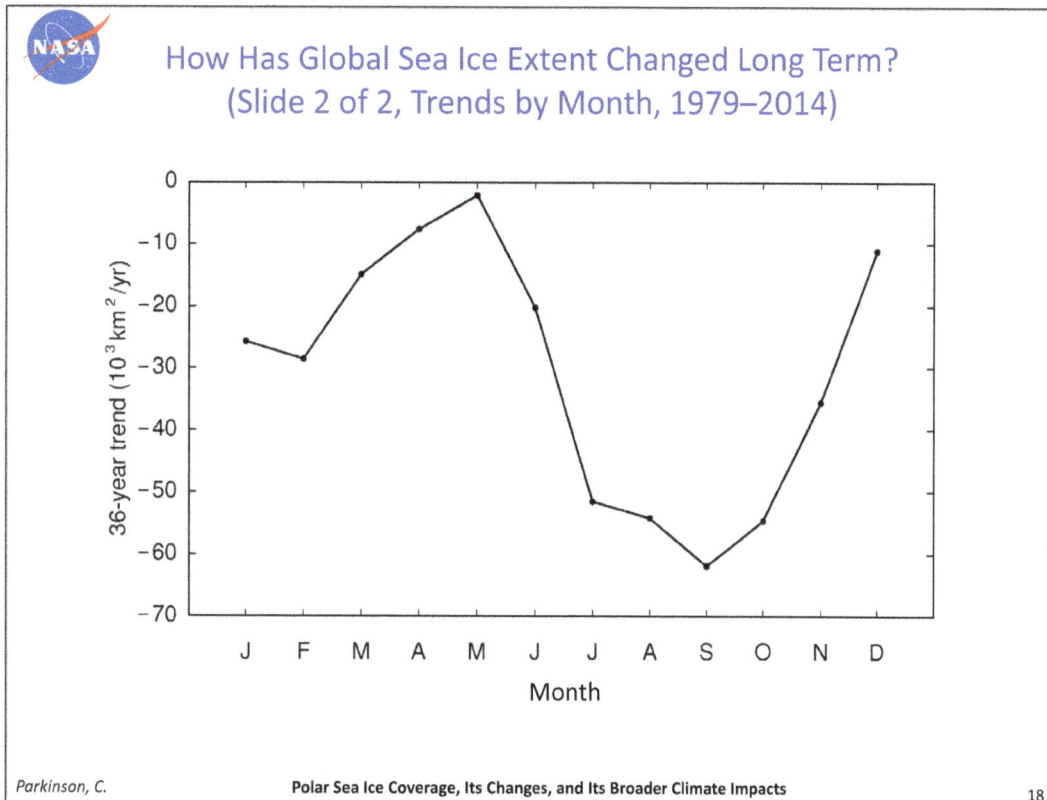

How Has Global Sea Ice Extent Changed Long Term?
(Slide 2 of 2, Trends by Month, 1979–2014)

Parkinson, C.   Polar Sea Ice Coverage, Its Changes, and Its Broader Climate Impacts   18

Slide 19

Summary

- Sea ice spreads over vast areas and has important effects on the rest of the climate system.
- Satellites allow the routine monitoring of the sea ice cover.
- The satellite data have shown prominent changes in polar sea ice coverage since the late 1970s. Over the 36-year period 1979–2014:
  - Arctic sea ice coverage decreased significantly, and this is true for all 12 months and all four seasons.
  - Antarctic sea ice coverage increased, and this is also true for all 12 months, although the Antarctic increases are not as large as the Arctic decreases.
  - Global sea ice coverage decreased, also in all 12 months.
- Much active research is ongoing to explain the sea ice changes and their impacts.
- Modeling studies, such as those pioneered by David Rind, provide a tool for projecting into the future and for quantitatively examining cause and effect.

Parkinson, C.   **Polar Sea Ice Coverage, Its Changes, and Its Broader Climate Impacts**   19

Slide 20

Acknowledgments and References

- Nick DiGirolamo of SSAI provided considerable assistance in the generation of many of the figures; funding for the work was provided by NASA.
- Several of the sea ice time series presented are updated through 2014 from time series in the following papers:
  - Parkinson, C.L., Cavalieri, D.J., Gloersen, P., Zwally, H.J. and Comiso, J.C. (1999). Arctic sea ice extents, areas, and trends, 1978–1996. *Journal of Geophysical Research*, 104 (C9), 20837–20856. doi:10.1029/1999JC900082.
  - Zwally, H.J., Comiso, J.C., Parkinson, C.L., Cavalieri, D.J. and Gloersen, P. (2002). Variability of Antarctic sea ice 1979–1998. *Journal of Geophysical Research*, 107 (C5), 3041. doi:10.1029/2000JC000733.
  - Parkinson, C.L. (2014a). Global sea ice coverage from satellite data: Annual cycle and 35-yr trends. *Journal of Climate*, 27 (24), 9377–9382. doi:10.1175/JCLI-D-14-00605.1.
- The length of the sea ice season results are updated from:
  - Parkinson, C.L. (2014b). Spatially mapped reductions in the length of the Arctic sea ice season. *Geophysical Research Letters*, 41, 4316–4322. doi:10.1002/2014GL060434.
- The plot of projected global warming is redrawn from:
  - Rind, D., Healy, R., Parkinson, C. and Martinson, D. (1995). The role of sea ice in 2 × CO$_2$ climate model sensitivity. Part 1: The total influence of sea ice thickness and extent. *Journal of Climate*, 8 (3), 449–463.

Parkinson, C.   **Polar Sea Ice Coverage, Its Changes, and Its Broader Climate Impacts**   20

Slide Notes

Slide 2   What Is Sea Ice?

Sea ice is quite distinct from icebergs, even though both sea ice and icebergs are examples of ice floating in the ocean. Whereas sea ice is formed from seawater, icebergs break off from land-based glacial ice formed from the compaction of snow. Sea ice is hence much saltier than icebergs and overwhelmingly thinner, typically not more than 5 m thick. Still, as long as a large sea ice floe is a meter or more in thickness, a small airplane can often land safely on it.

The photograph is from near the sea ice edge in the Bering Sea (between Alaska and Siberia) in March 1981 (photo by Claire L. Parkinson).

Slide 3   Where and How Extensive is Sea Ice in the Northern Hemisphere?

Sea ice coverage in the Northern Hemisphere has a prominent annual cycle, typically peaking in March (left image, averaged for the 36 years 1979–2014) and reaching a minimum in September (right image). *Sea ice concentration* is the percent areal coverage of sea ice at a given location (generally calculated for each grid element on a gridded map; for the maps in this presentation, the grid elements are approximately 25 km × 25 km). *Sea ice extent* is the summed area of all grid elements with at least 15% sea ice concentration. For comparison with the numbers on the slide, the area of the United States is 9.9×10^6 km^2 and the area of Canada is 10.0×10^6 km^2.

Data sources: The two maps on this slide (and the maps and plots later in this presentation) were created from data from the following sequence of satellite instruments: the Scanning Multichannel Microwave Radiometer (SMMR) on NASA's Nimbus 7 satellite, launched in October 1978; the Special Sensor Microwave Imager (SSMI) on the F8, F11, and F13 satellites of the Defense Meteorological Satellite Program (DMSP), first launched in June 1987; and the SSMI Sounder (SSMIS) on the DMSP F17 satellite, launched in November 2006. Throughout this presentation, SMMR data are used for the period from the late 1970s through mid-August 1987, SSMI data are used for mid-August 1987 through December 2007, and SSMIS data are used for January 2008 through December 2014. Each of the satellites carrying these instruments has a near-polar orbit. However, the orbits are not identical, and the differences in the orbits contribute to differences in how close the data coverage gets to the North Pole. The SMMR data reach poleward to 84.6°N; the SSMI data reach to 87.2°N; and the SSMIS data reach to 89.2°N. Hence, equatorward of 84.6°N, data on the maps are averaged for all 36 years 1979–2014, whereas for 84.6°N–87.2°N they are averaged for the SSMI and SSMIS years and for 87.2°N–89.2°N they are averaged only for the SSMIS years, 2008–2014, with no data poleward of 89.2°N.

Slide 4   Where and How Extensive is Sea Ice in the Southern Hemisphere?

As in the Northern Hemisphere, sea ice coverage in the Southern Hemisphere has a prominent annual cycle, although the seasons are reversed. Southern Hemisphere (or Antarctic) sea ice typically reaches its annual minimum in February (left image, averaged for 1979–2014) and its annual maximum in September (right image).

The larger annual range of sea ice extents in the Antarctic (this slide) versus the Arctic (previous slide) is in large part due to the different geographies: The central south polar region is covered by land ice rather than sea ice (limiting the extent of summer ice), and the Antarctic sea ice, in sharp contrast to Arctic sea ice, does not have land bounding its equatorward extension in winter.

Data sources: Satellite microwave data from SMMR, SSMI, and SSMIS.

Slide 5   Through What Mechanisms Does Sea Ice Impact the Rest of the Polar Climate System?

Comments beyond the answers on the slide: (1) The insulation effect of sea ice includes restricting the exchanges of mass and momentum as well as heat. For instance, the ice restricts the amount of water that gets evaporated into the atmosphere, thereby affecting the amount of precipitation (snowfall and/or rainfall) in surrounding regions; and the ice lessens the momentum transfer from the winds to the underlying water, frequently resulting in much calmer ocean conditions inside the ice cover than just outside it (greatly benefiting ship passengers subject to sea sickness). (2) The ocean circulation changes impacted by sea ice at times include overturning in the water column beneath the ice. (3) In addition to affecting climate, sea ice also affects polar ecosystems, including having impacts all the way up the food chain, from affecting microorganisms living within the ice floes to affecting such top predators as polar bears in the north polar region.

Slide 6   Do Sea Ice Impacts Extend Beyond the Polar Regions?

The Rind *et al.* (1995) study used the Goddard Institute for Space Studies (GISS) global climate model to quantify how the projected warming from doubled CO_2 is affected by sea ice, doing so by calculating the effects of doubled CO_2 in the GISS model both when including sea ice calculations and when not including them, in the latter case not allowing the sea ice annual cycle to change. As shown in the plot, the inclusion of sea ice calculations resulted in more warming at every latitude, with the greatest effect being in the vicinity of the sea ice. Overall, the simulated global warming was 4.17 K with sea ice calculations and 2.61 K without them, meaning that 37% of the simulated 4.17 K warming could be attributed to the inclusion of sea ice calculations (Rind *et al.*, 1995). Major excursions of cold Arctic air southward occur on occasion irrespective of the sea ice conditions, but suggestions have been made that the probability of such events could be higher now because of the changes in the Arctic sea ice cover. The southward excursions in January and February of 2015 contributed to record snowfalls in Boston, Massachusetts.

The image on the right is from the Atmospheric Infrared Sounder (AIRS) on the Aqua satellite and was provided by Ed Olsen of the AIRS Science Team.

Slide 7   How Can Sea Ice Be Monitored Routinely, Despite the Harsh Polar Conditions, Frequent Cloud Cover, and Darkness During the Polar Night?

Satellites in polar or near-polar orbits can observe the polar regions multiple times every day, totally avoiding the harsh polar conditions by flying hundreds of kilometers above the Earth. Using microwave instruments on board the satellites, sea ice can be distinguished from liquid water, and this distinction can be made irrespective of sunlight (as the microwave radiation is emitted from within the Earth system) or cloud coverage, for most cloud covers (as at selected wavelengths, the microwaves are not much affected by clouds).

The satellite pictured on the slide is NASA's Aqua satellite, as rendered by artist Marit Jentoft-Nilsen; and the microwave instrument pictured is the Advanced Microwave Scanning Radiometer for the Earth Observing System (AMSR-E), provided to the Aqua program by the Japan Aerospace Exploration Agency (JAXA). (AMSR-E photograph courtesy of JAXA and the Aqua spacecraft company Northrop Grumman.)

Slide 8   What Do Satellite Data Show About Sea Ice Changes in the Arctic? (Slide 1 of 4, Monthly Average Ice Extents, 11/1978–12/2014).

Each year the Arctic experiences a marked decrease in sea ice extent in spring and summer and a marked increase in sea ice extent in autumn and winter, with ice extents typically reaching a maximum in March (green dots) and a minimum in September (red dots), but with both the high values and the low values trending downward over the 36-year record. As mentioned on Slide 3, over the course of the 1979–2014 record, the average March ice extent was 15.2×10^6 km² and the average September ice extent was 6.3×10^6 km². For 2014 specifically, the numbers were markedly lower, at 14.6×10^6 km² in March and 5.3×10^6 km² in September. The next slide highlights the trends.

Data sources: Satellite microwave data from SMMR, SSMI, and SSMIS.

Slide 9   What Do Satellite Data Show About Sea Ice Changes in the Arctic? (Slide 2 of 4, Monthly Deviations and Yearly Averages, 1979–2014).

Yearly averages are calculated by averaging the ice extent values for each day of the year. Monthly deviations are calculated by subtracting from the value of the ice extent for an individual month the average of the ice extents for that month throughout the 36-year record. For instance, the first data point on the plot, for January 1979, is calculated by taking the ice extent for January 1979 and subtracting the average ice extent for the 36 Januaries 1979–2014. Both plots include the least squares fit line (or 'trend line') through the respective data points and the slope of the trend line, and both show the prominent downward trend in sea ice extents over the period 1979–2014. The trend of –53,100 km²/yr (which comes to –4.3%/decade) in the yearly average ice extents equals an ice extent loss each year of an area slightly exceeding the combined area of Massachusetts (27,300 km²) and Vermont (24,900 km²), or, alternatively, exceeding the area of Costa Rica (51,100 km²).

Data sources: Satellite microwave data from SMMR, SSMI, and SSMIS.

Slide 10   What Do Satellite Data Show About Sea Ice Changes in the Arctic? (Slide 3 of 4, Trends by Month, 1979–2014).

Over the 1979–2014 period, all 12 months show substantial decreases in Arctic sea ice extents. September has the greatest decreases, at –84,900 km²/year (–10.9%/decade), but even the month with the smallest decreases, May, has reductions of –30,100 km²/year (–2.2%/decade).

Data sources: Satellite microwave data from SMMR, SSMI, and SSMIS.

Slide 11   What Do Satellite Data Show About Sea Ice Changes in the Arctic? (Slide 4 of 4, Length of the Sea Ice Season).

Another way to examine changes in the sea ice cover is to consider changes in the length of the sea ice season, that is, in the number of days a year that sea ice is present. The 1979 and 2014 Season Length images show that a large area of the Central Arctic retains an ice cover throughout the year (or at least 360 days of it), but that much of the rest of the Arctic had a shorter ice season in 2014 than at the start of the record, in 1979. Particularly notable changes are in the Sea of Okhotsk and to the northeast of Scandinavia, where significant areas had ice coverage for 90 or more days in 1979 but were free of ice throughout 2014. When trend calculations are done on each grid element for the full 36-year record, 1979–2014, by far the majority of the changes show shortened ice seasons (oranges, browns, and reds) rather than lengthened ice seasons (blues and greens) (rightmost image), although this is not the case when considering only the first 10 years of the record, when lengthened sea ice seasons dominated the season length changes in the Western hemisphere and shortened sea ice seasons dominated in the Eastern hemisphere (lower left image). Hence, the results for the length of the sea ice season mirror those for the ice extent in terms of having trends toward lesser sea ice coverage that were far from convincing by the late 1980s but are quite convincing by 2015.

Data sources: Satellite microwave data from SMMR, SSMI, and SSMIS.

Slide 12   What Do Satellite Data Show About Sea Ice Changes in the Antarctic? (Slide 1 of 4, Monthly Average Ice Extents, 11/1978–12/2014).

As in the Arctic, each year sea ice extent in the Antarctic experiences a marked decrease in spring and summer and a marked increase in autumn and winter, although with the timing of the seasons reversed, Antarctic sea ice extents typically reaching a minimum in February (red dots) and a maximum in September (green dots). In the Antarctic case, however, the values are trending upward instead of downward, as shown more clearly on the next slide. As mentioned on Slide 4, over the course of the 1979–2014 record, the average February ice extent was 3.1×10^6 km^2 and the average September ice extent was 18.6×10^6 km^2. For 2014 specifically, the numbers were noticeably higher, at 3.8×10^6 km^2 in February and 19.8×10^6 km^2 in September.

Data sources: Satellite microwave data from SMMR, SSMI, and SSMIS.

Slide 13   What Do Satellite Data Show About Sea Ice Changes in the Antarctic? (Slide 2 of 4, Monthly Deviations and Yearly Averages, 1979–2014).

Both plots show that instead of having a prominent decrease in sea ice extent as in the Arctic, the Antarctic experienced an increase in sea ice extents over the period 1979–2014. The overall trend of 22,400 km^2/year in the yearly average ice extents is a 2.0%/decade increase in Antarctic sea ice extent. This equals an ice extent gain each year of an area slightly less than the area of New Jersey (22,600 km^2) or somewhat more than the area of El Salvador (21,000 km^2).

Data sources: Satellite microwave data from SMMR, SSMI, and SSMIS.

Slide 14   What Do Satellite Data Show About Sea Ice Changes in the Antarctic? (Slide 3 of 4, Trends by Month, 1979–2014).

Over the 1979–2014 period, all 12 months experienced an increase in Antarctic sea ice coverage, in direct contrast to the situation in the Arctic. December had the greatest increases, at 34,100 km^2/year (3.5%/decade), and February had the smallest increases, at 13,200 km^2/year (4.6%/decade). [February's higher %/decade value is due to the small amount of ice in February, allowing a lesser absolute decrease to constitute a higher percentage decrease.]

Data sources: Satellite microwave data from SMMR, SSMI, and SSMIS.

Slide 15   What Do Satellite Data Show About Sea Ice Changes in the Antarctic? (Slide 4 of 4, Regional Yearly Averages).

In contrast to some other portions of the Antarctic, the Antarctic Peninsula has a strong record of warming over the past several decades. It is, thus, not surprising that the region of the Bellingshausen and Amundsen Seas, to the immediate west of the Antarctic Peninsula, has experienced a downward trend in sea ice extents, despite the upward trend for the Antarctic as a whole. Scientists have been trying to understand the Antarctic sea ice increases; and the pattern of ice decreases in the Bellingshausen Sea and increases in the Ross Sea has helped identify possible shifts in atmospheric circulation (perhaps caused by the Antarctic ozone hole and perhaps including a stronger and/or more persistent low pressure system over the Amundsen Sea) as contributing factors in the 1979–2014 Antarctic sea ice expansion. Other possible contributing factors are changes in ocean circulation and/or melt and calving from the Antarctic ice sheet.

Data sources: Satellite microwave data from SMMR, SSMI, and SSMIS.

Slide 16   How Does Global Sea Ice Extent Vary Annually?

For many people who live in the Northern Hemisphere, it comes as a surprise that globally sea ice is least in February, in the midst of the Northern Hemisphere winter; but because of the much greater amplitude of the annual cycle in the Antarctic ice versus the Arctic ice, the annual cycle of global sea ice coverage looks more like that of the Antarctic than the Arctic. For global sea ice extents, the minimum monthly average was for February in each year of the 1979–2014 record and the maximum was for either October or November in every year except 1979, when the maximum was for July. On average over the course of the 36 years, the February ice extent was 18.2 × 10^6 km^2 and the monthly maximum ice extent was 26.6 × 10^6 km^2, for November.

Data sources: Satellite microwave data from SMMR, SSMI, and SSMIS.

Slide 17   How Has Global Sea Ice Extent Changed Long Term? (Slide 1 of 2, Monthly Deviations and Yearly Averages).

Both plots show that global sea ice extents decreased over the period 1979–2014. The overall trend of –30,600 km^2/yr in the yearly average ice extents equals an ice extent loss each year of an area slightly exceeding the area of Belgium (30,500 km²). In terms of percent per decade, the trend in yearly average ice extents is −1.3%/decade.

Data sources: Satellite microwave data from SMMR, SSMI, and SSMIS.

Slide 18   How Has Global Sea Ice Extent Changed Long Term? (Slide 2 of 2, Trends by Month, 1979–2014).

Over the 1979–2014 period, for every month the long term decreases in Arctic sea ice outweighed the increases in Antarctic sea ice, with the result that the Global ice extent shows decreases for each month. Some of these trends, however, are quite small, the lowest magnitude one being for May at −2200 km²/year. Others are significant, the highest magnitude one being for September at −61,900 km²/year (−2.4%/decade). [For people interested in statistical significance, all the trends presented on previous slides in this presentation are statistically significant at the 99% level. On this slide, in contrast, only the trends for February, July, August, September, October, and November are statistically significant at the 99% level.]

Data sources: Satellite microwave data from SMMR, SSMI, and SSMIS.

CLIMATE LECTURE 14

Arctic Sea Ice and Its Role in Global Change

Jiping Liu[1] *and Radley M. Horton*[2]

[1]*Department of Atmospheric and Environmental Sciences, University at Albany,*
State University of New York, Albany, NY, USA
[2]*Columbia University Center for Climate Systems Research, New York, NY, USA*

Jiping Liu is an Assistant Professor in the Department of Atmospheric and Environmental Sciences at the University of Albany. His research focuses on sea ice variability and change, including climate-cryosphere dynamics, feedbacks and modeling, atmosphere–sea ice–ocean interactions, and the application of remote sensing in ocean and sea ice.

Radley Horton is an Associate Research Scientist at the Center for Climate Systems Research at Columbia University and NASA Goddard Institute for Space Studies. His research focuses on the impact of climate change on the urban environment, including urban flooding, sea level rise, temperature extremes, and health effects.

Introduction

Sea ice is an important component of the global climate system. Sea ice forms, grows, and melts in the ocean. Sea ice grows during the fall and winter and melts during the spring and summer. Sea ice can melt completely in summer or survive multiple years. Sea ice can be classified by stages of development (thickness and age), that is, first-year sea ice (ice thickness typically <1.8 m) and multiyear sea ice (ice thickness typically >1.8 m). Sea ice occurs in both hemispheres. In the Northern Hemisphere, sea ice develops in the Arctic Ocean and surrounding bodies including Hudson and Baffin Bay, Gulf of St. Lawrence, the Greenland Sea, the Bering Sea, and the Sea of Okhotsk (sea ice can be observed as far south as Bohai Bay, China, ~38°N). In the Southern Hemisphere, sea ice only develops around Antarctica, reaching a maximum equatorward extension at around ~55°S).

During part of the year, sea ice covers about 25 million km². That means about 7% of the world's ocean is covered by sea ice. Arctic sea ice cover usually reaches its seasonal maximum in March, about 15–16 million km² and minimum in September, about 7 million km². Antarctic sea ice usually reaches its maximum extent in September, about 18–20 million km² and minimum extent in February, about 3 million km² (https://nsidc.org/cryosphere/sotc/sea_ice.html).

As a boundary between the ocean and the atmosphere, sea ice plays an important role in the global climate system and acts as an important indicator of global climate change. Firstly, sea ice has a high albedo, that is, bare ice has an albedo of about 0.5–0.6 and ice with snow has an albedo of about 0.8–0.9. This substantially reduces the incoming solar radiation absorbed by the surface. Secondly, sea ice has a strong insulating effect on the underlying ocean. This drastically reduces the exchange of heat, moisture, momentum, and gases (e.g., CO_2) between the ocean and the atmosphere. Thirdly, the formation and melting of sea ice strongly influences ocean thermohaline circulation through dense water formation (e.g., the North Atlantic Deep Water and the Antarctic Bottom Water). Finally, sea ice holds a large amount of relatively freshwater. This has a strong impact on the freshwater budget in the Arctic Ocean and the Southern Ocean. Because the albedo, insulating and density effects are so strong, even small changes in the amount of sea ice cover might be expected to drive large changes in the regional and ultimately global climate. Additionally, sea ice is important for polar marine ecosystem, for example, sea ice provides a habitat for algae and hunting ground for polar bears.

Since the beginning of the industrial revolution, CO_2 concentration in the atmosphere has increased about 40% (from 280 to 400 ppm). Currently, CO_2 emission from human activities is about 36 billion metric tons per year. There is an increasing concern about enhanced climate change in the polar regions due to the rising anthropogenic CO_2. The average surface temperature in the northern high latitude has risen at twice the rate as the rest of the world in the past decades. A number of positive feedbacks associated with sea ice have contributed to the amplification of climate change in the Arctic. The most important is the positive temperature-albedo feedback. That is, warming of the atmosphere or ocean causes a decrease of sea ice extent, the exposure of more open water (as well as more melt ponds over sea ice), and increased absorption of solar radiation at the surface, leading to further warming. Satellite observations show that Arctic sea ice has been declining at an astonishing rate associated with the amplified warming in the Arctic. Several emerging factors have been identified that are responsible for the observed Arctic sea ice loss. These include more open water and melt ponds, more soot and dust deposition on the ice pack, more penetration of solar radiation through sea ice, faster drift of the ice pack, increased vertical advective heat flux from the warm layer of Atlantic and Pacific origin water to the surface mixed layer, and changes in poleward heat transport. One of the most active areas in climate research today is observational and modeling studies of how Arctic sea ice may progress in the coming decades. To date, estimates of when the Arctic might become 'ice free' (often defined as having ice extent fall below 1 million km²) in September based on observational and modeling studies have ranged from this decade to the end of the century. Clearly, there is much we still need to learn about this rapidly changing region.

An even more active area of research concerns the possible links between Arctic sea ice loss and mid-latitude weather and climate. In early 2012, two research groups (Liu *et al.*, 2012; Francis and Vavrus, 2012) separately linked Arctic sea ice loss — via a weakened meridional temperature gradient and westerly winds — to a meandering jet stream that led to prolonged and more frequent extreme events (e.g., cold surges). A bevy of studies have expanded on, modified (e.g., proposed links between Arctic sea ice loss, enhanced fall snow cover in Siberia, and a meandering jet stream), and called into question (e.g., with arguments ranging from natural variability driving the extreme weather events to no significant change in the jet stream) the hypothesis, using a blend of observational and modeling studies. The explosion of publications and interest in this topic is evidence of just how far we are from a definitive answer. Ironically, by the time a link has been demonstrated to be true or false, the system may have transitioned to such a new low ice state as to render the prior analyses less relevant/applicable.

David Rind has been an important figure in advancing this topic. The response of sea ice to climate change affects the Earth's energy balance in ways that may lead to more climate change. In the mid-1990s, Rind examined the effects of sea ice changes in doubled-CO_2 climate model sensitivity experiments. He found that the total effect of sea ice on surface temperature changes (including cloud cover and water vapor feedbacks that arise associated with sea ice changes) accounts for 1.56°C of the 4.17°C global warming by analyzing doubled CO_2 simulations in an experiment with sea ice not allowed to change versus a control in which sea ice did change. Results were about four times larger than the sea ice impact when no feedbacks are allowed. These results pointed out the importance of properly constraining the sea ice response to climate perturbations, necessitating the use of more realistic sea ice models in climate system models (Rind et al., 1995, 1997).

In 2001, Rind found that El Niño (features warmer than normal sea-surface temperatures across the central and eastern equatorial Pacific) drives changes in the subtropical jet stream, which changes the path of storms that rearrange sea ice in the Pacific and Atlantic sides. This research showed how very disparate components of the climate system are connected, requiring that they all be understood before we can properly estimate future climate sensitivity (Rind et al., 2001).

In closing, in a rapidly changing Arctic, sea ice plays a growing role not only in a climate sense, but also via ecosystems, indigenous knowledge and use, economic development and geopolitics. As the Arctic continues to change and grow in importance, cross-disciplinary collaborations across a variety of interest groups will support effective and sustainable approaches to adaptation.

This lecture presents changes of Arctic sea ice in the context of global climate change, and the potential impacts of decreasing Arctic sea ice.

References

Francis, J. and Vavrus, S. (2012). Evidence linking Arctic amplification to extreme weather in mid-latitudes. *Geophysical Research Letters*, 39, L06801.

Liu, J., Curry, J., Wang, H., Song, M. and Horton, R. (2012). Impact of declining Arctic sea ice on winter snowfall. *Proceedings of National Academy Sciences*, 109, 4074–4079.

Rind, D., Chandler, M., Lerner, J., Martinson, D.G. and Yuan, X. (2001). Climate response to basin-specific changes in latitudinal temperature gradients and implications for sea ice variability. *Journal of Geophysical Research*, 106, 20161–20173.

Rind, D., Healy, R., Parkinson, C. and Martinson, D. (1997). The role of sea ice in 2 × CO_2 climate model sensitivity: Part II: Hemispheric dependencies. *Geophysical Research Letters*, 24, 1491–1494.

Rind, D., Healy, R., Parkinson, C. and Martinson, D. (1995). The role of sea ice in 2 × CO_2 climate model sensitivity. Part I: The total influence of sea ice thickness and extent. *Journal of Climate*, 8, 449–463.

Slide 1

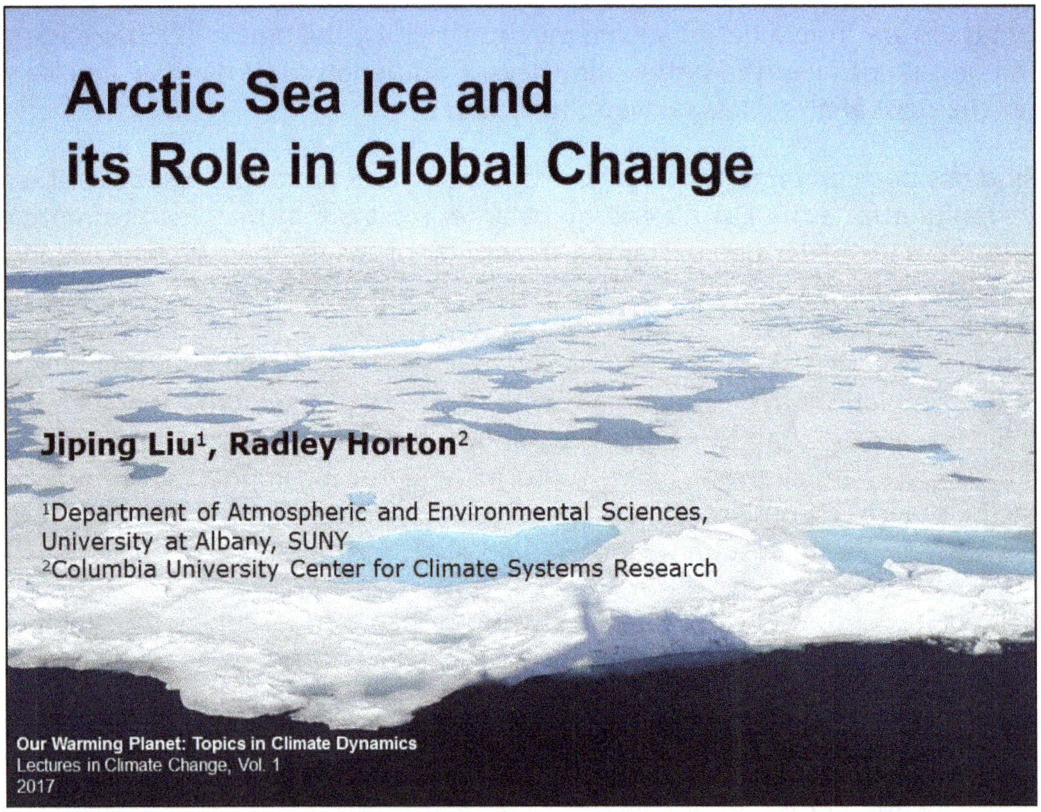

Slide 2

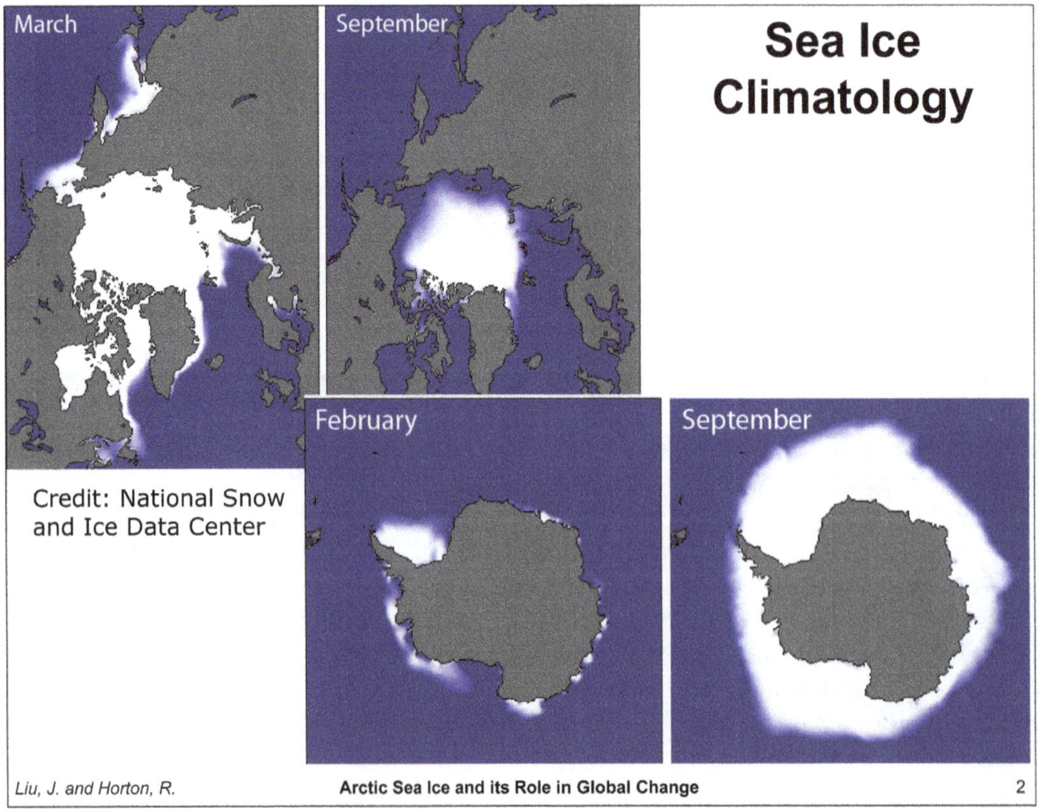

Slide 3

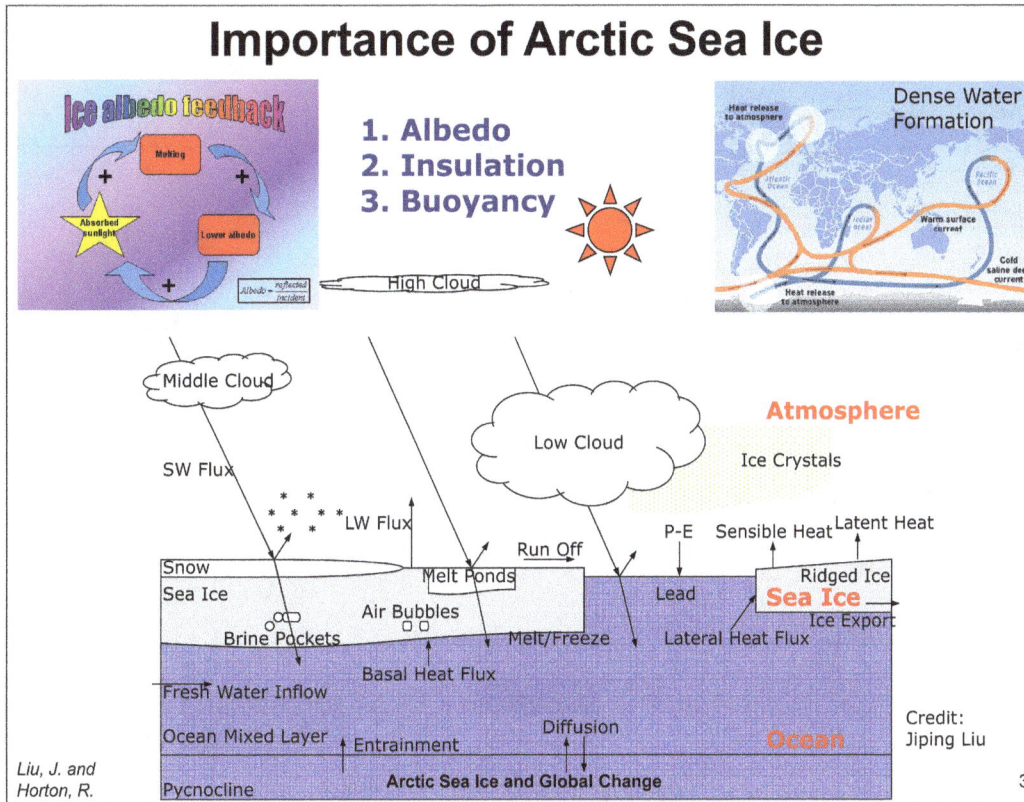

Slide 4

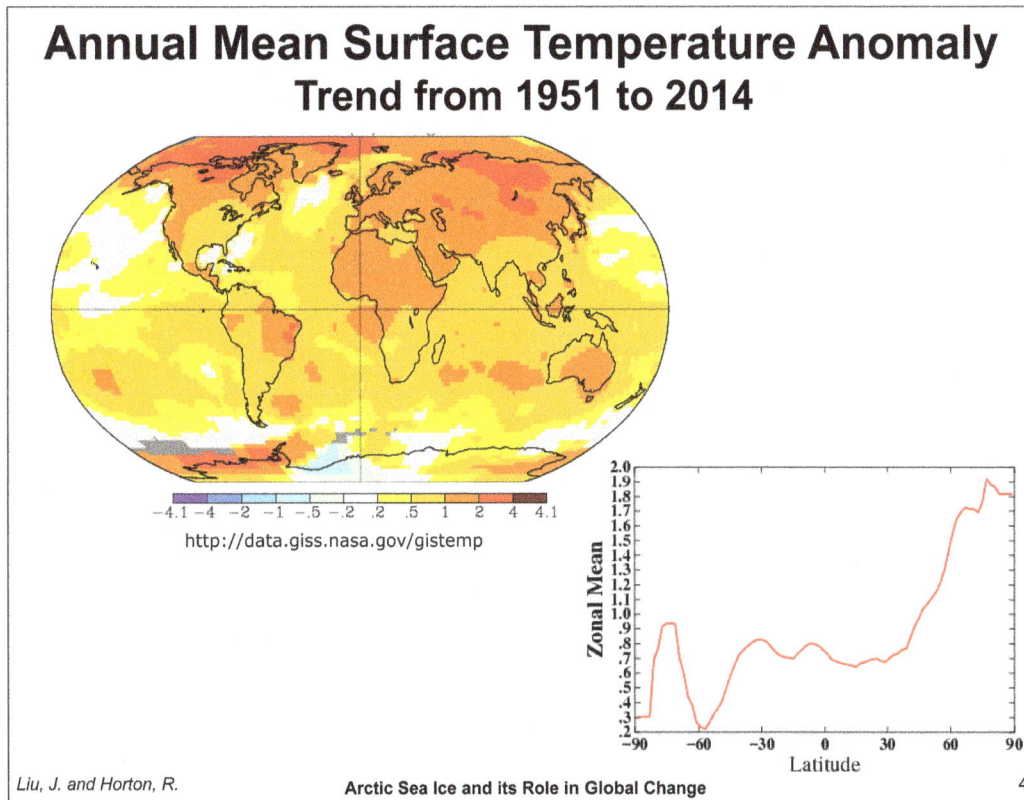

Slide 5

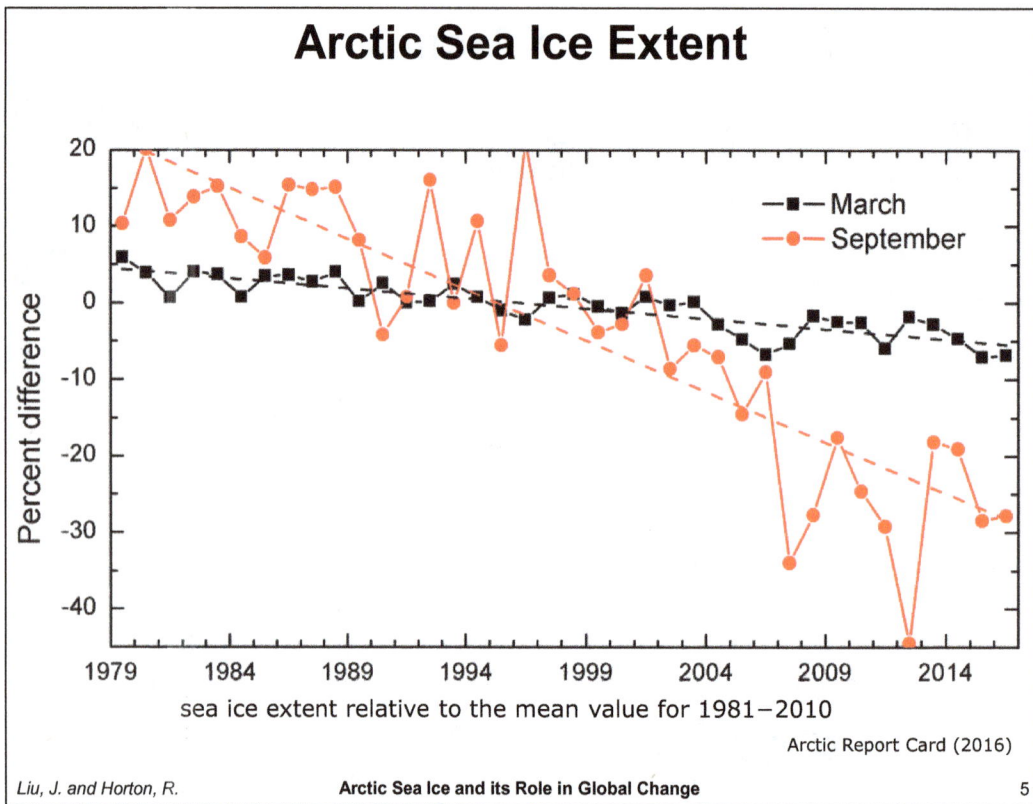

Slide 6

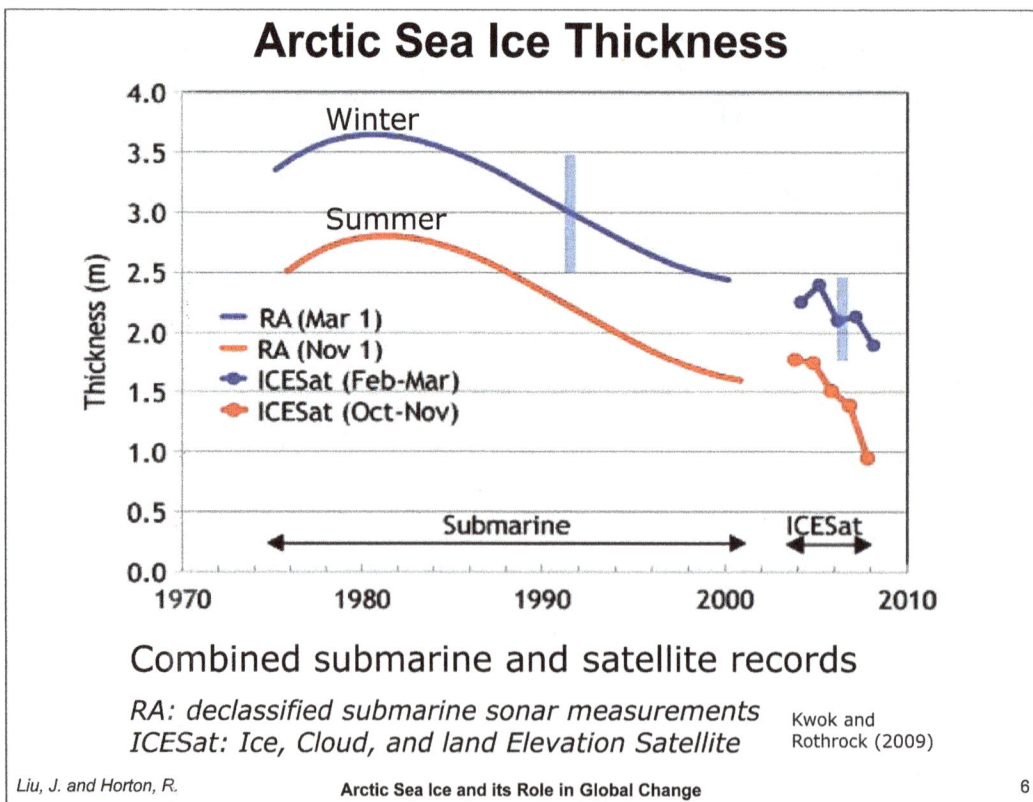

Slide 7

Arctic Sea Ice Age (End of Melt Season)

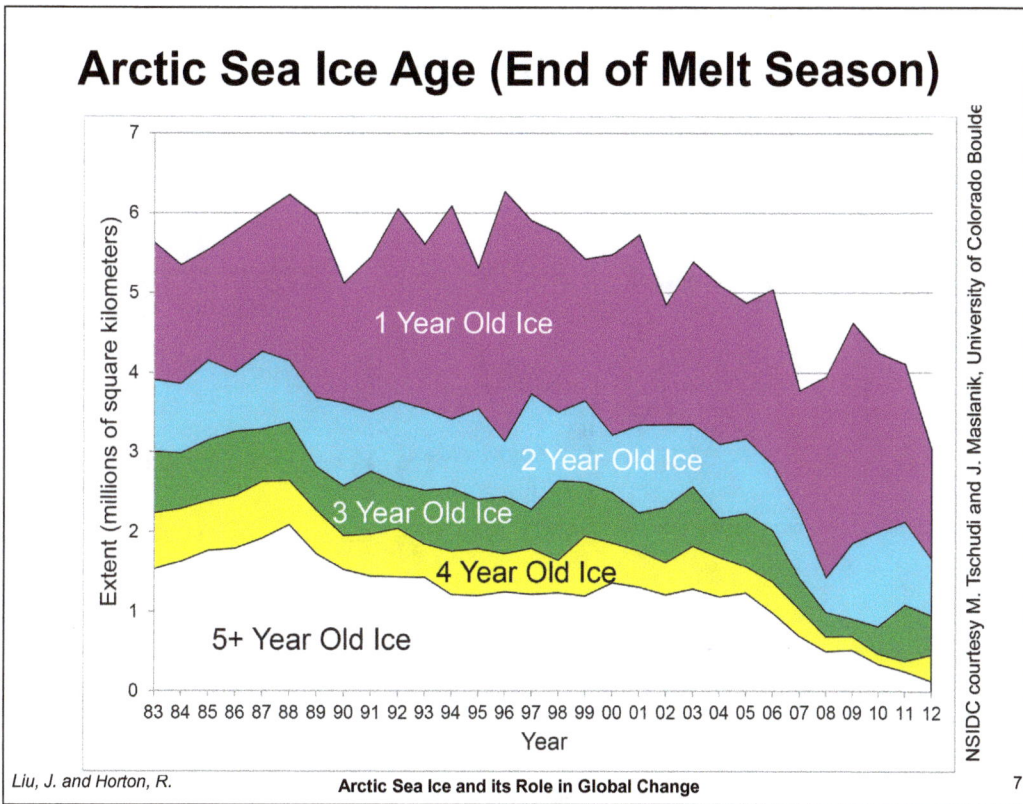

Slide 8

Arctic Sea Ice Drift Speed

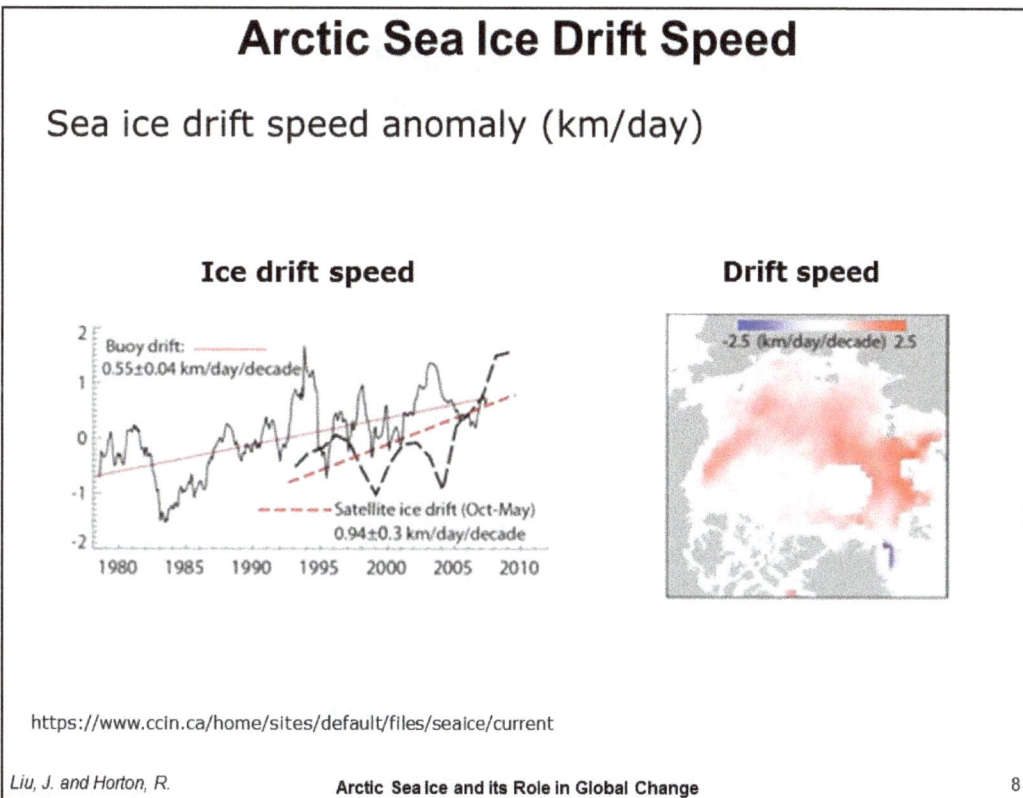

Slide 9

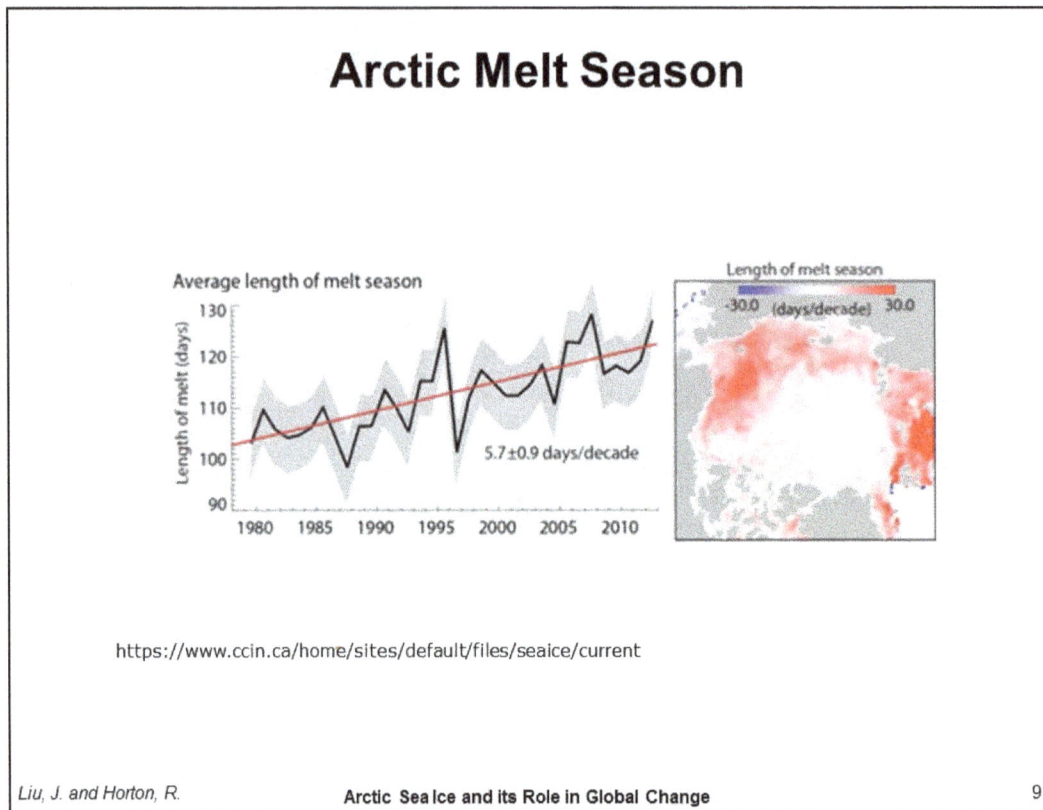

Slide 10

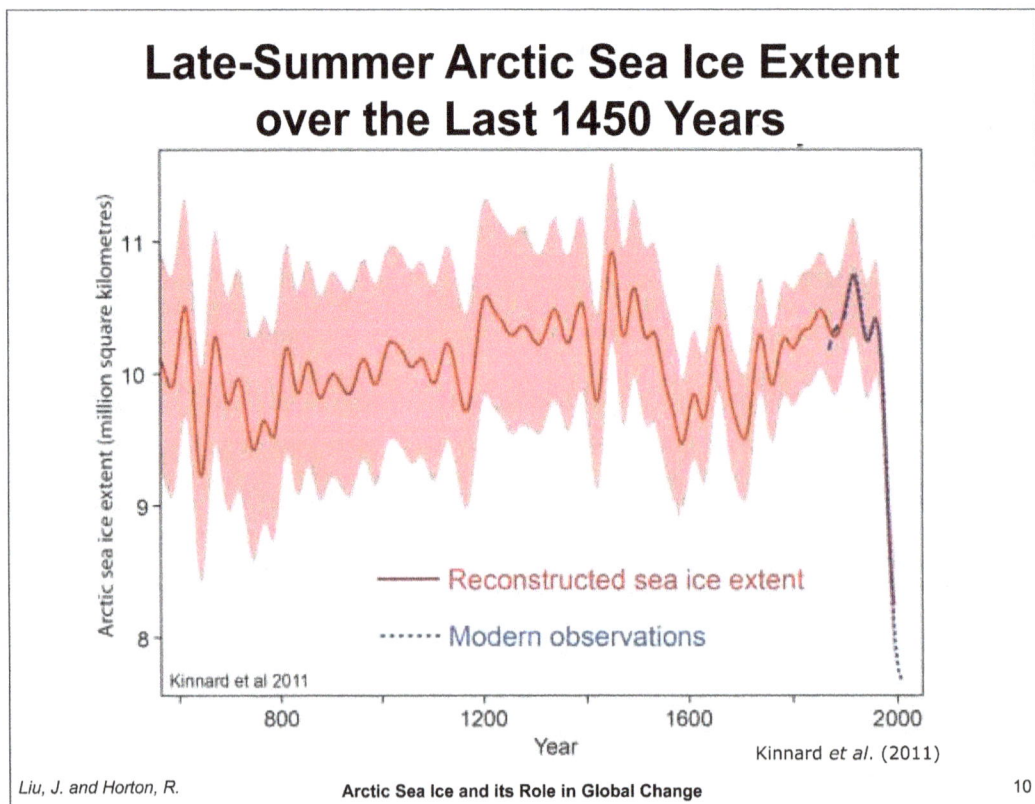

Slide 11

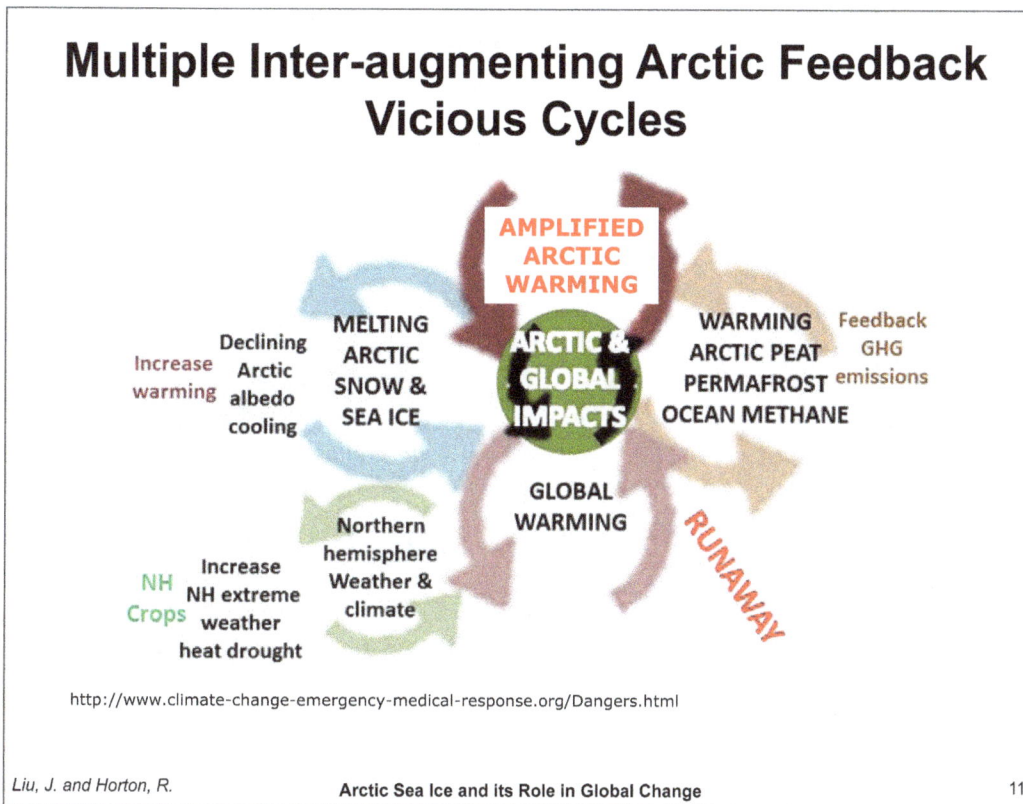

Multiple Inter-augmenting Arctic Feedback Vicious Cycles

AMPLIFIED ARCTIC WARMING

ARCTIC & GLOBAL IMPACTS

Increase warming

Declining Arctic albedo cooling

MELTING ARCTIC SNOW & SEA ICE

WARMING ARCTIC PEAT PERMAFROST OCEAN METHANE

Feedback GHG emissions

GLOBAL WARMING

RUNAWAY

Northern hemisphere Weather & climate

NH Crops

Increase NH extreme weather heat drought

http://www.climate-change-emergency-medical-response.org/Dangers.html

Liu, J. and Horton, R.   **Arctic Sea Ice and its Role in Global Change**   11

Slide 12

Weather and Climate

Loss of Arctic sea ice...

changes atmospheric pressure and winds...

potentially resulting in more severe winter storms in the eastern US and Eurasia

Potential for severe winters for eastern US

Warmer air & higher pressure surfaces over ice-free Arctic

Potential for severe winters for East Asia

Liu, J. and Horton, R.   **Arctic Sea Ice and its Role in Global Change**   12

Slide 13

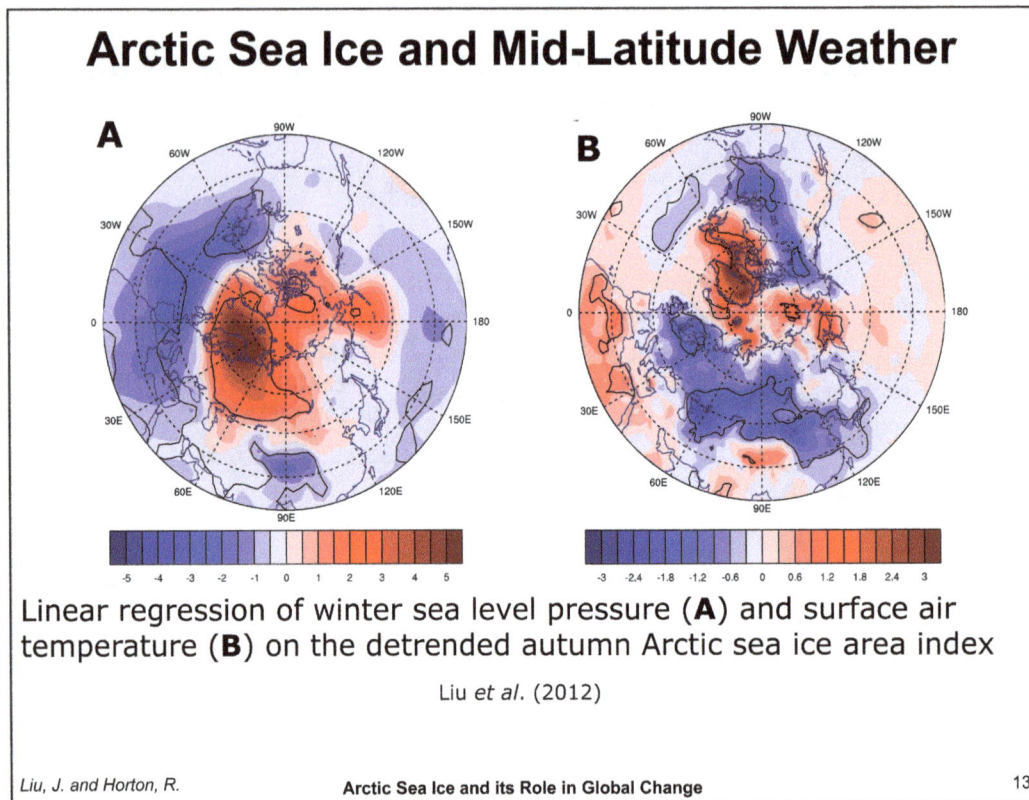

Arctic Sea Ice and Mid-Latitude Weather

Linear regression of winter sea level pressure (**A**) and surface air temperature (**B**) on the detrended autumn Arctic sea ice area index

Liu *et al.* (2012)

Slide 14

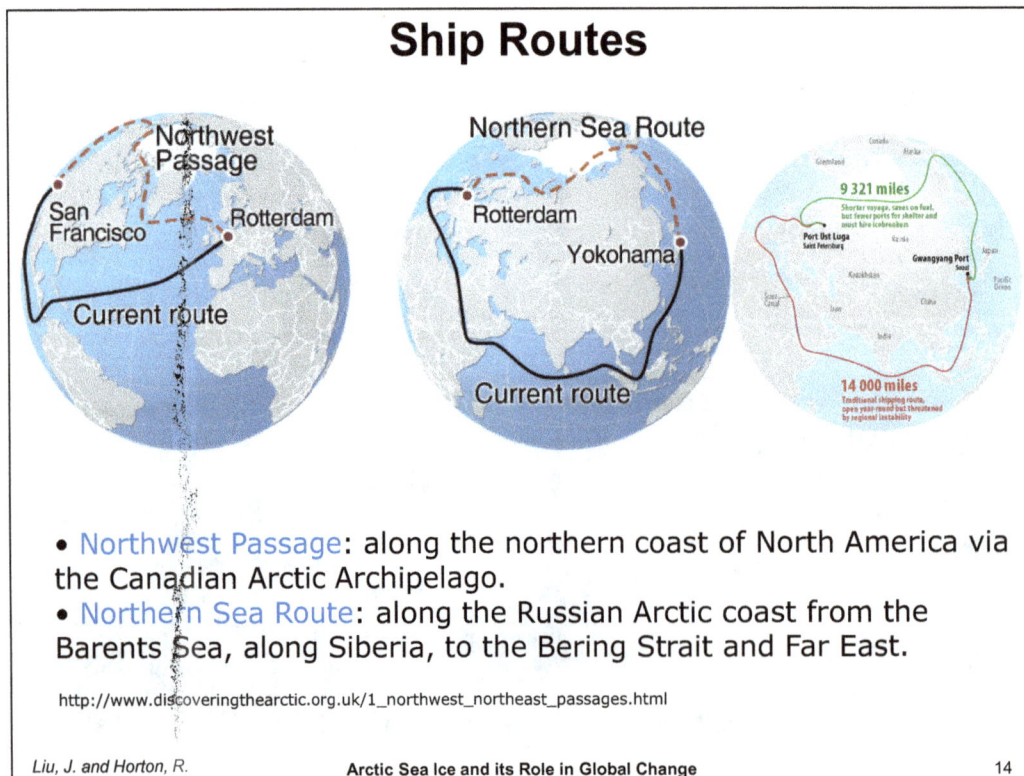

Ship Routes

- Northwest Passage: along the northern coast of North America via the Canadian Arctic Archipelago.
- Northern Sea Route: along the Russian Arctic coast from the Barents Sea, along Siberia, to the Bering Strait and Far East.

http://www.discoveringthearctic.org.uk/1_northwest_northeast_passages.html

Slide 15

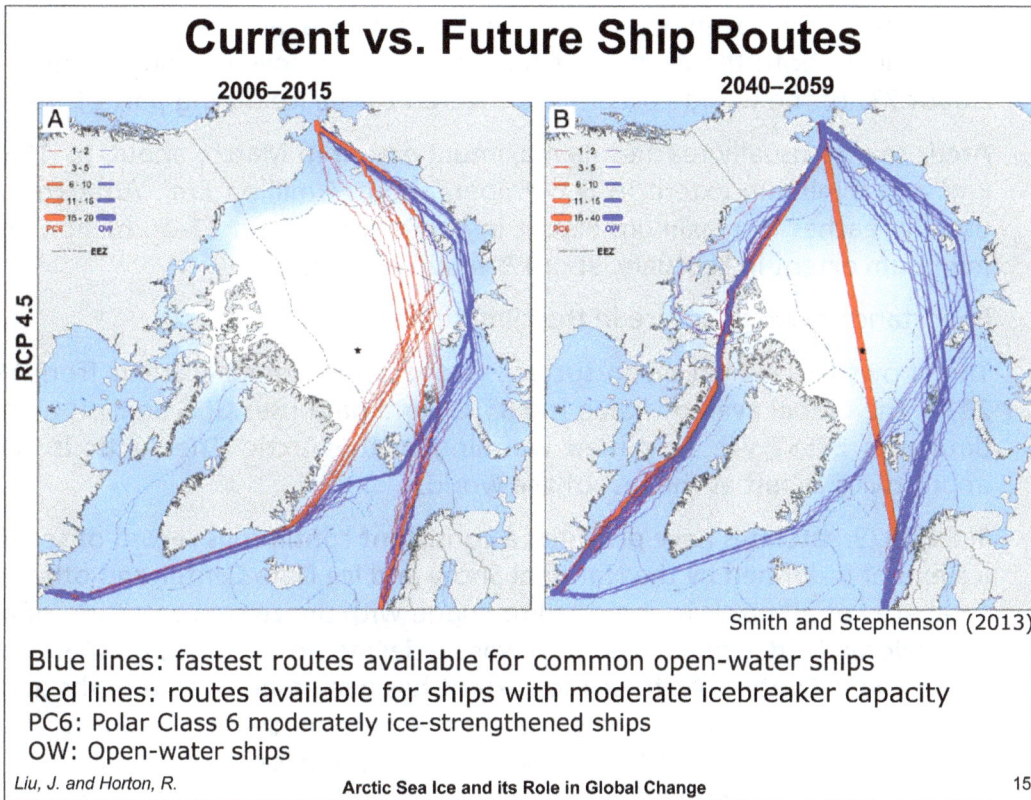

Current vs. Future Ship Routes

Blue lines: fastest routes available for common open-water ships
Red lines: routes available for ships with moderate icebreaker capacity
PC6: Polar Class 6 moderately ice-strengthened ships
OW: Open-water ships

Liu, J. and Horton, R.   **Arctic Sea Ice and its Role in Global Change**   15

Slide 16

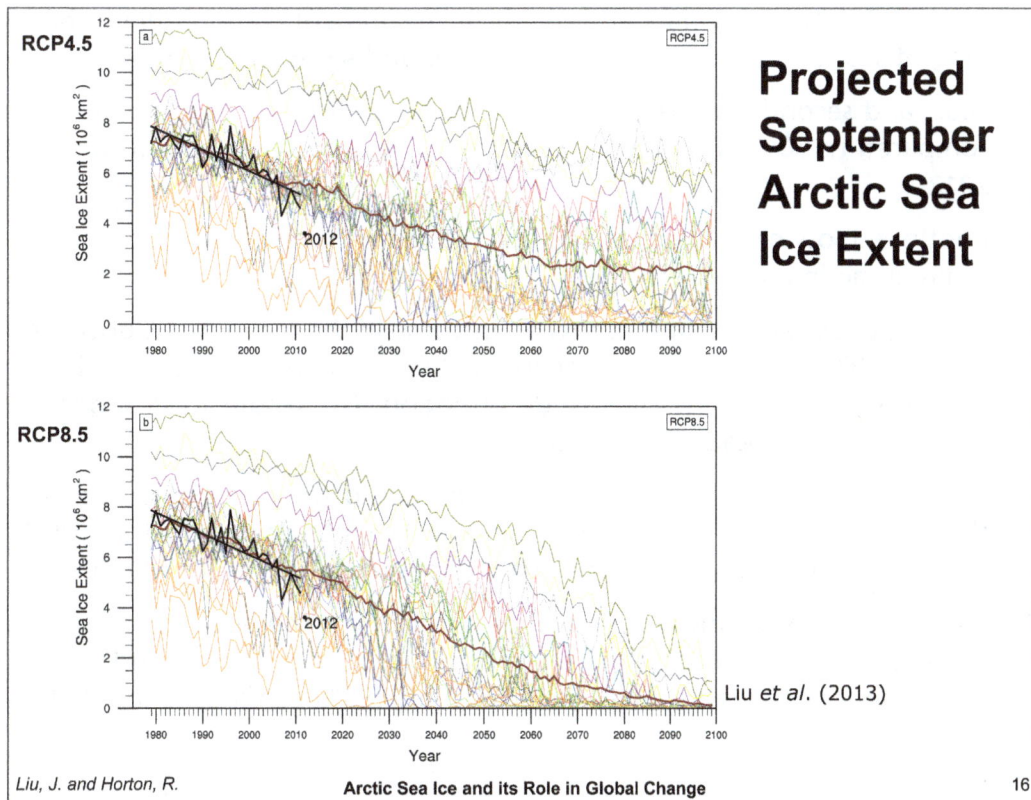

Projected September Arctic Sea Ice Extent

Liu *et al.* (2013)

Liu, J. and Horton, R.   **Arctic Sea Ice and its Role in Global Change**   16

Slide Notes

Slide 2   Sea ice is simply frozen ocean water. It forms, grows, and melts in the ocean. Sea ice occurs in both the Arctic and Antarctic. Sea ice covers about 25 million km^2. About 7% of the world's oceans are covered by sea ice during part of the year.

Arctic sea ice usually reaches its maximum extent in March, about 14–16 million km^2 and minimum extent in September, about 7 million km^2. Antarctic sea ice usually reaches its maximum extent in September, about 17–20 million km^2 and minimum extent in February, about 3–4 million km^2.

Slide 3   Importance of Arctic sea ice in the climate system.

Slide 4   This shows the annual mean surface temperature anomaly trend from 1951 to 2014. The global average value is 0.85°C (land data uses GISS analysis and ocean data uses ERSST_v4). Amplified warming in the Arctic. The Arctic is warming about twice as fast as the rest of the world.

Slide 5   Since 1979, satellites have provided a consistent continuous record of sea ice. Sea ice extent is defined by the National Snow and Ice Data Center and others as the total area of all pixels in the satellite image with an ice concentration of at least 15% (Note: the decision to use 15% was made back in the 1970s at NASA Goddard Space Flight Center). Arctic sea ice extent has decreased in all months and virtually all regions. As of 2014, the trend of September sea ice extent is –13% per decade relative to the 1981–2010 average. The trend is smaller in March (–3% per decade), but still decreasing and statistically significant. The last eight years have the eight smallest minimum extents since satellite observations.

Slide 6   Arctic sea ice is not only declining in extent but is also thinning dramatically since the late 1970s. The combined records from submarine and satellite show a thinning of sea ice thickness, ~1.75 m in winter and ~1.65 m in summer.

Slide 7   First- and second-year ice made up 70% of the ice cover in the Arctic in 2012, compared to 50% on in early 1980s. The oldest ice types almost disappeared in 2012.

Continued loss of thickest and oldest sea ice has prevented the recovery of the summer minimum.

Slide 8   Drifting buoys and satellites show an increase in the mean Arctic drift speed since the late 1970s. The increase in the ice drift speed tends to reduce sea ice mechanical strength and increase sea ice deformation. This results in stronger fracturing and more sea ice leads in the Arctic Ocean.

Slide 9   The amplified warming in the Arctic has caused the melt season to last longer, that is, melt season has increased by 5–6 days per decade in the last 30 years. A longer melt season allows the Arctic to absorb more heat to thin the ice.

Slide 10   This shows late-summer Arctic sea ice extent over the last 1450 years (red line), reconstructed from a combination of Arctic ice core, tree ring, and lake sediment data. The shaded area shows the uncertainty (95% confidence interval), and the blue dashed line shows modern observations. Arctic sea ice extent is currently lower than at any time in the last 1450 years.

Slide 11   GHG: greenhouse gases; NH: northern hemisphere.

Slide 12   Although individual extreme weather events cannot be directly linked to Arctic sea ice loss, recent observational data analysis and numerical model experiments suggest a link between decreasing Arctic sea ice and mid-latitude weather and climate, such as cold winter and increased snowfall.

Slide 13   For example, research suggested that decreasing autumn Arctic sea ice favors higher sea level pressure (SLP) and geopotential heights (GPH) over the Arctic and lower SLP and GPH in mid-latitudes. So, there is a weakening of the north–south temperature gradient and hence a weakening of westerly winds. The weakened westerly winds tend to enhance blocking circulations, favoring more frequent incursions of cold air mass from the Arctic into northern continents. With future loss of Arctic sea ice, cold conditions like winter 2009–2010 may happen more often.

However, the atmospheric dynamic is so variable, and the basic equations of motion on the time scale that we are discussing have so much inherent variability in them that you can have the same situations next time but everything is different. The only thing that will ultimately answer that is the durability of the effect. If, year after year, we are getting the same type of effect, that then looks like it is not a variability issue but a climate change issue.

Slide 14   Loss of Arctic sea ice means the opening up of the Arctic for commercial ships, offering a faster route for shipments between Europe and Asia, and holds the promise of increased trade for the High North of Arctic countries (Russia, Norway, and Canada). The number of ships using the route is on the rise.

Slide 15   By 2040–2059, loss of sea ice and changing ice conditions enable expanded September navigability for common open-water ships crossing the Arctic along the Northern Sea Route and the Northwest.

RCP: representative concentration pathway; RCP4.5 is a medium-mitigation emission scenario that stabilizes direct radiative forcing at 4.5 W/m^2 (~560 ppm CO_2 equivalent) at the end of the 21st century.

EEZ: the standard 200-nm Exclusive Economic Zone of the Russian Federation.

Slide 16   Time series of the simulated (colored lines) September sea ice extent from 1979 to 2100 for 30 CMIP5 models under the representative concentration pathway (RCP) 4.5 and 8.5 scenarios. The thick black line is the observations, the thick red line is the multimodel ensemble mean, and the black circle indicates September sea ice extent in 2012.

CLIMATE LECTURE 15

Antarctic Sea Ice and Global Warming

Douglas G. Martinson

Division of Ocean and Climate Physics, Lamont-Doherty Earth Observatory, and Department of Earth and Environmental Sciences, Columbia University, New York, NY, USA

Douglas Martinson is a Doherty Senior Research Scientist in the Division of Ocean and Climate Physics, and an Adjunct Professor in the Department of Earth and Environmental Sciences of Columbia University. His research has focused on how changes in the polar regions feedback to drive changes elsewhere in the world. The goal is to provide proper assessment of the impact of polar changes on global climate and to predict the consequence of the polar changes on those extra-polar regions sensitive to such change.

Introduction

Currently, there is considerable focus on the inexplicable increasing trend in time of the average Antarctic sea ice extent (Parkinson and Cavalieri, 2012), often noted in the context of the rapidly decreasing Arctic sea ice cover. While the increasing Antarctic sea ice trend was originally considered small, Simmonds (2015) has recently shown that it cannot be considered small anymore. Unfortunately, the social media uses this Arctic–Antarctic comparison in an attempt to further emphasize the paradox of the increasing Antarctic sea ice cover (often to emphasize the failings of global warming 'alarmists'), but such a comparison is unwarranted. The differences are so dramatic between the two poles that one is not justified in expecting the change in the Antarctic sea ice to echo that of the Arctic. For example, consider the geographic setting: the Arctic is a land-locked ocean that constrains the ice, whereas the Antarctic is a continent surrounded by ocean with primary winds blowing the ice floes to warmer waters where they rapidly melt. As a consequence, the Arctic until recently was largely covered by a relatively thick (~3 m) perennial sea ice cover, whereas the Antarctic polar oceans are covered by a large thin (~0.6 m) seasonal ice cover, with only small areas of perennial ice. Loss of Arctic highly reflective perennial ice in summer exposes the dark absorbing Arctic Ocean waters, driving the ice-albedo feedback and greatly enhancing the absorption of heat and enhanced melt back. In the Antarctic, the thin sea ice melts in spring, so the summer ocean water is always exposed, greatly limiting the ice-albedo feedback potential. But, even given these differences, we would still expect the Antarctic sea ice fields to decrease given a warming Earth.

The expanding Antarctic sea ice extent shows strong spatial variability. In the western Antarctic, the ice is declining, but elsewhere it is expanding to the extent that the expansion dominates the overall average. The Intergovernmental Panel on Climate Change (IPCC) assessment of climate includes a Coupled Model Intercomparison Project (CMIP) to determine how well the global climate models (GCMs) can simulate certain aspects of climate change under different forcing scenarios (e.g., doubled CO_2). The most recent of these, CMIP5 (http://cmip-pcmdi.llnl.gov/cmip5/), includes 49 coupled models that Shu, Song, and Qiao (2015) evaluated to determine how well they can simulate the changes in the polar sea ice fields. The models, in general, do a reasonable job capturing the declining Arctic sea ice, but only 8 of the 49 simulate increasing sea ice extent for the Antarctic (the others all show declining ice extent). Unfortunately, those eight did not capture the spatial variability in the changing ice extent. So, the models are not yet able to simulate the enigmatic increasing sea ice extent, though the multi-model ensemble (MME) mean does a fine job of capturing the 1979–2005 average (climatological) sea ice distribution and the climatological seasonal cycle.

Part of the failure of the models to capture the trend may lie in the fact that the Antarctic Ocean stratification leads to a very strong ice-ocean feedback. Specifically, the winter system, when ice grows, is well approximated by a simple linear system as shown in Slide 9, easily described by the system of equations shown on the slide (Martinson, 1990). This system shows a deep winter mixed layer (~100 m deep) at the freezing point (T_f) overlying the permanent pycnocline (the thermocline and halocline coincide in depth) leading to a fixed source of warm deep water — in this case, Upper Circumpolar Deep Water or a modified version thereof. Lateral advection resupplies any heat lost from the deep water through the pycnocline to the surface, thus maintaining the 'fixed' heat supply. Turbulent diffusion across the pycnocline leads to a very small heat flux (typically <5 W/m²). This sensible heat flux to the surface layer is not sufficient to satisfy the ocean–atmosphere heat loss of at least 35 W/m² in winter, so the remainder of the heat must be supplied by latent heat of fusion (sea ice growth). Sea ice rejects brine, which densifies the mixed layer and causes it to deepen to a statically stable depth by eroding into the pycnocline. In so doing, this drives a strong entrainment heat flux as the erosion of the pycnocline releases heat from the thermocline into the mixed layer. This heat flux is tremendous, typically well over the amount necessary to balance the original air–sea surface flux. In fact, in terms of ice growth, for every one unit of ice growth, the entrainment heat flux is strong enough to melt up to 16 units of ice! The net effect is that this feedback keeps the Antarctic sea ice cover very thin (typically ~60 cm).

The solutions to the system of equations lend themselves to prognostic scaling laws using simple to observe physical variables that describe the full solutions to better than 95%. These scaling laws have been integrated (Martinson and Iannuzzi, 1998), providing a few fundamental robust bulk properties that can be combined in simple ways to estimate key components of the system, such as the total ocean heat flux, bulk static stability, and ratio of ice production to ice melt from the negative feedback of the entrainment heat flux. These laws have been shown (Martinson and Iannuzzi, 1998) to capture well the ocean heat flux as measured during the ANZFLUX project in the Weddell Gyre (McPhee et al., 1996). Slide 12 shows their application to the Weddell Gyre, giving the ratio of how many units of ice would melt for each unit of ice growth. The ratio is tremendous (over 6 for much of the region, and reaching as high as 16) and clearly shows the strength of the feedback keeping the ice thin.

The feedback serves to self-regulate the ice thickness. That is, since any winter air–sea heat flux is stronger than the small turbulent diffusive flux, sea ice will grow, and no matter how much ice, the feedback will keep it thin, reducing the sensitivity of the ice thickness to changes in the air–sea heat flux. Simple experiments with the solutions to the system of equations in Slide 9 involving changing the winter heat flux by 40 W/m^2 lead to changes of less than 5 cm of ice thickness. This self-regulation is the consequence of detailed processes operating on small spatial scales, something very difficult to capture in large-scale GCMs. Most GCMs, ignoring the feedback, melt all Antarctic sea ice under a sufficiently warmed climate, as a thin sea ice cover is easy to remove without the feedback (similarly, a large ice extent is easy to alter). Rind, Healy, Parkinson, and Martinson (1995) tested the potential of this feedback by running some $2 \times CO_2$ simulations where the control run, without the feedback melted nearly all of the sea ice as in most models, but then the feedback was simulated by not allowing the annual cycle of the sea ice extent to change at all (an extreme consequence of the self-regulation). In the control run, the global warming from doubled CO_2 was 4.17°C, whereas in the feedback simulation the Earth warmed only 2.61°C. The fixed sea ice led to a reduction in the greenhouse warming by 37%, a staggering amount for a single process. During the early years of the CMIP projects (CMIP2), using still rather simple coupled models, evaluation of the result did not clearly reveal this sensitivity to thickness, but did find this effect strong with greater sea ice extent (Flato, 2004).

Later, continuing their experiments exploring the sensitivity of global warming to the treatment of the Antarctic sea ice fields, Rind and colleagues (Parkinson *et al.*, 2001) showed that errors in sea ice concentration (SIC) led to changes in the simulated temperatures. On average, each 1% decrease in SIC led to a yearly average increase of 0.0107°C in global warming.

Further, Rind *et al.* (2001) showed that changes in tropical sea surface temperature (SST) lead to changes in the meridional temperature gradients from the equator to pole and that this impacts *all* of the Antarctic polar basins. For example, the occurrence of El Niño conditions leads to a stronger SST gradient in the eastern Pacific basin, relaxing (spin down) the Pacific subpolar gyre. This same feature shifts the tropical Walker Cell leading to a reduction in the SST gradient in the Atlantic Ocean, leading to a spin up of the Weddell gyre. These results immediately explained the findings of Martinson and Iannuzzi (1998) showing the spin up of the Weddell Gyre during El Niño events. The opposite happens with La Niña, again as observed. They also explain why the response in the Atlantic sector is opposite to that in the Pacific sector, something observed in what is known as the Antarctic Dipole (Yuan and Martinson, 2001). The spin up or down of a polar gyre changes the surface divergence, which can lead to expansion or retraction of the sea ice cover.

While these experiments may not answer our questions regarding increasing sea ice extent, they clearly show how important it is to properly address details in the modeling of the sea ice, something that has been difficult to achieve.

In addition to these experiments, follow-on studies (e.g., Stammerjohn *et al.*, 2008; Lefebvre and Goosse, 2008) clearly show that different sectors of the Antarctic sea ice fields are dominated by different processes. This adds another layer of complexity, indicating that many processes affecting the sea ice must be considered if we expect to have any chance of properly capturing the observed trends through time.

Fortunately, with more powerful computers and clever numerical algorithms, excellent teams of modelers focusing on the sea ice components are now including a seemingly endless number of sea ice details at the scales required. These models are capturing the Arctic decreasing trend, though still fall short for the Antarctic (possibly due to the ice-ocean feedback, which is not active in the Arctic due to the presence of what is known as the Cold Halocline Water, so the entrainment of the thermocline is entraining cold water, not warm water as in the Antarctic).

Finally, regarding the overall fact that the Arctic sea ice is decreasing and the Antarctic increasing, Claire Parkinson provides excellent comments about this, as quoted in Slide 19. All regions do not respond the same, in fact some regions during global warming are cooling, and likewise, some regions show expanding ice instead of retreating ice. So, we are left with trying to explain why the Antarctic is not behaving as we expected. Antarctica is a last frontier, and this is just one more mystery surrounding it.

Acknowledgments

This work was supported by NASA Polar Programs Grant NAGW 1844, and NSF Grants OPP-08-23101 and PLR-1440435.

References

Flato, G. (2004). Sea-ice and its response to CO_2 forcing as simulated by global climate models. *Climate Dynamics*, 23, 229–241.

Lefebvre, W. and Goosse, H.(2008). An analysis of the atmospheric processes driving the large-scale winter sea ice variability in the Southern Ocean. *Journal of Geophysical Research*, 113, C02004. doi:10.1029/2006JC004032.

Martinson, D.G. (1990). Evolution of the Southern Antarctic mixed layer and sea-ice; Open ocean deep Water formation and ventilation. *Journal of Geophysical Research*, 95, 11641–11654.

Martinson, D.G. and Iannuzzi, R.A. (1998). Antarctic ocean–ice interaction: Implications from ocean bulk property distributions in the Weddell Gyre. In: *Antarctic Sea Ice: Physical Processes, Interactions and Variability*, Ant. Res. Ser., V74, AGU, Washington, DC, 243–271.

McPhee, M., Ackley, S., Guest, P., Huber, B., Martinson, D., Morison, J. and Stanton, T. (1996). The Antarctic cone flux experiment. *Bulletin of American Meteorological Society*, 77, 1221–1232.

Parkinson, C.L. and Cavalieri, D.J. (2012). Antarctic sea ice variability and trends 1979–2010. *The Cryosphere*, 6, 871–880. doi:10.5194/tc-6-871-2012.

Parkinson, C.L., Rind, D., Healy, R.J. and Martinson, D.G. (2001). The impact of sea ice concentration accuracies on climate model simulations. *Journal of Climate*, 14, 2606–2623.

Rind, D., Healy, R., Parkinson, C. and Martinson, D.G. (1995). The Role of Sea Ice in $2 \times CO_2$ climate model sensitivity Part I. The total influence of sea ice thickness and extent. *Journal of Climate*, 8, 449–463.

Rind, D., Chandler, M., Lerner, J., Martinson, D.G. and Yuan, X. (2001). Climate response to basin-specific changes in latitudinal temperature gradients and implications for sea ice variability. *Journal of Geophysical Research*, V106, 20161–20173.

Shu, Q., Song, Z. and Qiao, F. (2015). Assessment of sea ice simulations in the CMIP5 models, *The Cryosphere*, 9, 399–409. doi:10.5194/tc-9-399-2015.

Simmonds, I. (2015). Comparing and contrasting the behaviour of Arctic and Antarctic sea ice over the 35 year period 1979–2013. *Annals of Glaciology*, 56(69). doi:10.3189/2015AoG69A909.

Stammerjohn, S.E. Martinson, D.G., Smith, R.C., Yuan, X. and Rind. D. (2008). Trends in Antarctic annual sea ice retreat and advance and their relation to ENSO and southern annular mode variability. *Journal of Geophysical Research*, 113, C03S90. doi:10.1029/2007JC004269.

Yuan, X. and Martinson, D.G. (2001). The Antarctic dipole and its predictability.*Geophysical Research Letters*, 28, 3609–3612.

Slide 1

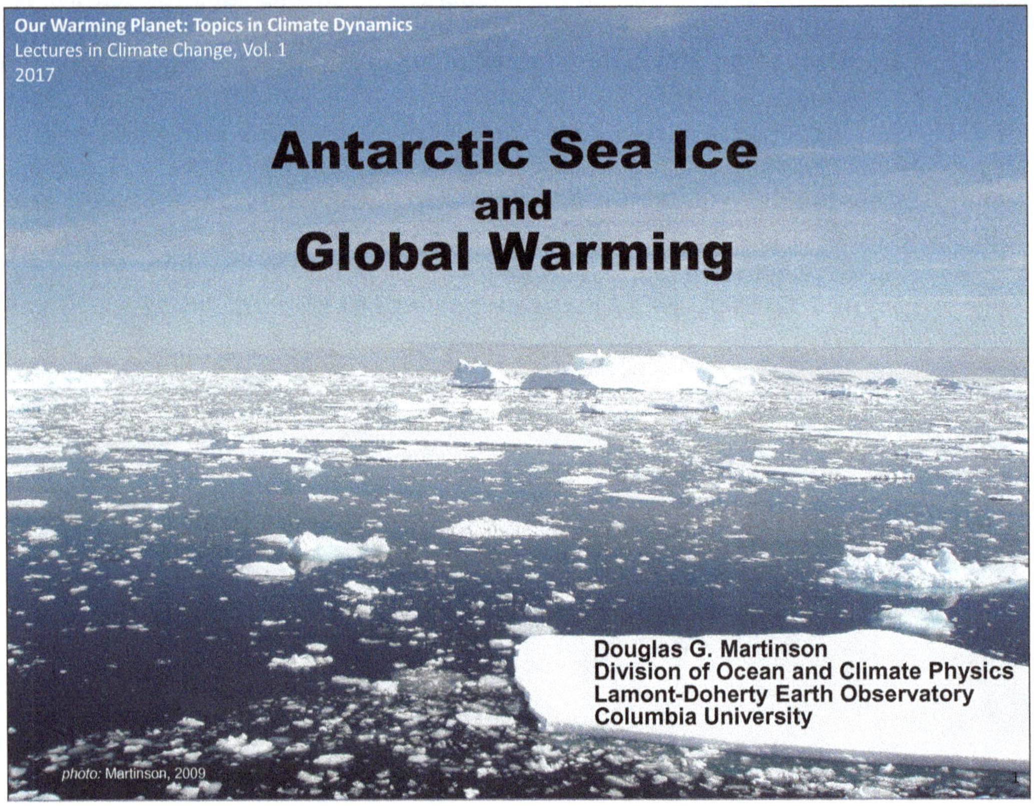

Slide 2

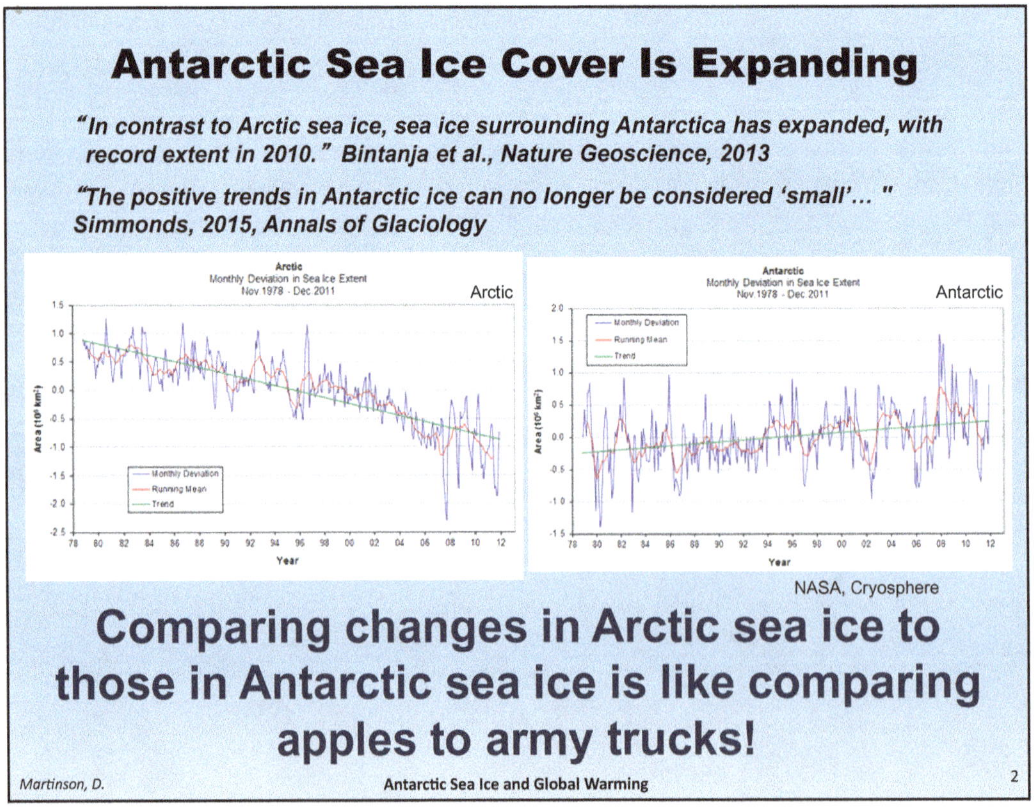

Slide 3

Differences Between Poles Are So Dramatic

One is not justified in expecting the change in
Antarctic sea ice to echo that of the Arctic.

Geographic setting

Arctic: land-locked ocean, limiting maximum sea ice cover

Antarctic: continent surrounded by ocean, ice drifts to warm waters; equatorward expansion limited by warm waters north of polar front (limits maximum ice cover)

Ice-albedo feedback

Arctic: covered by thick perennial sea ice cover, loss of reflective ice in summer results in absorption of heat into dark ocean melting more ice, absorbing more heat (=ice-albedo feedback);

Antarctic: polar oceans are covered by thin seasonal ice cover results in limited ice in summer, so limited ice-albedo feedback (minor feedback does occur where ice is north of polar circle).

Ocean-ice feedback

Arctic: stratification of surface ocean prevents deep ocean heat from affecting ice growth

Antarctic: stratification allows strong ocean-ice feedback heat flux, restricting ice thickness to a thin, easy-to-remove shell

Despite differences, we cannot explain why Antarctic sea ice is expanding!

Martinson, D.   Antarctic Sea Ice and Global Warming   photo: Martinson, 2009   3

Slide 4

Antarctic Sea Ice Is Seasonal
(Except For A Few Small Pockets)

Seasonal sea ice is less subject to ice-albedo feedback

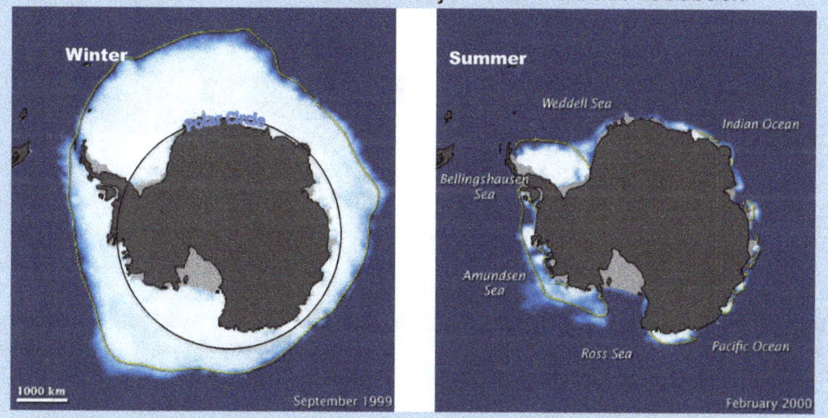

NASA, Earth Observatory

Martinson, D.   Antarctic Sea Ice and Global Warming   4

Slide 5

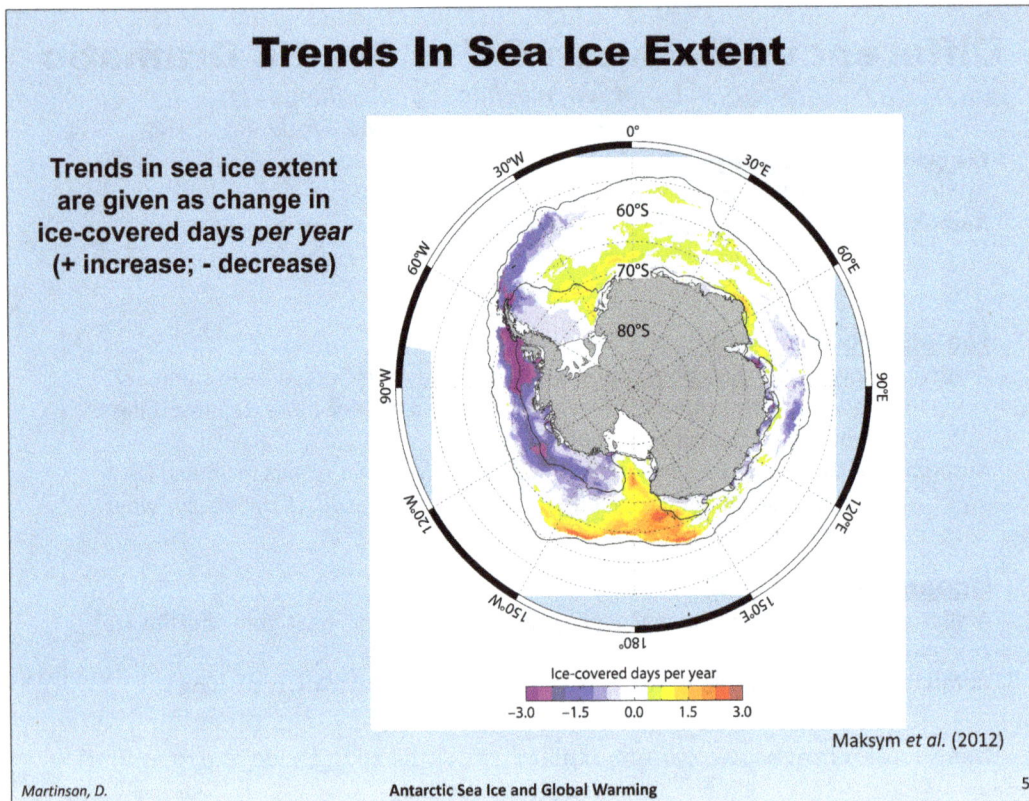

Slide 6

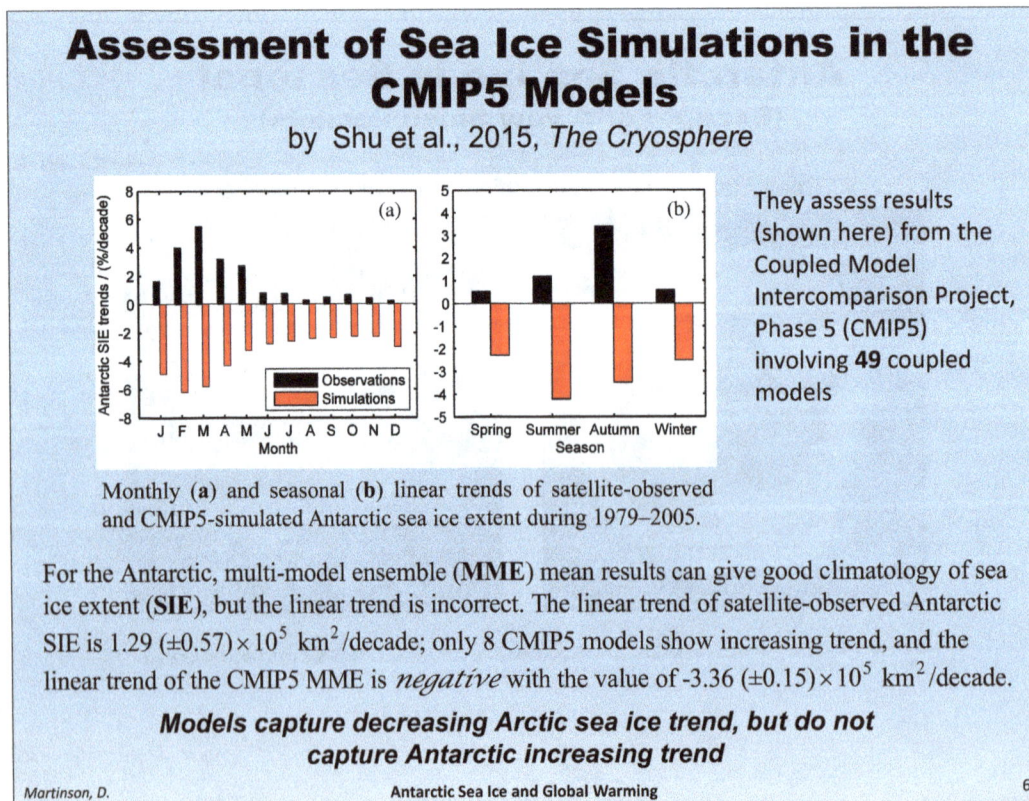

Monthly (**a**) and seasonal (**b**) linear trends of satellite-observed and CMIP5-simulated Antarctic sea ice extent during 1979–2005.

For the Antarctic, multi-model ensemble (**MME**) mean results can give good climatology of sea ice extent (**SIE**), but the linear trend is incorrect. The linear trend of satellite-observed Antarctic SIE is 1.29 ($\pm$0.57)$\times 10^5$ km^2/decade; only 8 CMIP5 models show increasing trend, and the linear trend of the CMIP5 MME is *negative* with the value of -3.36 ($\pm$0.15)$\times 10^5$ km^2/decade.

Models capture decreasing Arctic sea ice trend, but do not capture Antarctic increasing trend

Slide 7

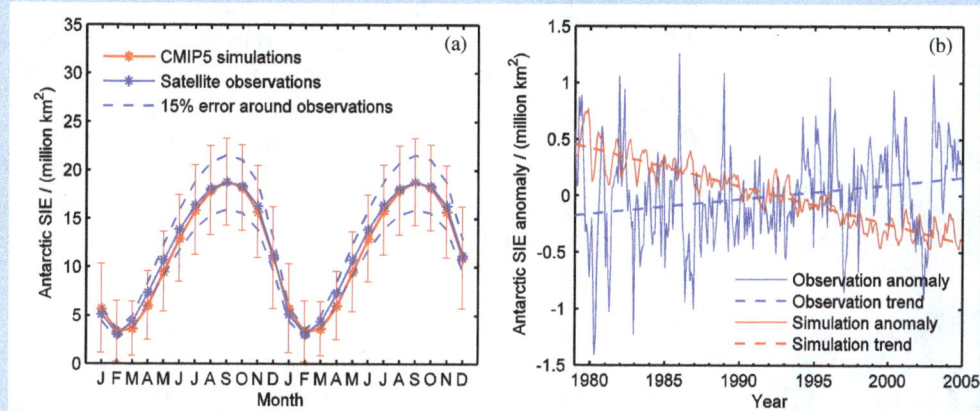

Assessment of Sea Ice Simulations in the CMIP5 Models

by Shu *et al.*, 2015, *The Cryosphere*
(continued)

Average of all simulations shown in red

Slide 8

Antarctic ocean–sea ice interaction demands detailed modeling of small spatial scale processes in surface ocean (difficult for large Global Climate Models: GCMs).

Slide 9

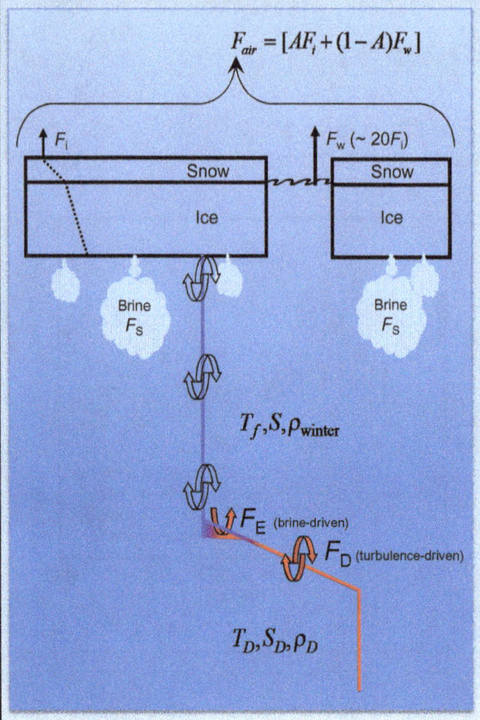

Winter Mixed Layer System

Equations can capture the effect that when ice grows it rejects brine, making mixed layer more dense. Mixed layer deepens as result, entraining warm thermocline water. This releases heat to the surface, melting ice that just grew or preventing an equivalent amount from growing, thus keeping ice cover thin. So much heat is released in Atlantic polar waters (Weddell gyre) that if one unit of ice grows, its brine entrains enough heat to melt 1–16 units of ice (this delays additional ice for many days).

It also moderates effect of changing surface heat loss, making ice more stable to changes in surface forcing. Possibly why more warming is not causing more ice loss. Process is difficult to model in GCMs because of detail required at fine scales.

Martinson, JGR, 1990

Martinson, D. **Antarctic Sea Ice and Global Warming** 9

Slide 10

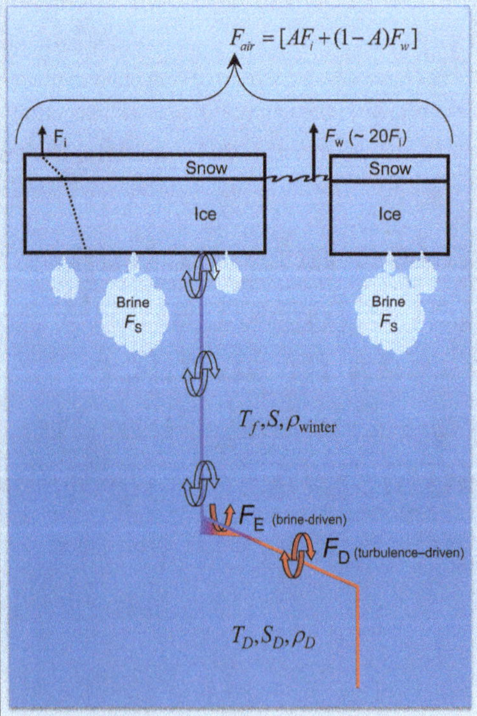

Winter Mixed Layer System (continued)

$$\frac{dT}{dt} = 0; \quad T = T_f$$

$$h\frac{\partial S}{\partial t} = w_e \Delta_z S + k_S \nabla S + F_{pe} - A\sigma_i \frac{\partial h_i}{\partial t}$$

$$\rho_i c_p \frac{\partial h_i}{\partial t} = \rho_w c_p \left(w_e \Delta_z T + k_S \nabla T \right) - F_{air}$$

$$\frac{dh_{ml}}{dt} = w_e + w$$

$$w_e = \frac{\partial \rho}{\partial t} \Big/ \nabla \rho$$

Martinson, JGR, 1990

Martinson, D. **Antarctic Sea Ice and Global Warming** 10

Slide 11

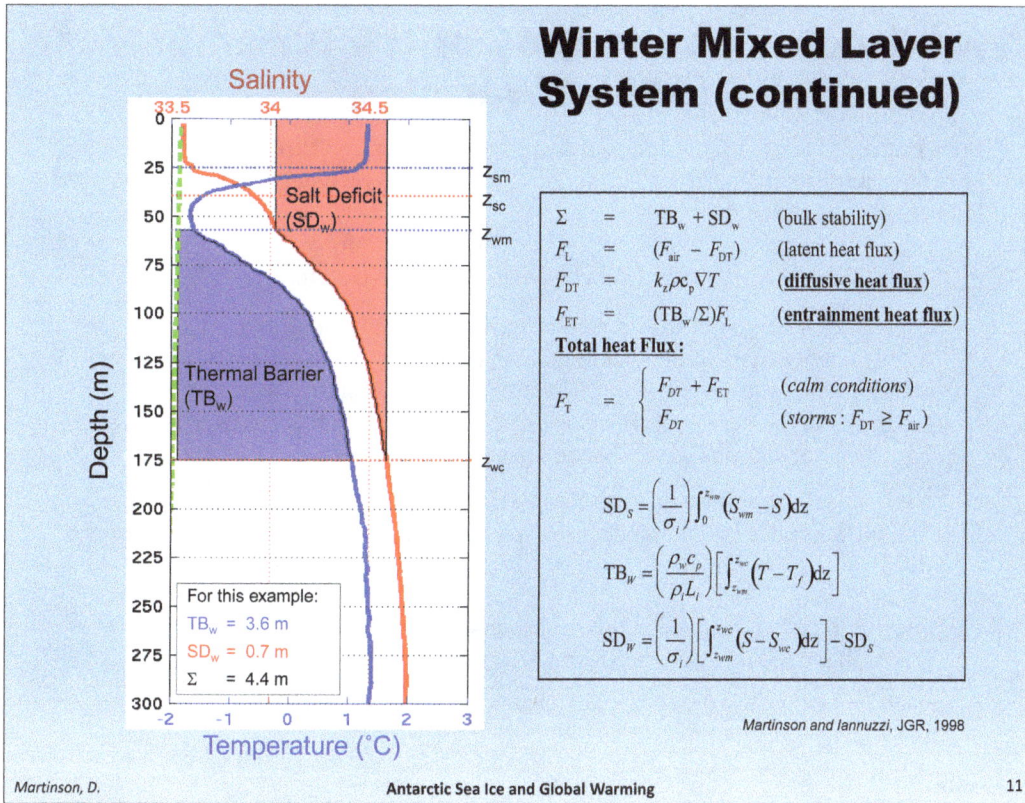

Slide 12

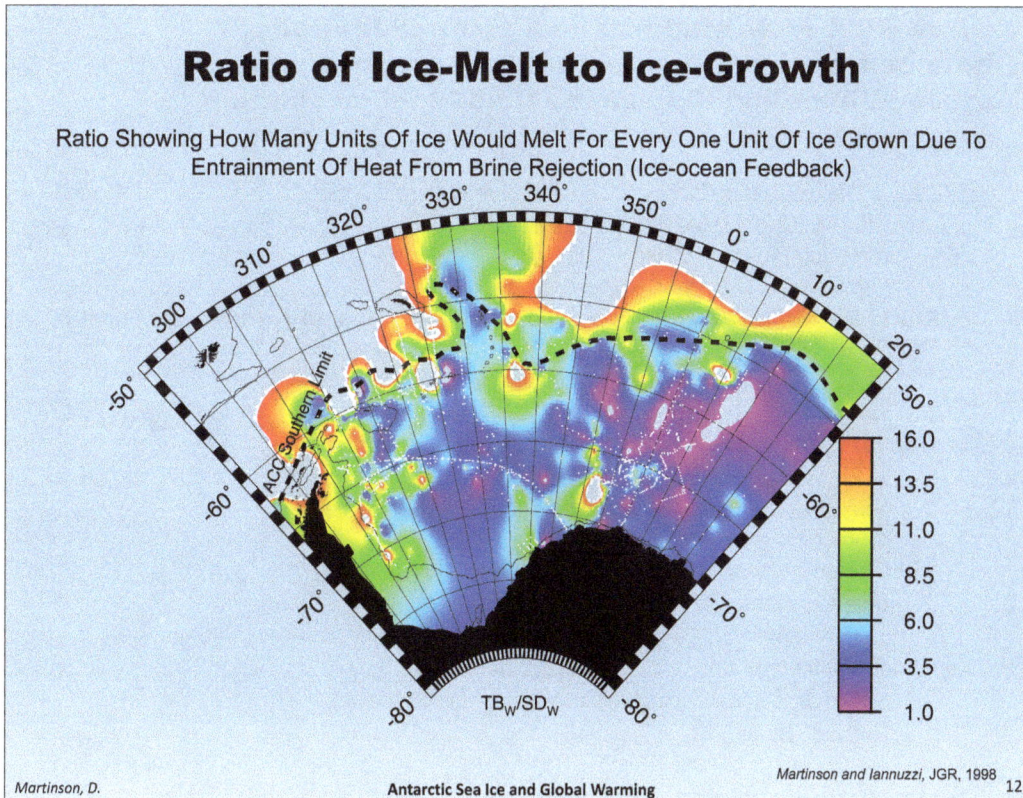

Slide 13

Two Studies on the Role of Ice Parameters in Global Warming

Numerous GCM tests of climate sensitivity to Antarctic sea ice distribution and thickness have been performed. In particular a series of studies by Rind and colleagues tested multiple aspects of this sensitivity.

Rind *et al.*, 1995 (varying ice thickness and extent with 2×CO_2 forcing):

Thinner ice and greater sea ice extent is easier to remove in GCMs, making results more sensitive than for thick ice. Simulating effect of ice-ocean feedback by not allowing sea ice to change in 2×CO_2 simulation, results in a reduction of climate sensitivity by 37%(!)—with changing ice, 2×CO_2 drives a global warming of 4.17˚C, but when not allowing ice to change, that amount was reduced by 1.56˚C (37%)

Note: initial coupled model runs (CMIP2) runs with interactive sea ice models, *do* find this response for greater ice extent, but not so clearly for ice thickness, *Flato*, 2004

Parkinson *et al.*, 2001 (exploring impact of changes in sea ice concentration as in Atmospheric Model Intercomparison project runs):

each 1% decrease in SIC causes yearly global warming of 0.0107˚C

Martinson, D. **Antarctic Sea Ice and Global Warming** 13

Slide 14

Other Rind *et al.* studies show communication between tropical climate and the Antarctic polar gyres influencing the sea ice fields and circulation.

Rind *et al.*, 2001:

Tropical Pacific warming anomaly (El Niño) enhances pole-equator temperature gradient (opposite for La Niña)

- Intensifies subtropical circulation
  - Subtropical jet (STJ) moves equatorward
- **STJ farther from available potential energy (APE) of cold Antarctic continent in Pacific**
  - Leads to less cyclogenesis in subpolar Pacific gyres
  - Thus, spin-down of subpolar gyre
- **Tropical warming anomaly perturbs Walker circulation**
  - Decreases tropical Atlantic subsidence
  - Slows down subtropical Atlantic circulation
    - SJT moves poleward
- **STJ closer to APE of cold Antarctic continent in Atlantic**
  - Leads to more cyclogenesis in subpolar Atlantic gyre
    - Thus, spin-up of Weddell gyre (perhaps expanding sea ice field to north)
- **Model results explain:**
  - Subpolar regional findings (Weddell spin-up/spin-down)
  - Antarctic Dipole

Martinson, D. **Antarctic Sea Ice and Global Warming** 14

Slide 15

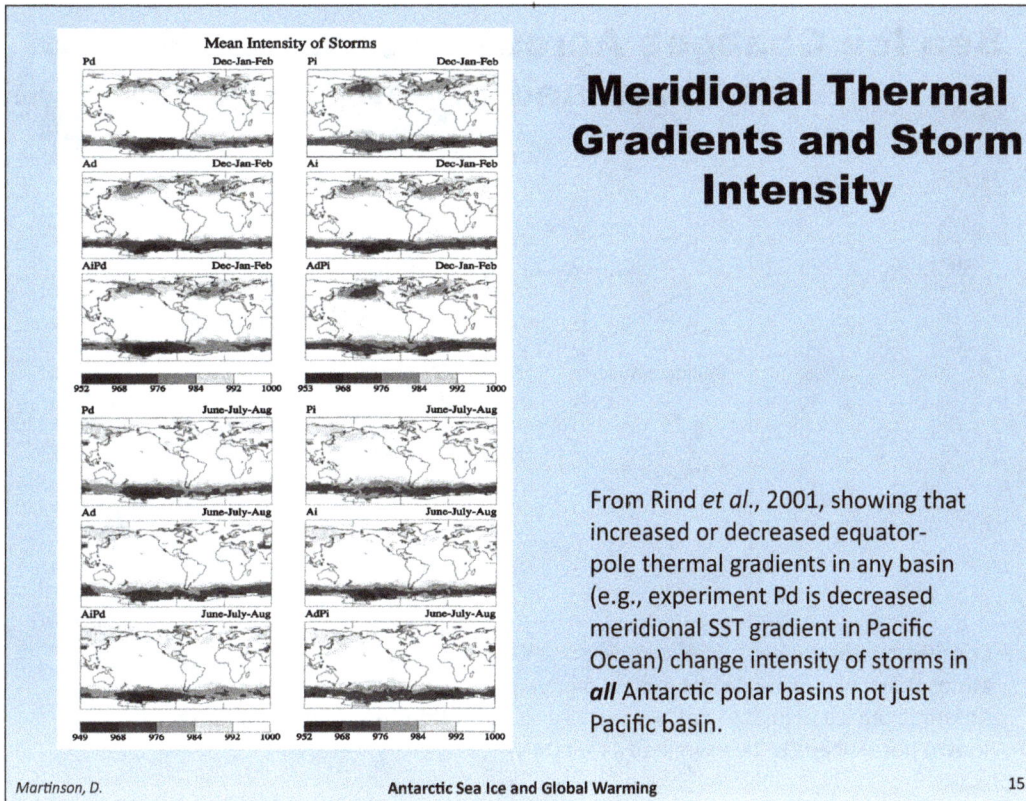

Meridional Thermal Gradients and Storm Intensity

From Rind *et al.*, 2001, showing that increased or decreased equator-pole thermal gradients in any basin (e.g., experiment Pd is decreased meridional SST gradient in Pacific Ocean) change intensity of storms in *all* Antarctic polar basins not just Pacific basin.

Martinson, D.   Antarctic Sea Ice and Global Warming   15

Slide 16

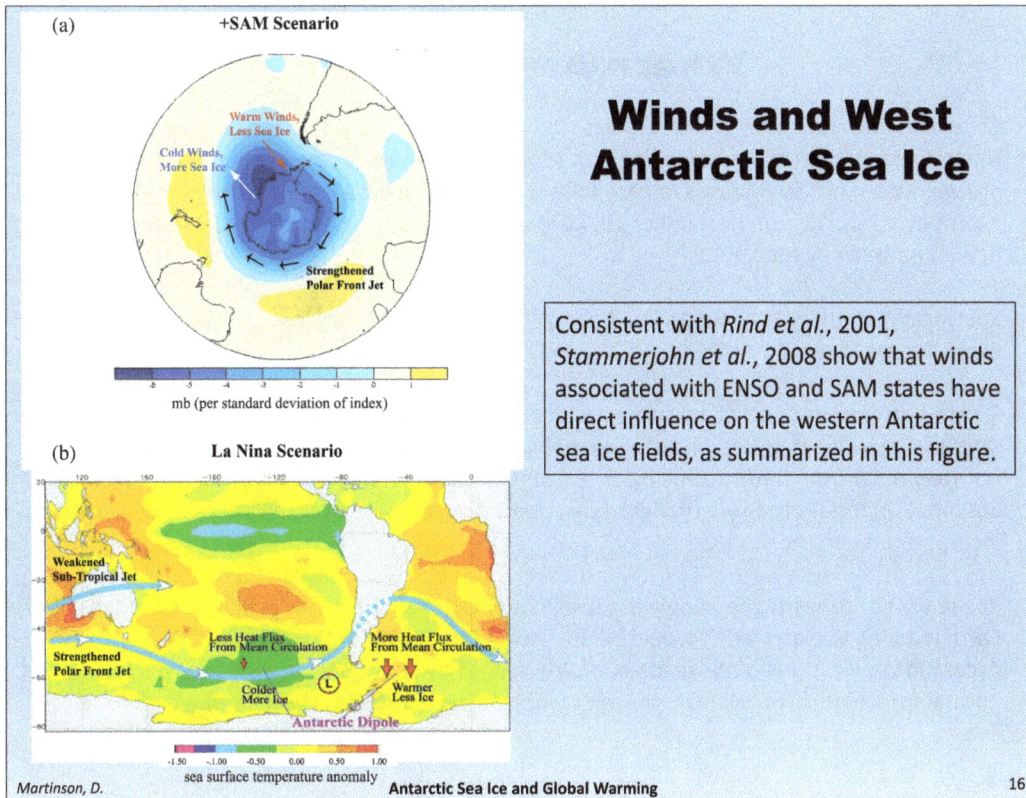

Winds and West Antarctic Sea Ice

Consistent with *Rind et al., 2001, Stammerjohn et al., 2008* show that winds associated with ENSO and SAM states have direct influence on the western Antarctic sea ice fields, as summarized in this figure.

Martinson, D.   Antarctic Sea Ice and Global Warming   16

Slide 17

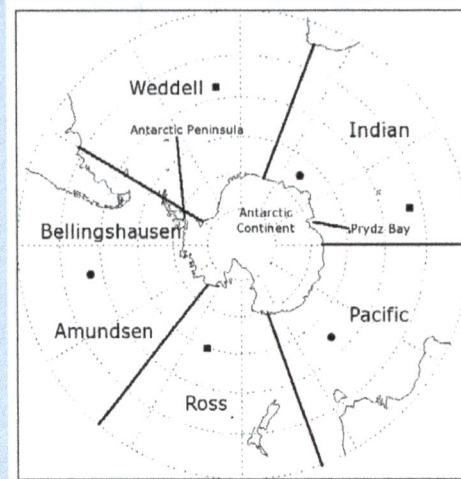

Sea Ice Changes Across Antarctica Need to be Examined Separately

Consistent with previous slide, *Lefebvre and Goosse* (2008) show that classic modes of atmospheric variability do not explain a large part of Antarctic winter sea ice extent. Antarctic sea ice field does not behave as a single entity—each sector (their sectors shown here) needs to be examined separately.

Slide 18

Where Are We Now?

Clearly, model and observational studies show that we do not yet understand why the Antarctic sea ice fields are expanding, but probably the most succinct summary of where we are was stated in a NASA press release following the Antarctic's largest maximum sea ice extent ever recorded on September 19, 2014 (shown here with the 1979–2014 mean maximum given by red line).

Specifically, quoting David Rind's longtime collaborator, Claire Parkinson at Goddard Space Flight Center:

'There hasn't been one explanation yet that I'd say has become a consensus, where people say, "We've nailed it, this is why it's happening,"' Parkinson said. 'Our models are improving, but they're far from perfect. One by one, scientists are figuring out that particular variables are more important than we thought years ago, and one by one those variables are getting incorporated into the models.'

'It's really not surprising to people in the climate field that not every location on the face of Earth is acting as expected–it would be amazing if everything did,' Parkinson said. 'The Antarctic sea ice is one of those areas where things have not gone entirely as expected. So it's natural for scientists to ask, "OK, this isn't what we expected, now how can we explain it?"'

Slide 19

Conclusions

This presentation focuses on early studies attempting to determine what role Antarctic sea ice parameters (e.g., thickness, concentration, extent) play in simulations of global warming. Most of these early studies are by David Rind and colleagues. Specifically, these studies show:

1. Feedbacks that lead to an ability of the sea ice to self-regulate (e.g., via the ice-driven ocean heat flux) can have a significant effect on the global temperature response to a doubling of CO_2. In the results shown here, not allowing the ice to disappear, as was common in nearly all such model runs, the global warming was reduced by 37% (i.e., a **HUGE** amount).
   Rind et al., 1995

2. Errors in sea ice concentration as small as -1% can lead to global temperature changes of +0.0107°C per reduction of 1% concentration.
   Parkinson et al., 2001

3. Changes in tropical–polar SST gradients in a single ocean basin, can influence conditions in all of the polar basins, not just the the perturbed basin.
   Rind et al., 2001

Point of these is not that they are the absolute answers, rather that they expose sensitivities in sea ice conditions not fully addressed in most GCMs

Martinson, D. **Antarctic Sea Ice and Global Warming** 19

Slide Notes

Slide 1   This presentation focuses on changes in Antarctic sea ice in global warming, particularly on our inability to model its increasing expansion when one would naturally assume it to shrink given global warming. However, it will be shown that while we donot know why it is expanding, the complex details of its ocean–sea ice interaction may present currently insurmountable difficulties in global climate models (GCMs, models that cannot include the details necessary to properly model the ocean–ice interaction. Early work by Rind focused directly on this problem, and showed a robustness to changed surface forcing due to the self-regulating nature of the ocean–ice interaction among other things, helping to focus the inadequacies of the GCMs in modeling sea ice.

Slide 2   First, we start by looking at the data for the Arctic and Antarctic sea ice. The Antarctic is often compared to the Arctic to show that there is either something wrong with the concept of global warming, or the Antarctic region must not be warming. This comparison is absolutely unwarranted! Bintanja, R., van Oldenborgh, G.J., Drijfhout, S.S., Wouters, B. and Katsman, C.A. (2013). Important role for ocean warming and increased ice-shelf melt in Antarctic sea-ice expansion.*Nature Geoscience*, **6**, 376–379.

Slide 3   First consider the most important, of many differences between the two regions, invalidating such a comparison. But, despite these differences, we still do not understand why the Antarctic sea ice cover is actually expanding.

Slide 4   With little sea ice in summer, when the sun is up, there can only be little loss of ice leading to enhanced absorption of solar radiation, because there is no ice to begin except for some very small pockets of perennial (year-round) ice. On the other hand, unlike the Arctic, much of the sea ice is equatorward of the Antarctic polar circle, meaning that there is some (low incident) sunlight in winter, and loss of that ice could lead to a minor ice-albedo feedback, but not like in the Arctic when in summer the sun is overhead and providing strong solar radiation into the dark ocean.

Slide 5   Expanding ice cover is not uniform, it shows strong spatial variations as shown here. Warm colors (red) show a strong increase in the Ross gyre (central Pacific) and less so in the Weddell gyre (Atlantic). A primary decreasing trend is clear in 'western' Antarctic (Antarctic peninsula to Ross Sea). Expanding regions exceed decreasing regions. Maksym, T., Stammerjohn, S.E., Ackley, S., Massom, R. (2012). Antarctic sea ice — A polar opposite? *Oceanography*, 25(3), 140–151.

Slide 6   The Intergovernmental Panel on Climate Change (IPCC) performs global climate model intercomparisons(CMIPs) to see how well the modeling community can simulate specific aspects of global climate change under different scenarios of greenhouse gas emissions.

Here, Shu *et al.* evaluate all 49 such models in the most recent (published in 2013; see http://cmip-pcmdi.llnl.gov for details) inter-model comparisons (CMIP5) to see their ability to capture the sea ice fields. The models seem to properly simulate the proper amount of long-term average ice (climatology), and capture the decreasing trend in the Arctic ice, but only a handful capture the increasing trend in the Antarctic sea ice cover. However, those that do capture the increasing trend do not have the increasing ice properly distributed spatially as shown in previous figure.

Slide 7   Another view of CMIP5 simulations of Antarctic sea ice. They do a decent job capturing the seasonal variability, but, again, the trend is opposite of that observed.

Slide 8   The Antarctic ocean–sea ice interactions are extremely important in the growth and regulation of the Antarctic sea ice cover. The details of these interactions we are about to examine are difficult to capture in the global climate models (GCMs). These GCMs have to model the Earth's oceans and atmosphere, and in order to cover such a large amount of space, and run for very long simulations, they cannot afford to model very small-scale processes (though the modelers are now coming up with efficient means for doing so, so there is hope).

Slide 9   Prose description of the all-important ice–ocean interactions of the water column.

1. Have a two-layer system (deep and surface water)
2. Surface layer is well mixed and intensely modified by surface interactions
   a. Cooling, freezing, mixing: Mixing is winddriven at the surface and turbulent across the pycnocline (interface between the layers)
   b. Winter mixed layer (called Winter Water, WW) is at freezing point
   c. Freezing drives ice formation, which releases a dense brine into water column, causing mixed layer to deepen, entraining warm thermocline water into mixed-layer melting ice that just formed, or preventing an equivalent amount of ice formation until heat if vented to atmosphere of what previous figure (with equations) is showing (and equations capturing). Can show this first, then equations (i.e., here is the system of equations describing those interactions).

Slide 10   This system is conveniently described by an analytic model, whose solutions are well described by scaling laws that will allow simple estimates of important ice-ocean interaction parameters (coming up).Equations only shown to emphasize how complicated system is but equations do exist to describe the important aspects of it.

Slide 11   The equations of the previous slides have full analytic solutions as well as scaling laws capturing the full solutions to better than 95% by simple combinations of easily observed variables. Those scaling laws have been integrated in depth to provide robust bulk properties that can be combined (as shown here), allowing easy diagnostic estimates from summer profiles of fundamental properties such as the net ocean heat loss via entrainment and via turbulent diffusion. Here, the system is shown as a predecessor to the following figure showing application of these laws to the Weddell gyre data giving how much ice will melt given a unit of ice growth (via the brine-induced entrainment term). Application to observations in the winter Weddell gyre gave excellent results.

Again, equations only show to demonstrate that system is complicated, but getting simpler from the easy to apply 'scaling laws'.

Slide 12   Figure shows effectiveness of heat released by the brine rejection during ice growth. The numbers show, for the Atlantic sector (Weddell gyre) of the Antarctic sea ice fields, the ratio of how many units of ice would melt from the entrainment of heat for one unit grown. The amount is staggering. It is difficult to imagine that any ice can grow, though it does grow in autumn. Specifically, when the sea ice melts in spring and summer, it creates a melt-layer of relatively fresh water (ice is almost fresh because it rejects the salt as a brine). This fresher is less dense than the water it melts into, so it 'floats' on the surface. In autumn, this layer cools to the freezing point and then ice starts to form with additional cooling. In winter, it is the brine rejection during this ice formation that makes the water denser, sinking into the underlying warm deep water releasing the heat to the surface. But, in autumn, it takes time before the rejected brine builds up sufficiently to make the surface layer as dense as the underlying water (typically 40–60 cm of ice). It is not until then that the surface water sinks into the warm deep water and initiates the ice-ocean feedback of releasing heat. Hence during the autumn ice grows without suffering through this negative feedback. Experiments with this system show that the feedback serves to self-regulate the ice, in the sense that changes in the surface forcing (stronger or weaker) result in little change of ice thickness. So, local warming does not necessarily imply thinner ice (as was tested in the GISS GCM).

Slide 13   Summary of findings from 2 of the more important early studies examining the role of various ice parameters (e.g., thickness, extent and concentration) in global warming simulations. First result tests previous model results (in preceding slides): specifically, because of the ocean-ice feedback, changes in the atmosphere do not really lead to changes in sea ice thickness, which is already being regulated by the entrainment heat flux. So for 1995 paper, the sea ice, which usually disappeared in global warming scenarios was fixed so as to not disappear. This reduced the global warming for $2\times CO_2$ by 37%. An enormous amount, showing how important such fine scale detailed processes must be considered.

Slide 14   Summary of one of the other very important papers showing how changes in tropical conditions. In this case, meridional SST gradients between tropics and Antarctic drive change in all polar basins, explaining some of the mysteries that had been observed in the observations (and not able to explain).

Specifically, the GCM shows the following:

*   Equatorial Pacific SST warm events (El Niño) lead to an increase in the Pacific pole-equator meridional temperature gradient.
*   This intensifies the subtropical cell resulting in an equatorward shift of the subtropical jet (STJ). This shift displaces the STJ farther from the source of available potential energy in the Antarctic which leads to a reduction in cyclogenesis and overall storm intensity influencing the driving of the south Pacific subpolar ocean circulation.

- The equatorial warming also alters the Walker cell circulation leading to a reduction in tropical Atlantic subsidence.
- This alters the Atlantic's vertical meridional cell which introduces changes comparable to that as if the tropical Atlantic cooled and initiated a meridional temperature gradient in the Atlantic of opposite sign to that of the Pacific.
- The result is a relaxation of the Atlantic meridional circulation, accompanied by a poleward shift of the STJ poleward, closer to the available potential energy.
- This increases storm intensity in the Atlantic (Weddell) sector of the polar gyre, invigorating the cyclonic gyre, is consistent with the findings here.
- This hemispheric mechanism also explains the anti-phasing of the Antarctic Dipole, found here and in Yuan and Martinson (2000), and is consistent with our broader understanding of Southern Hemispheric circulation and teleconnections.

Slide 15   Table 1. Description of Experiments

Label Description

Gi	increased gradient in all ocean basins
Gd	decreased gradient in all ocean basins
Gci	increased gradient in all basins plus 4°C cooling and increased sea ice
GWd	decreased gradient in all basins plus 4°C warming and reduced sea ice
Ai	increased gradient in Atlantic Ocean
Pi	increased gradient in Pacific Ocean
Ad	decreased gradient in Atlantic Ocean
Pd	decreased gradient in Pacific Ocean
AiPd	increased gradient in Atlantic, decreased gradient in Pacific
AdPi	decreased gradient in Atlantic, increased gradient in Pacific

Fundamental point is that any change in the meridional gradient in any basin does not isolate that basin for changes in storms, winds and other Antarctic impacts.

Slide 16   Results here show examination of specific regions, motivated by findings in Rind *et al.*(2001). The ENSO-Sam (El Niño-Southern Oscillation, ENSO, and Southern Annular Mode, SAM) connection can explain decreasing ice cover in the south eastern Pacific polar basin. Reflecting that spatial variation in sea ice extent trends (decreasing in western Antarctic, increasing elsewhere) may warrant examination of different processes (i.e., inclusion of many process in the GCMs).

Slide 17   Likewise, Lefebvre and Goosse explicitly note that the changes in the different sectors require different mechanisms to explain.

Slide 18   Some of the more important findings from the experiments described here.

Slide 19   Studies discussed are now old, but still have not been resolved. The statements in this slide follow the 2014 record-breaking ice extent, sum up where we are. Currently, there are superb sea ice modeling groups (e.g., Bitz, Holland, and Hunke) addressing these issues making everything including the kitchen sink interactive sea ice models in GCMs. They have had excellent results modeling the Arctic, but have not yet conquered the Antarctic problems.

SECTION 6

Paleocimate Perspective

CLIMATE LECTURE 16

The Importance of Understanding the Last Glacial Maximum for Climate Change

Dorothy Peteet

NASA Goddard Institute for Space Studies, USA, and the Lamont Doherty Earth Observatory, New York, NY, USA

Dorothy M. Peteet is a Senior Research Scientist at NASA/Goddard Institute for Space Studies and Adjunct Professor, Columbia University. She directs the Paleoecology Division of the New Core Lab. at Lamont Doherty Earth Observatory of Columbia and is studying the Late Pleistocene and Holocene archives of lakes and wetlands. She has documented past vegetational change using pollen and spores, and plant and animal macrofossils, among other paleo-indicators.

Introduction

The last glacial maximum (LGM) at approximately 23–18k (k—thousand calendar years) provides an important contrast to our present and pre-industrial climate in a warming world. Global observational datasets of LGM land and sea surface conditions have been synthesized and present some interesting challenges both for providing another scenario for understanding climate change and for climate sensitivity. These challenges are ongoing, as data increase and modeling improves. By definition, the LGM is defined as the time during the last glacial interval in which maximum ice was sequestered in ice sheets as visible in the marine isotopic records. Maximum cooling is visible from pollen and macrofossil records ^{14}C dated to this interval, and ice sheets and alpine glaciers are roughly at their maximum extent throughout the globe. The ice cores extracted from Greenland and Antarctica have given us high-resolution records of greenhouse gases, dust, and isotopes of hydrogen and oxygen which reveal the progression out of the LGM at 18k as climate warmed.

One of the first challenges addressed by D. Rind and D. Peteet (1985) was the disparity in temperature estimates they noted for the glacial using terrestrial vs. marine data. The disparity was very large, for while the marine sea surface temperature (SST) dataset produced by the Climate Long-range Investigation, Mapping, and Prediction group (CLIMAP, 1981) showed large areas of the subtropical ocean to be warmer than today at the LGM, both pollen data and glacial moraines indicated cold temperatures in the tropics and subtropics. The Rind and Peteet (1985) primary contribution was to show that modeling with the CLIMAP SST dataset did not produce enough surface air cooling to match the terrestrial data. In particular, regions with both pollen and glacial evidence for strong cooling such as East Africa, Indonesia, and Hawaii were not cool enough to generate such change. A companion experiment with SST's reduced everywhere by 2°C was in better agreement with the land data, but in some locations still not in agreement with the freezing line and pollen evidence of cooling. Similar modeling experiments by other groups produced similar results. One question posed was whether the lapse rate in tropical regions could have increased, but current understanding of atmospheric dynamics suggests that a strong divergence of the tropical LGM lapse rate from the moist adiabatic value is implausible. David Rind's research on atmospheric dynamics was central to this conclusion. Moist convection still represents the dominant vertical heat-transporting process at low latitudes (Webster and Streeten, 1978), as Rind and Peteet argued in this paper.

Subsequent significant efforts in both the marine and the terrestrial realm have been targeted at the question of LGM conditions in various parts of the globe, but the subtropics in particular is where the conflict in temperature estimates is large. Several new techniques for determining temperature shifts were employed, including determining the content of noble gases in groundwater. A decade later, Stute *et al.* (1995) presented a terrestrial dataset from Texas to lowland Brasil (40°N–40°S) including new groundwater samples which showed significant cooling (5–6°C) in the same region that faunal abundances from CLIMAP (1981) showed very little change (2°C). However, Sr/Ca ratios and oxygen isotope ratios in corals also showed a much stronger cooling, implying that the subtropics and tropics were cooler than CLIMAP reconstructions implied (Guilderson, Fairbanks,and Rubenstone, 1994). However, alkenone data and some isotopic data were still in conflict, and the debate continues.

A decade later, a series of Paleoclimate Model Intercomparison Project (PMIP2) experiments focused on the LGM with newer climate models that were state-of-the-art and included an interactive ocean. These experiments included coupled simulations with the thermohaline circulation (THC) shifts as well as changes in vegetation albedo feedbacks (Braconnot *et al.*, 2007). Some of these experiments included model results that had colder oceans than previously indicated by CLIMAP, and comparisons were made suggesting moisture shifts indicating more global aridity than previously modeled.

The 2007 Intergovernmental Panel on Climate Change (IPCC, 2007) showed a map of the difference in temperature between LGM and today, illustrating temperature shifts in the tropical oceans of not more than 2.5°C overall, but significantly colder than CLIMAP. Yet the radiative perturbation representing the forcings is presented with high scientific understanding of the forcings of orbital shift and greenhouse gases such as CO_2 and CH_4, but low understanding of the forcings of ice sheets, mineral dust, and vegetation. One of the benefits of numerous modeling efforts has been the understanding of how important the ocean circulation and SST forcing can be in influencing the global temperature. At the same time, it is clear that greenhouse gas forcings have been important but that shifts in vegetation, and feedbacks in albedo and the carbon cycle from those vegetational shifts can also be important. It is still unclear what the effects of the lowered carbon dioxide on vegetation really were, because experiments on vegetation with lowered CO_2 are difficult. It is also difficult to separate the effects of lowered temperature from moisture on vegetation, and more multidisciplinary studies are needed using a variety of proxies at one site (i.e., deuterium isotopes in leaf waxes, pollen, macrofossils, bryophytes, glaciers, etc.)

One of the modeling experiments (Crucifix, 2006) posed the question as to whether the LGM constrained climate sensitivity. A set of experiments with four different models showed large variation in short-wave cloud feedbacks at low latitudes both during the LGM and in the $2 \times CO_2$ experiment. For the latter, they get a climate sensitivity ranging from 2.1 to 3.9°C, while the sensitivity of the LGM, they concluded, cannot be calculated because the amount of actual forcing of cooling from greenhouse gases and the feedbacks is not known. Another modeling experiment used ensembles of 100 paired runs for pre-industrial and LGM boundary conditions and found global sensitivity was 4.3–9.8°C, due to changes in both vegetation and dust. Using a reconstructed tropical SST (30°N–30°S) cooling of 2.7 ± 1°C, they constrained the range to 5.8 +1.4°C, larger than most estimates (Schneider von Demling, Held, Ganopolski,and Rahmstorf, 2006). Recent attention to climate sensitivity has focused on the need to include fast feedbacks (such as climate-greenhouse gas) as well as slower feedbacks such as ice sheets and vegetation, although the name of this quest has shifted from 'climate sensitivity' to 'earth sensitivity' (Previdi et al., 2013). This change in our understanding of the importance of feedbacks for greenhouse gases and vegetation has large implications for understanding climate sensitivity of the LGM as well as for doubled CO_2.

Beyond looking at an equilibrium LGM climate, it is useful to understand the forcings that shifted the LGM towards warming. Questions concerning the triggers for warming focus on the very small orbital shifts and on ocean circulation as well as greenhouse gases. Comparison of Northern and Southern Hemispheres and the shift towards warming in each and the timing of this warming becomes critical to understanding how the climate system works. Rapid feedbacks such as shifts in the THC and sea ice must be important, as well as aerosols. Attention to detailed records of land, sea, and ice cores aid in piecing together this puzzle, which provides clues to how our climate system functions. Polar ice cores indicate dust deposition rates average 2–20 times greater during the LGM compared to the Late Holocene in some regions (Mahowald et al.,1999); and links to possible ocean fertilization and shifts in the strength and the change in the position of the Southern Hemisphere westerly winds has been used to explain climate fluctuations through the carbon cycle and greenhouse gas forcing (Anderson et al., 2009). Kohfeld et al. (2013) assembled datasets relevant to changes in the westerly winds and concluded that either the stronger or equatorward-shifted wind scenarios appear easier to tie to evidence for increased moisture on the west coast of continents, cooler temperatures and higher productivity in the southern ocean. However, they caution that many assumptions limit the certainty with which winds can be invoked as responsible for controlling glacial–interglacial conditions.

Defining the details of the termination of the LGM includes both the timing and the characteristics of the retreating ice sheet. The history of the retreat of the southeastern margin of the Laurentide Ice Sheet (LIS) is particularly relevant today as Greenland warms and meltwater enters the N. Atlantic. The conventionally accepted ages of the LGM retreat of the southeastern LIS are 26–21 calendar kyr (cal. kyr) (derived from bulk-sediment radiocarbon ages) (Stone and Borns, 1986) and 28–23 cal. kyr (varve estimates)(Ridge, 2004), but these estimates are problematic because they are so early.

Cosmogenic-nuclide exposure dating of the LGM requires the derivation of a local nuclide production rate. Balco and Schaefer (2006) first derived a production rate for 10Be using primarily radiocarbon dates of bulk sediment and/or varves, and derived an age of deglaciation of 22 cal. kyr. However, more recently Balco *et al.* (2009) utilized varve chronologies from New England to re-calculate the deglacial age for the New Jersey terminal moraine near 25 cal. kyr, and the Connecticut moraines at 21 cal. kyr (see arrows in Slide 19). Using accelerator mass spectrometry (AMS) 14C dating of initial macrofossils in 13 lake/bog inorganic clays, Peteet *et al.* (2012) find that vegetation first appeared on the landscape at 16 cal. kyr, suggesting that ice had not retreated until that time. The gap between previous age estimates is significant and has large implications for understanding the meltwater timing and links to shifts in THC. This new AMS chronology of LIS retreat is consistent with marine evidence of deglaciation from the North Atlantic, showing significant freshwater input and sea level rise only after 19 cal. kyr with a cold meltwater lid, perhaps delaying ice melt. It is also more consistent with the rise of greenhouse gas forcing beginning at 18k.

Key questions remain in our understanding of the LGM, its forcings, and climate sensitivity. These questions are relevant as we prepare for further warming with greenhouse gas increases and the feedbacks that accompany those increases. Hundreds of investigations have generated LGM data on land, ocean, and ice throughout the globe, and these are extremely valuable. Some questions that remain include the question of where the warming began (Northern or Southern Hemisphere) and whether the insolation increase was great enough to initiate Northern Hemisphere ice melt. If so, does this help explain the 100k cycle in paleorecords? Exactly, how strong are the forcings from the increases in carbon dioxide and methane throughout the deglaciation, and how strongly does ocean circulation control the CO_2 rise and then how important is the CH_4 forcing? As permafrost melts today, how important is the release of these greenhouse gases in changing our future temperature, and how much carbon will the ocean and northern peatlands sequester? How will Greenland meltwater affect the THC? Many other questions remain for us to investigate.

References

Anderson, R.F., Ali, S., Bradtmiller, L., Nielsen, S., Fleisher, M., Anderson, B. and Burckle, L.H. (2009). Wind-drivenupwelling in the Southern Ocean and the deglacial rise in atmospheric CO_2. *Science*, **323**, 1443–1448.

Balco, G. and Schaefer, J.M. (2006). Cosmogenic-nuclide and varve chronologies for the deglaciation of southern New England. *Quaternary Geochronology*, **1**, 15–28. doi:10.1016/j.quageo.2006.06.014.

Balco, G., Briner, J., Finkel, R.C., Rayburn, J.A., Ridge, J.C. and Schaefer, J.M. (2009). Regional beryllium-10 production rate calibration for late-glacial northeastern North America. *Quaternary Geochronology*, **4**, 93–107.doi:10.1016/j.quageo.2008.09.001.

Braconnot, P., Otto-Bliesner, B.S., Harrison, S., Joussaume, S., Peterchmitt, J.-Y., Abe-Ouchi, A., and Zhao, Y. (2007). Results of PMIP2 coupled simulations of the mid-holocene and last glacial maximum–Part 1: Experiments and large-scale features. *Climates of the Past*, **3**, 261–277.

CLIMAP Project Members. (1981). Seasonal reconstructions of the earth's surface at the last glacial maximum. *Geological Societyof America Map and Chart Series*, MC-36, 18.

Crucifix, M. (2006). Does the last glacial maximum constrain climate sensitivity? *Geophysical Research Letters,* **33**, L18701. doi:10.1029/2006GL027137.

Farrera, I., Harrison, S.P., Prentice, I.C., Ramstein G., Guiot, J. and Yu, G. (1999). Tropical paleoclimates at the Last Glacial Maximum: a new synthesis of terrestrial data. I. Vegetation, lake levels, and geochemistry. Climate Dynamics **15**(1), 823–856. doi:10.1007/s003820050317.

Guilderson, T., Fairbanks, R.G. and Rubenstone, J. (1994). Tropical temperature variations since 20,000 years ago: Modulating interhemispheric climate change. *Science,* 263, 663–664.

Intergovernmental Panel on Climate Change (IPCC).(2007).

Kohfeld, K., Graham, R.M., de Boer, A.M., Sime, L.C., Wolff, E.W., LeQuere, C. and Bopp, L. (2013). Southern hemisphere westerly wind changes during the last glacial maximum: Paleo-data synthesis. *Quaternary Science Reviews,* **68**, 76–95.

Mahowald, N., Kohfeld, K., Hansson, M., Balkanski, Y., Harrison, S., Prentice, C., Schulz, M. and Rodhe, H. (1999). Dust sources and deposition during the last glacial maximum and current climate: A comparison of model results with paleodata from ice cores and marine sediments. *Journal of Geophysical Research,* 1041, D13, 15895–15916.

Parrenin, F., Masson-Delmotte, V., Kohler, P., Raynaud, D., Paillard, D., Schwander, J., Barbante, C., Landais, A., Wegner, A. and Jouzel, J. (2013). Synchronous change of atmospheric CO_2 and Antarctic temperature during the last deglacial warming. *Science,* **339**, 1060–1063.

Peteet, D., Beh, M., Orr, C., Kurdyla, D., Nichols, J. and Guilderson, T. (2012). Delayed deglaciation or extreme Arctic conditions 21–16 cal. kyr at southeastern Laurentide Ice Sheet margin? *Geophysical Research Letters,* 39, L11706. doi:10.1029/2012GL051884.

Previdi, M., Liepert, B.G., Peteet, D., Hansen, J., Beerling, D., Broccoli, A.J., . . . Ramaswamy, V. (2013). Climate sensitivity in the Anthropocene. *Quarterly Journal of the Royal Meteorological Society.* doi:10.1002/qj.2165.

Ridge, J.C. (2004). The Quaternary glaciation of western New England with correlations to surrounding areas. In J. Ehlers and P.L. Gibbard (Eds.), *Quaternary Glaciations-Extent and Chronology Part II* (pp. 169–199), Amsterdam: Elsevier. doi:10.1016/S1571-0866(04)80196-9.

Rind, D. and Peteet, D.M. (1985). Terrestrial conditions at the last glacial maximum and CLIMAP sea surface temperature estimates: Are they consistent? *Quaternary Research,* **24**, 1–22.

Schneider von Deimling, T., Held, H., Ganopolski, A. and Rahmstorf, S. (2006). Climate sensitivity estimated from ensemble simulations of glacial climate. *Climate Dynamics,* **27**, 149–163. doi:10.1007/s00382-006-0126-8.

Stone, B. and Borns, H.W.J. (1986). Pleistocene glacial and interglacial stratigraphy of New England, Long Island, and adjacent Georges Bank and Gulf of Maine.*Quaternary Science Reviews,* **5**, 39–52.

Webster, P. and Streten, N. (1978). Late Quaternary ice age climates of tropical Australasia, interpretation and reconstruction. *Quaternary Research,* **10**, 279–309.

Slide 1

The Importance of Understanding the Last Glacial Maximum for Climate Change

Dorothy Peteet
NASA GISS and Lamont Doherty Earth Observatory

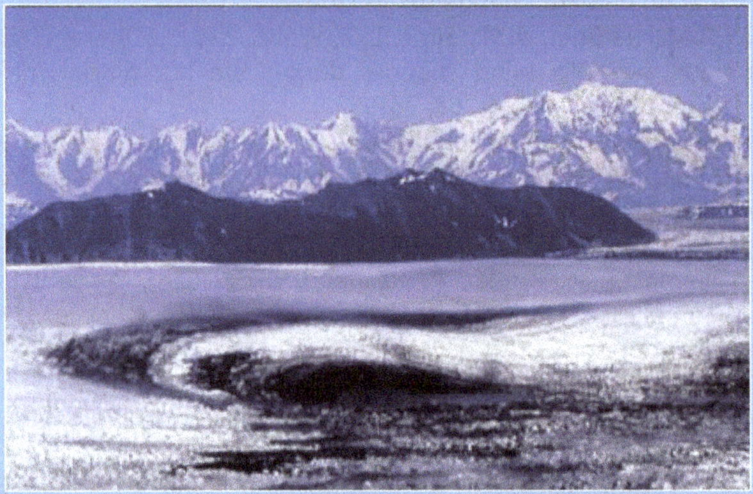

Bering Glacier, AK, 1987

Our Warming Planet: Topics in Climate Dynamics
Lectures in Climate Change, Vol. 1
2017

Slide 2

Definition of LGM

- Equilibrium climate 22-18k very different from today but with same continent configuration
- Climate processes same as today
- Vegetation same species as today
- Variation in orbital configuration, greenhouse gases, dust content
- C-14 and U-Th useful for dating

Peteet, D. The Importance of Understanding the Last Glacial Maximum for Climate Change 2

Slide 3

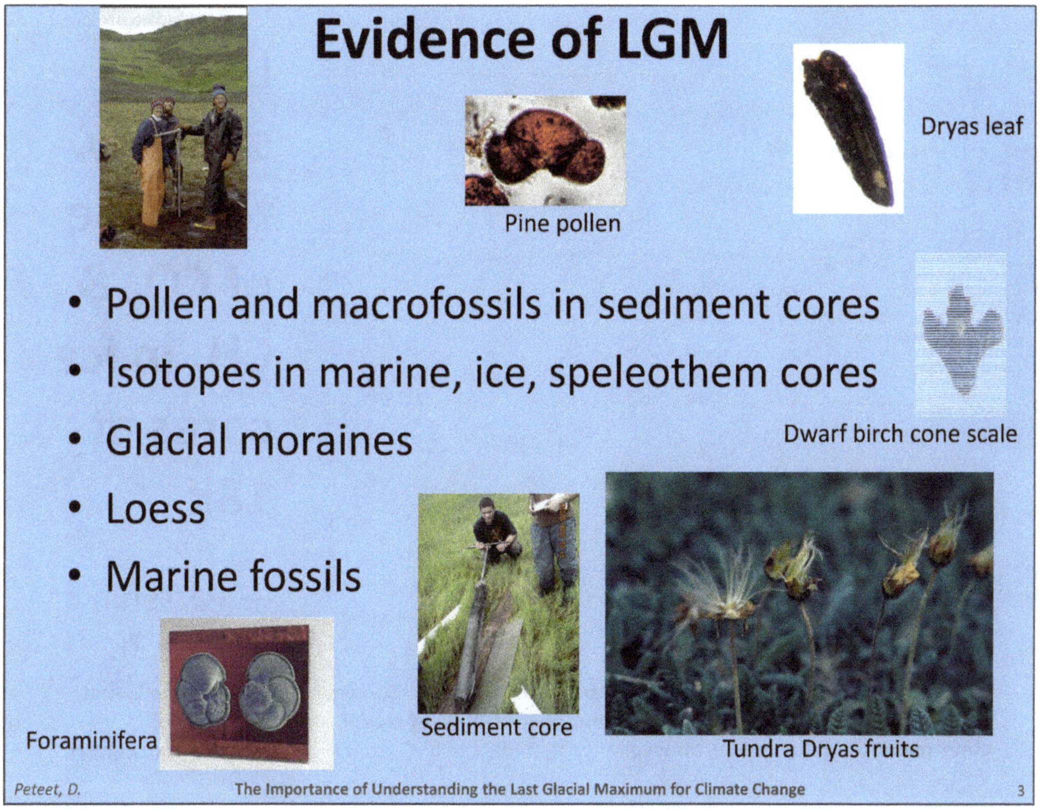

Slide 4

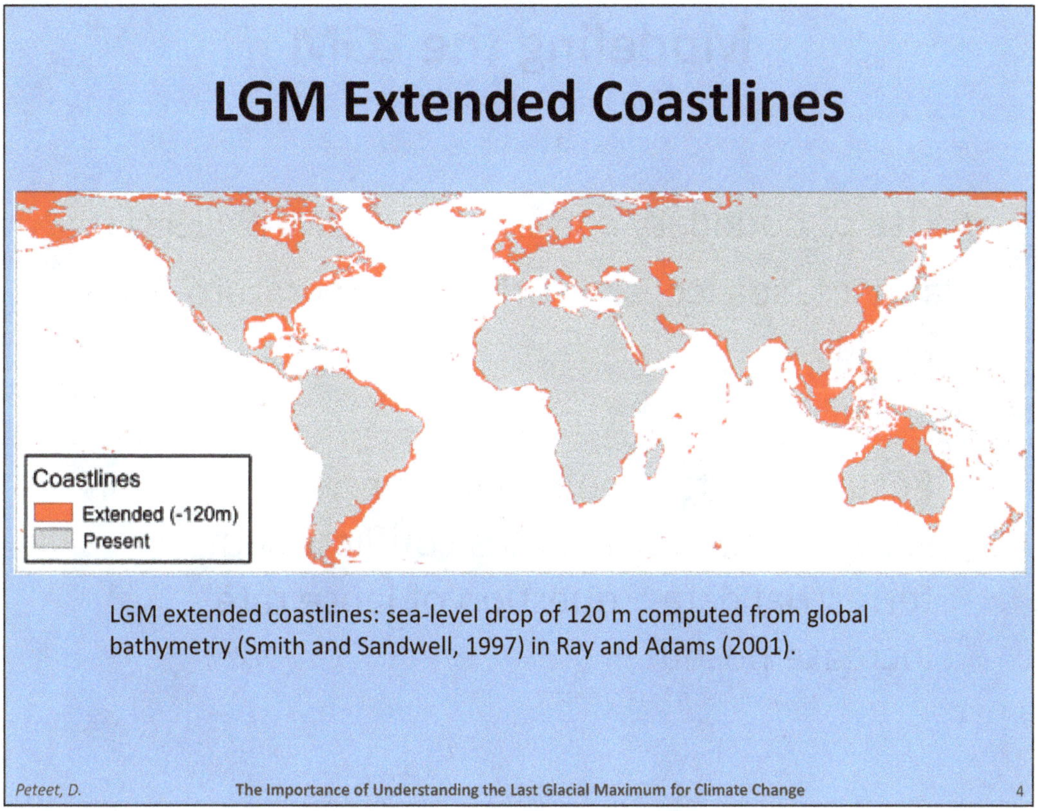

Slide 5

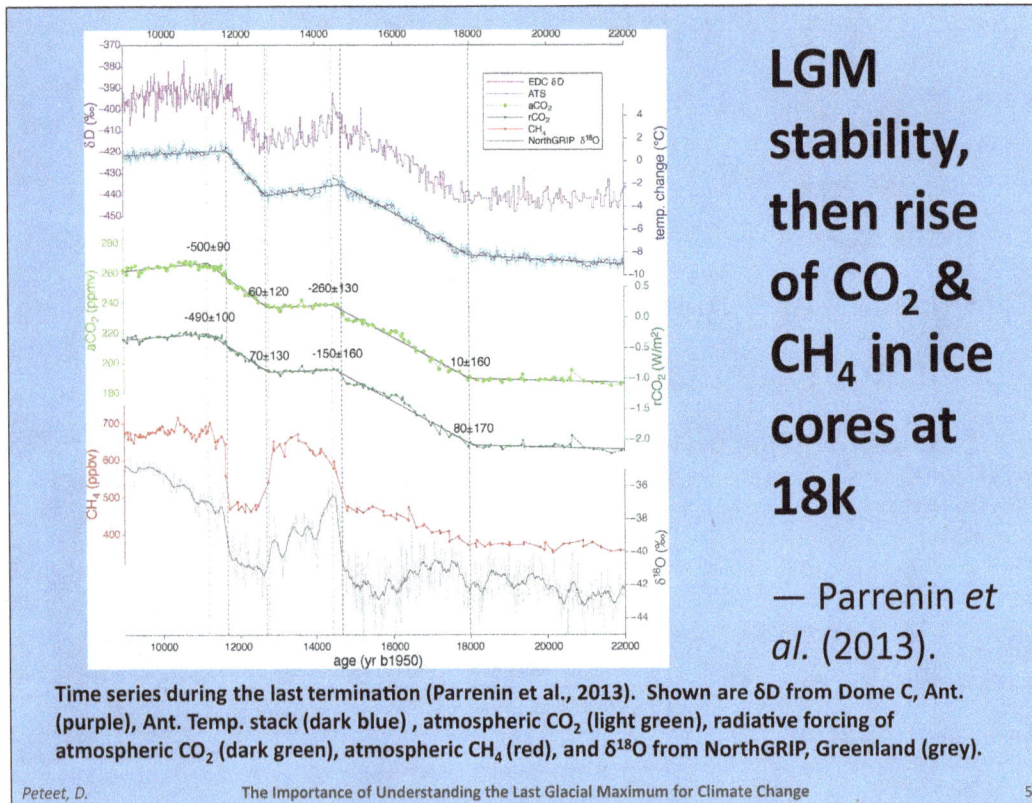

Time series during the last termination (Parrenin et al., 2013). Shown are δD from Dome C, Ant. (purple), Ant. Temp. stack (dark blue) , atmospheric CO_2 (light green), radiative forcing of atmospheric CO_2 (dark green), atmospheric CH_4 (red), and $δ^{18}O$ from NorthGRIP, Greenland (grey).

LGM stability, then rise of CO_2 & CH_4 in ice cores at 18k

— Parrenin *et al.* (2013).

Slide 6

Modeling the LGM
Rind and Peteet (1985)

* Use of boundary conditions—CLIMAP (1981) land ice, sea ice, sea surface temperature, land albedo

* Results: Parts of subtropics air temperature is warmer than today...thus conflicts with terrestrial data... question of lapse rate increase posed.

Slide 7

Glacial Minus Modern

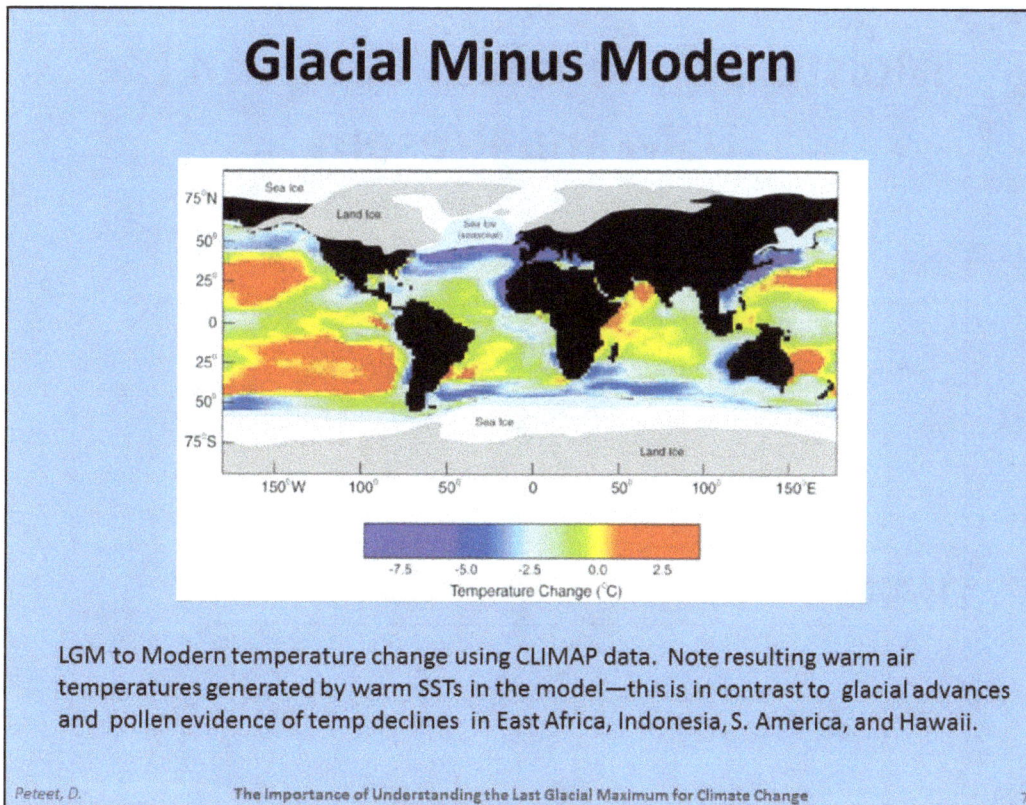

LGM to Modern temperature change using CLIMAP data. Note resulting warm air temperatures generated by warm SSTs in the model—this is in contrast to glacial advances and pollen evidence of temp declines in East Africa, Indonesia, S. America, and Hawaii.

Peteet, D.   The Importance of Understanding the Last Glacial Maximum for Climate Change   7

Slide 8

Data—LGM-Present Temp Difference

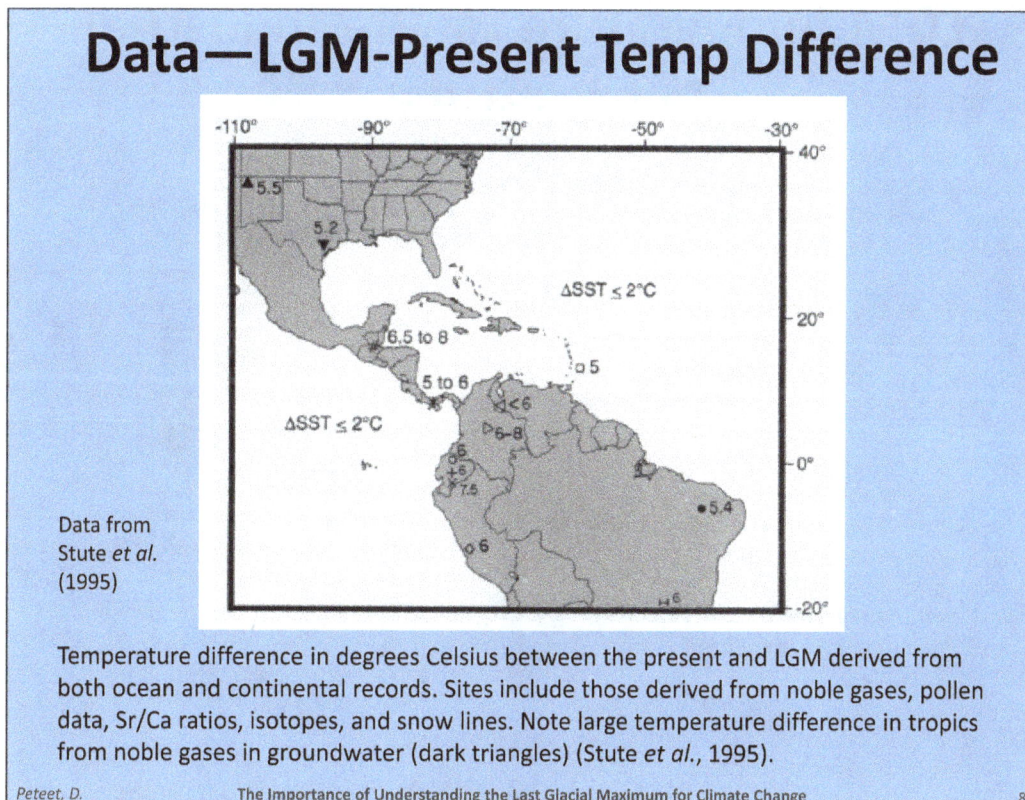

Data from Stute *et al.* (1995)

Temperature difference in degrees Celsius between the present and LGM derived from both ocean and continental records. Sites include those derived from noble gases, pollen data, Sr/Ca ratios, isotopes, and snow lines. Note large temperature difference in tropics from noble gases in groundwater (dark triangles) (Stute *et al.*, 1995).

Peteet, D.   The Importance of Understanding the Last Glacial Maximum for Climate Change   8

Slide 9

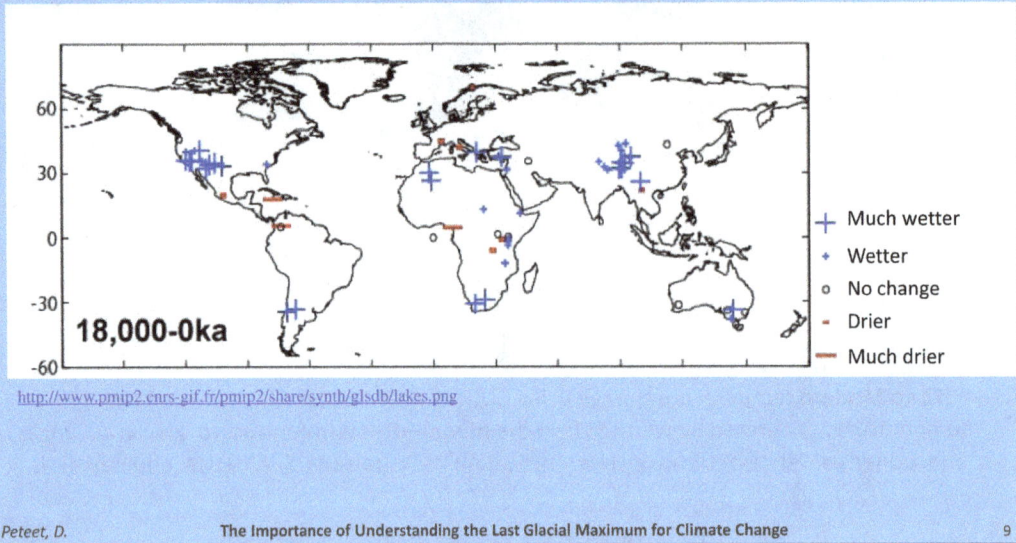

Moisture Differences Between the LGM and Present

18,000-0ka

Much wetter
Wetter
No change
Drier
Much drier

http://www.pmip2.cnrs-gif.fr/pmip2/share/synth/glsdb/lakes.png

Peteet, D.   The Importance of Understanding the Last Glacial Maximum for Climate Change   9

Slide 10

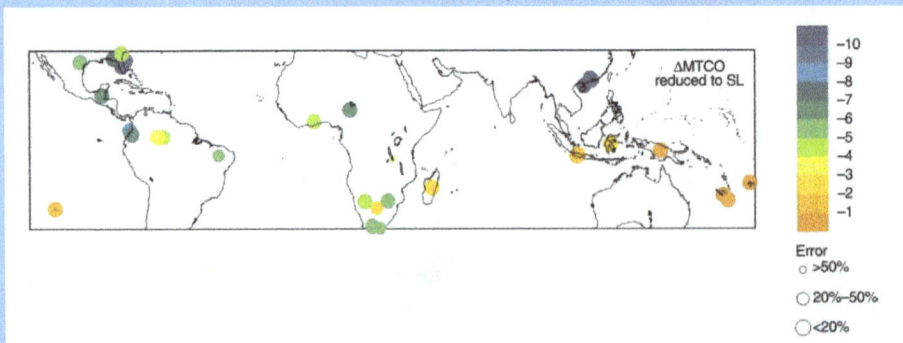

LGM Temperature vs. Pre-Industrial Temperature—
A Compilation from Data

Farerra *et al.* (1999)

ΔMTCO
reduced to SL

Error
o >50%
○ 20%–50%
○ <20%

Mean Temperature of coldest month...back to conundrum with glaciers...drier overall
But had moisture for glaciers in tropics...cold

Peteet, D.   The Importance of Understanding the Last Glacial Maximum for Climate Change   10

Slide 11

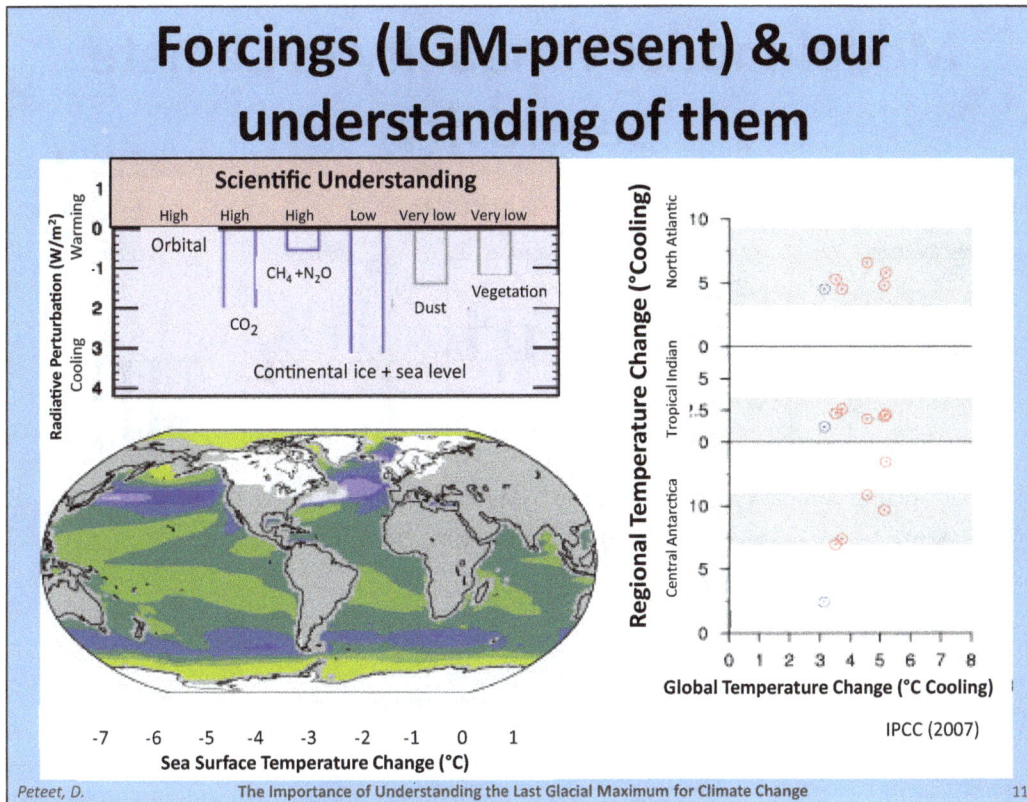

Forcings (LGM-present) & our understanding of them

Peteet, D.   The Importance of Understanding the Last Glacial Maximum for Climate Change   11

Slide 12

LGM Boundary Conditions for more Recent Modeling

- Large ice sheets at high latitudes and expanded mountain glaciers

- Sea level drop of 120 m

- Colder ocean temperatures throughout the globe ...but how cold, and what are transports?

- Vegetation shifts due to cooling, moisture change, albedo feedbacks, and the carbon cycle?

- Vegetation shifts due to lower CO_2?

- CO_2 and CH_4 at low values

Peteet, D.   The Importance of Understanding the Last Glacial Maximum for Climate Change   12

Slide 13

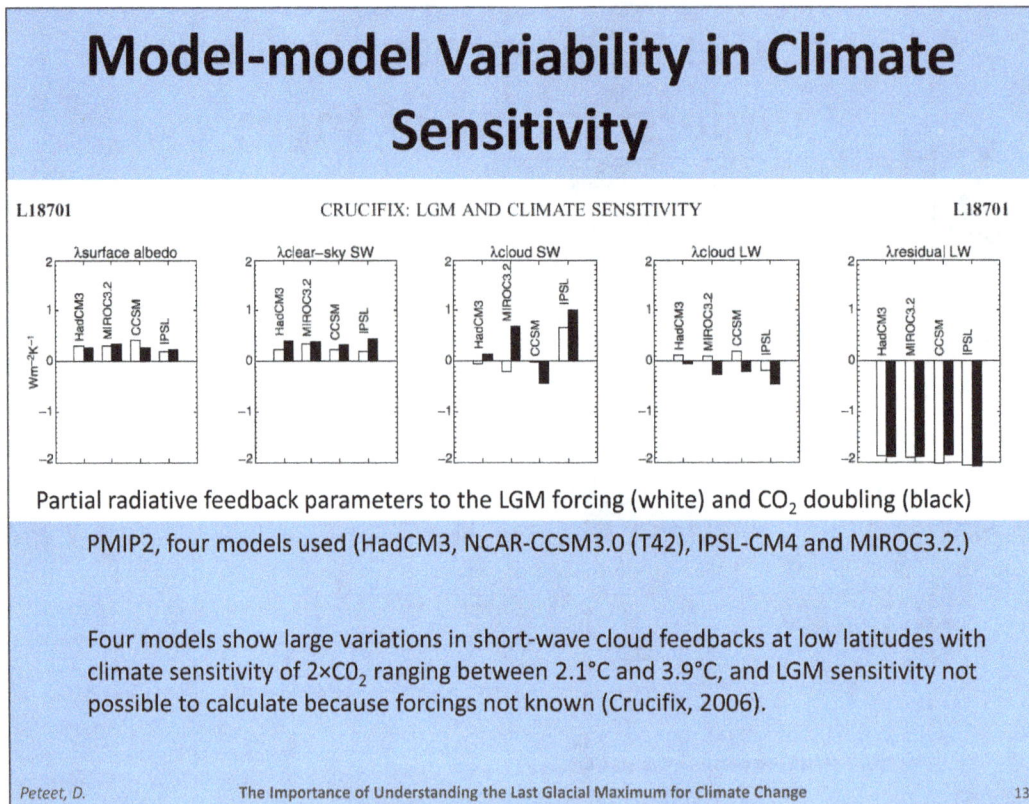

Model-model Variability in Climate Sensitivity

L18701   CRUCIFIX: LGM AND CLIMATE SENSITIVITY   L18701

Partial radiative feedback parameters to the LGM forcing (white) and CO_2 doubling (black)

PMIP2, four models used (HadCM3, NCAR-CCSM3.0 (T42), IPSL-CM4 and MIROC3.2.)

Four models show large variations in short-wave cloud feedbacks at low latitudes with climate sensitivity of 2×C0$_2$ ranging between 2.1°C and 3.9°C, and LGM sensitivity not possible to calculate because forcings not known (Crucifix, 2006).

Peteet, D.   The Importance of Understanding the Last Glacial Maximum for Climate Change   13

Slide 14

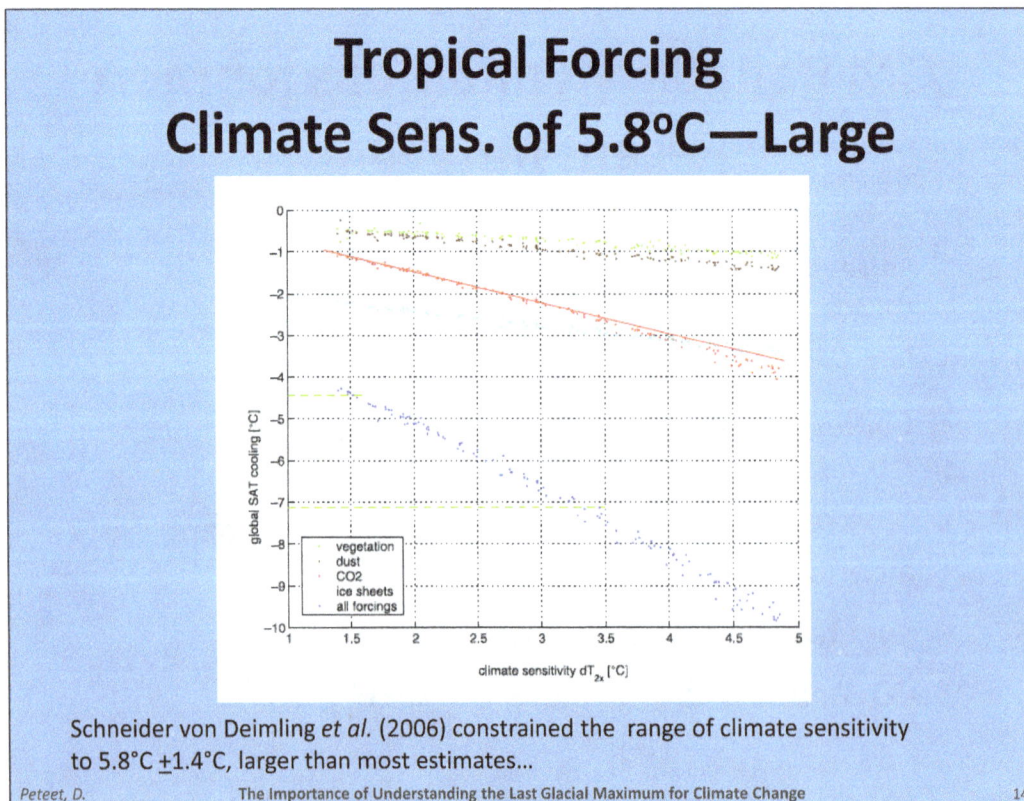

Tropical Forcing Climate Sens. of 5.8°C—Large

Schneider von Deimling *et al.* (2006) constrained the range of climate sensitivity to 5.8°C ±1.4°C, larger than most estimates...

Peteet, D.   The Importance of Understanding the Last Glacial Maximum for Climate Change   14

Slide 15

Climate Sensitivity to $2 \times CO_2$

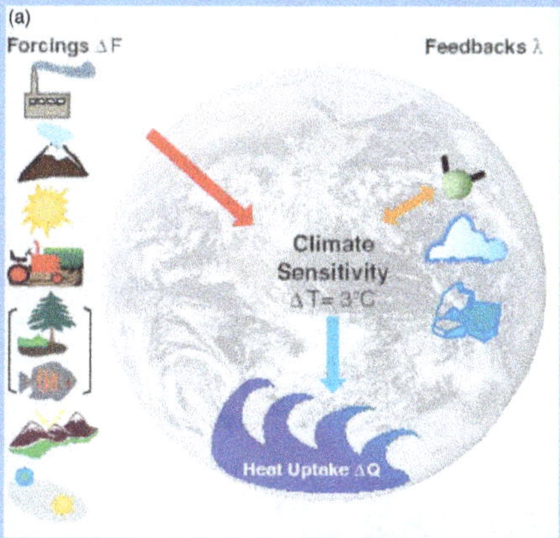

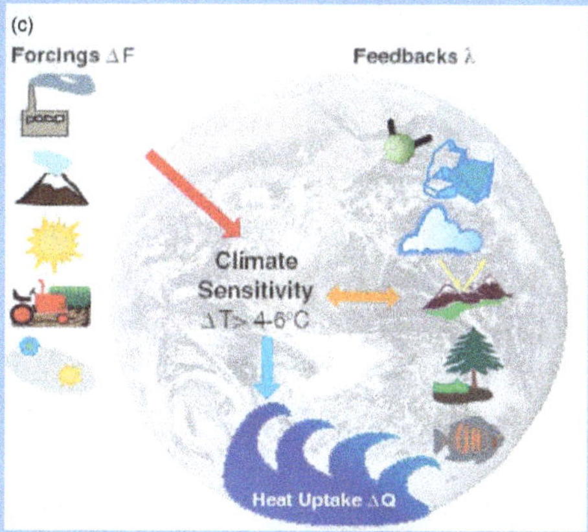

Previdi *et al.* (2013) show that if climate sensitivity from fixed forcings (on left) is dependent only on fast feedbacks due to water vapor, clouds and sea ice, sensitivity to doubled CO_2 is only 3°C.

In contrast, if changes in GHG feedbacks from climate-related terrestrial and marine C are included, the sensitivity to $2 \times CO_2$ becomes greater than 4°C–6°C .

Peteet, D. The Importance of Understanding the Last Glacial Maximum for Climate Change 15

Slide 16

Data Compilation of LGM – Holocene Shifts

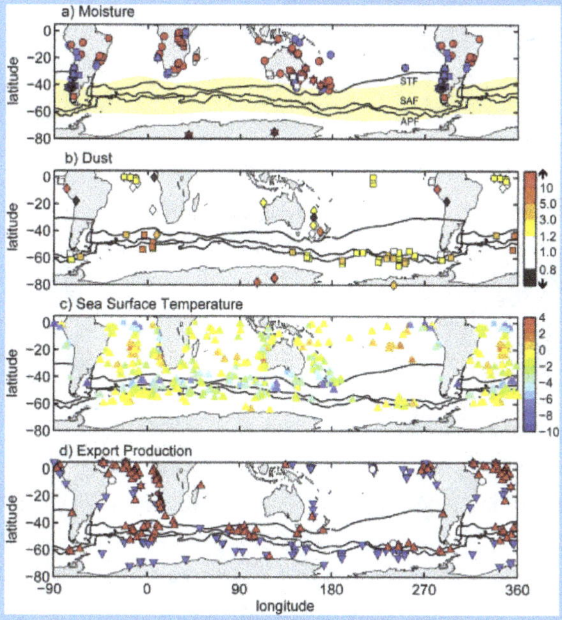

a) greater moisture on western side of continents,
b) abundant evidence of increases in dust,
c) SSTs warm in lower lats and cooler at high lats, and,
d) increase in export production at lower lats generally, with declines at higher latitudes.

Kohfeld *et al.* (2013)

Peteet, D. The Importance of Understanding the Last Glacial Maximum for Climate Change 16

Slide 17

LGM and Deglaciation

- Forcing of first warming?—abrupt shifts
- Land–ocean contrasts
- Trace gases
- N–S hemisphere contrast
- THC has power...affecting sea ice and winter temperature
- Aerosols?

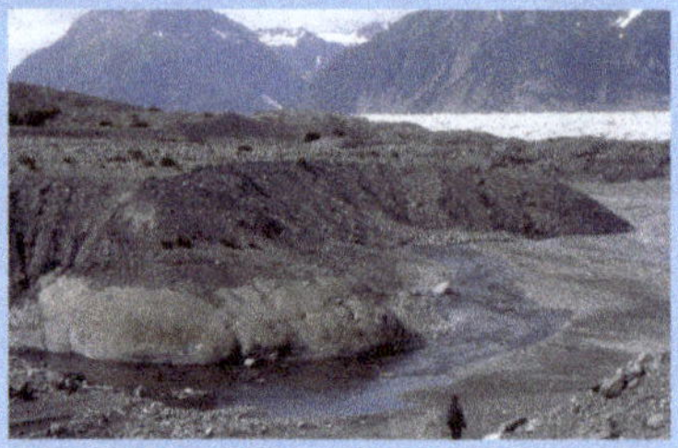

Peteet, D. The Importance of Understanding the Last Glacial Maximum for Climate Change 17

Slide 18

New Timing of Laurentide Deglaciation

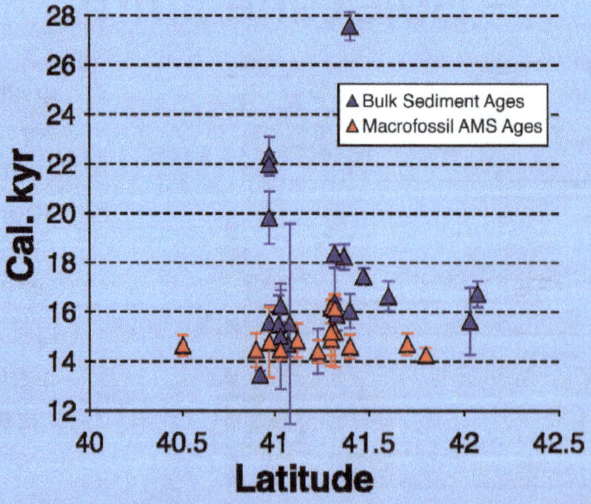

Suite of calendar ages (cal. kyr) calibrated from accelerator mass spectrometry (AMS) ^{14}C dates on plant macrofossils at 13 NE US sites, including 7 new lakes. Dates cluster at 14.5–15.6 cal. kyr, with 2 sites 16.1 and 16.7 cal. kyr. Bulk radiocarbon ages (blue), are generally significantly older. 2s error bars are given. Peteet *et al.* (2012)

Peteet, D. The Importance of Understanding the Last Glacial Maximum for Climate Change 18

Slide 19

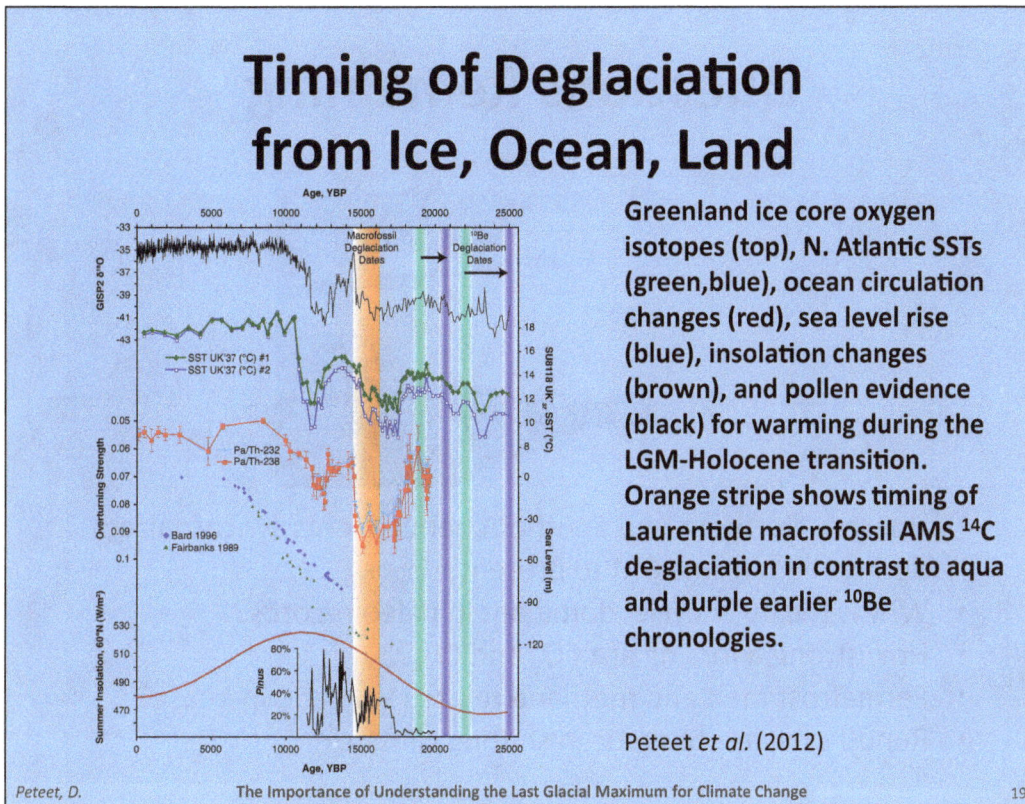

Timing of Deglaciation from Ice, Ocean, Land

Greenland ice core oxygen isotopes (top), N. Atlantic SSTs (green,blue), ocean circulation changes (red), sea level rise (blue), insolation changes (brown), and pollen evidence (black) for warming during the LGM-Holocene transition. Orange stripe shows timing of Laurentide macrofossil AMS ^{14}C de-glaciation in contrast to aqua and purple earlier [10]Be chronologies.

Peteet *et al.* (2012)

Peteet, D.   The Importance of Understanding the Last Glacial Maximum for Climate Change   19

Slide 20

Conclusions

- Change in our understanding of the importance of feedbacks for greenhouse gases and vegetation has large implications for understanding climate sensitivity of the LGM, deglaciation, and doubled CO_2.

- Many questions remain which we can investigate using multidisciplinary methods.

Peteet, D.   The Importance of Understanding the Last Glacial Maximum for Climate Change   20

Slide 21

Questions Remaining

- How are Northern and Southern hemispheres synchronized during shifts from LGM to present?
- Why is the 100k cycle dominant in paleo records?
- How strong a forcing are CO_2 and CH_4?
- Permafrost melt and methane release vs. net C increase?
- Runoff of water to arctic in shifting climate?

Peteet, D.   **The Importance of Understanding the Last Glacial Maximum for Climate Change**   21

Slide Notes

Slide 1 Photo of a large piedmont glacier, the Bering Glacier, AK, 1984, the size of Rhode Island and a possible diminutive modern analog for the Laurentide Ice Sheet.

Slide 2 Definition of the LGM between approximately 22–18k ago. It represents an equilibrium climate comparable to pre-industrial climate when our continents were in the same position as today. We assume that climatic processes and vegetation types were the same, but that variations in boundary conditions such as orbital configuration, greenhouse gas composition, and dust content of the atmosphere were different.

Slide 3 Evidence of the LGM is visible in both terrestrial and marine archives, including pollen and macrofossils from vegetation, glacial moraines left by ice sheets and alpine glaciers, loess deposits of windblown outwash near glaciated regions, and marine microfossils as well as oxygen and carbon isotopes recorded in marine and ice cores and speleothems.

Slide 4 Estimates of sea level lowering of 120 m are derived from several lines of evidence including corals, isotope records from benthic foraminifera, glacial isostasy, and ice margin reconstructions.

Slide 5 Ice core records from both poles show stability of low values of CH_4, CO_2, and oxygen and hydrogen isotopes at the LGM between 22 and 18k (figure from Parrenin et al., 2013).

Slide 6 Use of the dataset compiled by CLIMAP (1981) for GCM modeling included LGM estimates of land ice, sea ice, SST, and terrestrial surface albedo.

Slide 7 Results of GISS GCM modeling of the LGM indicated an inconsistency in air temperatures with the terrestrial (pollen and glacial) data in the subtropics (Rind and Peteet, 1985), and a very low climate sensitivity of only 3°C.

Slide 8 Stute et al. (1995) showed the consistent terrestrial dataset from Texas to lowland Brasil (40°N–40°S) including new groundwater samples which showed significant cooling (5–6°C) in the same region that faunal abundances from CLIMAP (1981) showed very little change (2°C).

Slide 9 Moisture differences between the LGM and today based upon paleodata (PMIP2).

Slide 10 Farrera et al. (1999) compiled the temperature shifts from both land and ocean throughout the tropics, and concluded that the tropics were colder than the CLIMAP reconstructions, and that lapse rates must have been steeper.

Slide 11 The 2007 IPCC (IPCC, 2007) showed a map of the difference in temperature between LGM and today, illustrating temperature shifts in the tropical oceans of not more than 2.5°C overall, but significantly colder than CLIMAP. Yet the radiative perturbation representing the forcings is presented with high scientific understanding of the forcings of orbital shift and greenhouse gases such as CO_2 and CH_4, but low understanding of the forcings of ice sheets, mineral dust, and vegetation.

Slide 12   More recent coupled GCM modeling experiments include questions about ice sheets, THC circulation, and feedbacks in the carbon cycle with varying boundary conditions. Questions that may help us understand the LGM include those that focus on the uncertainties in ocean circulation and SST, temperature gradients throughout the oceans, the role of the greenhouse gases in forcing temperature change in the de-glaciation just following the LGM, the climate sensitivity of the earth, and the timing and forcing of the last de-glaciation.

Slide 13   One of the modeling experiments (Crucifix, 2006) posed the question as to whether the LGM constrained climate sensitivity. A set of experiments with four-different models showed large variation in short-wave cloud feedbacks at low latitudes both during the LGM and in the $2 \times CO_2$ experiment. For the latter, they get a climate sensitivity ranging from 2.1 to 3.9°C, while the sensitivity of the LGM, they concluded, cannot be calculated because the amount of actual forcing of cooling from greenhouse gases and the feedbacks is not known.

Slide 14   Using a reconstructed tropical SST (30°N–30°S) cooling of 2.7 +1°C, these model-ers constrained the range to 5.8 +1.4°C, larger than most estimates (Schneider von Demling *et al.*, 2006).

Slide 15   In a study of climate sensitivity, Previdi *et al.* (2013) showed that climate or 'earth' sensitivity is increased greatly if feedbacks such as ice sheets and greenhouse gases are included with terrestrial change.

Slide 16   Kohfeld *et al.* (2013) assembled southern hemisphere datasets of (a) moisture, (b) dust, (c) SST, and export production which are relevant to changes in the west-erly winds. They conclude that either the stronger or equatorward-shifted wind scenarios appear easier to tie to evidence for increased moisture on the west coast of continents, cooler temperatures and higher productivity in the southern ocean, but caution that other interpretations are possible.

Slide 17   Understanding the termination of the LGM is important for learning how the climate system works. Key questions include examining the forcing for warming, the hemispheric responses in timing and magnitude, the greenhouse gas records and their relationship to surface air temperature on land and the vegetation, oceanic, and glacial responses. Aerosols such as dust may be very important.

Slide 18   Understanding linkages among ocean, atmosphere, and land is key to discern-ing causes of climate change. Careful attention to timing of de-glaciation has resulted in recent attention to the southeastern Laurentide margin, where AMS dating of macrofossils provides a suite of dates that define de-glaciation close to 16 cal. kyr = 16k (Peteet *et al.*, 2012).

Slide 19   This new AMS-dated timing is 5–9k later than the previously accepted age of 21–25k for de-glaciation as evidenced by varves and [10]Be dating of glacial moraines. However, it agrees with sea level rise and marine and ice core evidence of warming. The topic is controversial and needs further attention.

Slide 20   Key questions remain in the understanding of forcings and climate sensitivity. These questions are also very important in preparing for future climate shifts with global warming.

SECTION 7

Climate Change Impacts

Climate Lecture 17

Impacts of Sea Level Rise on Coastal Urban Areas

Vivien Gornitz

Center for Climate Systems Research, Columbia University and NASA Goddard Institute for Space Studies, New York, NY, USA

Vivien Gornitz is a Special Research Scientist at the Center for Climate Systems Research, Columbia University, and the NASA Goddard Institute for Space Studies. Her research focuses on the past, present, and future changes in sea levels, especially those associated with climate change.

Rising Seas and Coastal Hazards

Our planet is heating up largely because of increasing anthropogenic greenhouse gases from fossil fuel combustion and deforestation. Atmospheric levels of carbon dioxide and methane are now the highest in 800,000 years (Brook, 2008). Mean global temperature has increased by 0.7°C during the 20th century, likely surpassing temperatures of the past 1000 years (IPCC, 2013; Marcott *et al.*, 2013).

A major consequence of global warming is sea level rise. Between 1900 and 2010, global sea level rose 1.7 ± 0.2 mm/yr (IPCC, 2013). Since 1993, both satellites and tide gauges register a sea level rise of ~3.4 ± 0.4 mm/yr (CU Sea Level Rise Research Group, 2016; Nerem *et al.*, 2010). Since the mid-19th to early 20th century, modern sea level appears to have accelerated in two stages relative to the last few millennia: (1) the 20th-century sea level trend exceeds that of late Holocene by 1–2 mm/yr (Engelhart and Horton, 2012; Gehrels and Woodworth, 2013) and (2) since the 1990s, sea level is rising by more than 3 mm/yr.

On a centennial timescale, sea level rise is mainly controlled by thermal expansion of warming sea water and ice mass losses from glaciers and ice sheets. But the rise will not be globally uniform (Stammer *et al.*, 2013). Sea level varies on local to regional scales because of land motions due to ongoing glacial isostatic adjustments[1] stemming from the last deglaciation, neotectonics, and land subsidence due to excess subsurface fluid extraction (water, oil, and gas). Rates range from –17.6 mm/yr at Skagway, Alaska (glacial rebound from recent glacier retreat), to +9.1 mm/yr at Grand Isle, Louisiana (oil/gas extraction and sediment compaction). Cities, such as Bangkok, Manila, or Norfolk, Virginia, which likely pump groundwater to excess, show anomalously high rates. A weakened gravitational pull from a

---

[1] Adjustments of the Earth's crust to loading and unloading of ice masses during ice ages.

shrinking ice sheet lowers adjacent sea level. Water flows away from the ice sheet, while raising ocean levels far from the ice mass. Steric changes (ocean water temperature and salinity) may also affect ocean currents. For example, freshening of the North Atlantic Ocean due to increased ice loss from the Greenland Ice Sheet (and glaciers) may weaken the Atlantic Meridional Ocean Circulation (Hu *et al.*, 2011). This would lead to increased thermal expansion and an additional sea level rise of 0.4–0.55 m by 2100 along the northeast coast of North America (Hu *et al.*, 2011; Ezer, 2013), putting major cities such as Boston, New York City, and Washington, DC at greater risk. Coastal northeast United States is expected to become a regional sea level rise 'hot spot'. From the point of view of urban impacts, local to regional sea level changes will exert a much greater influence than global average changes.

Sea level rise will exacerbate numerous hazards facing coastal urban areas today (Estrada *et al.*, 2015; Hallegatte *et al.*, 2013; Mendelsohn *et al.*, 2012; Hanson *et al.*, 2011; Nicholls and Cazanave, 2010). These include storm surges, coastal and river flooding, shoreline erosion, saltwater intrusion, and land inundation. Flood exposure in coastal cities is increasing because of growing populations and assets in high-risk areas, and local land subsidence. Current and future flooding and storm damage frequency depends on various factors, such as rising sea level, surge, storm intensity, duration, and wave heights. These also significantly affect coastal erosion. The high winds, storm surges, and coastal flooding of tropical cyclones produce the greatest damages. Although the frequency, intensity, and geographical impacts of tropical cyclones have not changed substantially since the 1970s (Peduzzi *et al.*, 2012), the number of strongest North Atlantic cyclones has increased (Elsner *et al.*, 2008). Future changes in tropical cyclone characteristics remain uncertain, although the number of most intense cyclones is expected to grow (Knutson *et al.*, 2010). On the other hand, extra-tropical storms may show little change.

Many urban centers lie along tidal rivers (e.g., London, Rotterdam, New Orleans, Bangkok, Shanghai, and Haiphong). Heavy rainfall accompanying coastal storms can produce severe river flooding inland, which creates a dual threat for these cities from both high surges and overflowing rivers. Severe storms also generate high waves and/or water levels that lead to beach erosion and shoreline retreat. Sea level rise will generally increase erosion rates (IPCC, 2014, section 5.4.2.1; Bird, 2008). In urbanized areas, seawalls, groins, and other permanent structures can disrupt the natural flow of nearshore sediments, exacerbating beach erosion. Urban development prevents beaches, salt marshes, or mangroves from migrating inland, creating a 'coastal squeeze' that results in inundation of low-lying terrain by sea level rise and loss of natural ecosystems.

Another important impact of sea level rise is saltwater intrusion upstream and into coastal aquifers, potentially jeopardizing urban drinking water supplies and contaminating agricultural soils. Because of its low topography and unique geology, the City of Miami, Florida, faces multiple hazards that include coastal and inland flooding, storm surge inundation, beach erosion, and saltwater intrusion (Gornitz, 2016). The near surface, highly porous, permeable limestone bedrock — the Biscayne aquifer — is the city's major water resource. Saltwater infiltrates into the existing aquifers during heavy storms, or at times of low rainfall. Water management has become a major concern in Miami in light of the threat of increasing salinization.

Preparing for Rising Seas and Coastal Storms

Three major adaptation pathways exist to prepare for sea level rise. These include (1) shore-line protection, (2) accommodation, and (3) retreat (Gornitz, 2013). Shoreline protection can take 'hard' or 'soft' approaches. The former involves building engineered structures such as seawalls, bulkheads, boulder ramparts (revetments, riprap), groins, jetties, and breakwaters. These structures strengthen the existing shoreline and prevent slumping or erosion of poorly consolidated sediments, dampen wave action, and forestall flood damage. While resistant to average storm surges and wave heights, extreme events can still overtop them. Improper design or placement of these structures may, however, inten-sify erosion. Other structures such as dikes, tidal gates, and storm surge barriers protect against extreme floods or permanent inundation. Sea level rise will necessitate periodic strengthening, and raising of hard defenses. London, for example, plans to reinforce its defenses within the next 25–60 years, with the option to build a new barrier after that (Environment Agency, 2012). The Netherlands is already upgrading its sea defense system, adopting an integrated approach that includes 'building with nature' and allowing 'room for the river' (Delta Committee, 2008).

'Soft' defenses include beach nourishment, and rehabilitation of dunes and coastal wet-lands. Stable beaches and saltmarshes not only provide coastal ecosystem habitat, recrea-tional opportunities, but also act as buffers against storm surges and high waves. Restoration of the broadened beach and dune ridges, replanted with native vegetation between the Hague and Hoek van Holland, along the Delfland coast, has created a whole new nature district.

A second strategy entails accommodation to a rising sea. For example, stricter building codes will help strengthen structures and make them more storm-resilient. Buildings in flood-prone areas can be raised above the current (or projected) 100-year flood zone, con-structed on stilts or pilings, or ground floors used for non-residential purposes, such as busi-ness, parks, or recreation. Creation of more 'green infrastructure' (increased urban tree and grass plantings along sidewalks), or expanding parks will improve drainage and water infil-tration into the ground (e.g., Aerts et al., 2009). Another approach is to create neighbour-hoods of floating buildings or houseboats (e.g., Rotterdam, Seattle, Sausolito, and Bangkok). The Dutch have pioneered innovative multi-use flood defenses (such as dikes) which combine surge protection with housing, parking, parks, and commercial activities. Underground garages store excess water at times of high river or ocean levels (Rotterdam Climate Proof, 2010). Multi-purpose examples from Dordrecht, the Netherlands, Hamburg, Germany, and Toyko, Japan serve as models for other coastal cities (Stalenberg, 2012).

The ultimate option is retreat (Pilkey and Young, 2009). Defensive measures, even accom-modation, may offer a false short-term sense of security. Ultimately, it may become impos-sible to defend even the most heavily developed shorelines, particularly in high-risk areas. However, farsighted coastal management may avert many adverse consequences, well before many coastal areas become uninhabitable. For example, appropriate land use would limit housing density and building size in flood-prone areas, replacing them instead with natural 'buffer zones' of parks, restored wetlands, or recreational facilities.

Land-use policies that limit development in flood-prone areas, offer government acquisition of at-risk property, establish erosion setbacks and easements that create buffer zones for coastal wetlands or beaches, or *buyout programs* that reimburse shorefront land-owners for abandoning property in high-risk zones are just a few of the possible options. The most extreme choice is managed relocation in which financial incentives to rebuild after storms ends and a policy of coastal retreat is advocated (Pilkey and Young, 2009).

Case Study: New York City

New York City, with a population of 8.5 million, is the largest city in the United States and a major center of global finance and commerce. Significant economic assets and population exposed to coastal flooding lie along its 837 km of shoreline. Population within the 100-year flood zone[2] exceeds that of other vulnerable US coastal cities, such as Houston, New Orleans, and Miami (SIRR, 2014). Port facilities, major transportation routes, oil tanks and refineries, power stations, and wastewater treatment plants are some of the critical assets on the waterfront or within the 100-year flood zone. Coastal wetlands in Jamaica Bay buffer storm surges and waves and provide important ecosystem services. Marsh restoration is underway to counter the Bay deterioration that occurred during much of the 20th century (http://www.nycaudubon.org/jamaica-bay-project).

Winter cyclones and hurricanes have historically flooded parts of the city. Hurricane Sandy (a hybrid extra-tropical/tropical storm in NYC) in October, 2012, generated the highest recorded water level (3.38 m) at the Battery (the southern tip of Manhattan) in nearly 300 years, due to strong easterly winds, surge amplification,[3] maximum storm surge coinciding with high tide and full moon, plus historic sea level rise (0.44 m since 1856).

Following Hurricane Sandy, New York City has initiated new projects to reduce present and future climate-related risks, drawing upon experts from city and regional agencies, universities, the private sector, and the New York City Panel on Climate Change,[4] NPCC (NPCC, 2015; 2013; SIRR, 2013; NYC Hazard Mitigation Plan 2014). This case study highlights major new NPCC findings regarding future sea level rise, increased coastal flooding (NPCC, 2015), and some of the City's ongoing and planned responses.

Twentieth-century sea level rise in the New York City metropolitan area ranged between 2.4 and 4.1 cm per decade (NOAA, 2015), as compared to the global average rise of 1.7 cm per decade (IPCC, 2013). The NPCC (2015) finds that the mid-range[5] sea level at the Battery (southern tip of lower Manhattan) is projected to climb 28–53 cm by the 2050s and 46–99 cm

---

[2] The area flooded by a storm having an estimated 1% probability of occurrence per year (FEMA, 2013).
[3] Landward funneling of storm surge due to near right-angle geometry of New Jersey and Long Island shorelines.
[4] The NPCC, an advisory group originally convened in 2008 by former Mayor Michael Bloomberg, draws members from the academic, government, and private sectors.
[5] Future sea level rise projections for oceanic components derive from a suite of CMIP5 climate models and two greenhouse gas emission scenarios (RCP 4.5 and RCP 8.5 (IPCC, 2013)). Other sea level components are based on expert elicitation and literature surveys (NPCC, 2015). A 48-member model-based distribution is developed, in which the 10th percentile represents the low range; 25th to 75th percentiles the mid-range; and the 90th percentile a high-end estimate.

by the 2080s, relative to 2000–2004. In a worst-case, high-end scenario, sea level could reach 76 cm by the 2050s, 147 cm by the 2080s, and 190 cm by 2100.

Based on the above sea level rise projections, updated flood return curves (Federal Emergency Management Agency (FEMA, 2013), and assumed unchanged storm characteristics, flood heights for the 100-year storm (excluding waves) would increase from 3.4 m in the 2000s to 3.9–4.5 m by the 2080s (mid-range estimate). The annual flood likelihood would increase from 1 to 2.0–5.4% by the 2080s, up to 12.7%. The area potentially at risk to flooding would consequently expand in the future.

Adapting to Higher seas and Storm Surges

To strengthen its resiliency to climate change risks, New York City is undertaking initiatives to improve coastal flood mapping, continue collaboration with the NPCC to update and refine local climate projections, and strengthen coastal defenses using a variety of approaches tailored to specific neighborhood needs (SIRR, 2013; NYC, 2014). These include raising bulkheads, seawalls, enhancing beach nourishment, building local levees and storm surge barriers, implementing stricter building codes to reduce flood risk, increasing protection of critical infrastructure, including port facilities, utilities, telecommunications, sewer and drainage systems, and transportation arteries, and restoring wetlands and beach dunes. Another defense approach includes installation of several smaller, strategically placed, local storm surge barriers (SIRR, 2013).

Conclusions

Global warming in New York City is anticipated to increase temperatures, produce more frequent and prolonged heat waves, create more intense downpours, raise sea level, and lead to more damaging coastal flooding. In response, the city has embarked on a comprehensive long-term program to improve its climateresiliency, working closely with the NPCC, FEMA, and other state and national agencies. New York City's proposed strategies for enhancing coastal urban resiliency stand as a model for other urban coastal centers to prepare for climate change.

References

Aerts, J., Major, D.C., Bowman, M.J., Dircke, P. and Marfai, M.A. (2009). *Connecting Delta Cities: Coastal Cities, Flood Risk Management and Adaptation to Climate Change*. Amsterdam: Vrije Universiteit Press.

Bird, E. (2008). *Coastal Geomorphology: An Introduction* (2nd ed.). Chichester, UK: John Wiley & Sons, Inc.

Brook, E. (2008). Windows on the Greenhouse. *Nature, 453*, 291–292.

Colorado University Sea Level Rise Research Group, 2016. 2016_rel3: Global Mean Sea Level Time Series (seasonal signals removed) 6-30-2016. http://sealevel.colorado.edu

Delta Commissie (Committee) (2008). *Working Together With Water: A Living Land Builds for its Future*. http://www.deltacommissie.com/doc/deltareport_full.pdf

Engelhart, S.E. and Horton, B.P. (2012). Holocene sea level database for the Atlantic coast of the United States. *Quaternary Science Reviews* 54:12–25.

Environment Agency (2012). *Thames Estuary 2100: Managing Flood Risk Through London and the Thames Estuary: TE2100 Plan*. Strategic Environmental Assessment, Environmental Report Summary, London, UK. http://www.environment-agency.gov.uk

Elsner, J.B., Kossin, J.P. and Jagger, T.H. (2008). The increasing intensity of the strongest tropical cyclones. *Nature*, 455(7209), 92–95.

Estrada, F., Botzen, W.J. and Tol, R.S.J. (2015), Economic losses from US hurricanes consistent with an influence from climate change, *Nature Geoscience*, 8, 880-884.

Ezer, T. (2013). Sea level rise, spatially uneven and temporally unsteady: Why the U.S. East Coast, the global tide gauge record, and the global altimeter data show different trends. *Geophysical Research Letters*, 40, 5439–5444. doi:10.1002/2013GL057952.

Federal Emergency Management Agency (FEMA). (2013). *Flood Insurance Study: City of New York, New York, Bronx County, Richmond County, New York County, Queens County, Kings County.* Washington, DC: FEMA.

Gehrels, W.R. and Woodworth, P.L. (2013). When did modern rates of sea-level rise start? *Global and Planetary Change*, 1000:263–277.

Gornitz, V. (2013). *Rising Seas: Past Present, Future*. New York: Columbia University Press (Chapter 9 Coping with the rising waters).

Gornitz, V. (2017). Case Study 9.8. Coastal hazard and action plans in Miami, Florida, USA; Chapter 9. Coastal Zones in Urban Areas. Dawson, R.J., Shah Alam Khan, M., and V. Gornitz, Coordinating Lead Authors; Lead Authors: M. Fernanda, Lemos. Atkinson, L., Pullen, J., and Osorio, J. In C. Rosenzweig, W.D. Solecki, P. Romero-Lankao, S. Mehrotra, S. Dhakal, and S.A. Ibrahim (Eds.) *Climate Change and Cities: Second Assessment Report of the Urban Climate Change Research Network*. Cambridge, UK and New York, USA: Cambridge University Press, in press.

Gornitz, V. and D.C. Major (2017). Case Study 9.4. New York City: Preparing for Sea Level Rise and Coastal Storms, and other Climate Change Induced Hazards; Chapter 9. Coastal Zones in Urban Areas. Dawson, R.J., Shah Alam Khan, M., and V. Gornitz, Coordinating Lead Authors; Lead Authors: M. Fernanda, Lemos. Atkinson, L., Pullen, J., and Osorio, J. In C. Rosenzweig, W.D. Solecki, P. Romero-Lankao, S. Mehrotra, S. Dhakal, and S.A. Ibrahim (Eds.) *Climate Change and Cities: Second Assessment Report of the Urban Climate Change Research Network*. Cambridge, UK and New York, USA: Cambridge University Press, in press.

Hallegatte, S., Green, C., Nicholls, R.J. and Corfee-Morlot, J. (2013). Future flood losses in major coastal cities. *Nature Climate Change*, 3:802–806.

Hanson, S., Nicholls, R., Ranger, N., Hallegatte, S., Corfee-Morlot, J., Herweijer, C. and Chateau, J. (2011). A global ranking of port cities with high exposure to climate extremes. *Climatic Change* 104:89–111.

Hu, A., Meehl, G.A., Han, W. and Yin, J. (2011). Effect of the potential melting of the Greenland ice sheet on the meridional overturning circulation and global climate in the future. *Deep-Sea Research II*, 58:1914–1926.

IPCC (2013). Climate Change 2013. The Physical Science Basis. Contribution of Working Group 1 to the Fifth Assessment Report of the Intergovernmental Panel on Climate Change. Chapter 13: Sea Level Change, 1137–1216.

IPCC (2014). Climate Change 2014. Impacts, Adaptation, and Vulnerability. Contribution of Working Group II to the Fifth Assessment Report of the Intergovernmental Panel on Climate Change. Chapter 5: Coastal Systems and Low-Lying Areas, 361–409.

Knutson, T.R., McBride, J.L., Chan, J., Emanuel, K., Holland, G., Landsea, C. and Sugi, M. (2010). Tropical cyclones and climate change. *Nature Geoscience*, 3:157–163.

LafargeHolcim Foundation (2016). The Dryline. https://lafargeholcim-foundation.org/projects/the-dryline. (Accessed 7 Mar 2017).

Marcott, S.A., Shakun, J.D., Clark, P.U. and Mix, A.C. (2013). Reconstruction of regional and global temperature for the past 11,000 years. *Science,* 330:1198–1201.

Mendelsohn, R., Emanuel, K., Chonabayashi, S. and Bakkensen, L. (2012). The impact of climate change on global tropical cyclone damage. *Nature Climate Change*, 2:205–209.

Nicholls, R.J. and Cazanave, A. (2010). Sea-level rise and its impact on coastal zones. *Science,* **328**, 1517–1520.

Nerem, R.S., Chambers, D., Choe, C. and Mitchum, G.T. (2010). Estimating mean sea level change from the TOPEX and Jason altimeter missions. *Marine Geodesy,* **33**(1), (suppl. 1), 435.

NOAA TidesandCurrents. (2015). http://www.tidesandcurrents/noaa.gov

NPCC. (2015). New York Panel on Climate Change (2015). In C. Rosenzweig and W. Solecki (Eds.). *Building the Knowledge Base for Climate Resiliency* (vol. 1336, p. 150). New York: Annals of the New York Academy of Sciences.

NPCC. (2013). New York City Panel on Climate Change Climate Risk Information 2013. *Climate Risk Information: Observations, Climate Change Projections, and Maps.*

NYC Hazard Mitigation Plan.(2014). New York City Office of Emergency Management. http://www. nyc.gov/html/oem/html/planning_response/planning_hazard_mitigation_2014.shtml

Peduzzi, P., Chatenoux, B., Dao, H., De Bono, A., Herold, C., Kossin, J. and Nordbeck, O. (2012). Global trends in tropical cyclone risk. *Nature Climate Change*, 2:289–294.

Pilkey, O. and Young, R. (2009). *The Rising Sea.* Washington, DC: Island Press.

PlaNYC. (2013). *A Stronger, More Resilient New York* (p. 438). Mayor's Office of Long-Term Planning and Sustainability.

SIRR, PlaNYC, June (2013). *A Stronger, More Resilient New York.* The City of New York, Mayor Michael R. Bloomberg.

Stalenberg, B. (2012). Innovative flood defences in highly urbanized water cities. In J. Aerts, W. Botzen, M.J. Bowman, M.J., P.J.Ward and P. Dircke (Eds.), *Climate Adaptation and Flood Risk in Coastal Cities* (pp.145–164). London, New York: Earthscan.

Stammer, D., Cazenave, A., Ponte, R.M. and Tamisiea, M.E. (2013). Causes for contemporary regional sea level changes. *Annual Review of Marine Science,* **5**, 21–46.

Slide 1

Impacts of Sea Level Rise on Coastal Urban Areas

Vivien Gornitz

*Center for Climate Systems Research, Columbia University and
NASA Goddard Institute for Space Studies*

Our Warming Planet: Topics in Climate Dynamics
Lectures in Climate Change, Vol. 1
2017

Slide 2

Global Sea Level Change

Global Average Absolute Sea Level Change, 1880–2013

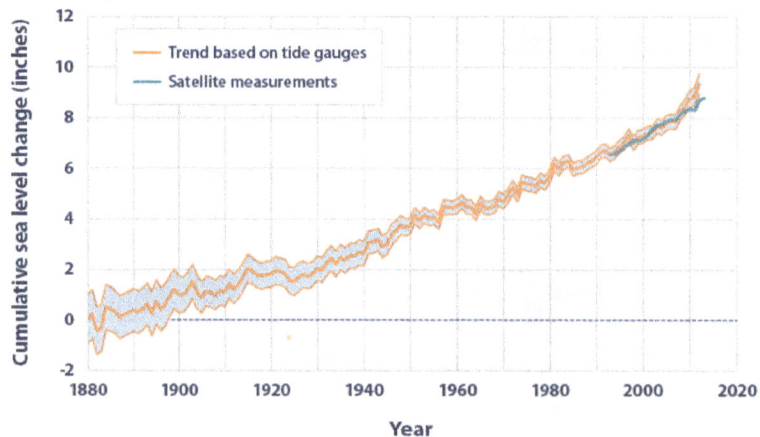

Data sources:
• CSIRO (Commonwealth Scientific and Industrial Research Organisation). 2013 update to data originally published in: Church, J.A., and N.J. White. 2011. Sea-level rise from the late 19th to the early 21st century. Surv. Geophys. 32:585–602.
• NOAA (National Oceanic and Atmospheric Administration). 2014. Laboratory for Satellite Altimetry: Sea level rise. Accessed April 2014. http://ibis.grdl.noaa.gov/SAT/SeaLevelRise/LSA_SLR_timeseries_global.php.

For more information, visit U.S. EPA's "Climate Change Indicators in the United States" at www.epa.gov/climatechange/indicators.

Slide 3

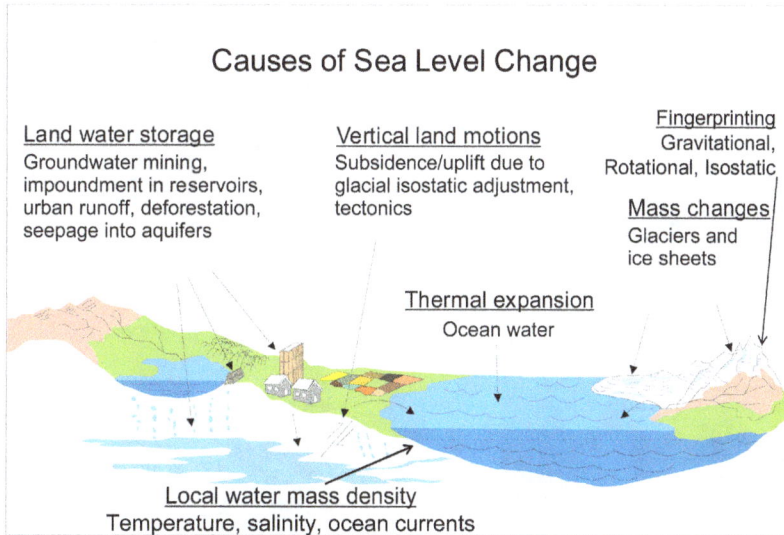

What Causes Sea Level Change?

Causes of Sea Level Change

Land water storage
Groundwater mining, impoundment in reservoirs, urban runoff, deforestation, seepage into aquifers

Vertical land motions
Subsidence/uplift due to glacial isostatic adjustment, tectonics

Fingerprinting
Gravitational, Rotational, Isostatic

Mass changes
Glaciers and ice sheets

Thermal expansion
Ocean water

Local water mass density
Temperature, salinity, ocean currents

Gornitz, V.   Impacts of Sea Level Rise on Coastal Urban Areas   3

Slide 4

Hurricane Katrina

(Photo: commander Mark Moran, Lt. Phil Eastman, and Lt. Dave Demmers, NOAA Aviation Weather Center).

Gornitz, V.   Impacts of Sea Level Rise on Coastal Urban Areas   4

Slide 5

Hurricane Sandy Flooding

a.   Waves Crashing Against Shore in Brooklyn

b.   Seaside Heights, NJ

Photo: Henry Zhang

c.   Water Cascading into WTC Site

d.   Hoboken, NJ

Photo: : John Minchillo/Associated Press

Gornitz, V.   Impacts of Sea Level Rise on Coastal Urban Areas   5

Slide 6

Historical Flooding in NYC

THE TOP 20 COASTAL STORM FLOODS
THE BATTERY—NEW YORK CITY—LAST 76 YEARS

STORM	DATE	WATER LEVEL (NAVD) FT	M
Hurricane Sandy	10/29/2012	11.1	3.38
Hurricane Donna	9/12/60	7.22	2.21
Nor'easter Dec. '92	12/11/92	6.92	2.11
Hurricane Irene	8/28/2011	6.72	2.05
Nor'easter	11/25/50	6.34	1.93
Ash Wednesday storm	3/6-7/62	6.14	1.87
Nor'easter	3/13-14/2010	6.06	1.85
Halloween ("Perfect Storm")	10/31/91	5.95	1.81
Blizzard of '84	3/29/84	5.75	1.75
Nor'easter	1/2/87	5.60	1.70
"Storm of the Century"	3/14/93	5.58	1.70
Nor'easter	11/12/68	5.58	1.70
Nor'easter	4/13/61	5.56	1.69
Nor'easter	2/19/60	5.54	1.68
Nor'easter	3/20/96	5.51	1.68
Nor'easter	10/19/96	5.49	1.67
Hurricane Gloria	9/27/85	5.45	1.66
Long Island Express	9/21/38	5.43	1.65
Hurricane of 1944	9/14/44	5.43	1.65
Nor'Ida	11/13-14/2009	4.79	1.46

Gornitz, V.   Impacts of Sea Level Rise on Coastal Urban Areas   6

Slide 7

Historical Sea Level Rise in NYC

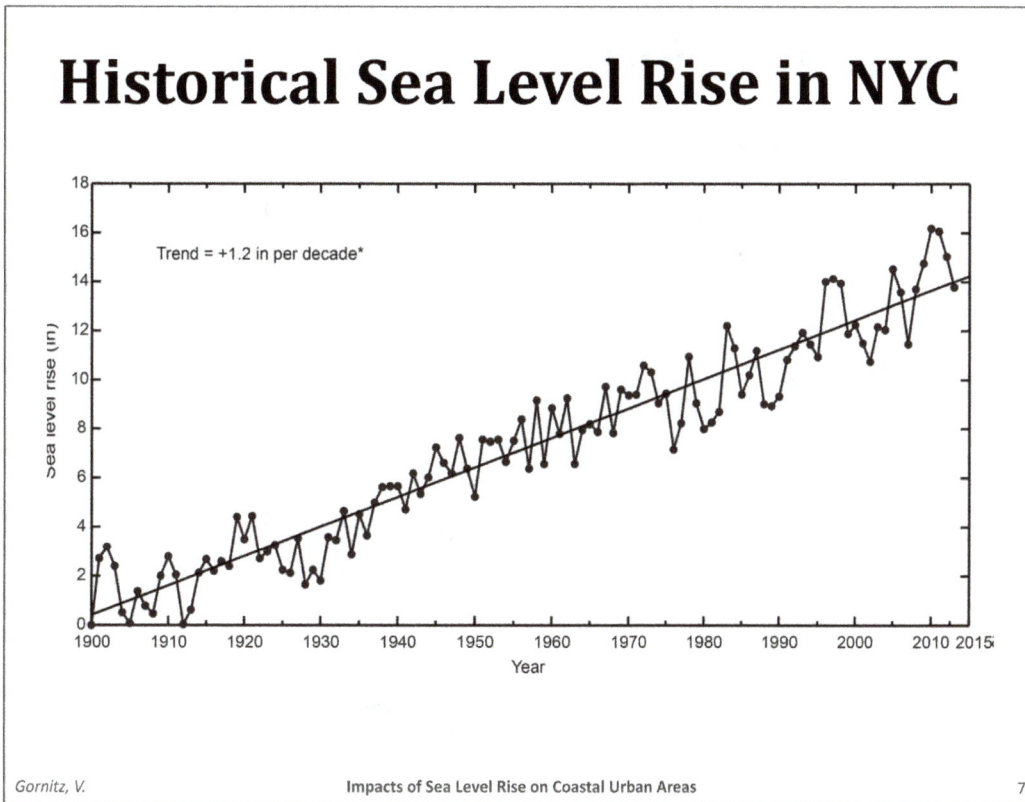

Trend = +1.2 in per decade*

Gornitz, V.   Impacts of Sea Level Rise on Coastal Urban Areas   7

Slide 8

New York City Sea Level Projections

Sea Level Rise Baseline (2000-2004)	Low-estimate (10th percentile)	Middle range (25th-75th percentile)	High-estimate (90th percentile)
2020s	+2 inches + 0.05 m	+4 to +8 inches +0.10 to +0.20 meters	+10 inches +0.25 meters
2050s	+8 inches +0.20 meters	+11 to +21 inches +0.28 to +0.53 meters	+30 inches +0.76 meters
2080s	+13 inches +0.33 meters	+18 to +39 inches +0.46 to +0.99 meters	+58 inches +1.47 meters
2100	+15 inches +0.38 meters	+22 to 50 inches +0.56 to +1.27 meters	+75 inches +1.91 meters

Based on 24 GCMs and two Representative Concentration Pathways. Shown are the low-estimate (10th percentile), middle range (25th percentile to 75th percentile), and high-estimate (90th percentile).

Gornitz, V.   Impacts of Sea Level Rise on Coastal Urban Areas   8

Slide 9

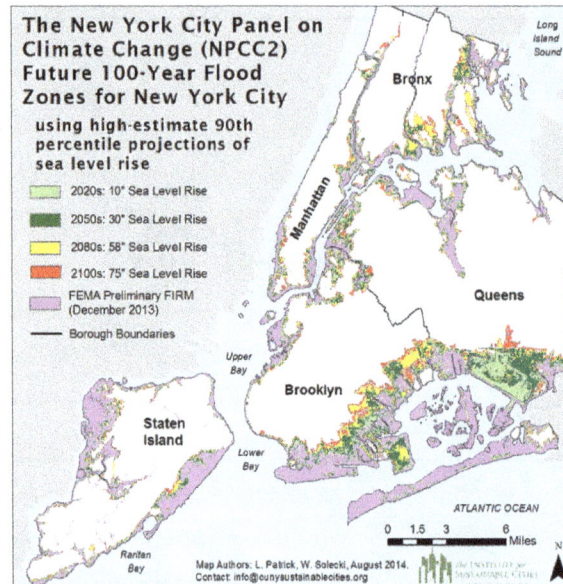

Slide 10

Increasing Urban Coastal Resilience

- **High resolution mapping** (i.e., LIDAR) to identify risk flood-prone areas

- Incorporate **sea level rise data into** latest 100-year **flood maps**

- Adapt existing storm emergency preparations to climate change

- **Improve coastal defenses**: strengthen and raise seawalls; build more dikes, levees, floodgates

- Raise land elevation, strengthen building codes, minimize new construction in flood-prone areas

- Create **"soft edges"** to dampen wave and tide energy—re-plant native vegetation; reduce land-sea slope

- Create series of **parks** along waterfront as **buffer zones**

- Restore or construct new wetlands and offshore reefs

- Widen beaches, rebuild and re-vegetate beach dunes.

Slide 11

'Hard' Coastal Defenses

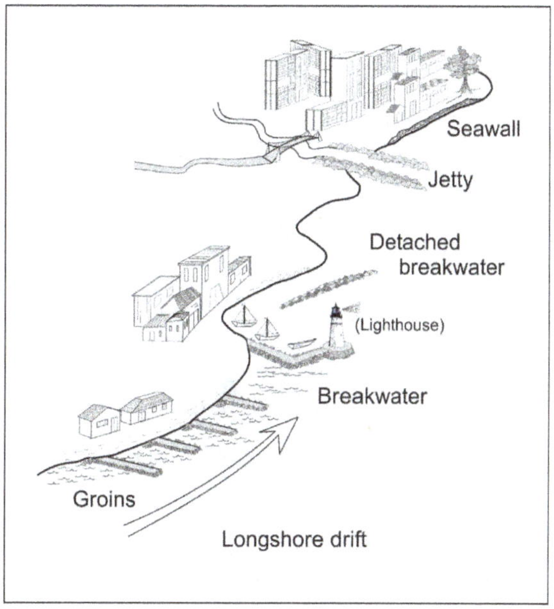
Seawall
Jetty
Detached
breakwater
(Lighthouse)
Breakwater
Groins
Longshore drift

Source: Gornitz (2013); *Rising Seas* Fig. 8.11

Gornitz, V. **Impacts of Sea Level Rise on Coastal Urban Areas** 11

Slide 12

Flood Barriers

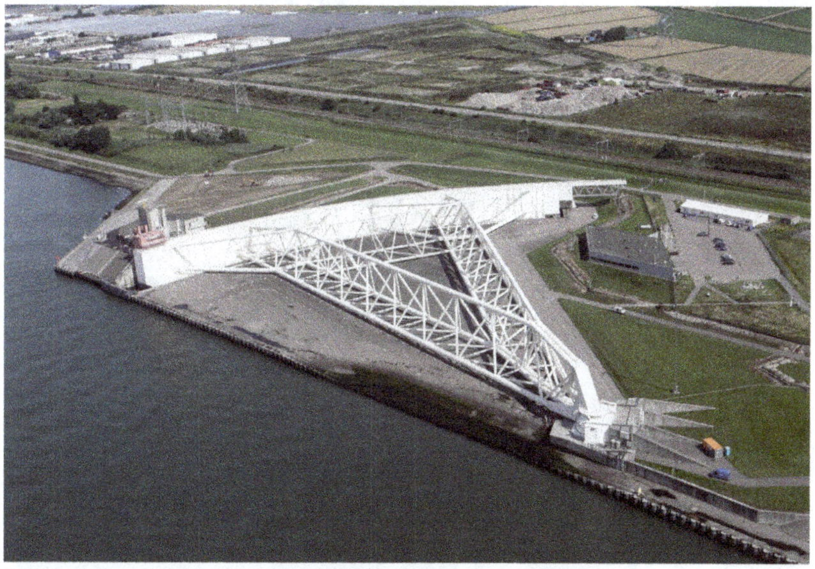

Northern half of the Maesland Barrier storm surge barrier, the Netherlands.
Photo: Joop van Houdt, Sept. 8, 2010. Wikimedia Commons

Gornitz, V. **Impacts of Sea Level Rise on Coastal Urban Areas** 12

Slide 13

'Soft' Coastal Defenses

Photo: V. Gornitz (Oct. 2010).

Slide 14

Multi-Use Development

Photo: V. Gornitz (2010).

Slide 15

Green Infrastructure

Source: Department of City Planning, City of New York City (2011).

Gornitz, V.   Impacts of Sea Level Rise on Coastal Urban Areas   15

Slide 16

Natural Protection Measures

Source: Galvin Brothers, Inc. http://chl.erdc.usace.mil/Articles/7/5/4/JamaicaBay.Grasses.jpg

Gornitz, V.   Impacts of Sea Level Rise on Coastal Urban Areas   16

Slide 17

Planned Berm and Park, Lower East Side of Manhattan

Source: LefargeHolcim Foundation, 2016; Bjarke Ingels Group

Gornitz, V. Impacts of Sea Level Rise on Coastal Urban Areas 17

Slide 18

Living with a Rising Sea Level

- By 2100, global sea level will rise up to 1 m and local sea level could rise even higher
- Sea level rise continues after 2100 because of persistence of atmospheric CO_2, continued global warming, thermal expansion, and ice sheet melting
- Sea level rise will increase land inundation, coastal storm flooding, saltwater intrusion, and coastal wetlands losses
- Cities can prepare by
  - improving existing coastal storm emergency preparations
  - strengthening coastal defenses including both "hard" engineered structures and "soft" natural protection measures (e.g., raised and re-vegetated dunes, restored wetlands, and park buffer zones)
- Strengthen building codes, revise land use zoning to avoid construction in flood-prone areas
- Adopt creative urban design, such as floating neighborhoods, houseboats, and canals
- As a last resort, plan for managed relocation

Gornitz, V. Impacts of Sea Level Rise on Coastal Urban Areas 18

Slide Notes

Slide 2   Global sea level rise, 1880–2013, based on tide gauge and satellite data. Between 1900 and 2010, global sea level rose 1.7 ± 0.2 mm/yr (IPCC, 2013). Since 1993, both satellites and tide gauges register a sea level rise of ~3.4 ± 0.4 mm/yr (CU Sea Level Rise Research Group, 2016; Nerem *et al.*, 2010) (Image source: EPA).

Since the mid-19th to early 20th century, modern sea level appears to have accelerated in two stages relative to the last few millennia: (1) the 20th-century sea level trend exceeds that of late Holocene by 1–2 mm/yr (Engelhart and Horton, 2012; Gehrels and Woodworth, 2013); and (2) since the 1990s, sea level is rising by more than 3 mm/yr.

Slide 3   Causes of sea level rise. On an annual to centennial time scale, thermal expansion of ocean water and ice mass losses from glaciers and ice sheets constitute the major contributors to global sea level rise. But the rise will not be globally uniform (Stammer *et al.*, 2013). Sea level varies on local to regional scales because of variations in ocean water density, land motions due to ongoing glacial isostatic adjustments (GIA: adjustments of the earth's crust to loading and unloading of ice masses) stemming from the last de-glaciation, neotectonics, and land subsidence due to excess subsurface fluid extraction (water, oil, and gas). Gravitational and associated geophysical changes that arise from ice mass losses will have geographically different effects on sea level change (a process sometimes referred to as 'fingerprinting').

Slide 4   Flooding of New Orleans after Hurricane Katrina in August, 2005. Tropical hurricanes are responsible from some of the worst natural disasters in US history.

Current and future flooding and storm damage frequency depends on various factors, such as rising sea level, surge, storm intensity, duration, and wave heights. These also significantly affect coastal erosion. The high winds, storm surges, and coastal flooding of tropical cyclones produce the greatest damages. Although the frequency, intensity, and geographical impacts of tropical cyclones have not changed substantially since the 1970s (Peduzzi *et al.*, 2012), the number of strongest North Atlantic cyclones has increased (Elsner *et al.*, 2008). Future changes in tropical cyclone characteristics remain uncertain, although the number of most intense cyclones is expected to grow (Knutson *et al.*, 2010). On the other hand, projected extra-tropical storm activity also remains uncertain.

Slide 5   Severe storms also generate high waves and/or water levels that lead to beach erosion and shoreline retreat. Sea level rise will generally increase erosion rates (IPCC, 2014, section 5.4.2.1; Bird, 2008). In urbanized areas, seawalls, groins, and other permanent structures can disrupt the natural flow of near-shore sediments, exacerbating beach erosion. Urban development prevents beaches, salt marshes, or mangroves from migrating inland, creating a 'coastal squeeze' that results in inundation of low-lying terrain by sea level rise and loss of natural ecosystems.

   a. Waves crashing against seawall in Brooklyn, New York City. Water overtopping the seawall is flooding the adjacent park during Hurricane Sandy on October

29, 2012. The park acts as a buffer zone absorbing the floodwater, which protects the private homes beyond.

b. Coastal flooding, Seaside Heights, New Jersey, during Hurricane Sandy in late October 2012. Seaside Heights is a small residential community situated on a narrow barrier island, roughly mid-way between Atlantic City, to the south, and Sandy Hook to the north. Barrier islands are highly dynamic environments, subject to wind and wave action and sediment mobility (Bird, 2008). In general, the barrier islands of New Jersey are eroding, in part due to historic sea level rise and in part due to the presence of hard structures. Storms such as Sandy produce extensive beach erosion.

c. Water cascading into the former World Trade Center site, lower Manhattan, New York City during Hurricane Sandy, October 29, 2012.

d. Water flooding the entrance to the PATH train in Hoboken, New Jersey during Hurricane Sandy. Hoboken lies across the Hudson River from New York City.

Slide 6 The top 20 coastal storm floods at the Battery, lower Manhattan, in New York City since 1938. Hurricane Sandy in October, 2012, tops the list, followed by a nor'easter in December 1992. Hurricane Sandy, technically a hybrid tropical–extra-tropical cyclone in NYC, caused severe damages due to the rare combination of strong easterly winds, surge amplification, maximum storm surge coinciding with high tide and full moon, plus historic sea level rise (0.44 m since 1856). Landward funneling of storm surge due to near right-angle geometry of New Jersey and Long Island shorelines.

Slide 7 Sea level rise will exacerbate numerous hazards facing coastal urban areas today (Estrada _et al._, 2015; Hallegatte _et al._, 2013; Mendelsohn _et al._, 2012; Hanson _et al._, 2011; Nicholls and Cazanave, 2010). These include storm surges, coastal and river flooding, shoreline erosion, saltwater intrusion, and land inundation. Flood exposure in coastal cities is increasing because of growing populations and assets in high-risk areas, and local land subsidence. Current and future flooding and storm damage frequency depends on various factors, such as rising sea level, surge, storm intensity, duration, and wave heights. These also significantly affect coastal erosion.

Coastal northeast United States is expected to become a regional sea level rise 'hot spot'. From the point of view of urban impacts, local to regional sea level changes will exert a much greater influence than global average changes.

Slide 8 Sea level rise projections for New York City (NPCC, 2015). The mid-range sea level at the Battery (southern tip of lower Manhattan) is projected to climb 28–53 cm by the 2050s and 46–99 cm by the 2080s, relative to 2000–2004. In a worst-case, high-end scenario, sea level could reach 76 cm by the 2050s, 147 cm by the 2080s, and 190 cm by 2100.

Future sea level rise projections for oceanic components derive from a suite of CMIP5 climate models and two greenhouse gas emission scenarios (RCP 4.5 and RCP 8.5 (IPCC, 2013). Other sea level components are based on expert elicitation and literature surveys (NPCC, 2015). A 48-member model-based distribution is developed, in which the 10th percentile represents the low range; 25th–75th percentiles the mid-range; and the 90th percentile a high-end estimate.

Slide 9   Areas potentially affected by the 100-year flood in the 2020s, 2050s, 2080s, and 2100 (NPCC, 2015). Projections use the high-end 90th-percentile sea level rise scenario.

(Note: This map is subject to limitations in accuracy due to uncertainties in the climate models, datasets, and methodology used in its development. Therefore, this map and data should not be used to assess actual coastal hazards, insurance requirements, or property values or be used in lieu of FIRMS issued by FEMA. The flood areas delineated above do not represent precise flood boundaries but rather illustrate: (1) areas exposed to the current 100-year flood that will continue to risk future flooding; (2) areas not currently within the 100-year flood zone, but are expected to be so, due to sea level rise; and (3) areas that do not currently flood and are unlikely to do so by 2100.)

Based on the above sea level rise projections, updated flood return curves (Federal Emergency Management Agency (FEMA), 2013), and assumed unchanged storm characteristics, flood heights for the 100-year storm (excluding waves) would increase from 3.4 m in the 2000s to 3.9–4.5 m by the 2080s (mid-range estimate). The annual flood likelihood would increase from 1 to 2.0–5.4% by the 2080s, up to 12.7%. The area potentially at risk to flooding would consequently expand in the future.

Slide 11   This image depicts a schematic diagram of 'hard' coastal defense structures designed to protect the coast from storms or erosion. If poorly designed, they may produce the opposite effect.

Three major adaptation pathways exist to prepare for sea level rise. These include (1) shoreline protection, (2) accommodation, and (3) retreat (Gornitz, 2013). Shoreline protection can take 'hard' or 'soft' approaches. The former involves building engineered structures such as seawalls, bulkheads, boulder ramparts (revetments, riprap), groins, jetties, and breakwaters. These structures strengthen the existing shoreline and prevent slumping or erosion of poorly consolidated sediments, dampen wave action, and forestall flood damage. While resistant to average storm surges and wave heights, extreme events can still overtop them. Improper design or placement of these structures may however intensify erosion.

Slide 12   The Maesland Barrier is an example of a movable barrier, which ordinarily remains open to shipping, but can be closed whenever a high storm surge threatens the city of Rotterdam, and the surrounding dike system upstream. In the future, rising sea levels will force more frequent closures of the barrier.

Structures such as dikes, tidal gates, and storm surge barriers protect against extreme floods or permanent inundation. Sea level rise will necessitate periodic strengthening, and raising of hard defenses, such as the Maesland Barrier. London, for example, plans to reinforce its defenses within the next 25–60 years, with the option to build a new barrier after that (Environment Agency, 2012). The Netherlands is already upgrading its sea defense system, adopting an integrated approach that includes 'building with nature' and allowing 'room for the river' (Delta Committee, 2008).

Slide 13 'Soft' defenses include beach nourishment, and rehabilitation of dunes and coastal wetlands. Stable beaches and saltmarshes not only provide coastal ecosystem habitat, recreational opportunities, but also act as buffers against storm surges and high waves.

Widening and extending the dune ridges seaward and replanting with native vegetation along the Delfland coast, the Netherlands, which has created a whole new nature district. The Dutch have recently placed a greater emphasis on use of 'soft' coastal defense measures, in addition to their traditional 'hard' engineering approach.

Slide 14 Multi-use development at Scheveningen, the Netherlands. The dikes are being raised and the beach widened. A boulevard and pedestrian walkways will run alongside the top of the dike. Note the housing developments on the top of the dike.

The Dutch have pioneered innovative multi-use flood defenses (such as dikes) which combine surge protection with housing, parking, parks, and commercial activities. Underground garages store excess water at times of high river or ocean levels (Rotterdam Climate Proof, 2010). Multi-purpose examples from Dordrecht, the Netherlands, Hamburg, Germany, and Tokyo, Japan, serve as models for other coastal cities (Stalenberg, 2012).

Slide 15 To strengthen its resiliency to climate change risks, New York City is undertaking initiatives to improve coastal flood mapping, continue collaboration with the NPCC to update and refine local climate projections, and strengthen of coastal defenses using a variety of approaches tailored to specific neighborhood needs (SIRR, 2013; NYC, 2014). These include raising bulkheads, seawalls, enhancing beach nourishment, building local levees and storm surge barriers, implementing stricter building codes to reduce flood risk, increasing protection of critical infrastructure, including port facilities, utilities, telecommunications, sewer and drainage systems, and transportation arteries, and restoring wetlands and beach dunes.

Creating a soft edge shoreline by recreating a natural wetland and installing protective structures to dampen waves and protect against storms, Brooklyn Bridge Park, New York City. In addition, the creation of more 'green infrastructure' (increased urban tree and grass plantings along sidewalks), or expanding parks will improve drainage and water infiltration into the ground (e.g., Aerts *et al.*, 2009).

Slide 16 Port facilities, major transportation routes, oil tanks and refineries, power stations, and wastewater treatment plants are some of the critical assets on the waterfront or within the 100-year flood zone in New York City. Coastal wetlands in Jamaica Bay buffer storm surges and waves and provide important ecosystem services. Marsh restoration is underway to counter the Bay deterioration that occurred during much of the 20th century.

The image shows salt marsh restoration Jamaica Bay. Native grasses are being transplanted by hand, Elders Point — West, Jamaica Bay, New York. Natural coastal vegetation helps to dampen destructive action of waves and storm surges, and also provides important habitat for wildlife.

Slide 17   Planned multi-use development in New York City. A section of the Dryline (formerly the Big U) — part of a New York City project to build berms and waterfront parks intended to shield lower Manhattan neighborhoods against future floods and higher sea levels. Source: LafargeHolcim Foundation (2016).

The berm and broad park zone will protect high rise housing developments and transportation arteries from present-day extreme storm surges, such as that produced by Hurricane Sandy in 2012, and from future sea level rise. In addition, the park will offer multiple recreational opportunities, including biking, jogging, picnicking, sports, and space for outdoor cultural events. Image credit: Bjarke Ingels Group.

CLIMATE LECTURE 18

Climate Change Challenges to Agriculture, Food Security, and Health

Cynthia Rosenzweig[1,2] and Daniel Hillel[2]

[1]*NASA Goddard Institute for Space Studies, New York, NY, USA*
[2]*Columbia University, New York, NY, USA*

Cynthia Rosenzweig is a Senior Research Scientist at the NASA Goddard Institute for Space Studies, where she heads the Climate Impacts Group, and is an adjunct Professor at Barnard College. She was a Coordinating Lead Author of the IPCC Working Group II Fourth Assessment Report. Her research focuses on the effects of climate change on agriculture.

Daniel Hillel has been a Senior Research Scientist at The Earth Institute at Columbia University. He is a water and soil scientist, focusing on improving water efficiency in light of climate change. He was awarded the World Food Prize in 2012.

'Food security exists when all people, at all times, have physical and economic access to sufficient, safe and nutritious food that meets their dietary needs and food preferences for an active and healthy life.'

— **Food and Agriculture Organization (FAO)**, Rome Declaration on World Food Security and World Food Summit Plan of Action, 1996.

We have learned from our brilliant and dedicated colleague, David Rind, that the already-evident and further-anticipated process of anthropogenic climate change raises worldwide concerns regarding agricultural production, food security, and public health. The four pillars of food security — its availability, accessibility, utilization, and stability — will all be affected by the expected changes in climate over the coming century. At both global and regional scales, the Intergovernmental Panel on Climate Change (IPCC) has demonstrated that the provision of food security and maintenance of ecosystem services are under threat from dangerous human interference in the Earth's climate (Porter *et al.*, 2014). The FAO emphasizes that the main cause of hunger and malnutrition is not the lack of food *per se*, but the inability of vulnerable groups to access food in many circumstances (FAO, 2014). Furthermore, the IPCC has determined that global food prices will likely increase by the year 2050 as a result of changes in temperature and precipitation (Porter *et al.*, 2014; Nelson *et al.*, 2014). At country scales, nations are concerned over potential damages that may arise in coming decades from climate change impacts on agriculture and the food system, as these are likely to affect their citizens' well-being, regional planning, resource use, trading patterns, and international policies.

While agroclimatic conditions, as well as land resources and their management, are key components of food production, current framing of agriculture expands beyond production to encompass the entire food system. The main food system components include the processes of (1) growth, harvesting, and production; (2) distribution; (3) food preparation, preservation, and processing; (4) marketing and retail; (5) food use and consumption; and (6) disposal of food waste (FAO, 2014; San Francisco Food Systems, 2005; Center for Ecoliteracy, 2015). This recognizes demand as well as supply, both of which are critically affected by distinct socioeconomic pressures, such as current and projected trends in population, income growth and distribution, as well as the availability and access to technology and development. Since 1970, for instance, average daily per capita intake has risen globally from 2400 to 2900 calories, spurred by economic growth, improved production systems, international trade, and globalization of food markets (WHO, 2015).

Feedbacks of such growth patterns on cultures and personal tastes, lifestyles, and demographic changes have, in turn, led to major dietary changes — mainly in developing countries, where shares of meat, fat, and sugar in total food intake have increased significantly (CGIAR, 2011). The increased availability of vegetable oils and caloric sweeteners, and a shift in diet towards animal-source foods away from legumes, coarse grains, and other vegetables has been observed in countries such as China, India, and Mexico, leading to a documented rise in the rates of obesity in developing nations (Popkin *et al.*, 2012). Further compounding the issue, so the FAO predicts, 60% more food will be needed worldwide by 2050 given the existing rate of food consumption (FAO *et al.*, 2012). Thus, the consequences of climate change on world food demand and supply will depend on many interactive dynamic processes.

Agriculture plays fundamental roles in human-driven climate change. On the one hand, it is a key sector that will be impacted by climate change over the coming decades, thus requiring adaptation measures. Growing periods will be altered, agroecological zones will shift, and productivity will be affected. On the other hand, agriculture is a major source of anthropogenic greenhouse gas emissions to the atmosphere, being responsible for about one-quarter of all net emissions when clearing of land for food production is taken into account (IPCC, 2014a).

A changing climate due to increasing anthropogenic emissions of greenhouse gases will induce changes in agricultural systems through a set of interactive processes. Both productivity and geographic distribution of crop species will be altered. The global production of wheat, rice, and maize, for example, are projected to be negatively affected by a 2°C warming over the coming century. Rising winter temperatures will affect vernalization of wheat, and thus its nutritional quality. However, individual locations may benefit from such temperature increases (IPCC, 2014b). The major climate factors contributing to these responses include increasing atmospheric carbon dioxide, rising temperature, and increasing frequency and severity of extreme events, especially droughts and floods. These factors, in turn, will influence water resources for agriculture, grazing lands, livestock, and associated agricultural pests. Effects will vary, depending on the degree of change in temperature and precipitation and on the particular management system in each location.

Elevated CO_2 at concentrations of about 550ppm in free-air CO_2 enrichment (FACE) experiments stimulated photosynthesis in C_3 crops (e.g., wheat, rice, and soybean) by about 10–45%, and in C_4 crops (e.g., corn) by about 0–10% (Kimball, 2011). The stimulation is reduced by low soil nitrogen. Biomass and yield were increased in FACE experiments in all C_3 species, but not in C_4 species, except when water was limiting and growth stimulation occurred through improved water use. There has been controversy about differences in results between FACE and enclosure (e.g., growth chamber) experiments. However, it appears unlikely that there is a significant difference in the response of C_3 crops to elevated CO_2 between the two experimental settings when the whole population of enclosure experiments is included, and their variability is taken into account (Kimball, 2011).

Studies have shown that recent climate change trends in some regions already appear to have discernible effects on some agricultural systems. Tao and Lin (2015) discuss how in China annual average temperatures have increased 1.2°C since 1960, the number of heat extremes has increased, the number of frost days has decreased significantly, and dry regions are experiencing less precipitation while wet areas are experiencing more precipitation (Tao and Lin, 2015). The study illustrated that these trends in China between 1980 and 2008 measurably affected crop yields, finding a decrease in wheat (1.27%), maize (1.73%), and soybeans (0.41%), and an increase in rice (0.56%), with varying intensities in these changes across different regions of the country (Tao and Lin, 2015; Fulu and Erda, 2014). The large spatial differences in the impacts on yields constitute a challenging aspect for food systems to manage and adapt to climate change (Rosenzweig and Hillel, 2015).

Two further challenges for agriculture have emerged in recent times — the need to deliver nutritious food for healthy diets and the need for environmental sustainability in the food system (Muller et al., 2014; Yadav et al., 2011). As climate changes as well as socioeconomic pressures will shape future demands for food, fiber, and energy, synergies must be identified between adaptation and mitigation strategies to allow for the development of robust options that meet both the climate and societal challenges in the coming decades.

Climate change will bring both challenges and opportunities to agriculture. Farmers and researchers are being called on to simultaneously and sustainably adapt to and mitigate climate change through their activities involving management practices, crop breeding, and new production systems. Some of these can be mutually reinforcing, especially in view of the projected increased climate variability under overall climate change. This is because most mitigation techniques currently considered in agriculture, including reduced tillage, were originally designed as 'best practice' management strategies aimed at enhancing the long-term stability and resilience of cropping systems in the face of climate variability and of increased production intensity.

By increasing the ability of soils to hold moisture and to avoid erosion, and by enriching ecosystem biodiversity through the establishment of more diversified cropping systems, mitigation techniques implemented locally for soil carbon sequestration can also help cropping systems to better withstand droughts and/or floods, both of which are projected to increase in frequency and severity in future warmer climates. These techniques can also help to enhance the quality of food. Agriculture will necessarily play a leading role in responding to both the challenges and opportunities presented by a dynamic environment, as the food system must deliver nutrition and health in a sustainable way under changing climate conditions.

If the emissions of greenhouse gases continue to grow, the greenhouse effect and surface warming will acquire ever-greater severity. The ultimate significance of the climate change issue is related to its global reach, affecting agricultural regions, food security, and health throughout the world in complex and interactive ways. Farmers and others in the agricultural sector are thus faced with the quadruple task of contributing to global reductions of carbon dioxide and other greenhouse gas emissions, coping with an already-changing climate, delivering healthy nutritious food, and sustainably managing soil and water resources. In short, agriculture is being called upon to ensure both human and planetary health.

References

Adiku, S. G. K., *et al.* (2015), Climate Change Impacts on West African Agriculture: An Integrated Regional Assessment (CIWARA), in *Handbook of Climate Change and Agroecosystems: The Agricultural Model Intercomparison and Improvement Project Integrated Crop and Economic Assessments, Part 2*, edited by C. Rosenzweig and D. Hillel, pp. 25–74, Imperial College Press.

Asseng, S., Ewert, F., Rosenzweig, C., Jones, J.W., Hatfield, J.L. and Wolf, J. (2013). Uncertainty in simulating wheat yields under climate change. *Nature Clim. Change*, 3(9), 827–832. doi: 10.1038/nclimate1916.

Center for Ecoliteracy. (2015). Food Systems, Climate Change, and Ecological Education. https://www.ecoliteracy.org/article/food-systems-climate-change-and-ecological-education.

Easterling, William E. (2011). "Guidelines for adapting agriculture to climate change." *Handbook of climate change and agroecosystems: impacts, adaptation, and mitigation*, 269–286. London, UK: Imperial College Press.

Ericksen, P.J. (2008). Conceptualizing food systems for global environmental change research. *Global Environmental Change*, 18(1), 234–245. doi:http://dx.doi.org/10.1016/j.gloenvcha.2007.09.002.

Ericksen, P., Stewart, B., Dixon, J., Barling, D., Loring, P., Anderson, M. and Ingram, J. (2010). The value of a food system approach. *Food Security and Global Environmental Change*, 25.

FAO (2014). Strengthening the enabling environment for food security and nutrition. Food and Agriculture Organization of the United Nations, Rome.

Fulu, T. and Erda, L. (2014). AgMIP Regional Activities in a Global Framework: The China Experience. In Rosenzweig, C. and Hillel, D. (Eds.), *Handbook of Climate Change and Agroecosystems* (Vol. 3, pp. 375–390). London, UK: Imperial College Press.

Gordo, O. and Sanz, J.J. (2005). Phenology and climate change: A long-term study in a Mediterranean locality. *Oecologia*, 146(3), 484–495. doi:10.1007/s00442-005-0240-z.

Gutierrez, A.P. and Ponti, L. (2011). Assessing the invasive potential of the Mediterranean fruit fly in California and Italy. *Biological Invasions*, 13(12), 2661–2676. doi:10.1007/s10530-011-9937-6.

Hairston, N. G., F. E. Smith, and L. B. Slobodkin (1960), Community Structure, Population Control, and Competition, *The American Naturalist*, 94(879), 421–425.

Hillel, D. and Rosenzweig, C. (2011). *Handbook of Climate Change and Agroecosystems*. London, UK: Imperial College Press.

Hillel, D. and Rosenzweig, C. (2015). *Handbook of Climate Change and Agroecosystems: The Agricultural Model Intercomparison and Improvement Project (AgMIP) Integrated Crop and Economic Assessments*. London, UK: Imperial College Press.

Howden, S.M., Crimp, S.J. and Nelson, R. (2010). Australian agriculture in a climate of change. Paper read at Managing climate change: papers from the Greenhouse 2009 conference.

Ingram, J. (2011). A food systems approach to researching food security and its interactions with global environmental change. *Food Security,* 3(4), 417–431. doi: 10.1007/s12571-011-0149-9.

IPCC. (2014a). Summary for Policymakers. In Edenhofer, O., Pichs-Madruga, R., Sokona, Y., Farahani, E., Kadner, S. and Minx, J.C. (Eds.), *Climate Change 2014, Mitigation of Climate Change. Contribution of Working Group III to the Fifth Assessment Report of the Intergovernmental Panel on Climate Change.* Cambridge, United Kingdom and New York: Cambridge University Press.

IPCC. (2014b). Summary for policymakers. In Field, C.B., Barros, V.R., Dokken, D.J., Mach, K.J., Mastrandrea, M.D. and White, L.L. (Eds.),*Climate Change 2014: Impacts, Adaptation, and Vulnerability. Part A: Global and Sectoral Aspects. Contribution of Working Group II to the Fifth Assessment Report of the Intergovernmental Panel on Climate Change* (pp. 1–32).Cambridge, United Kingdom and New York: Cambridge University Press.

Jones, J.W., Bartels, W.L., Fraisse, C., Boote, K.J., Ingram, K.T. and Hoogenboom, G. (2011). Use of Crop Models for Climate-Agricultural Decisions. In: Hillel D, Rosenzweig C (eds) *Handbook of Climate Change and Agroecosystems. ICP Series on Climate Change Impacts, Adaptation, and Mitigation,* vol 1. Imperial College Press, London, UK, pp. 131–157.

Kimball, B.A. (2011) Lessons from FACE: CO_2 effects and interactions with water, nitrogen and temperature. In D. Hillel and C. Rosenzweig (Eds.), *Handbook of Climate Change and Agroecosystems: Impacts, Adaptation, and Mitigation* (Chapter 5, pp. 87–107). London, UK: Imperial College Press.

Kong, A.Y.Y., Scow, K.M., Córdova-Kreylos, A.L., Holmes, W.E. and Six, J. (2011). Microbial community composition and carbon cycling within soil microenvironments of conventional, low-input, and organic cropping systems. *Soil Biology and Biochemistry,* 43(1), 20–30. doi:http://dx.doi.org/10.1016/j.soilbio.2010.09.005.

Muller, C., Elliott, J. and Levermann, A. (2014). Food security: Fertilizing hidden hunger. *Nature Climate Change,* 4(7), 540–541. doi:10.1038/nclimate2290.

Nelson, G.C., Valin, H., Sands, R.D., Havlík, P., Ahammad, H. and Willenbockel, D. (2014). Climate change effects on agriculture: Economic responses to biophysical shocks. *Proceedings of the National Academy of Sciences,* 111(9), 3274–3279. doi:10.1073/pnas.1222465110.

Popkin, B.M., Adair, L.S. and Ng, S.W. (2012). NOW AND THEN: The global nutrition transition: The pandemic of obesity in developing countries. *Nutrition Reviews,* 70(1), 3–21. *PMC.* Web. 18 Sept. 2015.

Porter, J.R., Xie, L., Challinor, A.J., Cochrane, K., Howden, S.M., Iqbal, M.M., Lobell, D.B. and Travasso, M.I. (2014). Food security and food production systems. In Field, C.B., Barros, V.R., Dokken, D.J., Mach, K.J., Mastrandrea, M.D., ... White, L.L. (Eds.), *Climate Change 2014: Impacts, Adaptation, and Vulnerability. Part A: Global and Sectoral Aspects. Contribution of Working Group II to the Fifth Assessment Report of the Intergovernmental Panel on Climate Change*(pp. 485–533). Cambridge, United Kingdom and New York: Cambridge University Press.

Powlson, D.S., Whitmore, A.P. and Goulding, K.W. (2011). Soil carbon sequestration for mitigating climate change: Distinguishing the genuine from the imaginary. In Hillel, D. and Rosenzweig, C., (Eds.), *Handbook of Climate Change and Agroecosystems: Impacts, Adaptation, and Mitigation,* 393–402. London, UK: Imperial College Press.

Rosenzweig, C., J. Antle, and J. Elliott (2016), Assessing impacts of climate change on food security worldwide, Eos, 97, doi:10.1029/2016EO047387. Published on 9 March 2016.

Rosenzweig, C., and D. Hillel (Eds.), 2015: Handbook of Climate Change and Agroecosystems: The Agricultural Model Intercomparison and Improvement Project (AgMIP) Integrated Crop and Economic Assessments. ICP Series on Climate Change Impacts, Adaptation, and Mitigation Vol. 3. Imperial College Press, doi:10.1142/p970.

Rosenzweig, C., Jones, J.W., Hatfield, J.L., Ruane, A.C., Boote, K.J., Thorburn, P. and Winter, J.M. (2013). The Agricultural Model Intercomparison and Improvement Project (AgMIP): Protocols and pilot studies. *Agricultural and Forest Meteorology,* 170, 166–182. doi:http://dx.doi.org/10.1016/j. agrformet.2012.09.011.

Ruane, A.C., Winter, J.M., McDermid, S.P. and Hudson, N.I. (2015). AgMIP Climate Data and Scenarios for Integrated Assessment. In Hillel, D. and Rosenzweig, C. (Eds.), *Handbook of Climate Change and Agroecosystems,* 45–78. London, UK: Imperial College Press.

San Francisco Food Systems. (2005). 2005 San Francisco Collaborative Food System Assessment (pp. 96). URL: http://www.sfgov3.org/Modules/ShowDocument.aspx?documentid=780.

Sinclair, T. R. (2011), Precipitation: The Thousand-Pound Gorilla in Crop Response to Climate Change, in *Handbook of Climate Change and Agroecosystems: Impacts, Adaptation, and Mitigation,* edited by C. Rosenzweig and D. Hillel, pp. 179-190, Imperial College Press.

Tao, F. and Lin, E. (2015). AgMIP Regional Activities in a Global Framework: The China Experience. In Hillel, D. and Rosenzweig, C. (Eds.), *Handbook of Climate Change and Agroecosystems: The Agricultural Model Intercomparison and Improvement Project Integrated Crop and Economic Assessments, Part 2,* 375–390. London, UK: Imperial College Press.

Thorburn, P.J., Boote, K.J., Hargreaves, J.N., Poulton, P.L. and Jones, J.W. (2015). Cropping Systems Modeling in AgMIP: A New Protocol-Driven Approach for Regional Integrated Assessments. In Hillel, D. and Rosenzweig, C. (Eds.), *Handbook of Climate Change and Agroecosystems: The Agricultural Model Intercomparison and Improvement Project (AgMIP) Integrated Crop and Economic Assessments — Joint Publication with American Society of Agronomy, Crop Science Society of America, and Soil Science Society of America (In 2 Parts), 3,* 79. London, UK: Imperial College Press.

WHO. (2015). Global and Regional Food Consumption Patterns and Trends. *Nutrition Health Topics.* Retrieved from World Health Organization website: http://www.who.int/nutrition/topics/3_foodconsumption/en/.

Valdivia, R.O., Antle, J.M., Rosenzweig, C., Ruane, A.C., Vervoort, J., Ashfaq, M. and Nhemachena, C. (2015). Representative agricultural pathways and scenarios for regional integrated assessment of climate change impact, vulnerability and adaptation. In Hillel, D. and Rosenzweig, C. (Eds.), *Handbook of Climate Change and Agroecosystems,* 101–145. London, UK: Imperial College Press.

Yadav, S.S., Redden, R., Hatfield, J.L., Lotze-Campen, H., and Hall, A.J. (2011). *Crop adaptation to climate change.* New Jersey, USA: John Wiley & Sons.

Slide 1

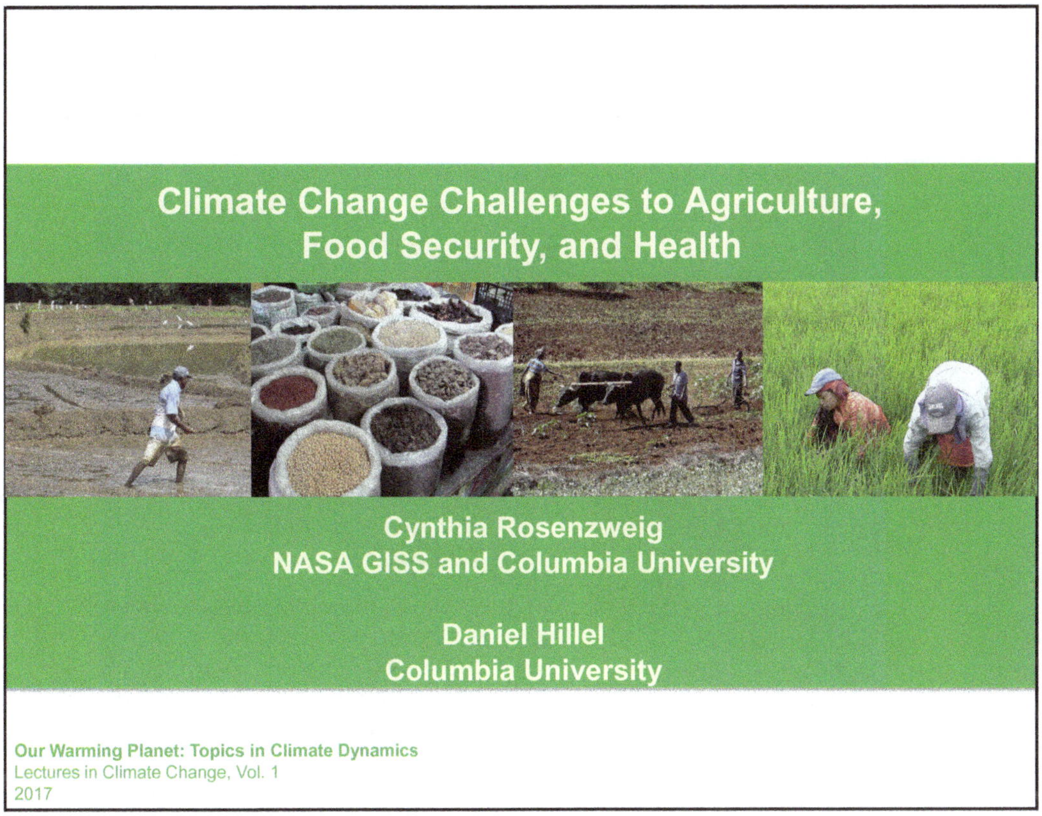
Climate Change Challenges to Agriculture,
Food Security, and Health
Cynthia Rosenzweig
NASA GISS and Columbia University
Daniel Hillel
Columbia University
Our Warming Planet: Topics in Climate Dynamics
Lectures in Climate Change, Vol. 1
2017

Slide 2

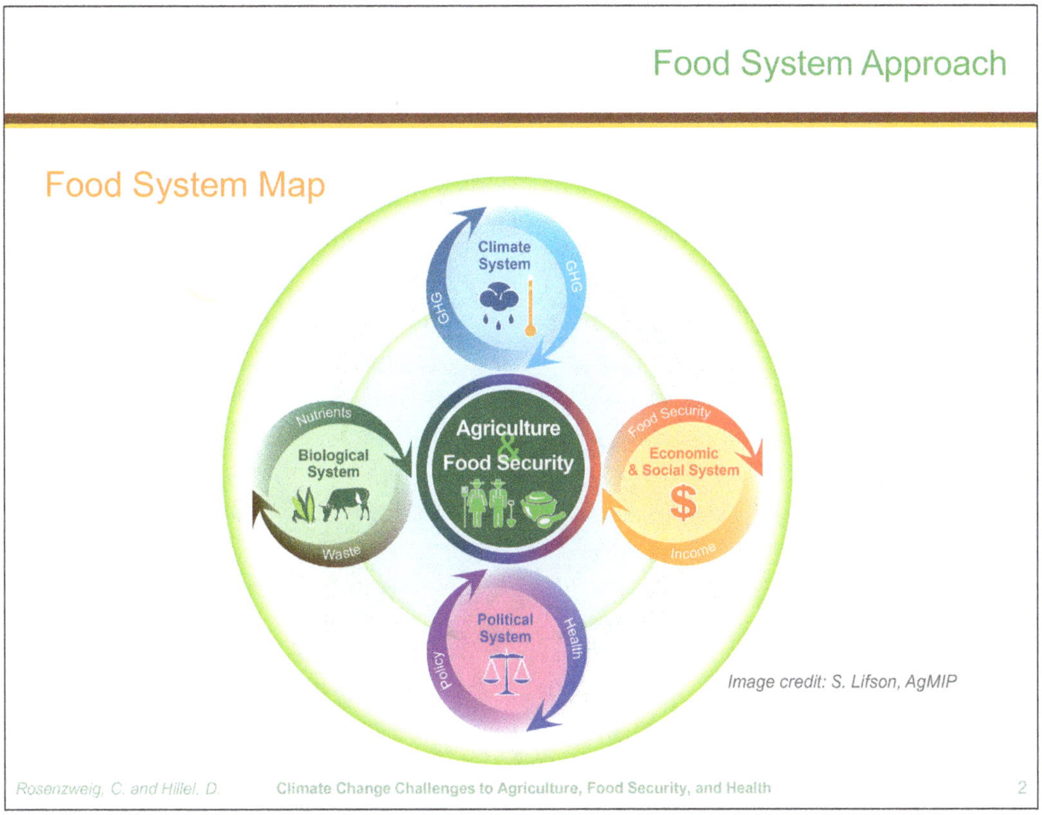
Food System Approach
Food System Map
Climate System
GHG
GHG
Nutrients
Biological System
Waste
Agriculture
&
Food Security
Food Security
Economic & Social System
$
Income
Political System
Policy
Health
Image credit: S. Lifson, AgMIP
Rosenzweig, C. and Hillel, D.
Climate Change Challenges to Agriculture, Food Security, and Health
2

Slide 3

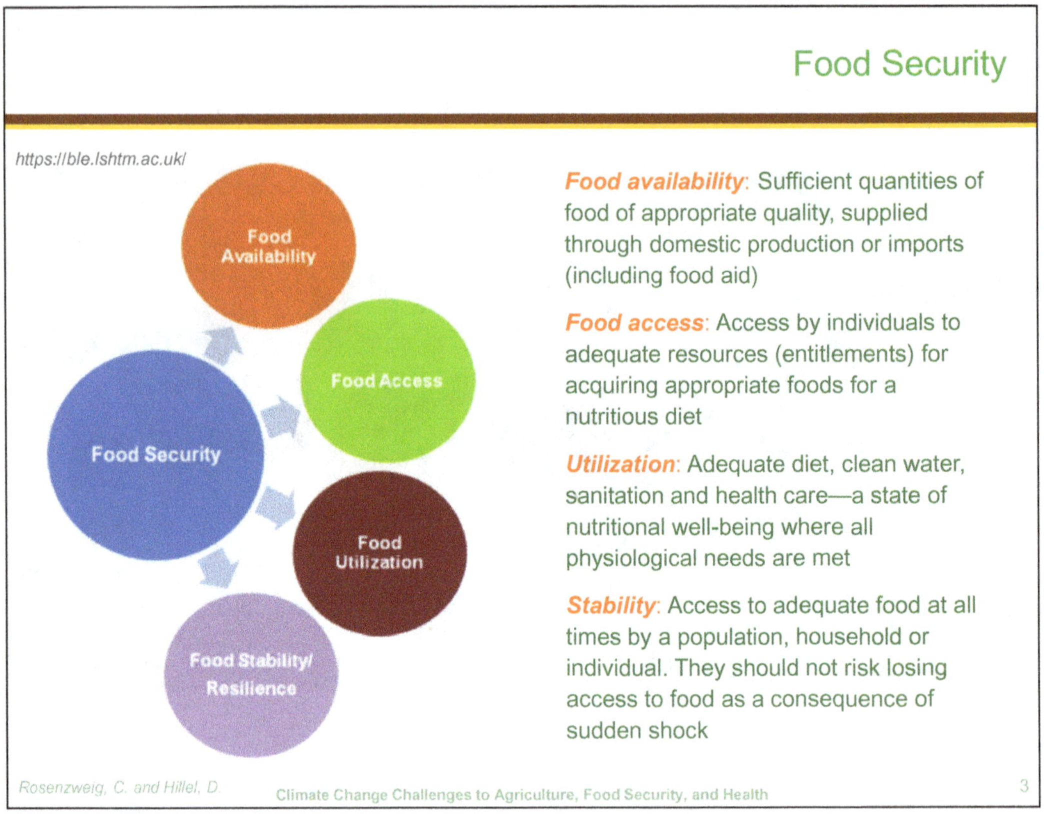
Food Security
https://ble.lshtm.ac.uk/
Food Availability
Food Access
Food Security
Food Utilization
Food Stability/ Resilience
Food availability: Sufficient quantities of food of appropriate quality, supplied through domestic production or imports (including food aid)
Food access: Access by individuals to adequate resources (entitlements) for acquiring appropriate foods for a nutritious diet
Utilization: Adequate diet, clean water, sanitation and health care—a state of nutritional well-being where all physiological needs are met
Stability: Access to adequate food at all times by a population, household or individual. They should not risk losing access to food as a consequence of sudden shock
Rosenzweig, C. and Hillel, D.
Climate Change Challenges to Agriculture, Food Security, and Health
3

Slide 4

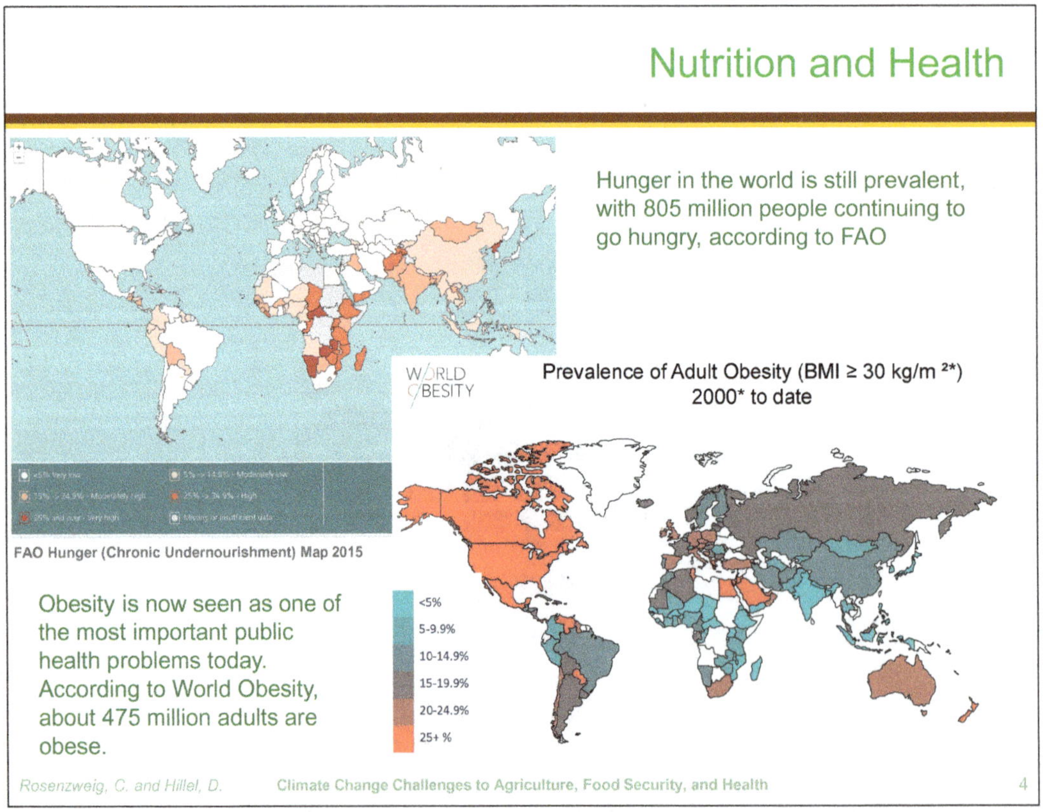
Nutrition and Health
Hunger in the world is still prevalent, with 805 million people continuing to go hungry, according to FAO
FAO Hunger (Chronic Undernourishment) Map 2015
WORLD OBESITY
Prevalence of Adult Obesity (BMI ≥ 30 kg/m 2*)
2000* to date
Obesity is now seen as one of the most important public health problems today. According to World Obesity, about 475 million adults are obese.
<5%
5-9.9%
10-14.9%
15-19.9%
20-24.9%
25+ %
Rosenzweig, C. and Hillel, D.
Climate Change Challenges to Agriculture, Food Security, and Health
4

Slide 5

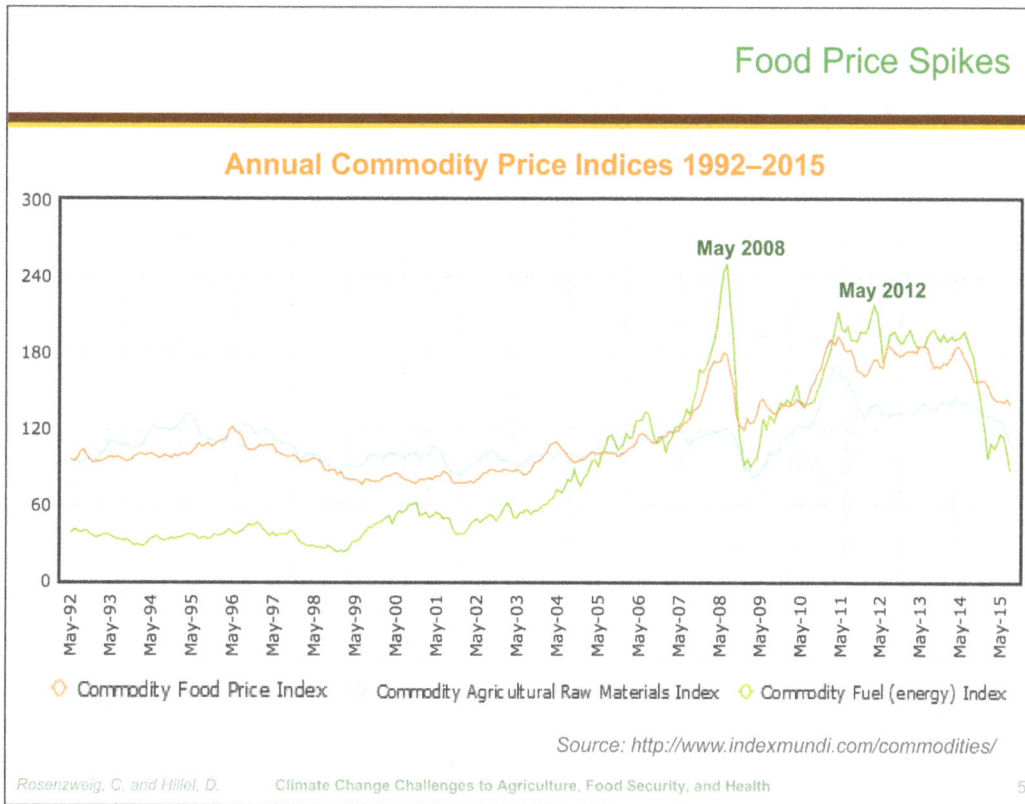

Food Price Spikes

Annual Commodity Price Indices 1992–2015

May 2008

May 2012

Commodity Food Price Index   Commodity Agricultural Raw Materials Index   Commodity Fuel (energy) Index

Source: http://www.indexmundi.com/commodities/

Rosenzweig, C. and Hillel, D.   Climate Change Challenges to Agriculture, Food Security, and Health   5

Slide 6

Current Climate Stresses

Corn shows the effect of the drought in Texas in August 2013
U.S. Department of Agriculture/Bob Nichols

California's accumulated precipitation 'deficit' from 2012 to 2014
NASA/Goddard Scientific Visualization Studio

Satellite image of flooding in Burma (Myanmar) August 2015
NASA Earth Observatory images by Joshua Stevens

Severe flooding in August 2015 in Pakistan wipes out harvest
Image:Rueters/London

Rosenzweig, C. and Hillel, D.   Climate Change Challenges to Agriculture, Food Security, and Health   6

Slide 7

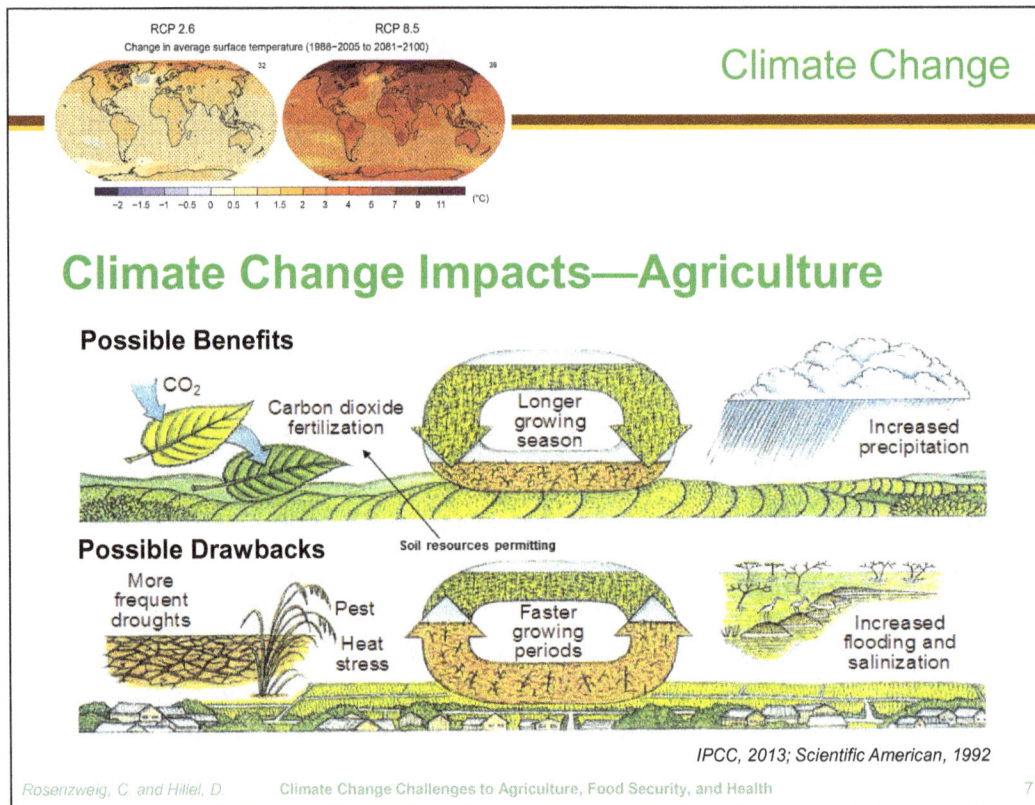

Slide 8

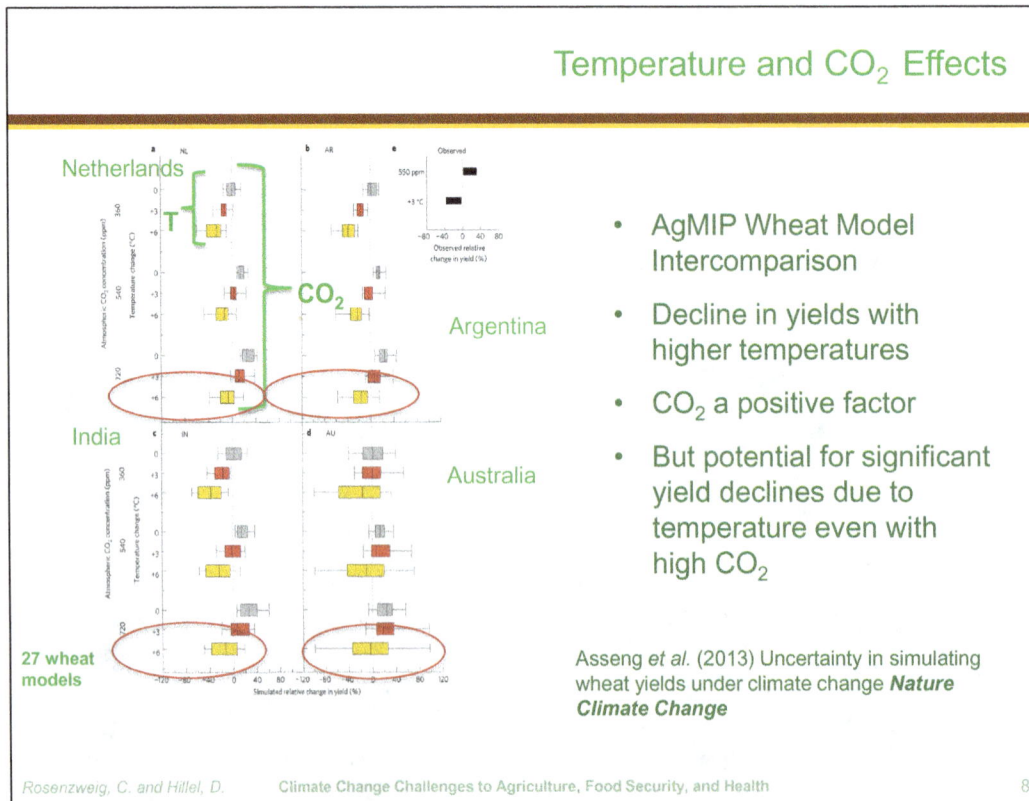

Slide 9

Yield Response to Water

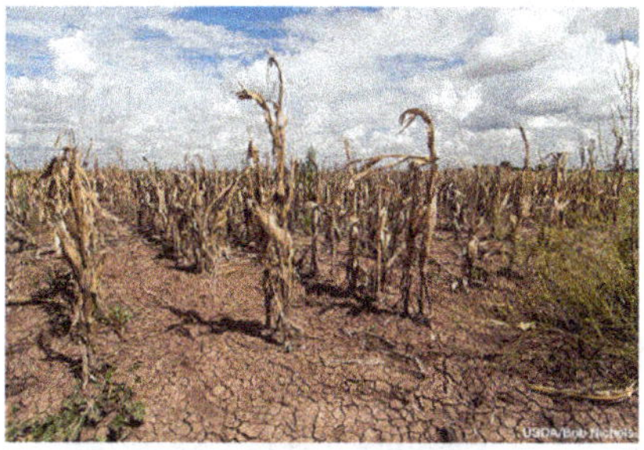

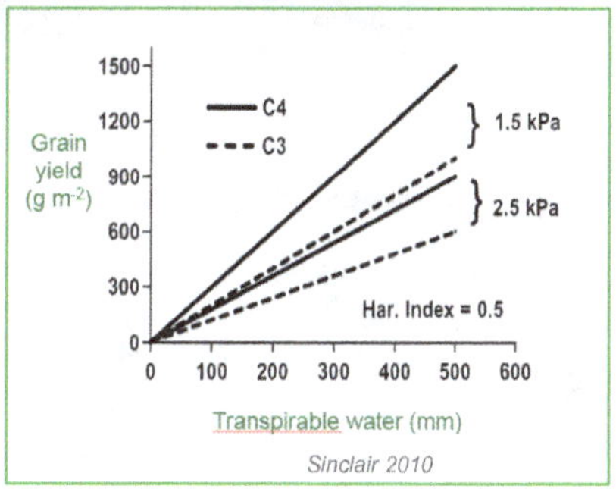

Maximum grain yield plotted as a function of the amount of transpirable soil water available through the growing season. Two vapor pressure deficit environments are presented.

- Crops need water—through precipitation or irrigation
- Drought stress affects yield during critical growth periods
- Excess water can be damaging as well
- C4 crops – e.g., corn and sorghum – favored at both higher and lower water stress

Slide 10

Effects on Soils

Carbon Cycling in Agroecosystems

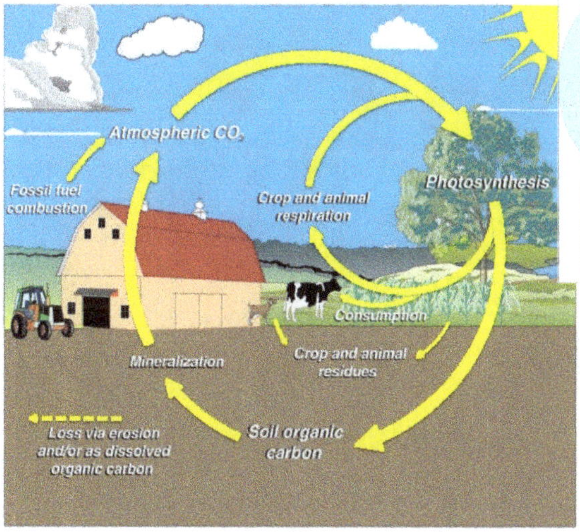

A.Y.Y. Kong et al.

Global Soil Regions

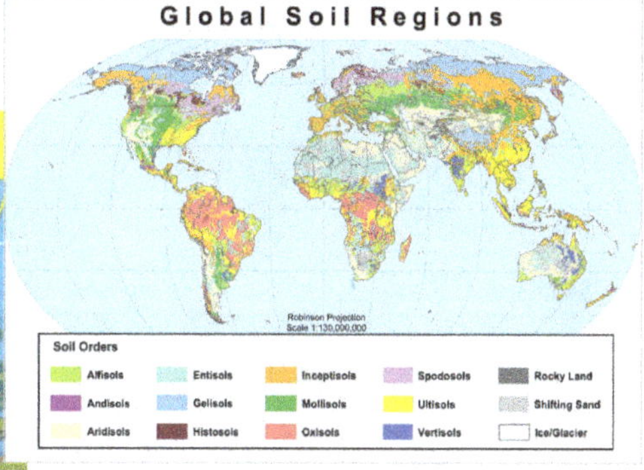

Source: USDA

- High temperature effects on soil biological and chemical processes
- Water-logging from increased heavy downpours

Slide 11

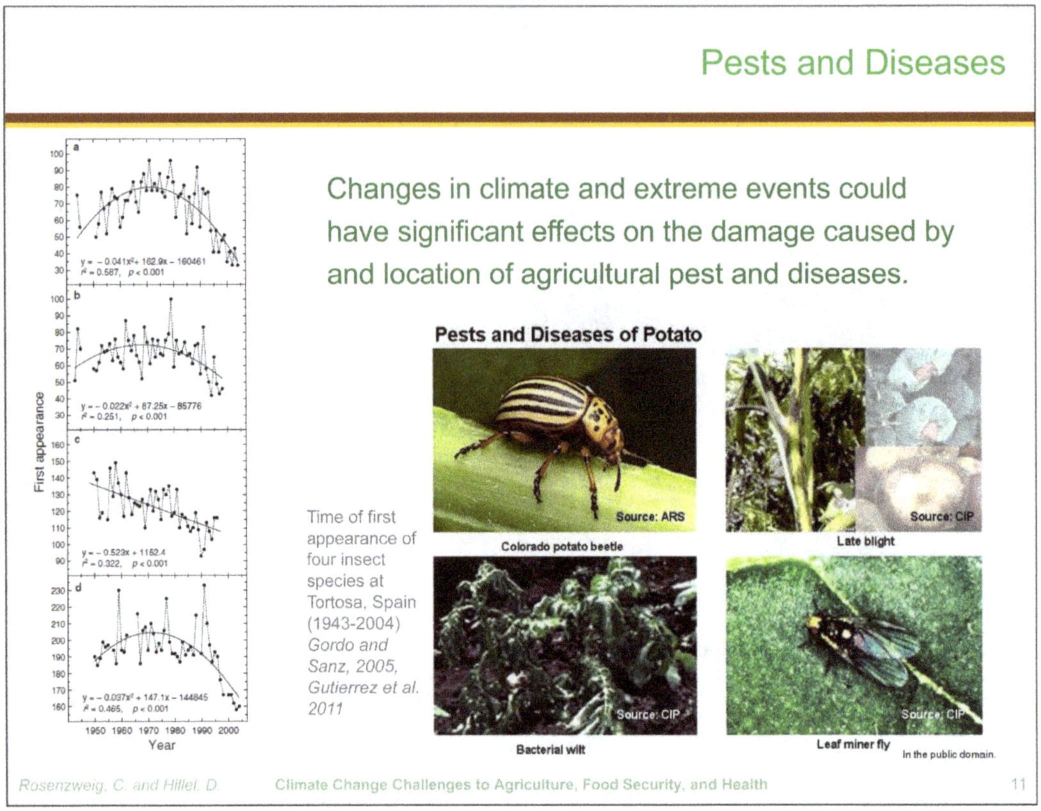

Slide 12

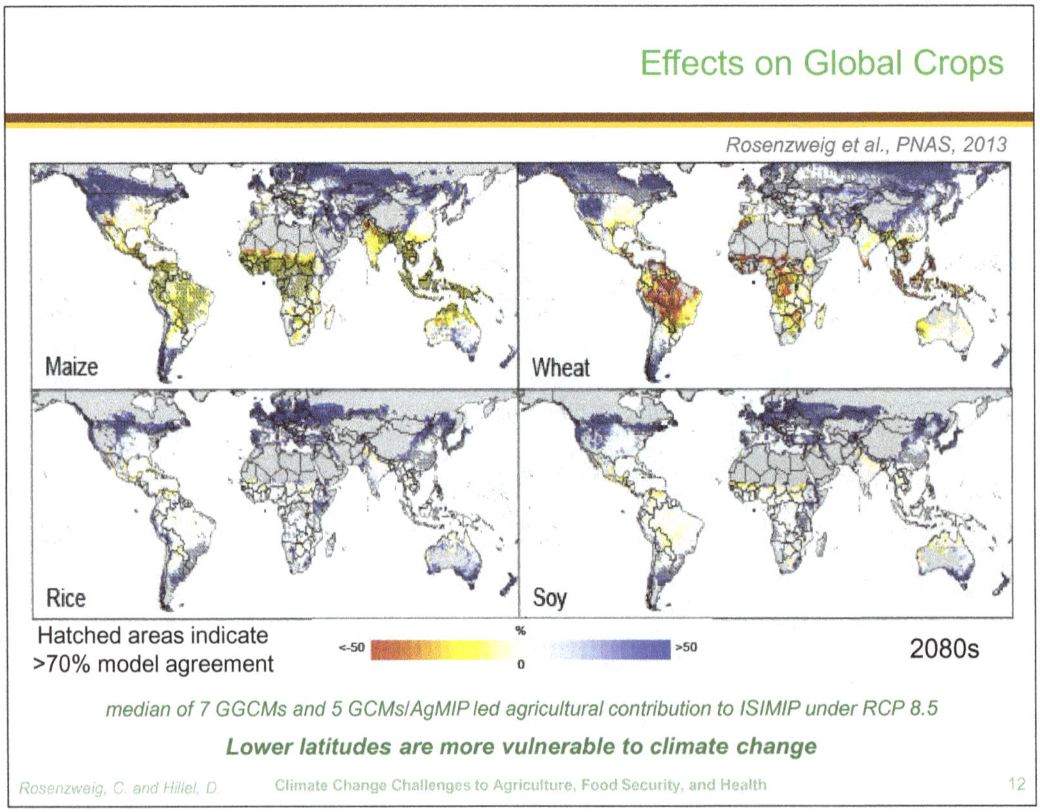

Slide 13

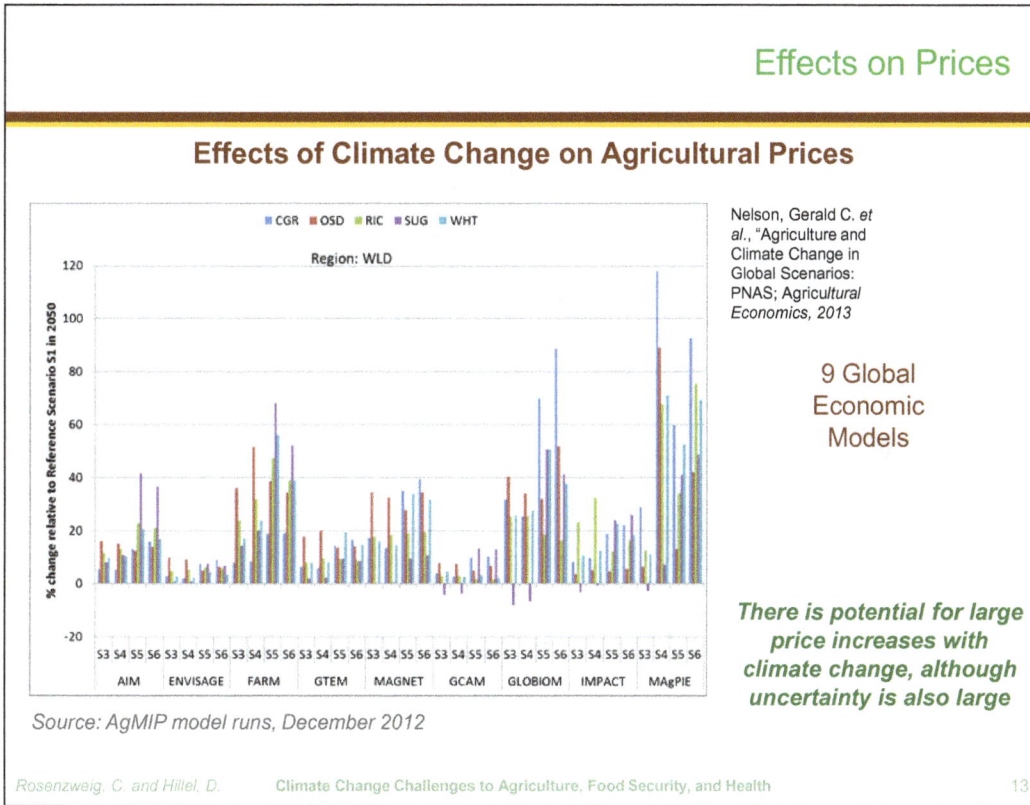

Effects on Prices

Effects of Climate Change on Agricultural Prices

Nelson, Gerald C. *et al.*, "Agriculture and Climate Change in Global Scenarios: PNAS; *Agricultural Economics, 2013*

9 Global Economic Models

There is potential for large price increases with climate change, although uncertainty is also large

Source: AgMIP model runs, December 2012

Rosenzweig, C. and Hillel, D.   Climate Change Challenges to Agriculture, Food Security, and Health   13

Slide 14

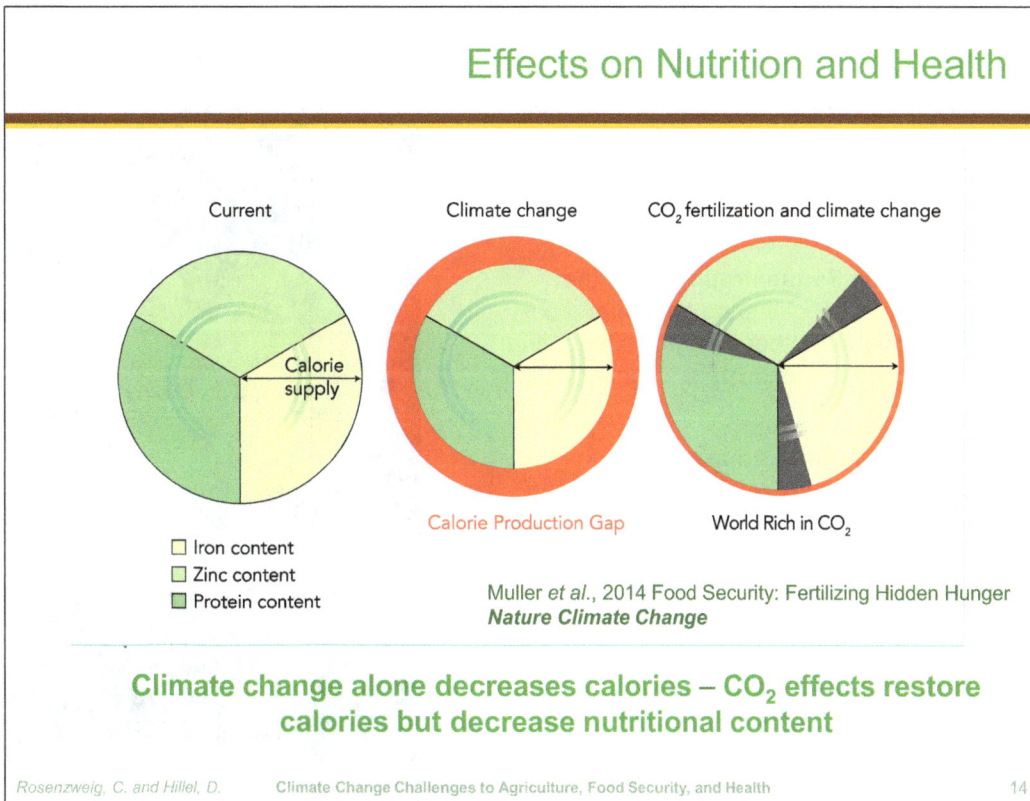

Effects on Nutrition and Health

Current

Climate change

CO_2 fertilization and climate change

Calorie supply

Calorie Production Gap

World Rich in CO_2

☐ Iron content
☐ Zinc content
☐ Protein content

Muller *et al.*, 2014 Food Security: Fertilizing Hidden Hunger
Nature Climate Change

Climate change alone decreases calories – CO_2 effects restore calories but decrease nutritional content

Rosenzweig, C. and Hillel, D.   Climate Change Challenges to Agriculture, Food Security, and Health   14

Slide 15

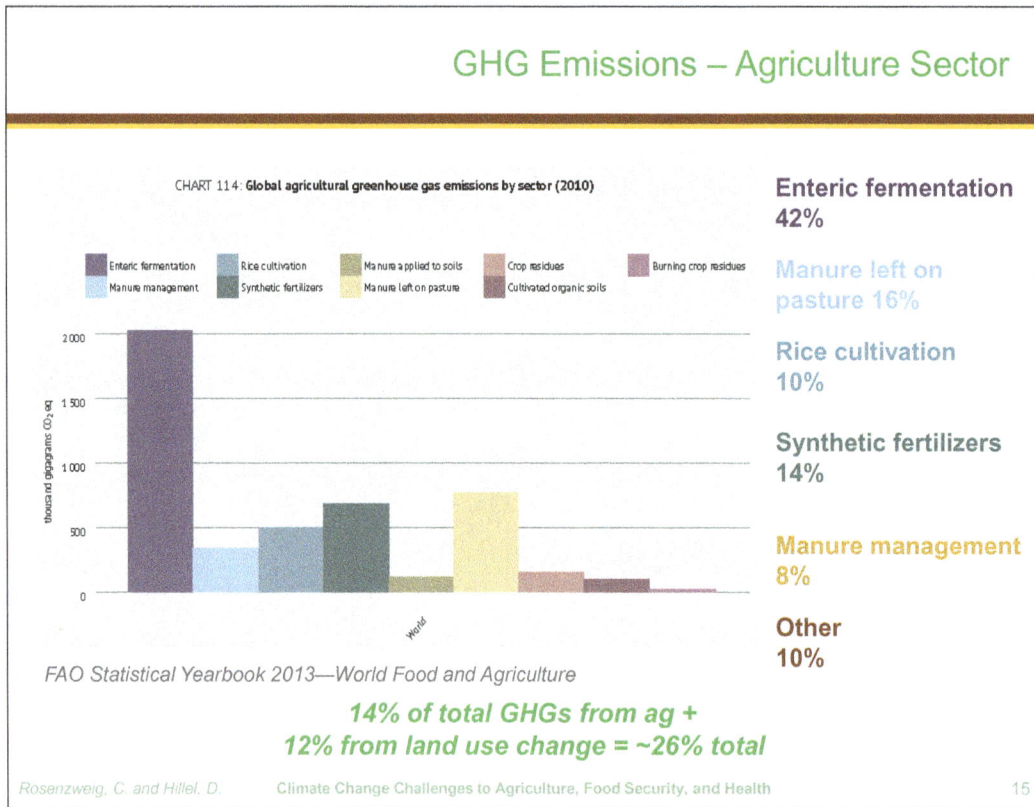

GHG Emissions – Agriculture Sector

CHART 11.4: **Global agricultural greenhouse gas emissions by sector (2010)**

- Enteric fermentation
- Manure management
- Rice cultivation
- Synthetic fertilizers
- Manure applied to soils
- Manure left on pasture
- Crop residues
- Cultivated organic soils
- Burning crop residues

Enteric fermentation 42%

Manure left on pasture 16%

Rice cultivation 10%

Synthetic fertilizers 14%

Manure management 8%

Other 10%

FAO Statistical Yearbook 2013—World Food and Agriculture

*14% of total GHGs from ag +
12% from land use change = ~26% total*

Rosenzweig, C. and Hillel, D.   Climate Change Challenges to Agriculture, Food Security, and Health   15

Slide 16

Climate Change Mitigation

Agricultural solutions for mitigating climate change

Soil Carbon Sequestration—has the possibility of sequestering carbon at a cost competitive with other methods of reducing atmospheric greenhouse gas concentrations. *(Center for Climate and Energy Solutions)*

Biomass Carbon Sequestration— Agroforestry can play a role in carbon sequestration and mitigating CO_2 emissions. *(USDA)*

Biofuels—With the exception of ethanol made from corn, use of biomass to produce heat, power, or fuels can result in zero or close to zero CO_2 emissions. *(Center for Climate and Energy Solutions)*

Rosenzweig, C. and Hillel, D.   Climate Change Challenges to Agriculture, Food Security, and Health   16

Slide 17

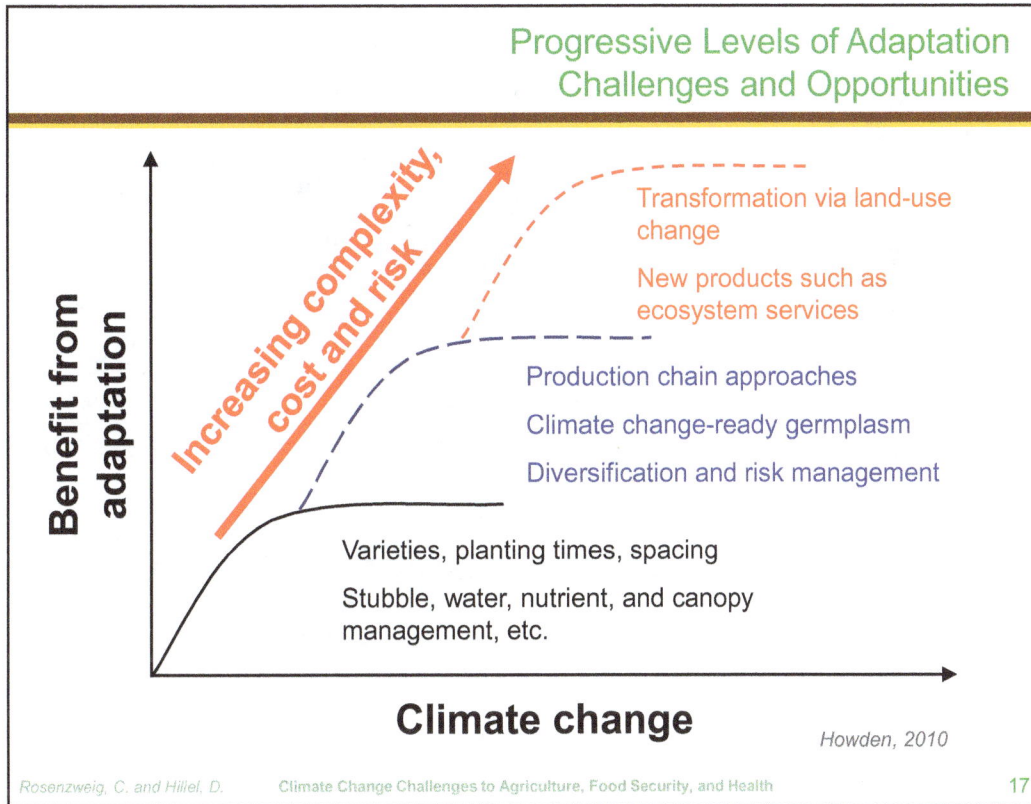

Slide 18

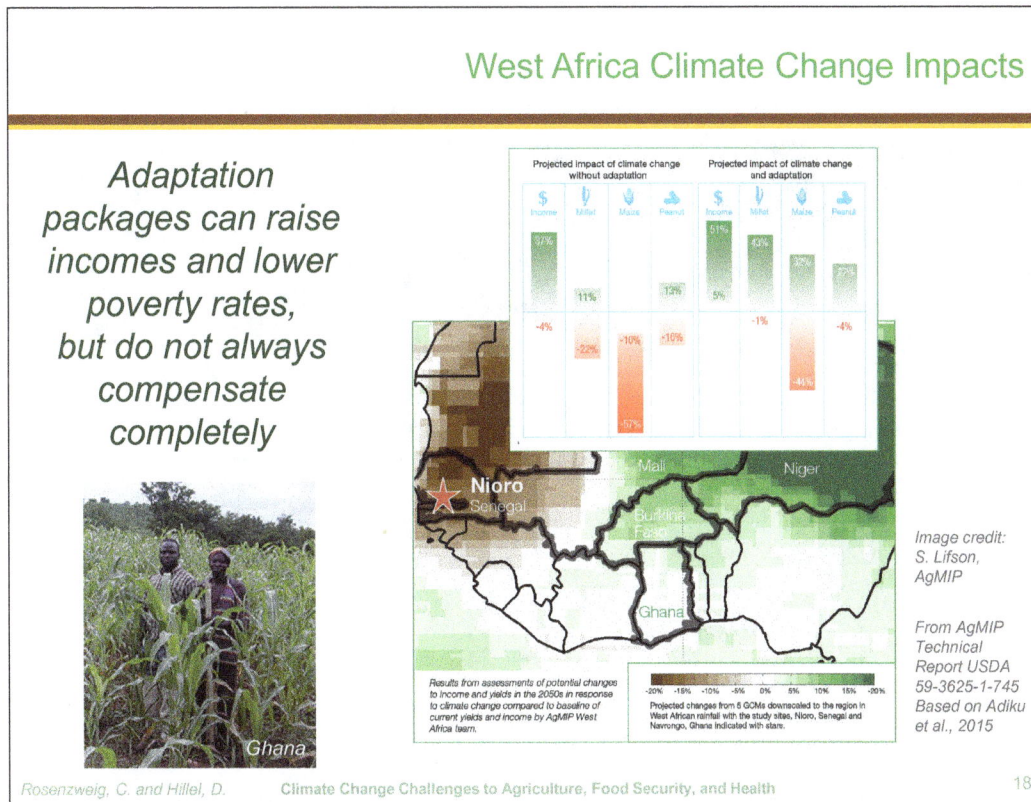

Slide 19

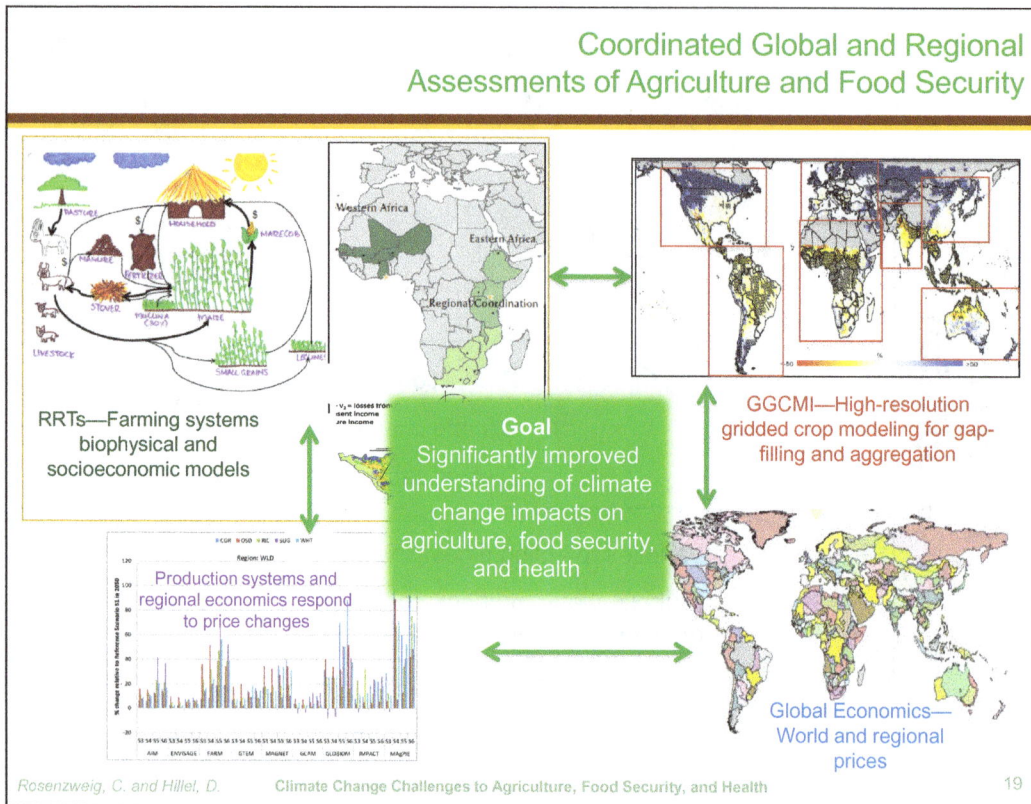

Slide 20

Slide Notes

Slide 1   We have learned from our brilliant and dedicated colleague, David Rind, that the already-evident and further-anticipated process of anthropogenic climate change raises worldwide concerns regarding agricultural production, food security, and public health.

Slide 2   The food system has four compass points: biophysical, social, environmental, and economic. The IPCC Fifth Assessment Report states that 'A food system is all processes and infrastructure involved in satisfying a population's food security, that is, the gathering/catching, growing, harvesting (production aspects), storing, processing, packaging, transporting, marketing, and consuming of food, and disposing of food waste (non-production aspects). It includes food security outcomes of these activities related to availability and utilization of, and access to, food as well as other socioeconomic and environmental factors' (IPCC, 2014b; Ericksen, 2008; Ericksen *et al.*, 2010; Ingram, 2011).

Within the food system, there are a number of intersecting subsystems. Supply is provided by farmers, with production affected by climate and soils as well as by genetics and management. Demand arises from consumers, and depends on dietary choices affected by social and cultural factors. The political system influences food via land use regulations, subsidies, and trade policies.

Slide 3   At the 1974 World Food Conference, the term 'food security' was defined with an emphasis on supply: *Food security* is the 'Availability at all times of adequate world food supplies of basic foodstuffs to sustain a steady expansion of food consumption and to offset fluctuations in production and prices.' The World Food Summit of 1996 defined food security as existing 'when all people at all times have access to sufficient, safe, nutritious food to maintain a healthy and active life'. This definition was again refined in The State of Food Insecurity 2001: 'Food security [is] a situation that exists when all people, at all times, have physical, social and economic access to sufficient, safe and nutritious food that meets their dietary needs and food preferences for an active and healthy life.'

Food availability: The availability of sufficient quantities of food of appropriate quality, supplied through domestic production or imports (including food aid).

Food access: Access by individuals to adequate resources (entitlements) for acquiring appropriate foods for a nutritious diet. Entitlements are defined as the set of all commodity bundles over which a person can establish command given the legal, political, economic, and social arrangements of the community in which they live (including traditional rights such as access to common resources).

Utilization: Utilization of food through adequate diet, clean water, sanitation, and health care to reach a state of nutritional well-being where all physiological needs are met. This brings out the importance of non-food inputs in food security.

Stability: To be food secure, a population, household or individual must have access to adequate food at all times. They should not risk losing access to food as a consequence of sudden shocks (e.g., an economic or climatic crisis) or cyclical events (e.g., seasonal food insecurity). The concept of stability can therefore refer to both the availability and access dimensions of food security.

Source: http://www.nourishlife.org/teach/food-system-tools/

Slide 4   Hunger in developing countries has improved since 1990 by 39%, according to the 2014 IFPRI Global Hunger Index. However, hunger in the world is still 'serious', with 805 million people continuing to go hungry, according to estimates by the Food and Agriculture Organization of the United Nations. Levels of hunger are 'extremely alarming' or 'alarming' in 16 countries, according to the 2014 GHI.

According to World Obesity, the epidemic of obesity is now recognized as one of the most important public health problems facing the world today. Adult obesity is more common globally than under-nutrition. There are around 475 million obese adults with over twice that number overweight — that means around 1.5 billion adults. Over 200 million school-age children are overweight, making this generation the first predicted to have a shorter lifespan than their parents.

http://www.worldobesity.org/aboutobesity/

Slide 5   There are food price shocks that have reverberations in many locations such as food riots and social disruption. Climate change may exacerbate the frequency and intensity of shocks to the food system.

Food price spikes in 2008 and to a lesser extent in 2011 and 2012 demonstrated that the steady decline in prices since the 1950s may not continue in the future. Sharp increases in food prices triggered riots in 30 countries. The food price spikes have been associated primarily with high energy prices and commodity trading, but climate factors played a role in some regions.

Slide 6   1.   In 2013, the state of Texas suffered one of the worst droughts in its history, causing significant economic damage to farmers and ranchers (top left).
2.   California's accumulated precipitation deficit from 2012 to 2014 shown as a percent change from the 17-year average based on NASA TRMM multi-satellite observations (top right).
3.   Satellite image of flooding in Burma (Myanmar) August 2015: excess water has harmed crops, displaced thousands of people, and led to deaths (bottom left).
4.   Flash flooding caused by torrential monsoon rains killed at least 28 in Pakistan in 2015. Ready-to-harvest major and minor crops, including cotton, paddy rice, and vegetables worth almost Rs85 billion, were completely washed out by rain and floods in Sindh (bottom right).

Farmers have dealt with climatic fluctuations since the advent of agriculture. Improving strategies for dealing with present climate extremes such as droughts, floods, and heat waves is an important way to prepare for climate change. Recent examples of droughts and floods that have affected agriculture include drought in California in 2015, drought in the US Mid-West in 2013–2014, floods in Europe in 2013, drought in Africa in 2016, floods in Thailand in 2011, and drought in Australia in 2006. A major food security concern is the concurrent occurrences of droughts and floods causing multiple breadbasket failures and major disruption to the world food system.

Recent flooding and heavy precipitation events in the US and worldwide have caused great damage to crop production. If the frequency of these weather extremes increase in the near future, as recent trends for the US indicate and as projected by global climate models, the costs of crop losses in the coming decades could rise dramatically (Nelson *et al.*, 2014).

Slide 7   Image sources: IPCC, 2013; Bongaarts, J., Scientific American, 1992.

Global studies done to date show that negative and positive effects will occur both within countries and over the world (Rosenzweig *et al.*, 2013). In large countries such as the United States, Russia, Brazil, and Australia, agricultural regions likely will be affected quite differently. Some regions will experience increases in production and some will experience declines. At the international level, this implies possible shifts in comparative advantage for production of export crops. It also implies that adaptive responses to climate change will necessarily be complex and varied.

In any one location, possible benefits may include increased production due to the physiological effects of carbon dioxide on crop growth and water use, longer growing seasons in high-latitude or high-elevation where crop production is now limited by low temperatures, and increased precipitation in currently dry areas. On the other hand, potential negative effects include more frequent droughts and floods, heat stress, increasing incidence of pests and diseases, speed-up of crop phenology due to higher temperatures, and extended salination in agricultural regions near the coast.

Climate change projections are fraught with uncertainty regarding both the rate and magnitude of temperature and precipitation variations in the coming decades. This uncertainty arises from a lack of precise knowledge of how climate system processes will change and of how population growth, economic and technological developments, and land-use patterns will evolve in the coming century.

To deal with these uncertainties, a scenario approach is often taken, which utilizes multiple climate models, multiple agricultural models, and several development pathways. The latter include representative agricultural pathways (RAPs). These are used for agricultural model intercomparison, improvement, and impact assessments and are consistent with the global pathways and scenarios used for a wide range of climate impact and vulnerability assessments (Valdivia *et al.*, 2015).

Slide 8   Simulated relative mean (30-year average, 1981–2010) grain yield change for increased temperatures (no change, gray; +3°C, red; +6°C, yellow) and elevated atmospheric CO_2 concentrations for the Netherlands (NL; **a**), Argentina (AR, **b**), India (IN; **c**), and Australia (AU; **d**). For each box plot, vertical lines represent, from left to right, the 10th percentile, 25th percentile, median, 75th percentile, and 90th percentile of simulations based on multiple models. Small graph shows observed range of yield impacts with elevated CO_2 and observed range of yields impacts with increased temperature (extrapolated, based on separate experiments with 40–345 ppm elevated CO_2 and 1.4–4.0°C temperature increases). Source: Asseng *et al.* (2013).

The AgMIP Wheat Model Intercomparison Team has tested the sensitivity of 27 wheat models to varying levels of temperature and carbon dioxide at four diverse sites around the world (Asseng *et al.*, 2013). They found a substantial decline in yields with higher temperature and an increase in yields with higher CO_2 levels. However, there were considerable intermode differences.

Slide 9   The graph shows maximum grain yield plotted as a function of the amount of transpirable soil water available through the growing season. The two vapor pressure deficit environments represent higher and lower water stress. C_4 crops are higher yielding in both environments.

To produce viable yields, crops need water either by precipitation or by irrigation. Drought stress affects yield during critical growth periods. Excess water can be damaging as well, through the processes of water-logging and lodging.

Slide 10   Sources: Map of Global Soil Regions, USDA, Hillel and Rosenzweig, 2011; Carbon Cycle in Agroecosystems, Kong *et al.*, 2011

The soils of the world, with the biota they support, are major absorbers, depositories, and releasers of organic carbon. Soils altogether contain an estimated 1700 Gt (billion metric tons) to a depth of 1 m and as much as 2400 Gt to a depth of 2 m (Hillel and Rosenzweig, 2011). The balance of soil carbon is greatly influenced by human management, including the clearing or restoration of natural vegetation and the patterns of land use. Cultivation spurs the microbial decomposition of soil organic matter while depriving it of replenishment, especially if the cropping system involves removal of plant matter (leaving but little organic residue in the field) and if the soil is fallowed (kept bare). Organic carbon is lost from soils both by oxidation and by erosion of topsoil.

Enrichment of the topsoil with organic matter makes it less prone to compaction, crust formation, and erosion; hence, it affects the quality of the environment. It also improves the function of the soil with respect to infiltration, aeration, seed germination, and plant nutrition. Conversion to no-till farming has been found to boost carbon storage in soils at rates varying from 0.1 to 0.7 Mg C ha^{-1} yr^{-1} (Hillel and Rosenzweig, 2011). However, such positive increments cannot be expected to continue indefinitely as any historically depleted soil will tend to approach its prior equilibrium (or C saturation) state within a few decades. Another issue is that the process is reversible. If the management practice that has led to an increase in soil organic carbon ceases, carbon will be released from the soil (Powlson, 2011).

A whole agroecosystem approach that considers the linked cycling of C, N, and P in soils is crucial to identifying soil management practices that truly contribute to mitigation of global climate change (Kong *et al.*, 2011).

Slide 11   Time of the first appearance of four insect species at Tortosa, Spain (1943–2004); Gordo and Sanz (2005); and Gutierrez *et al.* (2011).

Changes in temperature and rainfall patterns can alter pest species distribution, abundance, and damages in agricultural systems (Gutierrez *et al.*, 2011). Plant–herbivore–natural enemy interactions are likely to be affected. The effects on one trophic level may cascade to lower and/or higher levels in food chains and webs affecting system regulation and stability (Hairston *et al.*, 1960; Gutierrez *et al.*, 2011). Climate change could disrupt the synchrony between host plants and insect herbivores that have evolved to cope in current climate regimes (Gutierrez *et al.*, 2011).

Agricultural pests are responding to the climate changes that are already occurring. For example, Gordo and Sanz (2005) found that honey bees, cabbage white butterflies, potato beetle, and olive flies have been appearing earlier in the year in Spain since 1943.

Slide 12   Seven global gridded crop models were tested with five global climate model scenarios. For the first time, crop model uncertainty was characterized; hatched areas indicate where more than 70% of the model results were in agreement. Results showed that cropping regions in lower latitudes, where most of the developing countries are located, are more vulnerable to climate change than regions in higher latitudes where the more developed countries are situated (Rosenzweig *et al.*, 2013). Low-income countries have been shown to be most vulnerable, but there are risks for all agricultural regions. In contrast to previous assessments, AgMIP global gridded crop model results with realistic nitrogen fertilization showed steadily decreasing yields for wheat, maize, and soybean in mid- and high-latitude regions, even for small temperature increases.

When results are grouped by GGCMs with and without explicit nitrogen stress, results are substantially more negative with realistic nitrogen fertilization than without. GGCMs with realistic nitrogen fertilization capture more enhanced dynamics of crop growth and yield interactions with CO_2 fertilization. There is a great need for continuing rigorous model evaluation and improvement, and now over 15 global gridded crop models are testing historical climate sensitivity and climate change simulations.

Model agreement on the direction of yield changes is found in many major agricultural regions at both low and highlatitudes; however, reducing uncertainty in sign of response in mid-latitude regions remains a significant challenge.

Crop models simulate dynamic interactions among plant, soil, atmosphere, and management components. They are nonlinear, dynamic mathematical functions that describe the growth and yield of a crop and the changes in soil water and nutrients as affected by management and daily weather conditions (Jones *et al.*, 2011). Crop models represent the processes affecting growth and yield, in particular site factors (weather and soils), management (sowing date, fertilization, and cultivation), and crop genetics (crop cultivar) (Thorburn *et al.*, 2015).

The general mathematical form of a dynamic crop model is from Jones *et al.* (2011):

$\dot{x}(t)$ = $f(p_1, p_2, w(t), m(t), x(t))$, where
t = time, which goes from 0 at planting to T_F at crop maturity
$x(t)$ = vector of crop and soil state variables at time t
$\dot{x}(t)$ = vector of rates of changes of the state variables at time t
f = nonlinear set of equations defining dynamic physiological, physical, and chemical processes in plant and soil state variables

p_1 = vector of crop and cultivar-specific parameters of equations in f
p_2 = vector of soil physical and chemical parameters of equations in f
$w(t)$ = time-varying weather inputs to the model
$m(t)$ = time-varying management inputs to the model

Slide 13   Climate change is projected to exert upward pressure on agricultural prices, but with large uncertainty (Nelson *et al.*, 2014). Economic responses reduce yield loss, increase crop area, and reduce consumption. Economic models differ primarily in their assumptions regarding ease of land-use conversion, intensification, and trade.

S1 is the reference scenario, with no climate change. The S3–S6 scenarios are combinations of two GCMS with two crop models.

Slide 14 CO_2 levels will positively affect plant growth (especially in C_3 crops such as wheat, rice, and soy) and make up for some of the calorie lost due to climate change (red circles); it will also negatively affect the nutritional value of crops, reducing the amounts of essential minerals, such as iron and zinc, and protein (grey wedges).

Slide 15   Second only to energy generation, agriculture is a major source of greenhouse gas emissions, accounting for about 14% of the total. When land use changes primarily from land clearing for food production is added, the total represents about 26% of total greenhouse gas emissions. Of agricultural emission sources, enteric fermentation from ruminants accounts for 42%; manure left on pasture accounts for 16%; synthetic fertilizers, especially the production of nitrogen amendments accounts for 14%; release of methane (CH_4) from rice cultivation accounts for 10%; manure management in confined systems accounts for 8%; and other activities account for about 10%. Source: Muller *et al.*, 2014.

Slide 16   **Soil Carbon Sequestration** — Recent model studies of the economic potential for soil and forestry sequestration show that agriculture can sequester carbon at a cost competitive with other methods of reducing GHG emissions. http://www.c2es.org

Biomass Carbon Sequestration — Agroforestry can play a role in carbon sequestration and mitigating CO_2 emissions.

Biofuels — With the exception of ethanol made from corn, use of biomass to produce heat, power, or transportation fuels essentially results in zero or close-to-zero CO_2 emissions. However, using land for agroforestry and biofuels may compete with food production.

Slide 17   Three successive levels of agricultural adaptation may be distinguished (Howden, 2010). The first is often called 'autonomous adaptation', because no external interventions, such as new government policies or induced technologies, are needed (Easterling, 2011). Such first-level adaptations include changes in varieties, planting time, and spacing, as well as modifications to stubble, water, nutrient, and canopy management. The second tier of adaptations involves a broader set of interactions that may require new government policies or induced technologies. Second-tier adaptations include modifications to production chains, development of temperature and drought-tolerant climate-change ready germplasm, and diversification and risk management techniques such as new crop insurance programs. The third tier of adaptation is transformational, involving land-use change and/or development of new products such as ecosystem services.

Slide 18 Results from the regional integrated assessments in West Africa show that even with mixed rainfall projections and positive CO_2 effects, climate change adds pressure to many current smallholder agricultural systems (Adiku *et al.*, 2015; Hillel and Rosenzweig, 2015). Adaptation packages can raise incomes and lower poverty rates, but do not always compensate completely. Continued agricultural development, if manifested in sustainable trends, can help to ameliorate negative responses to climate-change stresses. Each smallholder farming system contains some vulnerable farmers and areas for which interventions will be necessary.

Slide 19 Source: Rosenzweig *et al.*, 2016. The AgMIP methods for regional integrated assessment take a farming systems approach that is transdisciplinary across physical, biophysical, and socioeconomic processes. Elements are simulated at field, farm, region, and global scales, utilizing multiple climate and crop models. Results are distributional, showing impacts on farmers' livelihoods and poverty rates.

Projections of future climate are uncertain due to lack of knowledge on how society and technology will develop, what climate change policies will be implemented, and how the climate system will evolve. Climate scenarios are useful tools for testing how agricultural systems may be affected by future climate change. Such tools developed by the Agricultural Model Intercomparison and Improvement Project (AgMIP) utilize common data formats, assemble time-series of quality-controlled historical climate data, and generate future climate scenarios combining observed site-based data with down-scaled global climate model simulation results (Ruane *et al.*, 2015).

Building blocks for the research community are now in place for coordinated global and regional assessments of future food security (Rosenzweig *et al.*, 2016). Crops, livestock, sites, and processes are important to full understanding and credibility of global simulations. Multi-model assessments have shown that focus on improving rigor of impact assessments is highly warranted. Regional integrated assessments of farming systems are needed to identify vulnerable groups and regions. At the global scale, crop, livestock, and economics models help to understand the implications of climate change in regard to mitigation, adaptation, nutrition, and livelihoods, and other significant parts of the global food system.

Slide 20 The ultimate significance of the climate change issue for agriculture is related to its global reach, affecting regions, food security, and health throughout the world in complex and interactive ways. Farmers and others in the agricultural sector are thus faced with the quadruple task of contributing to global reductions of carbon dioxide and other greenhouse gas emissions, adapting to an already-changing climate, delivering healthy nutritious food, and sustainably managing soil and water resources. In short, agriculture is called upon to deliver both human and planetary health.

CLIMATE LECTURE 19

Chemistry–Climate Interactions in a Changing Environment: Wildfire in the West and the US Warming Hole

Loretta J. Mickley

Harvard University, Cambridge, MA, USA

Loretta Mickley co-leads the Atmospheric Chemistry Modeling Group at Harvard. Her research focuses on chemistry–climate interactions in the troposphere. She seeks to understand how short-lived gases and particles affect climate and how climate, in turn, influences atmospheric composition.

Introduction

Tropospheric ozone and aerosols affect climate on both regional and hemispheric scales (Fiore *et al.*, 2012, Unger, 2012). Climate change, in turn, influences the distribution and lifetimes of these species, with consequences for human health and ecosystems (Jacob and Winner, 2009). Understanding the coupling between short-lived chemical species and climate change is a major challenge, given the uncertainties in chemical mechanisms and climate processes. For example, aerosols perturb regional climate directly, through absorption and scattering of incoming sunlight, and indirectly, through interactions with cloud droplets. Neither the optical properties of the mix of aerosols in the atmosphere nor the details of cloud-aerosol interactions have been wellcharacterized (Rosenfeld *et al.*, 2014; Chen *et al.*, 2015).

I summarize below two Harvard projects that shed light on chemistry–climate interactions. In the first project, we examined the impact of climate change on wildfire activity in the western United States and the implications for $PM_{2.5}$ air quality (Yue *et al.*, 2013, 2014). In the second project, we investigated the response of regional climate to trends in US aerosols.

Projections of Wildfires in the Western United States

Wildfire emissions can adversely affect air quality locally and downwind. Wildfire activity in North America is strongly related to weather conditions, such as temperature and humidity, and future climate change may exert a large influence on wildfire activity.

In Yue *et al.* (2013, 2014), we estimated future wildfire activity over the western United States using meteorological fields from an ensemble of climate models following the A1B scenario. We first developed an ecoregion-dependent fire prediction scheme by regressing local meteorological variables from the current and previous years together with fire indexes onto observed regional area burned. The regressions explain 25–60% of the variance in observed annual area burned in the West during 1980–2004, depending on the ecoregion. We also developed an alternative scheme by parameterizing daily area burned with temperature, precipitation, and relative humidity. Over California the parameterization includes the impact of Santa Ana wind and other geographical factors on wildfires. This approach explains as much as 64% in variance in area burned over southern California.

We then applied both these schemes to the archived meteorology from 15 IPCC climate models. Over the 2000–2050 time period, we calculated increases of 10–100% in area burned, depending on the ecoregion and the prediction method. Our projections are most robust in the southwestern desert, where all climate models predict significant ($p < 0.05$) meteorological changes. We found that the fire season lengthens by 23 days in the warmer and drier climate at midcentury. Using a chemical transport model, we calculated that the trend in wildfire activity increases summertime surface concentrations of organic carbon (OC) over the western United States by 46–70% at midcentury, relative to the present day. Black carbon (BC) increases by 20–27%. The smoke is most enhanced during extreme episodes: above the 84th percentile of concentrations, OC increases by ~90% and BC by ~50%, while visibility decreases from 130 km to 100 km in 32 Federal Class 1 areas in Rocky Mountains Forest. By relying on an ensemble of climate models and focusing on median results, we gain confidence in our results of increased fire activity and smoke in the western United States.

Climatic Effects of 1950–2050 Changes in US Anthropogenic Aerosols

Global mean surface temperatures have increased by about 0.89°C since 1880, as determined by a linear trend, with much of that increase likely due to greenhouse gases (IPCC, 2013). However, temperature trends on regional scales are more complicated. For example, the eastern United States experienced significant cooling between 1930 and 1990. The net cooling effect of anthropogenic aerosols is thought to have mitigated some global warming, but the importance of aerosol cooling on temperature trends in the United States has so far received little attention.

In Leibensperger *et al.* (2012a), we used the GEOS-Chem chemical transport model combined with the GISS Model 3 to calculate the aerosol direct and indirect radiative forcings from US anthropogenic sources over the 1950–2050 period. We used historical emission inventories and future projections from the IPCC A1B scenario. The aerosol simulation was evaluated with observed spatial distributions and 1980–2010 trends of aerosol concentrations and wet deposition. We found that the radiative forcing from US anthropogenic aerosols is strongly localized over the eastern United States. The forcing in the model peaks in 1970–1990, with values over the eastern United States (east of 100°W) of −2.0 W/m^2 for direct forcing, which includes contributions from sulfate (−2.0 W/m^2), nitrate (−0.2 W/m^2), OC (−0.2 W/m^2), and BC (+0.4 W/m^2). The aerosol indirect effect is of comparable magnitude to the direct forcing. The forcing declines sharply from 1990 to 2010 (by 0.8 W/m^2 direct and

1.0 W/m² indirect), mainly reflecting decreases in SO_2 emissions. Aerosol forcing in the model continues declining post-2010 but at a much slower rate since US SO_2 emissions have already declined by almost 60% from their peak.

In Leibensperger *et al.* (2012b), we investigated the climate response to changing US anthropogenic aerosol sources over the 1950–2050 period. We used the GISS Model 3 and compared to observed US temperature trends. Time-dependent aerosol distributions were generated as in Leibensperger *et al.* (2012a). We found that the regional radiative forcing from US anthropogenic aerosols elicits a strong regional climate response, cooling the central and eastern United States by 0.5–1.0°C on average during 1970–1990, with the strongest effects on maximum daytime temperatures in summer and fall. Aerosol cooling reflects comparable contributions from direct and indirect (cloud-mediated) radiative effects. Absorbing aerosol, mainly black carbon, has negligible warming effect. Aerosol cooling reduces surface evaporation and thus decreases precipitation along the US east coast, but also increases the southerly flow of moisture from the Gulf of Mexico, resulting in increased cloud cover and precipitation in the central United States. Observations over the eastern United States show a lack of warming in 1960–1980, a phenomenon known as the US 'warming hole'. This period is followed in the observations by very rapid warming, which we reproduce in Model 3 and attribute to trends in US anthropogenic aerosol sources. These results suggest that high aerosol loading may help explain the US warming hole. Present US aerosol concentrations are sufficiently low that future air quality improvements are projected to cause little further warming in the United States (0.1°C over 2010–2050). Our results suggest that most of the warming from aerosol reductions in the United States has already been realized over the 1980–2010 period.

Addendum: Notes about David Rind's Contributions

We began working with David in 1996, when NASA funded our multi-institutional project to study the effects of ozone and aerosols on climate (CACTUS: Chemistry, Aerosols, and Climate: Tropospheric Unified Simulation). The other CACTUS collaborators were atmospheric chemists; David was the only climate scientist on board. His patient, cheerful efforts to educate the rest of us on meteorological phenomena truly benefited our collaboration. After that, we kept working with David, through multiple other projects and acronyms. David led the early effort to use passive chemical tracers to validate model transport processes. In all our work with David, what I most valued was his gentle insistence on getting things right in the climate model and on interpreting model results with a mix of curiosity and appropriate caution.

References

Chen, Q., Heald, C.L., Jimenez, J.L., Canagaratna, M.R., Zhang, Q., He, L.Y., . . . Liggio, J. (2015). Elemental composition of organic aerosol: the gap between ambient and laboratory measurements. *Journal of Geophysical Research, 42,* 4182–4189.

Fiore, A.M., Naik, V., Spracklen, D.V., Steiner, A., Unger, N., ... Zeng, G. (2012). Global air quality and climate, *Chemical Society Reviews*, 41, 6663–6683.

IPCC. (2013). Summary for Policymakers. In Stocker, T.F., D. Qin, G.-K. Plattner, M. Tignor, S.K. Allen, J. Boschung, A. Nauels, Y. Xia, V. Bex and P.M. Midgley (Eds.), *Climate Change 2013: The*

Physical Science Basis. Contribution of Working Group I to the Fifth Assessment Report of the Intergovernmental Panel on Climate Change (p. 1535). Cambridge, United Kingdom and New York: Cambridge University Press.

Jacob, D.J. and Winner, D.A. (2009). Effect of climate change on air quality, *Atmospheric Environment,* 43, 51–63.

Leibensperger, E.M., Mickley, L.J., Jacob, D.J., Chen, W.-T., Seinfeld, J.H., Nenes, A.,. . . Rind, D. (2012a). Climatic effects of 1950–2050 changes in US anthropogenic aerosols — Part 1: Aerosol trends and radiative forcing, *Atmospheric Chemical Physics,* 12, 3333–3348.

Leibensperger, E.M., Mickley, L.J., Jacob, D.J., Chen, W.-T., Seinfeld, J.H., Nenes, A., . . . Rind, D. (2012b). Climatic effects of 1950–2050 changes in US anthropogenic aerosols – Part 2: Climate response, *Atmospheric Chemical Physics,* **12**, 3349–3362.

Rosenfeld, D., Sherwood, S., Wood, R. and Donner, L. (2014). Climate effects of aerosol–cloud interactions. *Science,* 343, 379–380.

Unger, N. (2012). Global climate forcing by criteria air pollutants. *Annual Review of Environment and Resources,* 37, 1–25.

Yue, X., Mickley, L.J., Logan, J.A. and Kaplan, J.O. (2013). Ensemble projections of wildfire activity and carbonaceous aerosol concentrations over the western United States in the mid-21st century. *Atmospheric Environment,* 77, 767–780.

Yue, X., Mickley, L.J. and Logan, J.A. (2014). Projection of wildfire activity in southern California in the mid-21st century. *Climate Dynamics,* 43, 1973–1991.

Slide 1

Chemistry–Climate Interactions in a Changing Environment: Wildfire in the West and the U.S. Warming Hole

2009 wildfire in Southern California

Loretta J. Mickley (Harvard)

with: Pattanun Achakulwisut, Becky Alexander, Thomas J. Breider, Daniel Cusworth, Rynda C. Hudman, Daniel J. Jacob, Jennifer A. Logan, Shannon N. Koplitz, Eric M. Leibensperger, Lee T. Murray, David Rind, Lu Shen, Dominick S. Spracklen, Amos P.K.A. Tai, Shiliang Wu, Xu Yue, and Lei Zhu.

Our Warming Planet: Topics in Climate Dynamics
Lectures in Climate Change, Vol. 1
2017

Slide 2

Two Key Issues in Chemistry–Climate Interactions in the Atmosphere

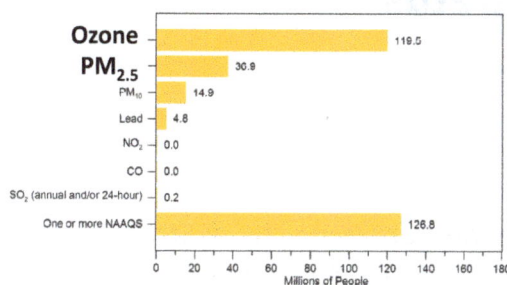

1. What is the climate penalty on air quality?
How will size of bars change with changing climate?

Millions of people in US living in areas in violation of the EPA standards.

2. How do trends in short-lived species affect global and regional climate?

Calculated trend in surface sulfate concentrations

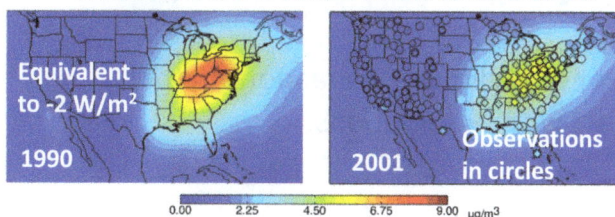

Regional forcings can be as large as global forcings from well-mixed GHGs.

Short-lived species can affect methane lifetime.

Leibensperger *et al.*, 2012

Slide 3

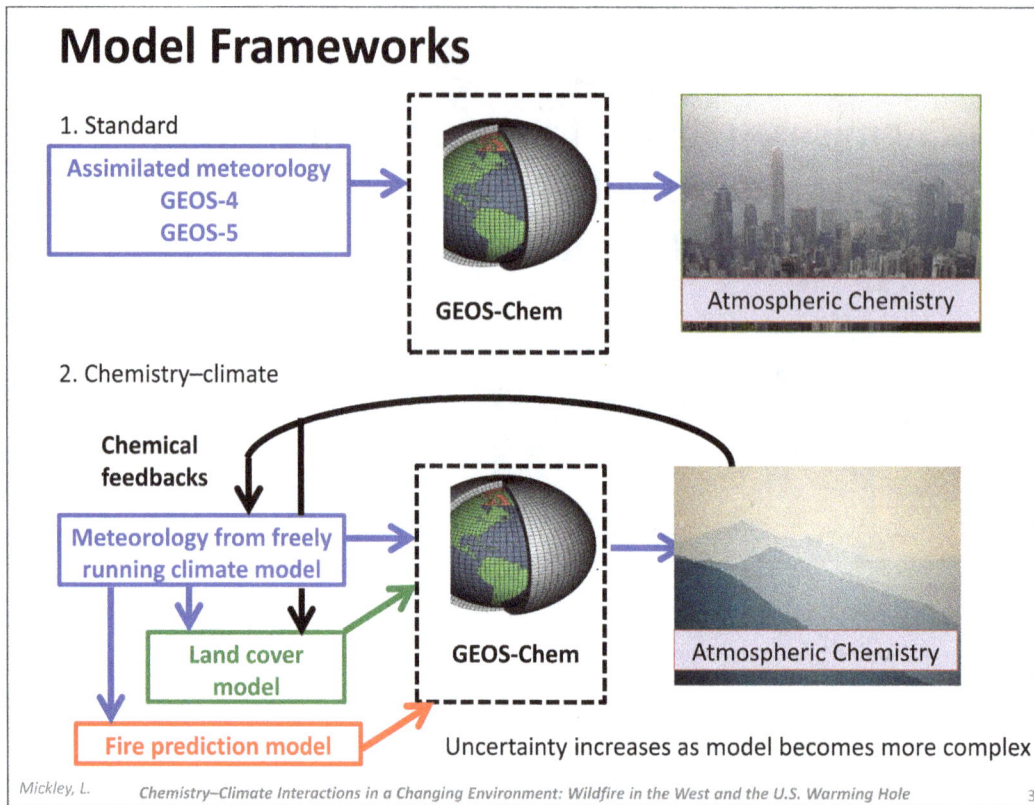

Slide 4

Methods to Deal with Uncertainties in Complex Model Apparatus

1. Build statistical models using observed sensitivities of target variables to meteorology.

2. Apply statistical model to meteorological fields from an ensemble of models such as CMIP5. Get robust result.

3. Compare results from Steps 1 to 2 with results from chemistry-climate models.

4. Benchmark chemistry–climate interactions in the recent + distant past with observations.

Mickley, L.   *Chemistry–Climate Interactions in a Changing Environment: Wildfire in the West and the U.S. Warming Hole*   4

Slide 5

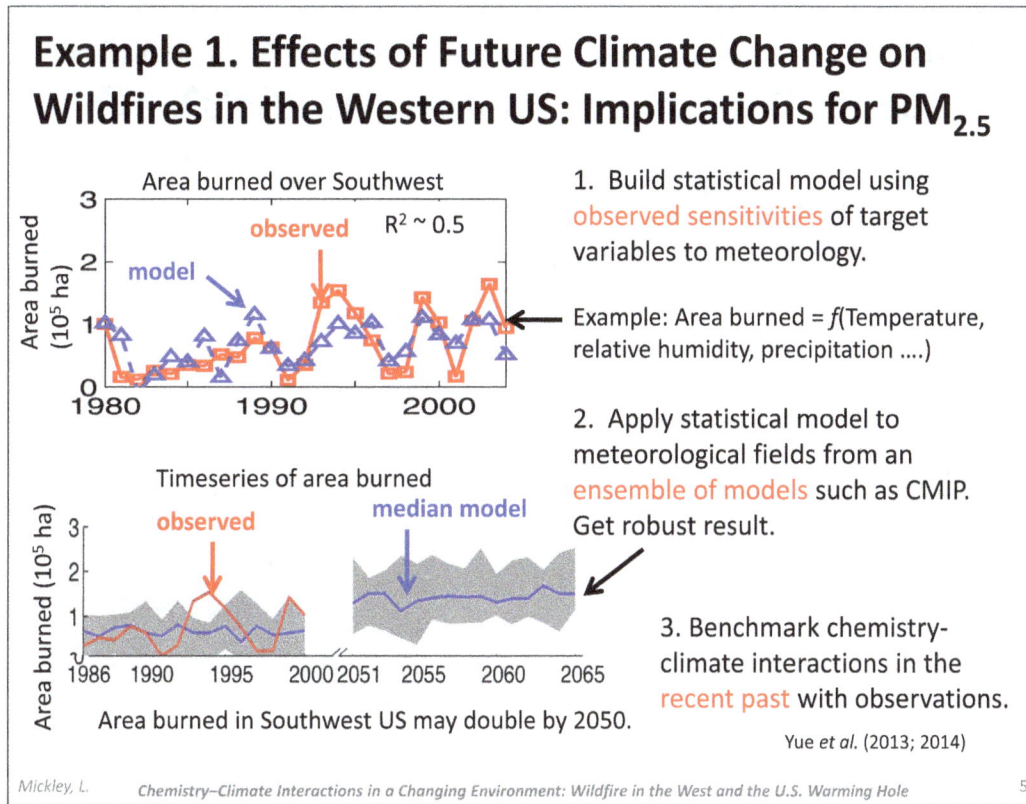

Example 1. Effects of Future Climate Change on Wildfires in the Western US: Implications for PM$_{2.5}$

Area burned over Southwest

observed   R^2 ~ 0.5

model

1. Build statistical model using observed sensitivities of target variables to meteorology.

Example: Area burned = f(Temperature, relative humidity, precipitation ....)

Timeseries of area burned

observed   median model

2. Apply statistical model to meteorological fields from an ensemble of models such as CMIP. Get robust result.

Area burned in Southwest US may double by 2050.

3. Benchmark chemistry-climate interactions in the recent past with observations.

Yue *et al.* (2013; 2014)

Mickley, L.   *Chemistry–Climate Interactions in a Changing Environment: Wildfire in the West and the U.S. Warming Hole*   5

Slide 6

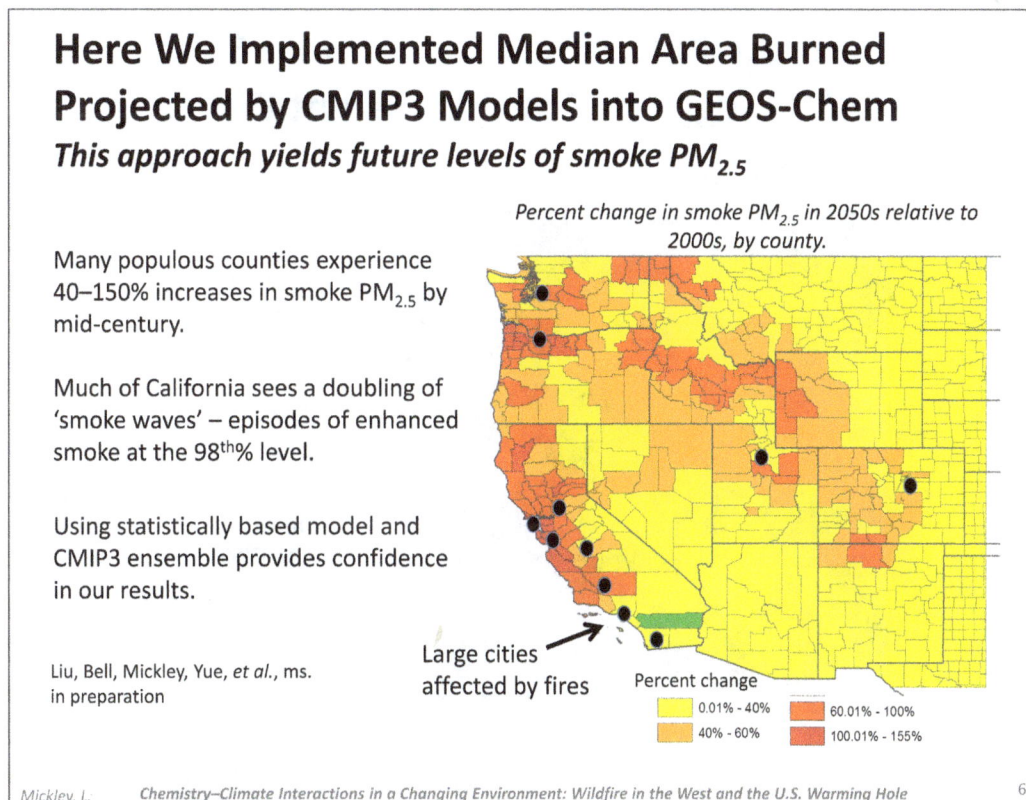

Here We Implemented Median Area Burned Projected by CMIP3 Models into GEOS-Chem
This approach yields future levels of smoke PM$_{2.5}$

Many populous counties experience 40–150% increases in smoke PM$_{2.5}$ by mid-century.

Much of California sees a doubling of 'smoke waves' – episodes of enhanced smoke at the 98th% level.

Using statistically based model and CMIP3 ensemble provides confidence in our results.

Liu, Bell, Mickley, Yue, *et al.*, ms. in preparation

Percent change in smoke PM$_{2.5}$ in 2050s relative to 2000s, by county.

Large cities affected by fires

Percent change

0.01% - 40%	60.01% - 100%
40% - 60%	100.01% - 155%

Mickley, L.   *Chemistry–Climate Interactions in a Changing Environment: Wildfire in the West and the U.S. Warming Hole*   6

Slide 7

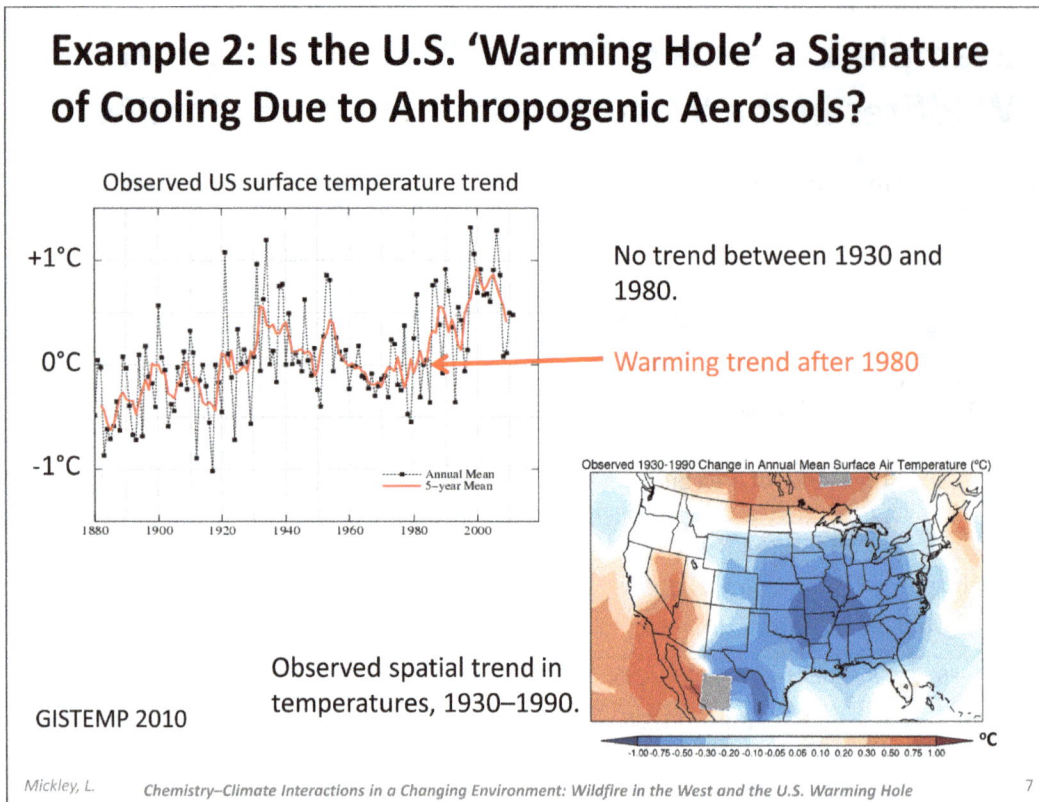

Example 2: Is the U.S. 'Warming Hole' a Signature of Cooling Due to Anthropogenic Aerosols?

Observed US surface temperature trend

No trend between 1930 and 1980.

Warming trend after 1980

Observed spatial trend in temperatures, 1930–1990.

Observed 1930-1990 Change in Annual Mean Surface Air Temperature (°C)

GISTEMP 2010

Mickley, L.   *Chemistry–Climate Interactions in a Changing Environment: Wildfire in the West and the U.S. Warming Hole*   7

Slide 8

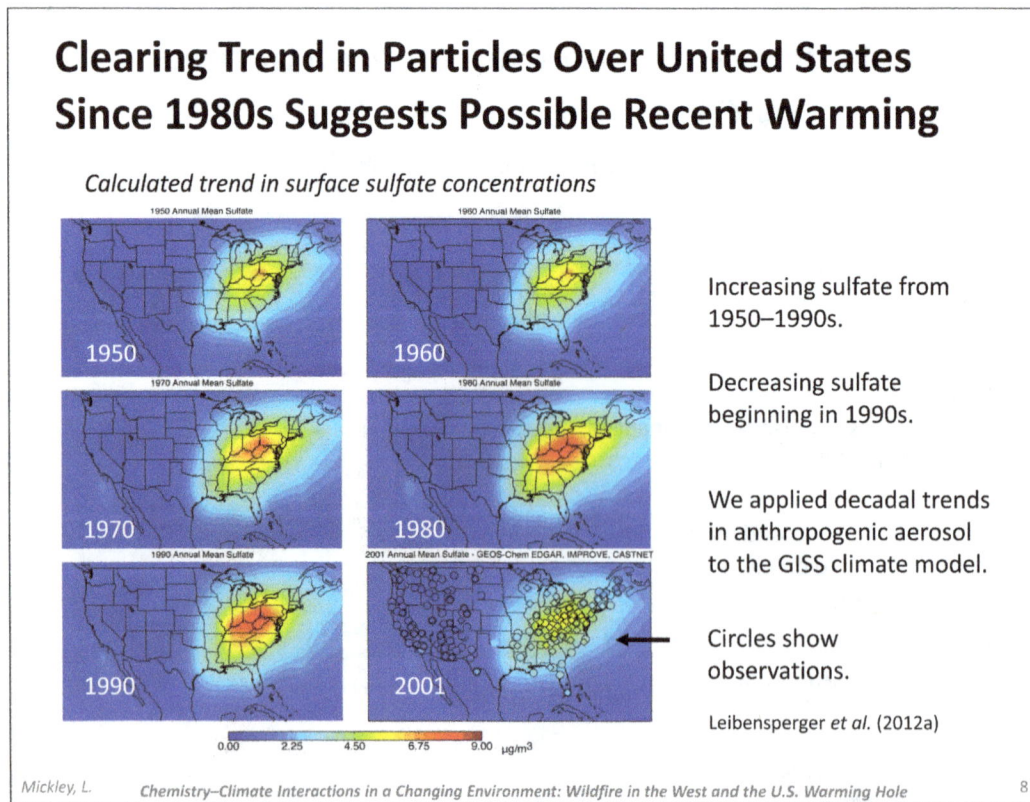

Clearing Trend in Particles Over United States Since 1980s Suggests Possible Recent Warming

Calculated trend in surface sulfate concentrations

1950 1960 1970 1980 1990 2001

Increasing sulfate from 1950–1990s.

Decreasing sulfate beginning in 1990s.

We applied decadal trends in anthropogenic aerosol to the GISS climate model.

Circles show observations.

Leibensperger *et al.* (2012a)

Mickley, L.   *Chemistry–Climate Interactions in a Changing Environment: Wildfire in the West and the U.S. Warming Hole*   8

Slide 9

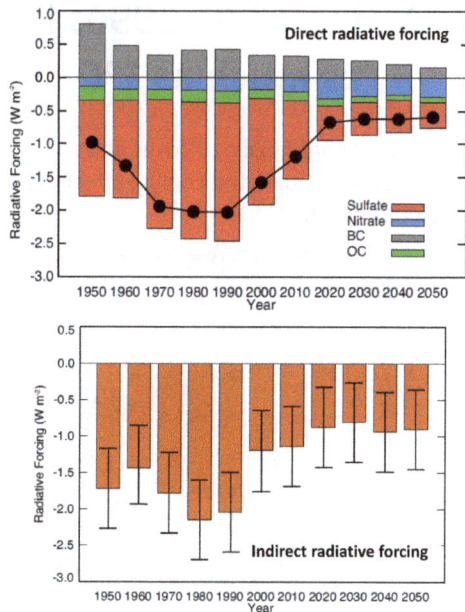

Slide 10

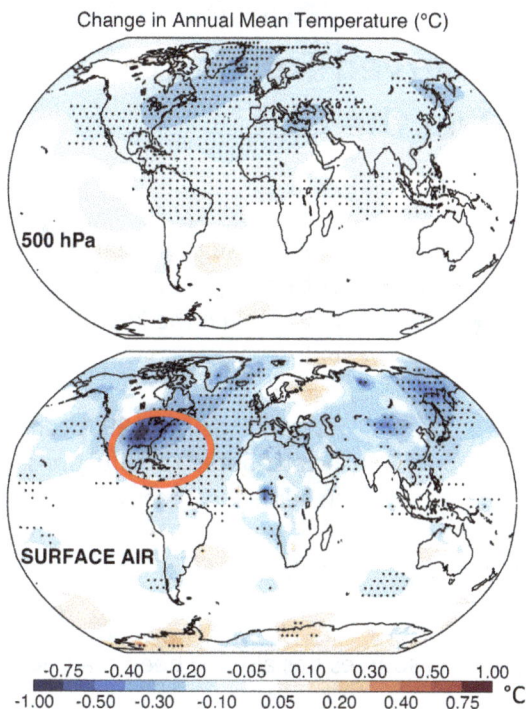

Slide 11

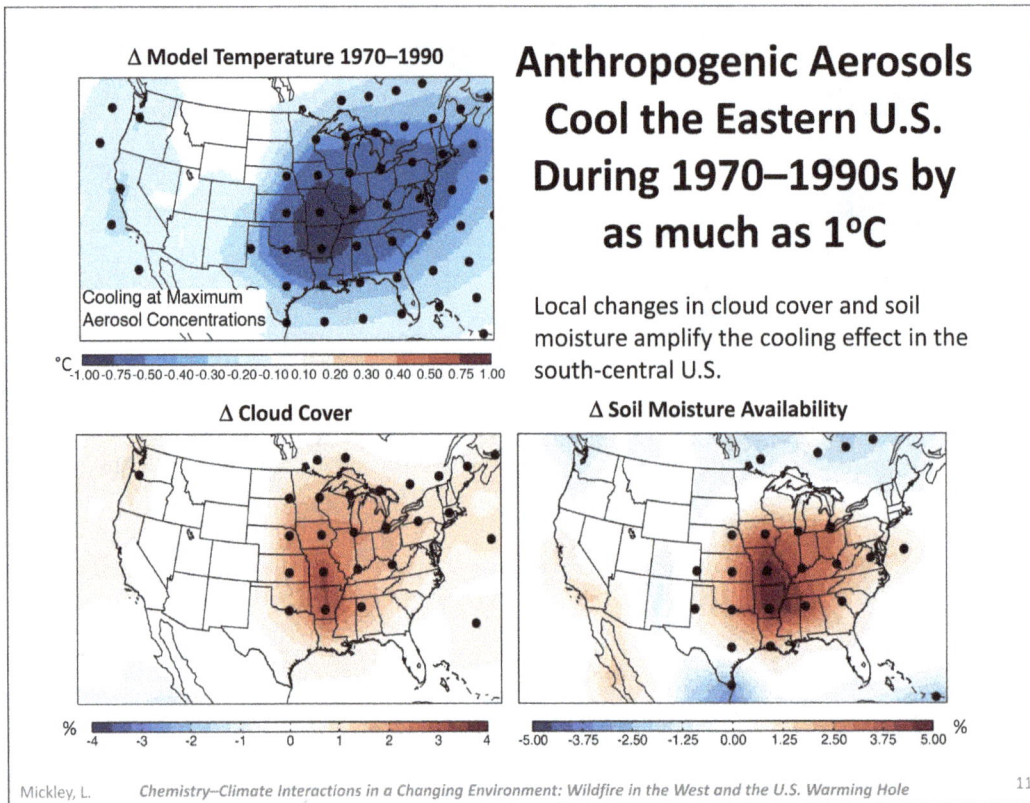

Mickley, L.   *Chemistry–Climate Interactions in a Changing Environment: Wildfire in the West and the U.S. Warming Hole*   11

Slide 12

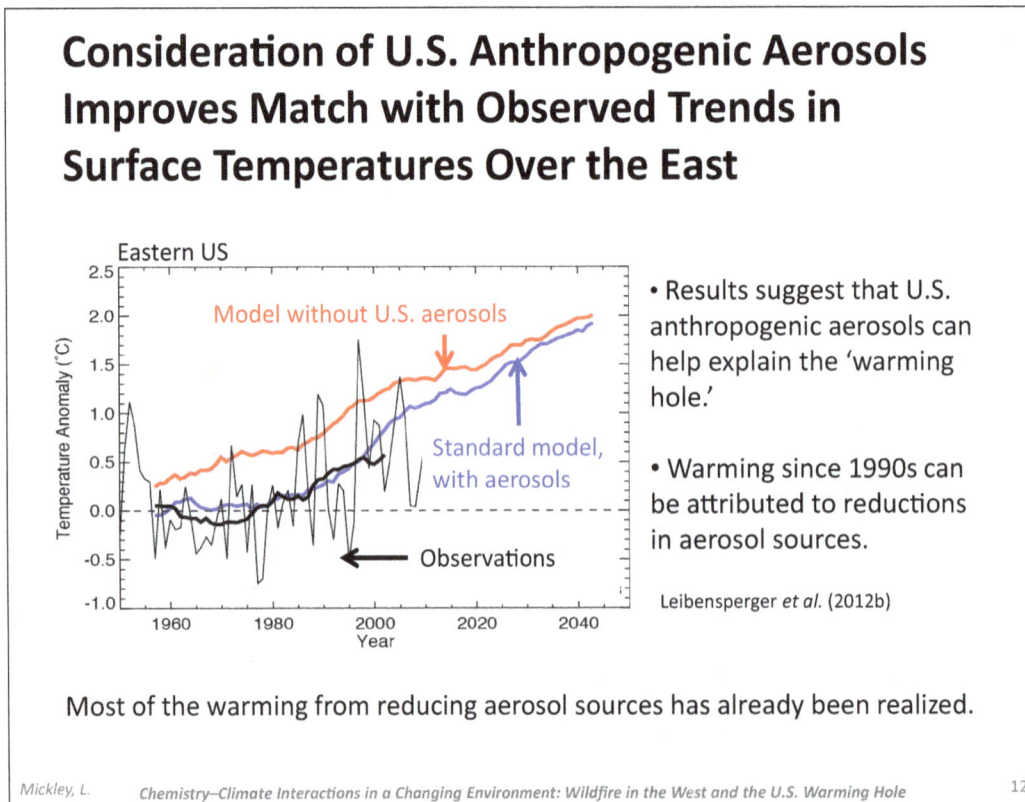

Mickley, L.   *Chemistry–Climate Interactions in a Changing Environment: Wildfire in the West and the U.S. Warming Hole*   12

Slide 13

Key Conclusions

• Air quality depends on weather. Global climate change may impose a "climate penalty" on U.S. air quality.

• We find that a warming climate will lengthen the fire season in the West by 23 days by the 2050s and increase summertime surface concentrations of organic carbon (OC) by 46–70% and of black carbon (BC) by 20–27%.

• Decreases in anthropogenic aerosol due to air quality regulations may impose large regional forcings and inadvertently lead to surface warming.

• We find that the anthropogenic aerosols during the 1970s–1980s cooled the eastern U.S. during 1970–1990s by 0.5–1 °C. This cooling likely contributed to the observed U.S. warming hole.

Slide 14

Global Climate in the 21st Century
BY DAVID RIND

Ambio, Vol. 13, No. 3, 1985

Rising atmospheric carbon dioxide might bring dust-bowl conditions to the US-Canadian grain belt. Photo: Earthscan/USIS.

It was a typical summer's day in the latter part of the 21st century in Dallas, Texas. The temperature was again above 100°F (38°C); it would get this hot, or hotter, seventy times during the summer. . .

An example from the mid-1980s of David's early efforts to spread the word on climate change.

Thank you David!

Slide Notes

Slide 2   Ozone and aerosols are profoundly influenced by meteorology, and climate change could impose a 'climate penalty' on regional air quality. Both ozone and aerosols can, in turn, affect climate on regional or global scales.

Slide 3   In the standard model framework for a chemical transport model (CTM), assimilated meteorology is applied to the model, yielding atmospheric composition. In chemistry–climate models, there are one- or two-way effects amongmeteorology, chemistry, land cover, and other variables.

Slide 4   Chemistry–climate models are complex and prone to large uncertainty. Here are listed some methods to deal with uncertainties.

Slide 5   In Yue *et al.* (2013, 2014), we built a fire prediction scheme using observed sensitivities of area burned to meteorological variables. We then applied the scheme to archived meteorological fields from CMIP3.

Slide 6   We found significant increases in smoke PM2.5 in the western US in the warmer, drier climate of the 2050s.

We are working with epidemiologists to determine the health impacts of the increased smoke.

Slide 7   The US warming hole of the 20th century has not been wellexplained and may be simply due to chance.

We were curious if anthropogenic aerosols played a role in the cooling.

Slide 8   Most of aerosols over the eastern US is composed of sulfate, which is highly scattering.

Using GEOS-Chem, we calculated a strong decrease insulfate from 1950 to the 2000s.

Slide 9   Using the GISS Model 3, we calculated both the direct and indirect forcings from aerosols over the eastern US from 1950 to 2050. For the future years, we followed the A1B scenario.

Slide 10   We found strong cooling from US anthropogenic aerosol during the 1970–1990 timeframe relative to the entire 1950–2050 timeframe. Dots show those gridboxes where cooling is statistically significant.

Slide 11   We find that feedbacks involving soil moisture and cloud cover are especially strong in the warming-hole region.

Slide 12   We are continuing to investigate the aerosol effects on regional climate.

SECTION 8

Educational Perspective

CLIMATE LECTURE 20

The Educational Global Climate Model (EdGCM)

Mark A. Chandler

Columbia University, Center for Climate Systems Research, Goddard Institute for Space Studies, New York, NY, USA

Mark Chandler (PhD in Geoscience from Columbia University, 1992) is a paleoclimatologist at Columbia University and works at NASA's Goddard Institute for Space Studies, which is part of Columbia's Morningside Campus in New York. His research involves the study of extreme climate change in Earth's geologic history, with a particular focus on past periods of global warmth. In addition, he has worked for many years to improve educational access to NASA's global climate models (GCMs) through a project he directs called the Educational Global Climate Modeling Project (EdGCM). The EdGCM Project develops software and educational materials that make it possible for teachers and students to operate a GCM on desktop computers. He encourages educational experiences that immerse people in realistic global climate change research and hundreds of universities, schools, and research institutions currently use EdGCM as a means of lowering the barriers to participation in the science of climate modeling.

Introduction

Climate change will profoundly impact Earth's environmental health as well as the world's economic and geopolitical landscape over the coming decades. The impacts of climate change are, in fact, already beginning to be experienced and have the potential to affect every living plant and animal on Earth within decades (IPCC, 2014). Given this reality, every citizen of this planet should have the right to knowledge about the Earth's climate system and have the option to adapt to, or help mitigate the profound changes that are coming. In addition, a portion of the workforce needs to be capable of interpreting and analyzing climate information because, since the impacts of climate change will be widespread, pervasive, and continue to change over time, more professions will be interacting with climate data. We are already at, or past, the point where educators and their students require access to the scientific and technological resources — computer models, data, and visualization tools — that scientists use daily in the study of climate change.

Although scientists use many methods to study Earth's climate system, global climate models (GCMs) have become the primary tools for exploring the complex interactions between components of the entire system: atmosphere, oceans, and land. GCMs are used to make projections of future climate change, to simulate climates of the past, and even to help scientists look for life on other planets (Flato *et al.*, 2013; Way *et al.*, 2017). Like any model, a

GCM can help people evaluate actions before they are taken. Like Business Intelligence software, they are Climate Intelligence tools. Unfortunately, GCMs are black boxes to most people. A previous chapter in this book by Gary Russell, entitled *Building a Climate Model*, is one example of the growing body of literature aimed at the general public describing the inner workings of GCMs. This literature goes a long way toward explaining climate model fundamentals. However, it will not be enough to alleviate their black-box nature unless people are afforded hands-on, authentic learning experiences as well.

Twenty years ago it was thought that GCMs would become commonplace in schools and perhaps might even be running on the desktops of congressional staffers (Randall, 1996). But today very few schools, including institutions of higher learning, have hands-on access to GCMs. Most GCMs still require supercomputing resources and skilled programmers to operate them. It might surprise many people that this lack of access extends to many research scientists as well, because GCMs are developed primarily at national labs where security and cybersecurity concerns have made such facilities less accessible over the past two decades. In some sense, the limited accessibility of GCMs has been beneficial, since these immensely complex models have been developed and applied almost entirely by highly trained scientists who are positioned to understand the strengths and limitations of the models. Experts who are better able to avoid misuse and misinterpretation of simulation results. Some skeptics have painted this as nefarious, but it is simply a best practice of most professions to have trained professionals operating complex tools — whether that be in engineering, medicine, law, economics, or any other vocation. Nevertheless, the resulting lack of familiarity with climate models and the techniques used to project future climate change has oftentimes engendered public distrust of important scientific findings based on modeling methodologies.

The Educational Global Climate Modeling Project (EdGCM), discussed in this lecture, makes one of NASA's GCMs accessible through a user-friendly interface so educators can run it using desktop or laptop computers, and without the need for scientific programming skills. EdGCM affords students the opportunity to explore the subject of climate change in the same manner as research scientists, and in the process become knowledgeable about a topic that is affecting their lives. It will better prepare them to grapple with the myriad of complex climate issues of the coming decades.

The EdGCM Project: Objectives

The overarching objective of the EdGCM Project is to improve the quality of teaching and learning of climate change science through broader access to GCMs. But, also to provide appropriate technology and materials to help educators use these models effectively. The design of EdGCM encourages students to learn about the scientific process while performing authentic investigations and real research projects that follow this process. The tools also provide a mechanism for linking classrooms with the research at national labs and universities and, ultimately, all of these things will demystify the science of climate change.

Teaching the Scientific Method Via Climate Modeling

Guiding introductory-level students through the scientific process is best accomplished by laying out a pathway where discrete steps are clearly and simply identified (Bush *et al.*, 2017).

With classroom time at a premium, it is most efficient to combine using the tools — in this case climate models — with learning about climate science, and simultaneously conducting authentic research to experience the scientific method in action. The steps presented here engage participants in the fundamentals of the modeling process, emphasizing a simple step-by-step approach, as well as the basic analysis of climate model results through data processing and scientific visualization. Although this simple step-by-step approach is certainly commonplace in climate research, it is also common that climate scientists will iterate through subsets of these steps many times during the course of a typical research project. While EdGCM usage mirrors the scientific process as viewed through the lens of a climate modeler, we are mindful that there is no universal scientific method. With EdGCM projects students participate in five basic steps: (1) experiment design, (2) simulation setup, (3) running experiments, (4) analyzing output using data-processing and scientific visualization methods, and (5) communicating results.

(1) *Experiment Design*: Hundreds of simulations have been conducted using NASA GCMs (and extensively published in scientific journals) so students can explore the typical design of climate modeling research using pre-existing, classic examples. EdGCM employs a variety of classic simulations in its design so teachers and their students can quickly begin working with authentic research topics using the method of 'rediscovery'. Rediscovery being the ability to re-examine previous research using similar methods to the original work. Using EdGCM in this fashion to begin means there is little upfront instruction required in the use of the software. The classic simulations are pre-packaged and can be run with the click of a 'Play' button. These simulations cover a range of topics and are chosen because they are highly cited in the scientific literature and can be used to examine fundamental climate science issues. There are climate simulations that depict simply the modern climate ('modern' here meaning the mid-20th century, before the impacts of anthropogenic warming become substantial). There are 'business-as-usual' greenhouse gas scenarios to examine future climate change, as well as ice age scenarios that contrast the climates of today with those of the last glacial maximum. All of these simulations make it possible for students to apply the actual GCM, and perform analyses, in a nearly identical fashion to those originally published by NASA scientists. This alone can provide many courses with a capstone moment; the ability to compare student-generated results to peer-reviewed scientific publications.

(2) *Simulation Setup:* Once students are familiar with the basic design of classic climate change experiments the next step is to allow them to design and setup their own unique simulations. Within the archetypal climate modeling lab this step would require experience in scientific programming and knowledge of climate modeling jargon. EdGCM reduces this need using a graphical point-and-click interface to accomplish most of GCM setup. The graphical interface is a form, called Setup Simulations, which is sub-divided into several logical sections. When fully expanded the form allows students to modify nearly every feature of the GCM, short of changes that require recompilation of model code, however, each section can be shown or hidden, allowing the teacher to focus students' attention on specific topics. Simulation Setup forms can be duplicated easily, because it is actually quite rare to set up GCM simulations from scratch, with most scientists instead duplicating previous experiments and modify only those inputs that are required to generate a unique experiment.

Completing a General Information section begins the task of setting up a climate simulation. This information is used to uniquely identify each simulation and store it as metadata in a relational database. Supplying a short Run ID and a modest descriptive title are the most basic information, as are the name of an experiment's owner and the start date of the project. But, this is also where users define the length, in calendar years, of the simulation. The existence of the EdGCM database also means that students can manipulate the GCM's input files (initial and boundary conditions) without needing to understand the details of complex file formats, and although pro-users have the option to select, or even create, individual input files, students can easily select directories that contain sets of input files that are compatible with one another. For example, a set of modern boundary condition files may be used for present day or future climate experiments, a set of ice age files could be selected to simulate the climate effects of large ice sheets, or a Cretaceous set could be chosen to simulate climate in the age of the dinosaurs.

After the basic information and input files are selected, students can assign values for a variety of climate forcings, which can be combined in a multitude of ways to create unique climate scenarios. These forcings include the level of greenhouse gases (carbon dioxide, methane, nitrous oxides, and chlorofluorocarbons), the luminosity of the sun, and the physical parameters of the Earth's orbit. Each forcing type may also be assigned a trend, which can take the form of simple or complex functions. Simple functions, like linear or exponential trends and step functions, are easily assigned with a few clicks of a button, while more complex trends, such as those incorporating episodically increasing and decreasing forcings, can be created and loaded as datasets, with only a bit more effort. With Setup Simulations educators can easily explain to students the fundamental details of how climate models are controlled and how, with a few simple commands, nearly limitless experiments are possible. Students can quickly design simulations to examine past, present, and future climates, or a myriad of 'thought experiments'.

(3) *Running Experiments*: In the context of climate modeling studies, a research project may be composed of one, often two, or sometimes several computer simulations. Ordinarily, it would take months to complete such a series of simulations with a cutting-edge GCM, even using a supercomputer. EdGCM uses a coarser resolution GCM (i.e., fewer grid cells define the 3-D geography of the atmosphere, land, and oceans) making it feasible to simulate several hundred years of the Earth's climate on a laptop in less than a day (Hansen *et al.*, 1983). Once a new simulation is set up, therefore, a click of the Play button will start the model running and any curriculum that is devoting 1 or 2 weeks to climate topics would easily be able to incorporate some aspect of climate modeling. However, there are clearly many educational settings that would only be able to devote a few days of time to the subject of climate change. Anticipating such situations, EdGCM also makes it possible for teachers to pre-compute GCM simulations and then upload them to lab computers or to a web-based utility, called EzGCM, that can mimic in a matter of minutes, a GCM simulating decades of climate change (The EdGCM Project, 2017). Even with this method, students can still be asked to perform some of the tasks of running the GCM, including the selection of specific simulations, choosing which climate variables to monitor, and using graphical controls to launch the climate model run. While the simulation is underway, text-based information is presented, offering

the student feedback — status messages generated by the GCM. More importantly, animated line plots convey the progression, and change, through time of the state of forcings and climate variables (e.g., CO_2, temperature, cloud cover, and precipitation). There are numerous climate variables to explore, since the number is limited only by the plethora of diagnostics calculated by the GCM itself. These displays of climate forcings and climate variables, along with the ability to simultaneously compare time series from multiple climate change experiments is an immersive experience for students even if schedules limit them to only a class period or two for the subject of climate change.

(4) *Analyzing Model Output — Post-Processing and Scientific Visualization*: Raw computer climate model output, the data of climate simulations, is generally written out in a binary format that must be prepared for direct scientific analysis. This form of data preparation is often described by the generic term 'post-processing' and it is actually the first part of the analysis step in the scientific process. Post-processing involves making scientific decisions as well as a substantial amount of programming, since the raw binary output must be reformatted using computational coding methods.

A graphical user interface can be used to automate a number of the most-used post-processing techniques, such as averaging across temporal and spatial domains and extracting individual variables from larger datasets. We believe that these post-processing steps can and should be run by the students as part of their learning about the scientific method for several reasons: (1) unlike GCM runs, post-processing can be computationally executed in a matter of seconds rather than hours or days; (2) processing the data interactively, and in real time, empowers users to understand that raw computer model output demands a degree of analysis prior to visualization, something that is probably not obvious to students who are used to CGI-laden movies and video gaming; (3) by requiring an interactive step involving post-processing the learner begins to recognize the importance of making analytical choices prior to detailed examination of individual variables; and (4) all of this effort conveys to users the significance of decisions that are made throughout the scientific process. We think that having students perform these steps is also necessary because smartphones and tablets have hidden from average users the crucial steps that lie between a complex computer operation and the visual display of information. In EdGCM, therefore, users make a series of choices that advance them through required post-processing steps, to end up with a resulting data file that is prepared for visual analysis.

After post-processing is complete, the next step is to represent the climate data visually. Scientific visualization involving plotting, mapping, and other 2-D images condenses complex information into a manageable form. It also makes it possible to compare multiple variables to one another, or to compare the same variable across time or geographic locations. A further, very common means of condensing climate model results can be accomplished by differencing 2-D arrays and then graphing the resulting anomaly array. These types of scientific visualization strategies are commonplace for earth science researchers, but students would normally have little access to such capabilities because such software generally involves some level of scripting and programming (or extraordinarily high costs). EdGCM comes with its own visualization software, called Panoply, that is easy for almost any student to use — and it is probably the most fun part of the process since color maps and cross-sections of the GCM output make the simulation results come to life (Schmunk, 2017).

(5) *Communicating Results*: No scientific study is complete without communicating the results to others. Whether the audience is the general public, other scientific specialists, classmates, or the teacher, it is crucial that students understand this is the culmination of their research and a fundamental part of the scientific process. EdGCM's eJournal feature is a tool that makes it simple to create reports that integrate text with in-line figures. This teaches students to prepare a written report of their results in a style similar to a publishable scientific manuscript.

An individual eJournal report can contain up to 20 sections that accommodate either text or images. Sections may be added or rearranged and images are easily inserted as needed. Once inserted into an eJournal an image becomes a figure. A figure in an eJournal manuscript is not just a graphic, but has an associated text field that becomes the figure caption. Figure captions are editable and students learn to use them to emphasize specific information in the image that is pertinent to their analyses. If a figure is moved within the eJournal its figure caption follows.

EdGCM also incorporates a tool called the Image Browser, from which the figures for eJournals are derived. The Image Browser is a window into an image library of photos, graphs, maps, and plots that are stored in the EdGCM folder on the computer's hard drive. Thumbnails of the images, along with other metadata, are stored in the EdGCM database. The image library is fully editable by a teacher (and by students, if they have permission) and can contain images in every major graphics file format. Even images created by the students themselves (possibly using EdGCM's scientific visualization applications) can be saved to the library and displayed for use with eJournal reports.

When an eJournal report is finished it can be converted to html through a single button click and it will automatically be previewed in the user's default web browser. Once the report is finalized (edited, reviewed, and graded) another click generates a unique eJournal directory, which includes the text of the report, the original images and their thumbnails, the associated figure captions, and the html that defines the layout. This eJournal directory can be uploaded to a school web server or emailed to the teacher, at which point the report is considered 'published' and in final form. eJournal reports can even make grading more efficient, since they have the consistent style of an organized scientific manuscript.

Concluding Remarks

Lessons and exercises that guide students through the methods outlined above deliver authentic research experiences, which teach students to follow a multi-step approach that is similar to the scientific research process that climate modelers use. The steps emphasized by EdGCM include: setting up and running simulations, post-processing computer model output, using scientific visualization to analyze simulation output, and communicating results. When students participate fully in this approach, they can quickly and effectively gain experience in (1) the scientific process, (2) climate science content, and (3) the use of computer models. This engages students and can help educators teach this important subject in a 3D fashion. This can excite students about science, make them proud of their own scientific efforts, and get them communicating through their own social avenues.

References

Bush, D., Sieber, R., Seiler, G., & Chandler, M. (2017). University-level teaching of Anthropogenic Global Climate Change (AGCC) via student inquiry. *Studies in Science Education*, 1–24. doi: 10.1080/03057267.2017.1319632

Flato, G., Marotzke, J., Abiodun, B., Braconnot, P., Chou, S. C., Collins, W. J., Cox, P., Driouech, F., Emori, S. & Eyring, V. (2013). Evaluation of Climate Models. In T. F. Stocker, D. Qin, G.-K. Plattner, M. Tignor, J. B. S.K. Allen, A. Nauels, Y. Xia, V. Bex & P. M. Midgley (Eds.), *Climate Change 2013: The Physical Science Basis. Contribution of Working Group I to the Fifth Assessment Report of the Intergovernmental Panel on Climate Change* (pp. 741–866). Cambridge, United Kingdom and New York, NY, USA: Cambridge University Press.

Hansen, J., Russell, G., Rind, D., Stone, P., Lacis, A., Lebedeff, S., Ruedy, R., & Travis, L. (1983). Efficient Three-Dimensional Global Models for Climate Studies: Models I and II. *Monthly Weather Review*, 111(4), 609–662.

IPCC. (2014). Climate Change 2014: Synthesis Report. In [Core Writing Team RK Pachauri and LA Meyer] (Ed.), *Contribution of Working Groups I, II and III to the Fifth Assessment Report of the Intergovernmental Panel on Climate Change*, (pp. 151). Geneva, Switzerland: IPCC.

Randall, D. A. (1996). A University Perspective on Global Climate Modeling. *Bull. Am. Meteor. Soc.*, 77(11), 2685–2690.

Schmunk, R. B. (2017). Panoply netCDF, HDF and GRIB Data Viewer (Version 4.7): NASA Goddard Institute for Space Studies. Retrieved from https://www.giss.nasa.gov/tools/panoply/

The EdGCM Project. (2017). EzGCM. Retrieved May, 2017, from http://ezgcm.com/

Way, M. J., Aleinov, I., Amundsen, D. S., Chandler, M. A., Clune, T., Del Genio, A. D., Fujii, Y., Kelley, M., Kiang, N. Y., Sohl, L., & Tsigaridis, K. (2017, In press). Resolving Orbital and Climate Keys of Earth and Extraterrestrial Environments with Dynamics 1.0: A general circulation model for simulating the climates of rocky planets. *Astrophys. J. Supp. Series*. doi: arXiv:1701.02360

Slide 1

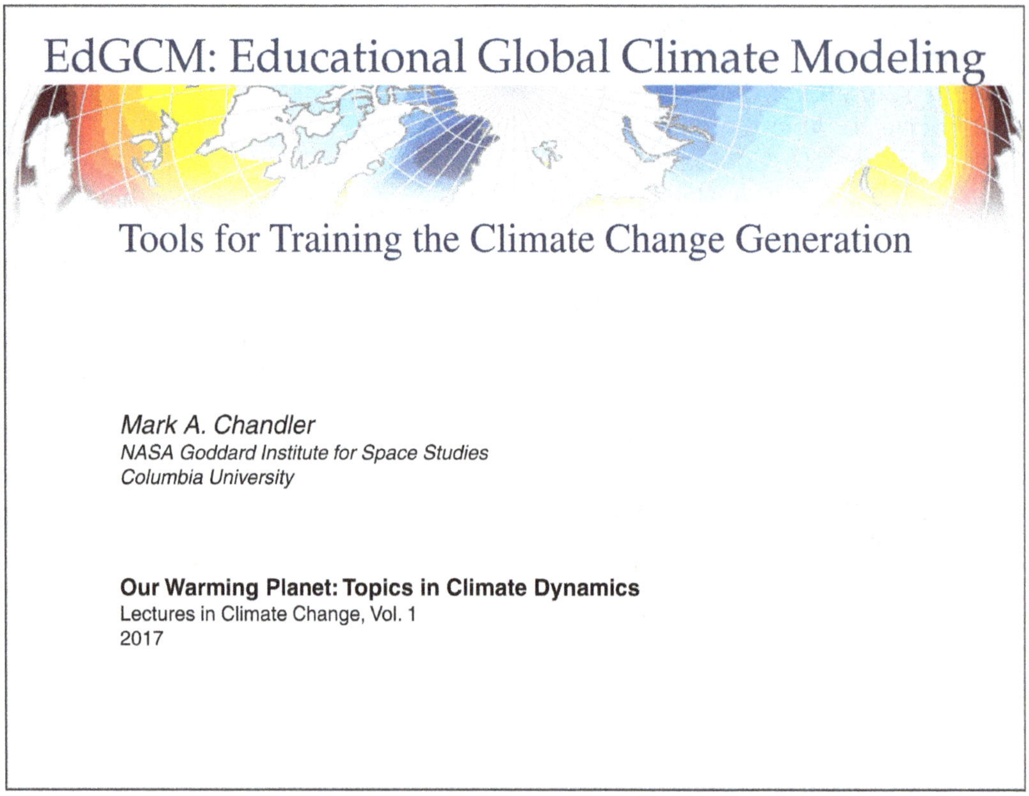

Slide 2

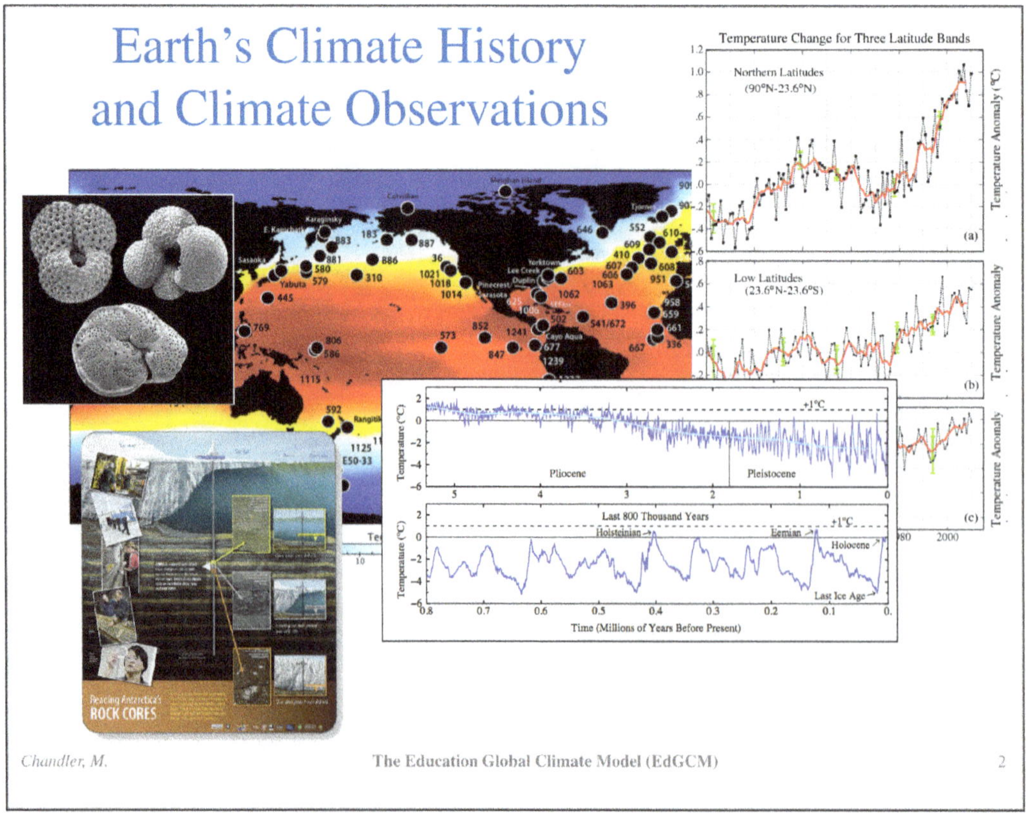

Slide 3

Climate Change Intelligence

Like all models, climate models help people evaluate actions before they are taken, whether those actions be business plans, mitigation scenarios, or adaptation strategies.

Chandler, M.   The Education Global Climate Model (EdGCM)   3

Slide 4

What is a GCM?

Atmosphere

Oceans

Cryosphere

Land Surface

Chandler, M.   The Education Global Climate Model (EdGCM)   4

Slide 5

GCMs Were Expected to be in Common Use by Now…

'…Very soon it will be possible to run a GCM on a laptop computer.'

'GCMs will begin running on workstations in high schools, and possibly elementary schools. They may even be running in the offices of congressman.'

Dr. David Randall
Bulletin of the American Meteorological Society, 1996

Slide 6

What is *Ed*GCM?

- **A GLOBAL CLIMATE MODEL**
  The NASA/GISS GCM Model II

- **A GRAPHICAL USER INTERFACE**
  Wrapped around the global climate model

NASA-GISS
www.giss.nasa.gov

Columbia University
edgcm.columbia.edu

Slide 7

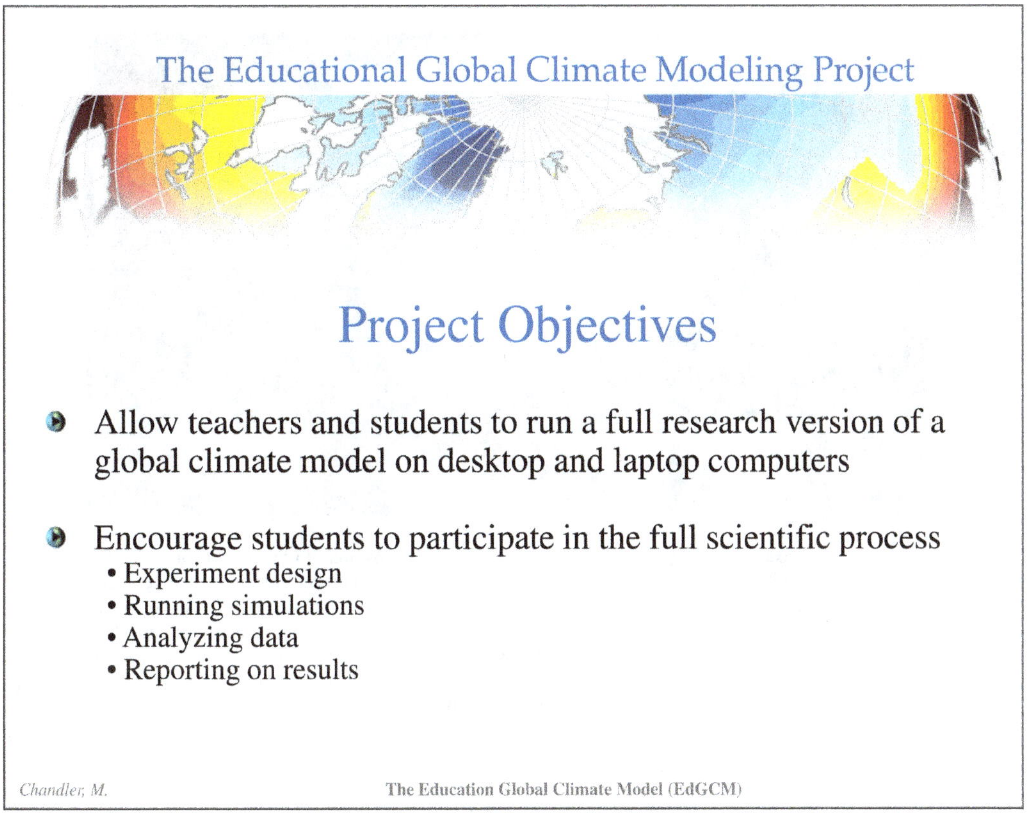

Slide 8

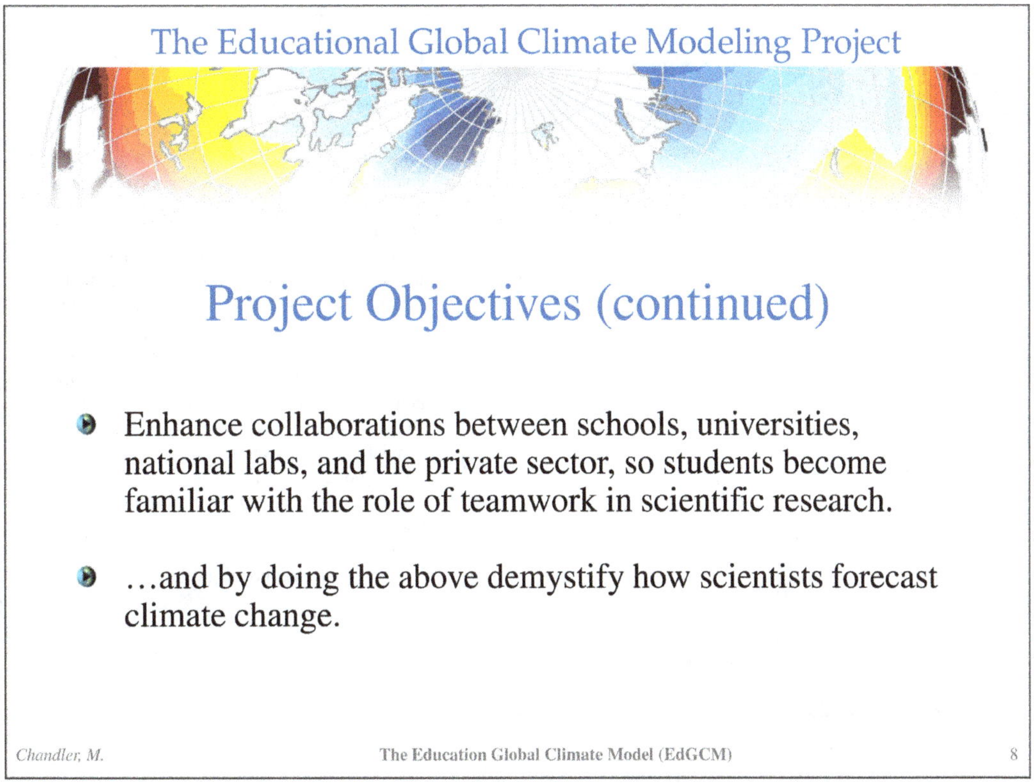

Slide 9

The scientific process often begins with an examination of prior research, in this case exploring how climate scientists have designed classic research projects that used computer models to study climate change. Materials that present the essential background of a climate science topic and present a hypothesis for testing can act as a guide to learning how climate modeling has been employed by other researchers to produce scientific results. The EdGCM Project provides short summaries of a number of such topics, and also supplies a web-based template that allows teachers or students to create summaries of their own research projects, which can then be added to the online collection.

Slide 10

Experiments are designed using the "Setup Simulations" form. This form acts as an interface for manipulating the GCM itself, with well-defined fields for entering the names of input files, such as initial and boundary conditions. The interface is divided into logical sections, each of which the educator can show or hide, depending upon which components of the GCM they want students to focus. Point-and-click controls select model options (e.g., number of simulated years per experiment, levels of greenhouse gases in the atmosphere, orbital parameters, and paleocontinental configurations). With the guidance of the Setup Simulations tool, students can easily run experiments that simulate a wide variety of climates, from future global warming to past ice ages. It is equally possible to run "thought experiments" that allow students to test the sensitivity of the climate model to hypothetical changes in solar luminosity or greenhouse gas levels.

Slide 11

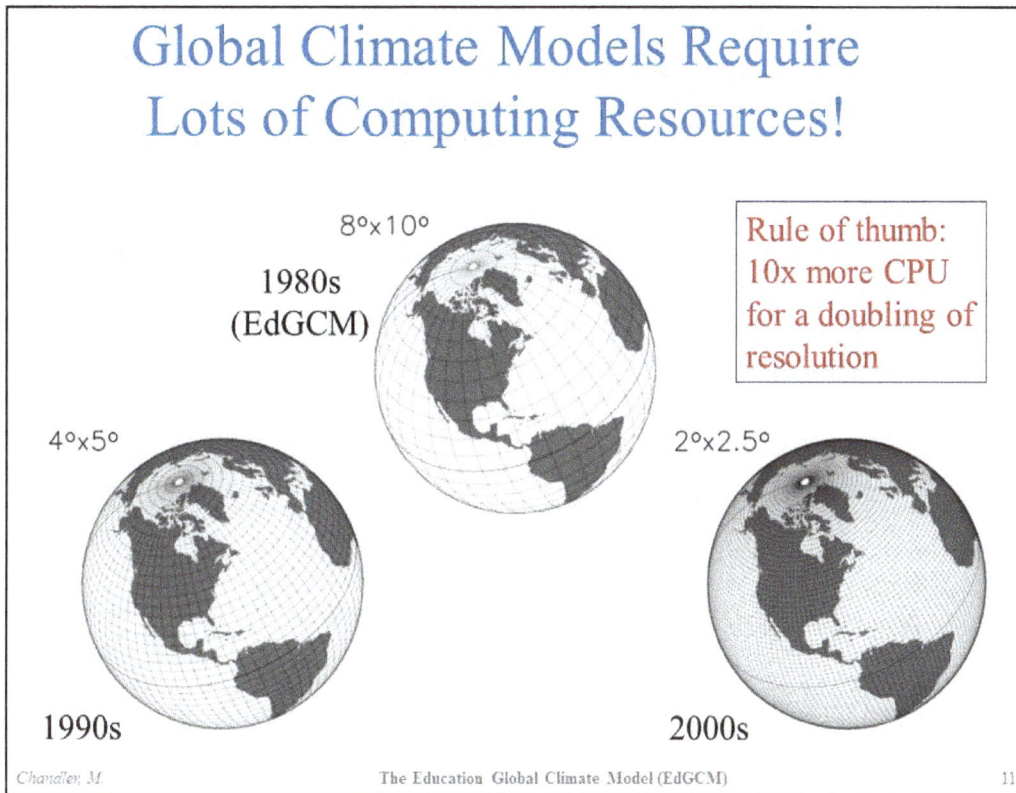

Slide 12

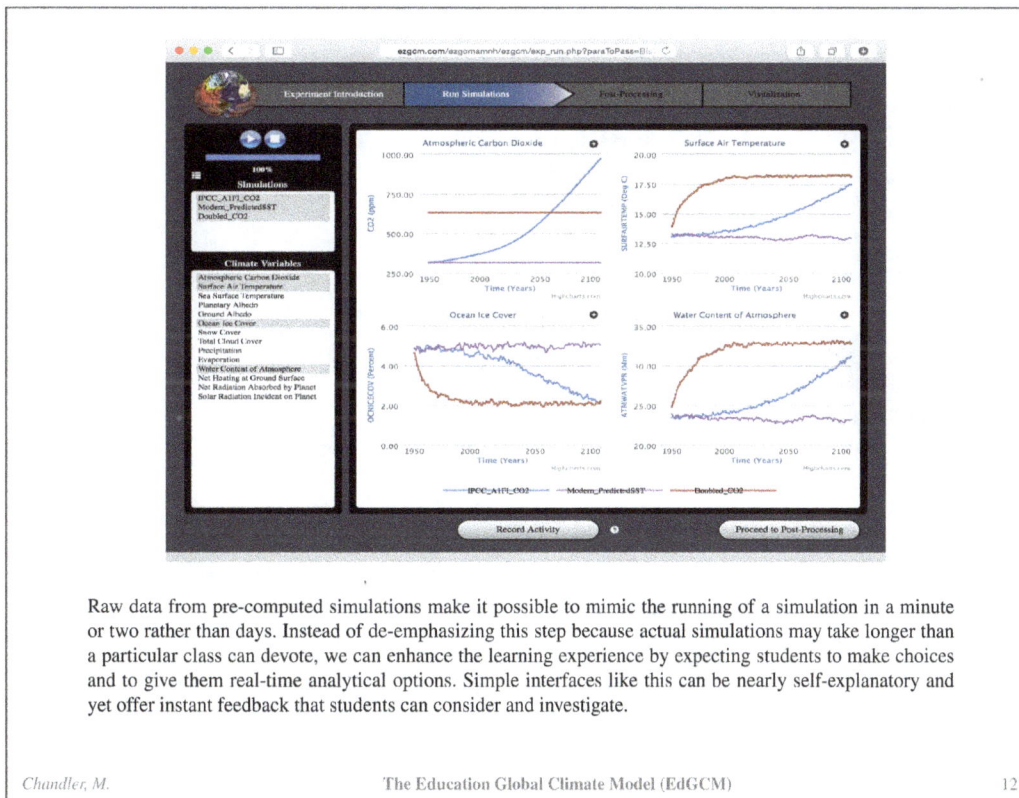

Raw data from pre-computed simulations make it possible to mimic the running of a simulation in a minute or two rather than days. Instead of de-emphasizing this step because actual simulations may take longer than a particular class can devote, we can enhance the learning experience by expecting students to make choices and to give them real-time analytical options. Simple interfaces like this can be nearly self-explanatory and yet offer instant feedback that students can consider and investigate.

Slide 13

The post-processing of model output is typically performed by dedicated systems analysts or by scientists who are also skilled programmers. This can be a limiting factor on the amount of analysis that can be performed in many research programs. However, it is more akin to an insurmountable obstacle for most schools and educational institutions. To solve this problem EdGCM automates the most-used post-processing operations, including:

- The production of summary tables for all variables produced by the global climate model (over 400), averaged over months, seasons, and years for any subset of the length of a simulation.

- The creation of global maps, vertical atmospheric slices, global time series, and zonal averages for approximately 80 different climate variables.

Slide 14

EdGCM comes installed with two scientific visualization applications (EVA and Panoply) for analyzing and interpreting climate model output. Students can create both line plots and global maps from the GCM's 1-D and 2-D output. Continental overlays, latitude by longitude grids, as well as a wide variety of map projections and colorbars can be applied. The scaling of the data is also easily changed, which helps to highlight various features of interest within the data without altering the data values.

Slide 15

Communicating the results of your climate modeling research project is what we define as the final step of the scientific process. The professional communication of science results generally requires that reports be in the form of a scientific manuscript, containing not only text, but images and figure captions (like this one). EdGCM makes this much simpler for students to accomplish through a tool called eJournal, which encourages report writing in sections, with in-line images and figure captions. EdGCM also creates html code that lays out the manuscript and makes it convenient and efficient to publish student manuscripts to school or personal websites.

Slide 16

A tweet of an image created using EdGCM showing the results of a rediscovery experiment. The map shows the surface air temperature anomaly as simulated by the GISS GCM in much the same manner that the climate model was used by NASA in its earliest 3-D global warming simulations, The original simulations were published nearly 30 years ago by GISS scientists, including David Rind, and played a significant role in raising the level of concern about the potential planetary-scale side effects of burning fossil fuels.

Slide 17

Slide 18

Slide Notes

Slide 1   Global Climate Modeling in the classroom — presentation at AAAS Meeting, 2012

AAAS Meeting, Vancouver, British Columbia, Canada, February 17, 2012
Authors: Mark Chandler and Linda Sohl, NASA Goddard Institute for Space Studies at Columbia University
Jian Zhou and Renee Sieber, McGill University, Department of Geography

Slide 2   Climate scientists use many methods to learn about Earth's climate and Earth's CHANGING climate:

- Paleoclimate data (from Hansen new article)
- Current Observations (from GISTEMP)
- Current Impacts (Svante Scientific Inc. slides)
- GCMs to examine and further inform...but, GCMs really do provide the only tools that can lead us toward detailed projections that can help us prepare for the consequence of continuing climate change and help us assess mitigation proposals.

Slide 3   Computer simulations are the primary tools scientists use and policy-makers rely on to assess the consequences of climate change. Like all models, climate models help people evaluate actions before they are taken, whether those actions be mitigation strategies or adaptation plans...or do nothing.

Slide 4   What is a GCM? The acronym 'GCM' originally stood for general circulation model. These are numerical computer models of the atmosphere (and now the oceans too) — essentially the circulating fluid components of the Earth. However, more recently the acronym 'GCM' has come to refer to a class of 3-D models of the entire Earth climate system known as global climate models. These models include a general circulation model as their core component, but they also simulate many additional components of the Earth's climate system, including ATMOSPHERE, OCEANS, CRYOSHPERE, and the LAND SURFACE. Cryosphere includes snow and sea ice, as well as the ice sheets on Greenland and Antarctica. The land surface component includes simple parameterizations of vegetation, ground hydrology, and river drainage. Some components, such as ice sheets, are not yet fully coupled with the climate system in a two-way fashion. In otherwords, they exist in the GCM and they can impact the atmosphere and oceans, but they are not themselves affected in return by changes in the atmosphere or oceans. This means that the climate change impacts on some features, such as ice sheets, must be calculated separately. As the GCMs become more sophisticated and as climate scientists learn more about how all the components operate, more and more of these 'offline' components are becoming dynamically coupled to the rest of the simulated climate system in GCMs.

Slide 5   GCMs were once predicted to be ubiquitous. This never happened — largely because the IPCC process has dominated the efforts of climate modeling groups who see the need for much higher resolution GCMs with ever more complex physics and chemistry.

Slide 6   EdGCM is a suite of software wrapped around a GCM to make it easier to operate, analyze output, and organize large amounts of data/output.

It is ultimately a joint project between NASA/GISS and Columbia University, CCSR. Funding has also been received from NSF and many individual contributions from our users.

Slide 7   We need to provide access to advanced capabilities (in this case GCMs) for those outside of the large government labs (NASA, NCAR, NOAA, and DOE). But, this brings with it many new requirements that are not so easily met.

Slide 8   We need to provide access to advanced capabilities (in this case GCMs) for those outside of the large government labs (NASA, NCAR, NOAA, and DOE). But, this brings with it many new requirements that are not so easily met.

Slide 11   The computing power required by a GCM is an order of magnitude higher each time the resolution is doubled. This occurs about once every decade, based somewhat on available supercomputing technology of at the time. In reality, the computing resources required increase somewhat more rapidly, since scientists have also been adding more complex physics over the decades as well.

Slide 17   • An accessible global climate model
• More than 50,000 downloads
• Over 150 institutions already using EdGCM
• Used successfully in science competitions

Slide 18   Change the world at http://edgcm.columbia.edu (using EdGCM to do climate change experiments).

About the Editors

Co-Editors

Cynthia Rosenzweig is a leader in the field of climate change impacts. She is a Senior Research Scientist at Columbia University's Center for Climate Systems Research and a professor in the Department of Environmental Science at Barnard College. She is also a senior research scientist at the NASA Goddard Institute for Space Studies, where she heads the Climate Impacts Group. She is the co-founder of the Agricultural Model Intercomparison and Improvement Project (AgMIP), a major international collaboration to improve global agricultural modeling, understand climate impacts on the agricultural sector, and enhance adaptation capacity in developing and developed countries. She is Co-Chair of the Urban Climate Change Research Network (UCCRN) and Co-Chair of the New York City Panel on Climate Change (NPCC). She was a coordinating lead author on observed climate change impacts for the IPCC Working Group II Fourth Assessment. She was named as one of *Nature*'s "Ten People Who Mattered in 2012". A recipient of a Guggenheim Fellowship, she joins impact models with climate models to project future outcomes under altered climate conditions.

Co-Editors

David Rind is a Senior NASA Climate Researchers Emeritus at the Goddard Institute for Space Studies. For more than 30 years he was a climate research scientist for NASA, as well as an adjunct professor at Columbia University, teaching graduate-level courses in climate dynamics and atmospheric dynamics. During this time he had the pleasure of working with a number of graduate students, as well as many close colleagues. He has more than 300 publications relating to climate and climate change, is a fellow of the American Geophysical Union, and a recipient of many awards, including the 2007 Nobel Peace Prize as a lead author on the Intergovermental Panel on Climate Change (IPCC).

Andrew Lacis (Ph.D. in Physics from University of Iowa, 1970) is a senior research scientist at the NASA Goddard Institute for Space Studies (GISS) in New York City. Dr. Lacis has been a member of the GISS climate modeling group since the mid-1970s. His area of expertise is development of fast and accurate radiative transfer techniques to model solar and thermal radiation with applications to the study of global climate change and the remote sensing of planetary atmospheres. He is a fellow of the American Geophysical Union, and is the principal architect of the radiation modeling methodology that is used in the general purpose GISS coupled atmosphere-ocean Global Climate Model (GCM).

Danielle Manley is a Staff Associate of Research at Columbia University's Center for Climate Systems Research (CCSR), where her primary role is in communicating climate risk information to stakeholders for local decision-making and resilience planning. She serves as a project manager for the New York City Panel on Climate Change (NPCC). She is a researcher for the Consortium for Climate Risk in the Urban Northeast, a NOAA-funded Regional Integrated Sciences and Assessments Program, as well as for the Adaptation for Development and Conservation (ADVANCE) partnership between CCSR and the World Wildlife Fund. She has a Bachelor's of Science in Environmental Science from the University at Buffalo, and a Master's Degree in Climate and Society from Columbia University.

Appendix — Supplementary Material

Included with this publication are downloadable PowerPoint slides of each lecture for students and teachers around the world to be better able to understand various aspects of climate change. These supplementary material may be retrieved by using the following Access Token:

http://www.worldscientific.com/r/10256-SUPP

The Token may be either clicked on (for eBook only), or manually copied/typed into the address bar of your internet browser.

If you purchased the eBook directly from the World Scientific website, your access token would have been activated under your username.

Users will be directed to the sign in page, and prompted to either sign in or register an account with World Scientific.

Upon successful login, you will be redirected to the book's page; click on the 'Supplementary' tab (third tab from the left, in the screenshot below) to locate the .zip file containing all 20 lecture slide sets. Click on the 'Powerpoint Slides (271 MB)' hyperlink to commence downloading.

For subsequent access to the supplementary materials, simply login with the same username.

For any enquiries on access issues, please email sales@wspc.com.sg.